All Source Positioning, Navigation, and Timing

For a listing of recent titles in the
Artech House *GNSS Series*,
turn to the back of this book.

All Source Positioning, Navigation, and Timing

Rongsheng (Ken) Li

ARTECH
HOUSE

BOSTON | LONDON
artechhouse.com

Library of Congress Cataloging-in-Publication Data
A catalog record for this book is available from the U.S. Library of Congress

British Library Cataloguing in Publication Data
A catalog record for this book is available from the British Library.

ISBN 13: 978-1-63081-703-9

Cover design by Andy Meaden

Artech House
685 Canton Street
Norwood, MA 02062

10 9 8 7 6 5 4 3 2 1

To the mathematicians and scientists before me, who laid a solid foundation for modern-day engineering

To the young talents after me, who will stand on my shoulders and make our society better

Praise to the All Mighty who predestined and blessed all these activities

Contents

CHAPTER 14
Signal of Opportunity PNT 365

CHAPTER 15
Real-Time Systems and the Software Development Approach and Applications to PNT Systems 393

CHAPTER 1

Introduction and Overview

1.1 Why All Source Positioning, Navigation, and Timing?

The problem of positioning, navigation, and timing (PNT) is one that was solved by Global Positioning Systems (GPS) or more generally by Global Navigation Satellite Systems (GNSS), which today includes GPS built by the United States, GLONASS by USSR/Russia, Galileo within Europe, and BeiDou by China, as well as a number of satellite-based augmentation systems. Today, every cell phone has a GNSS receiver that costs probably less than $3 to manufacture, and there are most likely more GNSS receivers than the number of people in most civilized countries. Indeed, GNSS has been playing a critical role in the operations of military, civil, and commercial systems, as well as consumer products. It is not an overstatement that GNSS is one of the most important inventions of human society.

However, for military as well as safety and mission-critical commercial applications, GNSS has a fundamental disadvantage: its vulnerability. Specifically, we are talking about both the availability and the integrity that can be seriously impaired under the following four scenarios:

1. The GNSS signal can be easily jammed by even a low-power jammer in a local operational area.
2. Civilian GNSS signals can be easily spoofed.
3. GNSS signals can be blocked by mountains and buildings, leading to significantly reduced availability when GNSS satellites at low elevation angles are not visible by users who are in a natural canyon or in an urban canyon (a canyon between high-rise buildings).
4. The GNSS constellation could be among the first things that will be taken out in a major strategic war in the future.

As will be elaborated upon in the next section, there are many alternative or complementary PNT technologies that can potentially provide PNT solutions when GNSS is denied or challenged. However, as we will find, none of these technologies will be able to provide sufficient performance, availability, and integrity under all operational or application scenarios. It is the consensus of the various governments and the industry that there is not a single magic solution that can replace GNSS when GNSS is denied or challenged. The solution will necessarily be based on a combination and integration of many different sensors and information sources, consequently, the All Source PNT (i.e., PNT by integration of information from all sources available).

1.2 Complementary PNT Technologies and All Source PNT

In this section, we will provide an overview of a number of alternative or complementary PNT technologies in terms of their mechanism and operational principles, typical performance, advantages, and limitations. A summary comparison of the various complementary PNT technologies is provided in Table 1.1.

The most inexpensive complementary PNT technology is the multiconstellation or multifrequency GNSS receivers, which exist in almost all cell phones today. When we use all the possible signals from each of the GNSS constellations including GPS, GLONASS, Galileo, Beidou, as well as regionally available GNSS and

Table 1.1 Comparison of Complementary PNT Technologies

	Complementary PNT Technologies	*Typical Performance and Mechanism*	*Advantages*	*Disadvantages*
1	Multiconstellation or multifrequency GNSS	0.01 to 20m; use all GNSS to improve availability and integrity by using a large number of signals and by cross-checking	Potentially solve canyon and urban canyon problem; greatly improved integrity	Can still be jammed; does not help if GNSS constellation is gone; not trusted by military applications
2	Signal of opportunity	1 to 1,000m; leverage large amount of strong signals available from terrestrial and space sources	Large amount of signals, some of them millions of times stronger than GPS; very hard to jam; cross-checking allows improved integrity	Requires a reference station and communication from it to users that can be jammed or made unavailable; not trusted by many mission-critical applications; receivers must be able to handle large amount of different types of signals; need signal source location or ephemeris data
3	Celestial navigation	10 to 200m; use stars and resident space objects (RSO) to determine attitude and position; typically integrated with INS; short-wave infrared (SWIR) camera provides day time capabilities	Stars and RSOs are always available; not concerned with cloud below	Need to deal with cloud above; midnight gap: some of the low Earth orbit (LEO) RSOs are not illuminated during midnight; performance may not be sufficient for some purposes; may need gimbals or beam directors to track RSOs
4	Vision-based navigation	0.1 to 100m; matching image against database image or simply use measured motion in a sequence of images; 0.5% to 5% visual odometry	Complementary to celestial: not concerned with cloud above	Need to deal with cloud below; rain, snow, seasonal changes, and constructions can cause mismatches and reduce availability; not available above featureless ocean
5	Terrain referenced navigation (TRN)	0.1 to 100m; using radar or laser altimeter to measure altitude to match against terrain database	Weatherproof; cloud not an issue; sensors can be low cost and low SWAP	Detectable active signals; not available over flat land or flat ocean

Table 1.1 Comparison of Complementary PNT Technologies *(Cont.)*

	Complementary PNT Technologies	*Typical Performance and Mechanism*	*Advantages*	*Disadvantages*
6	Magnetic anomaly referenced navigation (MARN)	10 to 1,000m; using magnetometer to measure earth magnetic field to match against map/database	Mature, high accuracy, low SWAP sensor; very hard to jam or spoof	Lack of high-accuracy magnetic anomaly map or database; must be able to accurately calibrate out active field and hard or soft iron effects on the vehicle
7	Gravity gradient anomaly referenced navigation (GGARN)	10 to 1,000m; using gravity gradient instrument (GGI) to measure gravity gradient to match against map or database	Hard to jam or spoof; less impacted by the environment (than MARN)	High-accuracy GGI still in development; high-accuracy gravity maps or database with terrain effects included are not generally available
8	Gravity anomaly referenced navigation (GARN)	100 to 1,000m; using gravimeter (high-accuracy accelerometer) to measure gravity to match against map or database	Same as above; may use existing high-accuracy accelerometers	Must filter out the accelerations due to motion; only applicable to stable or long journey vehicles such as submarines
9	Network collaborative PNT	1 to 1,000m; perform communication, time transfer, ranging, and even measuring line-of-sight (LOS) vector through comm. links to achieve better performance for the entire network	Easy augmentation on the existing communication links; expand the benefits of expensive and less deployed PNT assets to the network partners	Communications can still be vulnerable and are subject to jamming and spoofing

GNSS augmentation systems such as the Wide Area Augmentation System (WAAS), Quasi-Zenith Satellite System (QZSS), and the European Geostationary Navigation Overlay Service (EGNOS), there are in total a few hundred high-quality signals that can be tracked for PNT purposes. At any location, there will be easily more than 50 signals visible to a multiconstellation or multifrequency receiver. This provides the opportunity for greatly improved availability even in natural canyon and urban canyon scenarios.

The large amount of signals also allows highly reliable cross-checking to eliminate spoofing, to detect faults in the data, and to reject multipath signals, greatly improving the integrity of the solution. By using differential GNSS augmentations and by using carrier-phase tracking and ambiguity resolutions, one can achieve centimeter-level positioning performance.

Unfortunately, all these great benefits of multiconstellation, multifrequency GNSS can still be easily denied by a low-power jammer due to the very low power of the GNSS signals received by a ground user. During the time of a strategic war, all the GNSS constellations will be the target of attack.

Signal of opportunity (SOOP) PNT takes advantage of the very large amount of radio frequency (RF) signals that are available from both terrestrial sources (such as

AM/FM radios, TV stations, cell phone towers) and space sources (such as satellite radio, satellite TVs, satellite-based mobile phone services). Most of these sources have very high power levels (typically 30 to 70 dB above the GNSS signal) and they are hard to be completely jammed or destroyed even during the war.

The biggest issue of SOOP PNT is that all these signal sources are not necessarily synchronized in any fashion and, consequently, it is often necessary to use a reference station that receives the same signals as the user to allow time difference of arrival (TDOA)-based positioning and time synchronization with the reference station. The reference station not only solves the problem of time synchronization, but also provides the benefits of improved performance in a similar way as differential GPS (DGPS). However, the use of a reference station introduces a major element of vulnerability. It can become a target of attack, and the communication between the reference station and the user receiver is subject to jamming and spoofing.

Celestial navigation is an old technology that was used in ship and aircraft systems before. The modernization of celestial navigation includes the development of SWIR star-trackers that are capable of tracking stars and satellites during the day and the use of not only stars but also what are generally called RSOs for improved navigation performance.

The use of stars will allow a star-tracker aided inertial navigation system to maintain very good attitude accuracy with velocity error and position error bounded and Schuler-tuned (see Chapter 10 for details). The use of RSOs, which are of finite distances to the navigator with the LOS directions being sensitive to both attitude and position, will help to further improve the position and velocity performance. Even with both stars and RSOs used, the performance of celestial navigation falls into the categories of en route navigation. It will be extremely hard to improve the performance further.

Celestial navigation is subject to availability reduction caused by cloud covers above the navigator. Cloud is much less a problem for aircraft flying at a high altitude. At a low altitude, especially for marine and ground vehicles, there is a high probability of cloud blocking the views of stars and satellites. Even with cloud mapping and cloud avoidance designs, there is still a significant chance that stars and RSOs are not visible when we need the performance.

Vision-based navigation is complementary to celestial navigation in several ways. Celestial navigation looks up to see stars and RSOs. Vision-based navigation looks down to ground to see the ground image when we are navigating on an aircraft. In this case, one will need to be concerned about clouds below the vehicle, which becomes more a problem for high-altitude vehicles than low-altitude vehicles.

Vision-based navigation for a ground user will need to use images around the navigator, and typically a three-dimensional (3-D) database will be required to support such operations.

In any case, there are fundamentally two different approaches available for vision-based navigation. The first approach is to compare the camera image against a database of geographically tagged images and use the registration errors to determine attitude and position. In many cases, the database does not have to store the images. Instead, features of images can be stored. In this case, the camera images will be processed first to extract features in order to match features in the database.

The second approach does not use a database. Instead, it uses the motion of images as indicated by a sequence of images to derive information on the change of attitude and change of position. This is visual odometry, which will allow us to perform relative navigation.

There is also something in between. That is SLAM (Simultaneous Localization And Mapping), which is fundamentally the same as the database-based approach except that the database or database contents are updated or created as we discover new scenes and features in the environment.

Both celestial navigation and vision-based navigation are based on passive optical images of the environment that cannot be jammed or spoofed. Neither approach needs to generate any detectable signals.

Vision-based navigation does have the capability to achieve scalable performance in the sense that when we are closer to the image that we are using, higher accuracy can be achieved. One of the biggest issues of vision-based navigation using a database is that the images can be changing due to snow, rain, season, and construction. Although things can be done to manage the match of an image against the image in the database in such a way that not every part of the image has to be matched, these issues do reduce availability and performance.

TRN, MARN, GGARN, and GARN all belong to a larger class of wide-sense terrain referenced navigation. The general principle for all of these types of navigation methods is the same. We measure physical variables, such as altitude from the terrain, the magnetic field anomaly, the gravity field anomaly, and the gravity gradient anomaly, which are functions of location, using the appropriate types of sensors. We then match the measured variables against the predicted values based on our location or attitude and the map or database. The difference allows us to estimate navigation and sensor errors and to make corrections to our navigation state variables.

Classic TRN uses radar or laser altimeter to measure the altitude above the terrain and uses that to match against the database. Classic TRN has to use active signals for the radar altimeter, which has some levels of detectability and some levels of vulnerability to be jammed. The other disadvantage of TRN is that it will not work over flat land or flat ocean.

Magnetic anomaly and gravity anomaly only use passive sensors and it will be extremely difficult if not completely impossible to jam or spoof the gravity field or magnetic field from some distance. Both magnetic anomaly and gravity anomaly are also available everywhere and can be used by aircraft, surface vehicles, and underwater vehicles.

High-precision, low SWAP, reasonably priced magnetometers are available today to support magnetic anomaly referenced navigation. The current hurdle is mostly the lack of the high-accuracy magnetic anomaly map and the limited capabilities to model and calibrate the magnetic field in the nearby environment.

However, high-precision, low SWAP, reasonably priced GGIs are still under development. In addition, gravity gradient referenced navigation also suffers from the lack of a high-precision map or database.

GARN uses gravity directly instead of gravity gradient. That immediately poses the need to eliminate the impact of the acceleration due to vehicle motions. Consequently, GARN is only applicable to special classes of low dynamic vehicles such as submarine or marine vehicles when long time averaging and filtering can reduce the impact of motion to acceptable levels.

Network collaborative PNT leverages a connected world by network communications. The network partners will exchange information through the network and more importantly perform additional network-based measurements such as time-transfer measurements (time of transmit (TOT) and time of arrival (TOA)), the range measurements, and the LOS vector measurements. These measurements will allow network partners to determine their position and time based on the position and time of their collaborative partners. One important benefit of such an approach is that it will allow expensive resources such as atomic clocks, celestial navigation devices, or resources that enjoy more availability and integrity of GNSS to propagate their availability and integrity to network partners that may not be as lucky or as resourceful. It is not hard to appreciate that network collaborative PNT cannot work alone as a PNT technology. There must be nodes in the network that have other PNT technology-based equipment available that provide the network with anchoring information about position and time.

As can be seen, none of these complementary PNT technologies will be a silver bullet to provide consistent performance, availability, and integrity in all operational and application scenarios. That is exactly the reason that the consensus in the industry and the various government communities is that the only solution is to integrate these technologies together, hence the All Source PNT.

References [1–4] show the U.S. government's initiative in the development of All Source PNT technology for which DARPA used the term All Source Positioning and Navigation (ASPN). References [5–67] provide examples of publications relevant to alternative or complementary PNT and All Source PNT technologies developed by the industry or the academia that can be integrated into traditional PNT technologies such as inertial navigation [68–86], using various types of Kalman filtering-based integration techniques [87–91].

1.3 Generic All Source PNT Architecture

Most of the readers of this book probably have some experience with integrated GPS/Inertial Navigation System (INS), which integrate GPS with INS. All Source PNT, similar to GPS/INS, is a problem of sensor fusion and requires similar tools and architecture. The following are the most important architectural aspects of All Source PNT:

1. The All Source PNT system must integrate multiple sensor and information sources.
2. When multiple sensors and information sources are used, measurements from each of the sensor and information sources must be able to be time-tagged with respect to the same time reference or clock (time synchronization/time matching).
3. Similarly, measurements from each of the sensors and information sources must be able to be transformed to the same spatial coordinate frames or references (spatial alignment/spatial match).

4. An algorithm must be used to combine all the information together and must support the information coming asynchronously from different sources. Extended Kalman filter or unscented Kalman filter is typically what we use to achieve this purpose.
5. There must be ways to detect, contain, isolate, and respond to various types of faults and threats.

Consequently, Figure 1.1 is naturally a typical architecture of an All Source PNT system.

An astute reader will realize that the architecture described in Figure 1.1 is motivated by the general architecture of the Kalman filter [87–91]. Indeed, this is reflected by the following key aspects of the architecture:

1. The propagation solutions;
2. Time-matching buffer;
3. Measurement receiving, time-tagging, and processing;
4. Measurement prediction;
5. Narrow-sense Kalman filter.

In an All Source PNT system, there are typically two types of solutions that get fused together. The propagation-based solution propagates the solution forward using numerical integration algorithms, or difference equations, or simply by hardware. The measurement-based solution is where the sensor measures part of the PNT

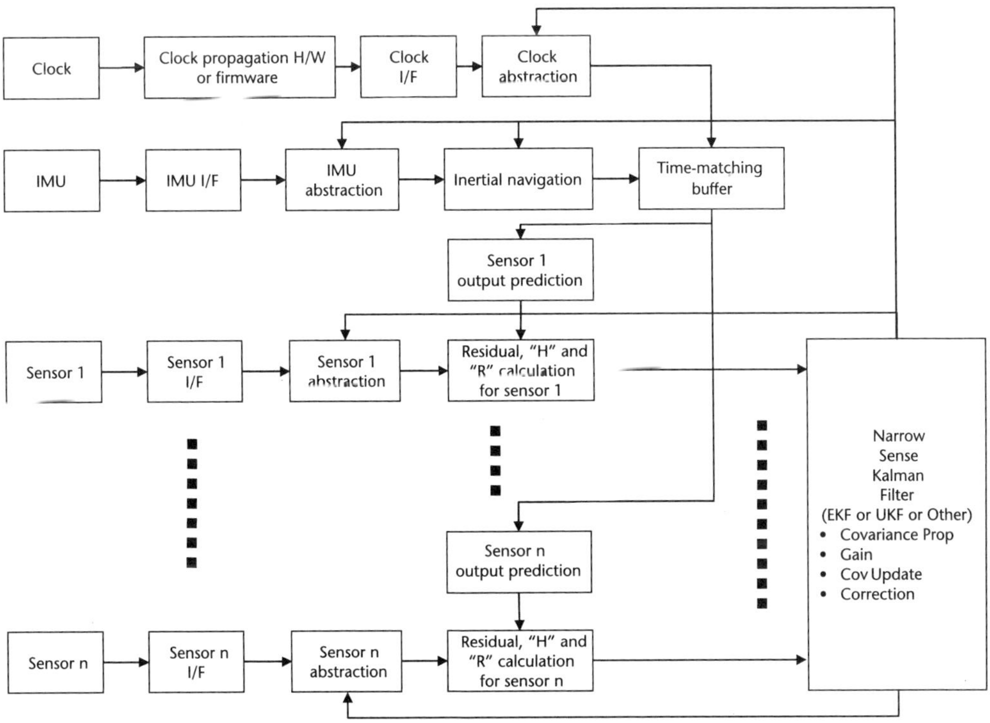

Figure 1.1 All Source PNT functional architecture.

state directly through an algebraic or static transformation (versus the differential or difference equations for the propagation-based solution).

Typical examples of propagation solutions are the propagation of time based on some oscillator output; the propagation of attitude, velocity, and position based on strapdown inertial measurement unit (IMU) measurements; and the propagation of position based on velocity measurement derived from Doppler radars or odometers. Another example of propagation solution is when the propagation is not done based on any sensor data. Instead, it simply uses a mathematical model such as the vehicle kinematics model or the vehicle dynamic model or a combination of both.

Typical examples of measurements are the measurements of position and velocity by a GPS receiver; the measurements of pseudo-ranges, delta ranges, and carrier phase by a GPS receiver; the measurements of TOA/pseudo-ranges, or TDOA by a signal of opportunity receiver; the measurements of star position and RSO position in a star-tracker field of view (FOV); the measurements of a ground image by a camera; the measurements of altitude by an altimeter; the measurements of magnetic field by a magnetometer; the measurements of specific force by a gravimeter; the measurements of gravity gradients by a GGI; the measurements of ranges between two operational nodes in a network; and the measurements of LOS vectors pointing from one node to another node.

All these examples of measurements can be used as updates to a Kalman filter, which will then produce corrections to the propagation-based solution and corrections to error sources of the measurements themselves. It can be expected that all the different sources of sensor information, either a source for propagation solution or a source for measurement-based solution, will be operating in general in an asynchronous fashion.

The time-matching buffer in the architecture is designed specifically to deal with the fact that the sensor inputs come in an asynchronous manner. We will typically save the propagation-based solution (time and INS solution, for example) into a time-matching buffer that is typically organized as a circular buffer with each of the records containing the time tag of validity and the state variables of the propagation solution.

When a measurement from any one of the measurement sensor comes in, the time tag of the measurement will be used to retrieve a sample of the stored states in the time matching buffer. Since the sensors operate in an asynchronous manner, the time tag will in general not match with the time tag of a particular sample in the time matching buffer exactly. Instead, it will be found to be somewhere between two samples. Interpolations will be performed to create the best approximation of the state using the two (or more) samples to create a sample that matches with the time tag provided.

The time-matched state created by the time-matching buffer will then be used by the measurement prediction component to predict the measurement at the time tag of the measurement based on the measurement model. The measurement and the time-matched predicted measurement will then be differenced to generate the most important variable in a Kalman filter-based architecture, the Kalman filter residual or the innovation. (Note that, in some literature, the residuals and innovations are given different meanings, but in most of the literature, including this book, we mean exactly the same thing by innovation and residual.)

Since it is really the entire system that forms the whole Kalman filter, we will use the term narrow-sense Kalman filter to mean the part of system that does the covariance propagation, the gain computation, the covariance update, and the generation of the corrections by multiplying the gain matrix with the residual. By definition, the narrow-sense Kalman filter will then generate the corrections that can be applied back to both the propagation solution and the measurement models.

Obviously, in order for the architecture described above to work correctly, each of the devices in the system must use the same time reference. This can be achieved by a number of time transfer mechanisms such as time transfer by Network Timing Protocol (NTP) or Precision Timing Protocol (PTP), time transfer using 1 pulse per second (1PPS), or time transfer by correlations (see later chapters).

Another thing that is important is the alignment of different coordinate frames used by different sensor sources. In general, we will need to maintain sensor misalignment states for each of the sensors as part of the Kalman filter states.

1.4 Structure of This Book

This book is centered on the architecture described by Figure 1.1. After this introduction and overview, we use Chapter 2 to collect the most important basic concepts and tools to be used throughout the book. Among them are the definitions of coordinate frames, the transformation between frames, various representation of attitude, the gravity and magnetic field models, and the most important tool for sensor fusion: Kalman filtering and its variations.

Chapters 3 through 5 can be considered as the first iteration of attacking the problem of All Source PNT. The idea is simple. In a Kalman filter-based system, we blend two types of solutions: the propagation-based solutions and the measurement-based solution. In Chapter 3, we describe various types of propagation-based solutions for PNT. In Chapter 4, we describe various types of measurement solutions for PNT, and in Chapter 5, we discuss in general how these solutions are integrated or fused together.

Chapters 6 through 14 can be considered as the second iteration of attacking the problem of All Source PNT. In these chapters, we discuss some of the promising and practically useful complementary PNT technologies in such a way that each of them can be included into the common architecture of All Source PNT. Each of these chapters is intended to serve two purposes: it can be used as a tutorial for a specific type of PNT application, and it can be used as part of the overall All Source PNT solution.

In particular, Chapter 6 is dedicated to integrated clock. Chapter 7 focuses on attitude determination. Although it is titled "Spacecraft Attitude Determination," most of the contents are generally applicable to any attitude determination problems. Chapter 8 discusses the spacecraft orbit determination problem.

Chapter 9 discusses TRN. Chapter 10 discusses modern celestial navigation. Chapter 11 is dedicated to image and vision-based navigation where we discuss three aspects of the technology including the geo-database referenced method, simultaneous localization and mapping (SLAM) and visual odometry.

Chapter 12 focuses on gravity anomaly and MARN, which is an extension of Chapter 9. Chapter 13 covers collaborative PNT and Chapter 14 is on SOOP PNT.

Finally, we use Chapter 15 to cover the less covered contents: the general hardware and real-time software architecture and their development process. In particular, we will address the issue of the open architecture standard.

References

[1] DARPA, "Broad Agency Announcement: All-Source Positioning and Navigation (ASPN)," Strategic Technology Office, DARPA-BAA-11-14, 2010.

[2] https://www.transportation.gov/pnt/national-positioning-navigation-and-timing-pnt-architecture.

[3] https://www.darpa.mil/program/adaptable-navigation-systems.

[4] Owano, N., "DARPA Seeks New Positioning, Navigation, Timing Solutions," *TecXplore*, March 28, 2015, https://techxplore.com/news/2015-03-darpa-positioning-solutions.html.

[5] Groves, P. D., "The Complexity Problem in Future Multi-Sensor Navigation and Positioning Systems: A Modular Solution," *The Journal of Navigation*, Vol. 67, 2014, pp. 311–326.

[6] Si, S., et al., "A New Underwater All Source Positioning and Navigation (ASPN) Algorithm Based on Factor Graph," *2019 Chinese Control and Decision Conference (CCDC)*, 2019, https://ieeexplore.ieee.org/document/8832700.

[7] Groves, P. D., *Principles of GNSS, Inertial, and Multisensor Integrated Navigation Systems*, 2nd ed., Norwood, MA: Artech House, 2013.

[8] Penn, T. R., "All Source Sensor Integration Using an Extended Kalman Filter," Air Force Institute of Technology, Master's thesis, March 2012.

[9] Pullen, S., T. Walter, and P. Enge, "Integrity for Non-Aviation Users: Moving Away from Specific Risk," *GPS World*, July 2011, pp. 28–36.

[10] Thomas, M., et al., *Global Navigation Space Systems: Reliance and Vulnerabilities*, Royal Academy of Engineering, 2011.

[11] Wang, L., and S. Li, "Enhanced Multi-Sensor Data Fusion Methodology Based on Multiple Model Estimation for Integrated Navigation System," *Int. J. Control Autom. Syst.*, Vol. 16, 2018, pp. 295–305.

[12] Gong, X., J. Zhang, and J. Fang, "A Modified Nonlinear Two-Filter Smoothing for High-Precision Airborne Integrated GPS and Inertial Navigation," *IEEE Transactions on Instrumentation and Measurement*, Vol. 64, 2015, pp. 3315–3322.

[13] Gao, W., Y. Zhang, and J. Wang, "A Strapdown Interial Navigation System/Beidou/Doppler Velocity Log Integrated Navigation Algorithm Based on a Cubature Kalman Filter," *Sensors*, Vol. 14, 2014, pp. 1511–1527.

[14] Abdolkarimi, E. S., G. Abaei, and M. R. Mosavi, "A Wavelet-Extreme Learning Machine for Low-Cost INS/GPS Navigation System in High-Speed Applications," *GPS Solutions*, Vol. 22, 2018, p. 15.

[15] Chang, L., K. Li, and B. Hu, "Huber's M-Estimation-Based Process Uncertainty Robust Filter for Integrated INS/GPS," *IEEE Sens. J.*, Vol. 15, 2015, pp. 3367–3374.

[16] Giorgi, G., and C. Narduzzi, "Performance Analysis of Kalman-Filter-Based Clock Synchronization in IEEE 1588 Networks," *IEEE Transactions on Instrumentation and Measurement*, Vol. 60, No. 8, August 2011.

[17] Stein, S. R., and R. L. Filler, "Kalman Filter Analysis for Real Time Applications of Clocks and Oscillators," *Proceedings of the 42nd Annual Frequency Control Symposium*, 1988.

[18] Kim, H., et al., "Tracking Low-Precision Clocks with Time-Varying Drifts Using Kalman Filtering," *Journal IEEE/ACM Transactions on Networking (TON)*, Vol. 20, No. 1, February 2012, pp. 257–270.

[19] Wang, B., et al., "Kalman Filter Simulation and Characterization of BDS Satellite Clock," *2018 European Frequency and Time Forum (EFTF)*, April 10–12, 2018.

[20] https://en.wikipedia.org/wiki/Precision_Time_Protocol.

[21] Whelan, D., G. Gutt, and P. Enge, "Boeing Timing & Location: An Indoor Capable Time Transfer and Geolocation System," *Proceedings of Fifth Stanford University, Symposium on Position, Navigation, and Timing*, Menlo Park, CA, 2011.

[22] Tan, Z., et al., "New Method for Positioning Using IRIDIUM Satellite Signals of Opportunity," *IEEE Access*, Vol. 7, June 24, 2019, pp. 83412–83423.

[23] Chowdhary, G., et al., "GPS-Denied Indoor and Outdoor Monocular Vision Aided Navigation and Control of Unmanned Aircraft," *Journal of Field Robotics*, Vol. 30, No. 3, pp. 415–438.

[24] Won, D. H., et al., "Integration of Limited GNSS Signals with Monocular Vision Based Navigation," *12th International Conference on Control, Automation and Systems (ICCAS)*, pp. 2012, 1364–1367.

[25] George, M., and S. Sukkarieh, "Camera Aided Inertial Navigation in Poor GPS Environments," *2007 IEEE Aerospace Conference*, 2007, pp. 1–12.

[26] Giebner, M. G., "Tightly-Coupled Image-Aided Inertial Navigation System Via a Kalman Filter," M.S. thesis, AFIT/GE/ENG/03-10, Graduate School of Engineering and Management, Air Force Institute of Technology (AU), Wright-Patterson AFB OH, ADA415246, March 2003.

[27] Hartley, R., and A. Zisserman, *Multiple View Geometry in Computer Vision*, 2nd ed., New York: Cambridge University Press, 2003.

[28] Kauffman, K., et al., "Simulation Study of UWB-OFDM SAR for Navigation with INS Integration," *Proceedings of the 2011 International Technical Meeting of the Institute of Navigation*, 2011.

[29] Veth, M. J., "Fusion of Imaging and Inertial Sensors for Navigation," Ph.D. dissertation. Air Force Institute of Technology (AU), Wright-Patterson AFB OH, ADA468382, September 2006.

[30] Chiu, H. -P., et al., "Precise Vision-Aided Aerial Navigation," *IEEE/RSJ International Conference on Intelligent Robots and Systems (IROS)*, 2014.

[31] Chiu, H. -P., et al., "Constrained Optimal Selection for Multi-Sensor Robot Navigation Using Plug-and-Play Factor Graphs," *IEEE International Conference on Robotics and Automation (ICRA)*, 2014.

[32] Chiu, H. -P., et al., "Robust Vision-Aided Navigation Using Sliding-Window Factor Graphs," *IEEE International Conference on Robotics and Automation (ICRA)*, 2013.

[33] Trinh, N., et al., "A General Approach to Online Sensor Calibration for Navigation Using Factor Graphs," *ION International Technical Meeting*, 2013.

[34] Kumar, R., et al., "Plug-and-Play Navigation Algorithms Using Factor Graphs," *Joint Navigation Conference (JNC)*, 2012.

[35] Kaess, M., et al., "iSAM2: Incremental Smoothing and Mapping Using the Bayes Tree," *Intl. J. of Robotics Research*, Vol. 31, February 2012, pp. 217–236.

[36] Lindsten, F., et al., "Geo-Referencing for UAV Navigation Using Environment Classification," *Proc. IEEE Intl. Conf. on Robotics and Automation (ICRA)*, 2010.

[37] Patterson, T., et al., "Utilizing Geographic Information System Data for Unmanned Aerial Vehicle Position Estimation," *Proc. of Canadian Conf. on Computer and Robot Vision*, 2011.

[38] Sim, D., et al., "Integrated Position Estimation Using Aerial Image Sequences," *IEEE Transactions on Pattern Analysis and Machine Intelligence*, Vol. 24, No. 1, 2002.

[39] Conte, G., and P. Doherty, "Vision-Based Unmanned Aerial Vehicle Navigation Using Geo-Referenced Information," *EURASIP Journal of Advances in Signal Processing*, 2009.

[40] Kschischang, F., B. Fey, and H. Loeliger, "Factor Graphs and the Sum Product Algorithm," *IEEE Transactions on Information Theory*, Vol. 47, No. 2, February 2001.

[41] Mourikis, A., et al., "Vision-Aided Inertial Navigation for Spacecraft Entry, Descent, and Landing," *IEEE Transactions on Robotics*, Vol. 25, No. 2, April 2009.

[42] Chiu, H., et al., "Robust Vision-Aided Navigation Using Sliding-Window Factor Graphs," *Proc. IEEE Intl. Conf. on Robotics and Automation (ICRA)*, 2013.

[43] Kumar, R., et al., "Registration of Video to Geo-Referenced Imagery," *Proc. IEEE Intl. Conf. on Pattern Recognition (ICPR)*, 1998.

[44] Mourikis, A., and S. Roumeliotis, "A Multi-State Constraint Kalman Filter for Vision-Aided Inertial Navigation," *Proc. IEEE Intl. Conf. on Robotics and Automation (ICRA)*, 2007.

[45] Sinko, J., "RTK Performance in Highway and Racetrack Experiments," *Navigation*, Vol. 50, No. 4, 2003, pp. 265–275.

[46] Hollowell, J., "Heli/SITAN: A Terrain Referenced Navigation Algorithm for Helicopters," *IEEE Symposium on Position Location and Navigation. A Decade of Excellence in the Navigation Sciences*, March 1990.

[47] McGuffin, J. T., "Terrain Referenced Navigation Electromagnetic-Gravitational Correlation," U.S. Patent 5272639A, 1992 (2019-11-14 Application status is Expired—Lifetime).

[48] Fellerhoff, J. R., "SITAN Implementation in the Saint System," *IEEE Journal*, 1986, pp. 89–95.

[49] Baird, C. A., and M. R. Abramson, "A Comparison of Several Digital Map-Aided Navigation Techniques," *IEEE Journal*, 1984, pp. 286–293.

[50] Hostetler, L. D., and R. D. Andreas, "Nonlinear Kalman Filtering Techniques for Terrain-Aided Navigation," *IEEE Transactions on Automatic Control*, Vol. AC-28, No. 3, March 1983, pp. 315–323.

[51] Priestley, N., "Terrain Referenced Navigation," *IEEE Symposium on Position Location and Navigation, A Decade of Excellence in the Navigation Sciences*, March 20, 1990.

[52] Henley, A. J., "Terrain Aided Navigation: Current Status, Techniques for Flat Terrain and Reference Data Requirements," *IEEE Symposium on Position Location and Navigation. A Decade of Excellence in the Navigation Sciences*, March 20, 1990.

[53] Alving, B., et al., "Terrain Referenced Navigation of AUVs and Submarines Using Multi-Beam Echo Sounders," Norwegian Defense Research, 2004, http://citeseerx.ist.psu.edu/viewdoc/download?doi=10.1.1.459.3521&rep=rep1&type=pdf.

[54] Metzger, J., "Sigma-Point Filter for Terrain Referenced Navigation," *AIAA Guidance, Navigation, and Control Conference and Exhibit*, San Francisco, CA, August 15–18, 2005.

[55] Richeson, J. A., "Gravity Gradiometer Aided Inertial Navigation Within Non-GNSS Environments," Ph.D. Dissertation, University of Maryland, College Park, 2008.

[56] Rogers, M. M., "An Investigation into the Feasibility of Using a Modern Gravity Gradient Instrument for Passive Aircraft Navigation and Terrain Avoidance," Master's Thesis, Air Force Institute of Technology, 2009.

[57] Wu, X., "Gravity Gradient Survey with a Mobile Atom Interferometer," Ph.D. Dissertation, Stanford University, 2009.

[58] Metzger, E. H., and A. Jircitano, "Inertial Navigation Performance Improvement Using Gravity Gradient Matching Techniques," *J. Spacecraft*, June 1976.

[59] Meryl, J., H. Rice, and S. Shinners, "The Universal Gravity Module for Enhanced Submarine Navigation," *IEEE 1998 Position Location and Navigation Symposium*, Palm Springs, CA, 1998.

[60] Wu, L., et al., "Performance Evaluation and Analysis for Gravity Matching Aided Navigation," *Sensors*, April 2017.

[61] Wang, H., et al., "Location Accuracy of INS/Gravity-Integrated Navigation System on the Basis of Ocean Experiment and Simulation," *Sensors*, December 2017.

[62] Wang, B., et al., "A Particle Filter-Based Matching Algorithm with Gravity Sample Vector for Underwater Gravity Aided Navigation," *IEEE/ASME Transactions on Mechatronics*, Vol. 21, No. 3, June 2016.

[63] Morales, G., "Magnetometer Aided Navigation Filters for Improved Observability and Estimation on Ground Vehicles," Master's Thesis, Auburn University, 2016.

[64] Han, K., et al., "UAV Vision: Feature Based Accurate Ground Target Localization Through Propagated Initializations and Interframe Hymnographies," *Proc. IEEE Intl Conf. on Robotics and Automation (ICRA)*, 2012.

[65] Kumar, R., et al., "Registration of Highly-Oblique and Zoomed in Aerial Video to Reference Imagery," *Proc. IEEE Intl Conf. on Pattern Recognition (ICPR)*, 2000.

[66] Kumar, R., et al., "Aerial Video Surveillance and Exploitation," *Proc. of the IEEE*, Vol. 89, No. 10, October 2001.

[67] Wiles, R., et al., "Video Geo-Registration: Algorithms and Quantitative Evaluation," *Proc. IEEE Intl Conf. on Computer Vision (ICCV)*, 2001

[68] Savage, P. G., "Strapdown Inertial Navigation Integration Algorithm Design Part 1: Attitude Algorithms," *Journal of Guidance, Control, and Dynamics*, Vol. 21, No. 1, 1998, pp. 19–28.

[69] Savage, P. G., "Strapdown Inertial Navigation Integration Algorithm Design Part 2: Velocity and Position Algorithms," *Journal of Guidance, Control, and Dynamics*, Vol. 21, No. 2, March-April 1998.

[70] Titterton, D. and J. Weston, *Strapdown Inertial Navigation Technology*, Second Edition, The American Institute of Aeronautics and Astronautics, 2004.

[71] Roetenberg, D., "Inertial and Magnetic Sensing of Human Motion," Ph.D. thesis, Universiteit Twente, 2006.

[72] Jordan, J. W., "An Accurate Strapdown Direction Cosine Algorithm," NASA TN-D-5384, September 1969.

[73] Bortz, J. E., "A New Mathematical Formulation for Strapdown Inertial Navigation," *IEEE Transactions on Aerospace and Electronic Systems*, Vol. AES-7, No. 1, 1971, pp. 61–66.

[74] Ignagni, M. B., "Optimal Strapdown Attitude Integration Algorithms," *Journal of Guidance, Control, and Dynamics*, Vol. 13, No. 2, 1990, pp. 363–369.

[75] Ignagni, M. B., "Efficient Class of Optimized Coning Compensation Algorithms," *Journal of Guidance, Control, and Dynamics*, Vol. 19, No. 2, 1996, pp. 424–429.

[76] Mark, J. G., and D. A. Tazartes, "On Sculling Algorithms," *3rd St. Petersburg International Conf. on Integrated Navigation Systems*, St. Petersburg, Russia, May 1996.

[77] Miller, R., "A New Strapdown Attitude Algorithm," *Journal of Guidance, Control, and Dynamics*, Vol. 6, No. 4, 1983, pp. 287–291.

[78] Wu, M., et al., "Optimization-Based Alignment for Inertial Navigation Systems: Theory and Algorithm," *Aerosp. Sci. Technol.*, Vol. 15, 2011, pp. 1–17.

[79] Chang, L., J. Li, and S. Chen, "Initial Alignment by Attitude Estimation for Strapdown Inertial Navigation Systems," *IEEE Transactions on Instrumentation and Measurement*, Vol. 64, 2015, pp. 784–794.

[80] Silson, P. M. G., "Coarse Alignment of a Ship's Strapdown Inertial Attitude Reference System Using Velocity Loci," *IEEE Transactions on Instrumentation and Measurement*, Vol. 60, 2011, pp. 1930–1941.

[81] Wu, Y., and X. Pan, "Velocity/Position Integration Formula Part I: Application to In-Flight Coarse Alignment," *IEEE Transactions on Aerospace and Electronic Systems*, Vol. 49, 2012, pp. 1006–1023.

[82] Fang, J., and S. Yang, "Study on Innovation Adaptive EKF for In-Flight Alignment of Airborne POS," *IEEE Transactions on Instrumentation and Measurement*, Vol. 60, 2011, pp. 1378–1388.

[83] El-Sheimy, N., S. Nassar, and A. Noureldin, "Wavelet De-Noising for IMU Alignment," *IEEE Aerospace and Electronic Systems Magazine*, Vol. 19, 2004, pp. 32–39.

[84] Xu, X., et al., "A Kalman Filter for SINS Self-Alignment Based on Vector Observation," *Sensors*, Vol. 17, 2017, p. 264.

[85] Liang, Y., and Y. Jia, "A Nonlinear Quaternion-Based Fault-Tolerant SINS/GNSS Integrated Navigation Method for Autonomous UAVs," *Aerosp. Sci. Technol.*, Vol. 40, 2015, pp. 191–199.

[86] Chang, L., J. Li, and K. Li, "Optimization-Based Alignment for Strapdown Inertial Navigation System: Comparison and Extension," *IEEE Transactions on Aerospace and Electronic Systems*, Vol. 52, 2014, pp. 1697–1713.

[87] Gelb, M., *Applied Optimal Estimation*, Cambridge, MA: MIT Press, 1974.

[88] Brown, R. G., and P. Hwang, *Introduction to Random Signals and Applied Kalman Filtering*, New York: John Wiley & Sons, 1996.

[89] Li, R., and A. Wu, "Kalman Filtering & Applications," *UCLA Lecture Notes*, 2005–2019.

[90] Julier S. J., J. K. Uhlmann, and H. Durrant-Whyte, "A New Approach for Filtering Nonlinear System," *Proceedings of the American Control Conference*, 1995, pp. 1628–1632.

[91] Julier, S. J., and J. K. Uhlmann, *General Method for Approximating Nonlinear Transformation of Probability Distributions*, Technical Report RRG Department of Engineering Science, University of Oxford, November 1996.

CHAPTER 2

Foundations

2.1 Introduction

In this chapter, we will review some of the key definitions, concepts, symbols, equations, and algorithms associated with the coordinate frames and coordinate transformations. We will also review Earth models and gravity and magnetic field models and summarize the most important tool to support All Source PNT: the Kalman filter, its architecture, algorithms, and variations, as well as how we should apply Kalman filtering in PNT applications.

There are many books and articles available that cover some aspects of this chapter [1–8]. Therefore, the purposes of this chapter are not to derive or elaborate these facts in detail. Instead, we will summarize the most useful facts in one place, define the symbols that we will be using in the rest of the book, and present certain topics in an unique way in the hope of bringing additional benefit to the readers.

2.2 Coordinate Frames

We will be using many different coordinate frames; among them, the following are used constantly:

- Body frame (B): Of an entity such as vehicle, a platform, or a sensor.
- Local-level navigation frame (N): A coordinate frame that translationally moves with the body frame but remains locally level (the z-axis is aligned with the local vertical). The azimuth of this frame can be constrained in many different ways. The most popular is the wander-azimuth frame, which remains defined at the North and South Poles of the Earth. In addition, we will also use two other frequently used local level frames: NED (North-East-down) and ENU (East-North-up). Both NED and ENU are right-handed, local-level frames. For NED, the z-axis points down along the local vertical, the x-axis points North, and the y-axis points East. For ENU, the z-axis points up, the x-axis points East, and the y-axis points North.
- Earth frame (E): We typically use an Earth-centered, Earth-fixed (ECEF) frame, which is defined as:
  - Origin: Earth's center of mass (typically per WGS84 ellipsoid);
  - z-axis: The direction of the International Earth Rotation and Reference

Systems Service (IERS) Reference Pole (IRP) (rotational axis of the WGS 84 ellipsoid);
  - x-axis: Intersection of the IERS Reference Meridian (IRM) and the plane passing through the origin and normal to the z-axis;
  - y-axis: Completes a right-handed, ECEF orthogonal coordinate system.
- Inertial frame (I): The inertial frame in Newtonian sense. The accelerometers and the gyros measure the delta velocity/specific force and delta angle/rotation rate of the sensor body frame with respect to the inertial frame. For the inertial frame, most of the time we do not care about the specific frame definition other than the fact the frame itself is inertial. However, the definition of the inertial frame becomes important if we deal with heavenly bodies such as stars, planets, and satellites.

A stable Earth-centered inertial (ECI) coordinate system of convenience may be defined as being coincident with the ECEF coordinate system at a given time. As will be discussed later, the Earth axis of rotation is actually not inertially fixed. As the result, we want to introduce the concept of fixed epoch ECI frame. Without fixing the epoch, the instantaneous ECI frame is actually moving. In order to transform ECI frame to ECEF frame, we must precisely model the Earth nutation and precession motions. This will be further discussed in Section 2.4.

2.3 Mathematical Representation of Attitude and Coordinate Transformations

2.3.1 Three Most Frequently Used Attitude Representations

Used Attitude RepresentationsAttitude represents the pose of an object as well as the coordinate transformation from one frame to another frame. There are many different types of representations of attitude, but we typically will have to use three most important representations almost all the time: the Euler angles, the directional cosine matrix (DCM), and the quaternion. Table 2.1 summarizes the key features of the three different types of attitude representations.

2.3.2 Inner and Cross-Product-Related Symbols

In the following, we will introduce a number of frequently used symbols.

Given 3×1 vectors, $\vec{e}$ and $\vec{f}$, the inner product of the two vectors can be written as

$$\vec{e} \cdot \vec{f} = \left\langle \vec{e}, \vec{f} \right\rangle = \left(\vec{e}, \vec{f} \right) = \vec{e}^T \vec{f}$$

and it is well-known that

$$\vec{e} \cdot \vec{f} = \vec{f} \cdot \vec{e}$$

The cross-product of two vectors can be written as

$$\vec{e} \times \vec{f} = [\vec{e} \times] \vec{f} \tag{2.1}$$

Table 2.1 Comparison of Three Types of Attitude Representations

	Euler Angles	*DCM*	*Quaternion*
Number of Parameters	3	3 × 3	4
Essence	Sequence of three single-axis rotations	Representation of unit vectors of the frame axes in the other frame	Eigen-axis rotation
Singularity	Depending on the rotational sequence, the Euler angles may not be defined for certainty attitude	No singularity	No singularity
Other pros and cons	Minimum number of parameters; must be used to relate to platform gimbal angles; intuitive and visualizable; often convenient for analytical expressions; singularity; sequence is often a source of confusion and mistakes	Convenient and almost always the choice for performing coordinate transformations; most number of parameters	Good compromise between singularity and number of parameters; no trig functions for transformations; not as convenient as DCM for vector transformations

where we have introduced the frequently used symbol

$$[\vec{e}\,\times] = \begin{bmatrix} 0 & -e_3 & e_2 \\ e_3 & 0 & -e_1 \\ -e_2 & e_1 & 0 \end{bmatrix} \tag{2.2}$$

and it is well known that

$$\vec{f} \times \vec{e} = -\vec{e} \times \vec{f} \tag{2.3}$$

2.3.3 DCM

Given the unit vectors along the three axes of the frame A

$$\vec{x}_A,\ \vec{y}_A,\ \vec{z}_A$$

Their coordinates in frame A are simply

$$^{A}\vec{x}_A = \begin{bmatrix} 1 \\ 0 \\ 0 \end{bmatrix},\quad ^{A}\vec{y}_A = \begin{bmatrix} 0 \\ 1 \\ 0 \end{bmatrix},\quad ^{A}\vec{z}_A = \begin{bmatrix} 0 \\ 0 \\ 1 \end{bmatrix}$$

Let the coordinates of $\vec{x}_A$, $\vec{y}_A$, $\vec{z}_A$ in frame B be given by

$$^{B}\vec{x}_A,\ ^{B}\vec{y}_A,\ ^{B}\vec{z}_A$$

then the DCM $C_A^B = C_{B\_A}$ is defined to be

$$C_A^B = C_{B\_A} = \left[\,^{B}\vec{x}_A,\ ^{B}\vec{y}_A,\ ^{B}\vec{z}_A\right] \tag{2.4}$$

Let the vector v and its coordinate vectors in frames A and B be given by ${}^{A}\vec{v}$, ${}^{B}\vec{v}$; then

$${}^{B}\vec{v} = C_A^B\,{}^{A}\vec{v} \tag{2.5}$$

Note that, for the above equations to be true, the frames do not have to be orthogonal or right-handed and the DCM does not need to be orthonormal in general.

The DCM has the following frequently used properties:

1. The columns of DCM are the representation of the unit vectors of the source frame in the destination frame

$$\begin{aligned} {}^{B}\vec{x}_A &= C_A^B(:,1) \\ {}^{B}\vec{y}_A &= C_A^B(:,2) \\ {}^{B}\vec{z}_A &= C_A^B(:,3) \end{aligned} \tag{2.6}$$

2. The DCM is a matrix consists of a matrix of inner products or the directional cosines. This is true for nonorthogonal frames as well

$$\cos\alpha_{ij} = \left\langle u_{Bi}, v_{Aj} \right\rangle = C_A^B(i,j) \tag{2.7}$$

where u_{Bi} is the unit vector of frame B's ith axis, u_{Aj} is the unit vector of frame A's jth axis, and α_{ij} is the angle between u_{Bi} and u_{Aj}.

3. Multiple transformation chain rule

$$C_A^C = C_B^C C_A^B \tag{2.8}$$

4. For orthogonal frames only, we have

$$\left(C_A^B\right)^{-1} = \left(C_A^B\right)^T \tag{2.9}$$

5. For two orthogonal frames A and B, there exists a vector v such that

$$v = C_A^B v \tag{2.10}$$

This is the mathematical version of Euler rotation theorem. Mathematically, this equation means that the orthonormal matrix has at least 1 eigenvalue that equals to 1. Geometrically, it means that a coordinate transformation can be represented by a single-axis rotation. The axis is the eigen-axis, which remains the same before and after the rotation.

2.3.4 Euler Angles

Euler angles represent attitude by a sequence of three rotations that rotate one frame to another. There are three basic rotations: around the x-axis, around the y-axis, and around the z-axis.

Each one of the Euler angle rotations can be represented by one of the following three DCMs:

$$C_{\text{OLD}}^{\text{New}} = R_x(\varphi) = \begin{bmatrix} 1 & 0 & 0 \\ 0 & \cos\varphi & \sin\varphi \\ 0 & -\sin\varphi & \cos\varphi \end{bmatrix} \tag{2.11}$$

$$C_{\text{Old}}^{\text{New}} = R_y(\theta) = \begin{bmatrix} \cos\theta & 0 & -\sin\theta \\ 0 & 1 & 0 \\ \sin\theta & 0 & \cos\theta \end{bmatrix} \tag{2.12}$$

$$C_{\text{Old}}^{\text{New}} = R_z(\psi) = \begin{bmatrix} \cos\psi & \sin\psi & 0 \\ -\sin\psi & \cos\psi & 0 \\ 0 & 0 & 1 \end{bmatrix} \tag{2.13}$$

There are two types of Euler angles in terms of sequence of rotations:

- *Type 1:* Rotations around each axis sequentially. Singularity happens when the second rotation is +/–90°. There are 6 different sequences: 123, 132, 231, 213, 312, and 321.
- *Type 2:* The first and the third rotations are around the same axis. Singularity happens at 0° and 180° for the second rotation. There are six different sequences: 121, 131, 212, 232, 313, and 323.

Examples

$$R('321',\psi,\theta,\varphi) = R_x(\varphi)R_y(\theta)R_z(\psi)$$
$$R('313',\psi,\theta,\varphi) = R_z(\varphi)R_x(\theta)R_z(\psi)$$

The singularity associated with the Euler angle is due to the fact that, for certainty attitude, there is more than one sequence of Euler angles to achieve the same attitude.

For example, the following two sequences achieve the same results:

- 0° yaw, 90° pitch, 45° roll;
- 45° yaw, 90° pitch, 0° roll

With pitch = 90°, one cannot distinguish roll and yaw. Figure 2.1 provides an illustration.

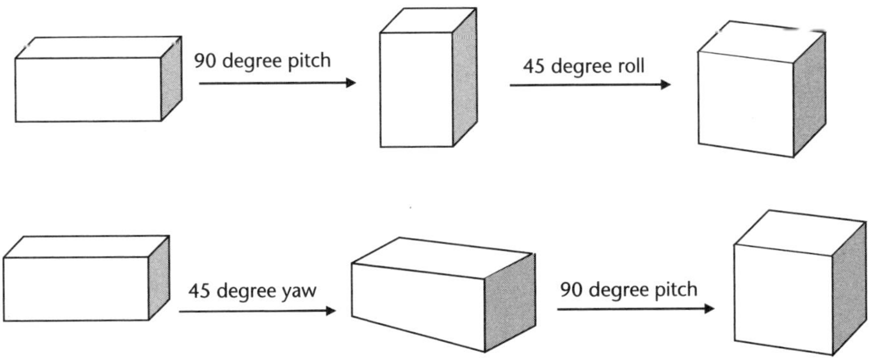

Figure 2.1 Two different Euler angle sequences achieving the same end result.

2.3.5 Rotation Around an Arbitrary Axis

Given an axis of rotation as illustrated in Figure 2.2, frames A and B, a unit vector representing the axis of rotation

$$\vec{e} = \begin{bmatrix} e_1 \\ e_2 \\ e_3 \end{bmatrix}$$

and the angle of the rotation, θ_A^B

The DCM from A to B can then be written as

$$C_A^B = R_{\vec{e}}\left(\theta_A^B\right) = \left(\cos\theta_A^B\right)I + \left(1 - \cos\theta_A^B\right)\vec{e}\vec{e}^T - \sin\theta_A^B\left[\vec{e}\,\times\right] \tag{2.14}$$

The following is an example to show the trivial case of a single axis rotation around x-axis.

$$e = \begin{bmatrix} 1 \\ 0 \\ 0 \end{bmatrix},\ \vec{e}\vec{e}^T = \begin{bmatrix} 1 & 0 & 0 \\ 0 & 0 & 0 \\ 0 & 0 & 0 \end{bmatrix},\ \left[\vec{e}\,\times\right] = \begin{bmatrix} 0 & 0 & 0 \\ 0 & 0 & -1 \\ 0 & 1 & 0 \end{bmatrix},$$

$$C_A^B = \left(\cos\theta_A^B\right)\begin{bmatrix} 1 & 0 & 0 \\ 0 & 1 & 0 \\ 0 & 0 & 1 \end{bmatrix} + \left(1 - \cos\theta_A^B\right)\begin{bmatrix} 1 & 0 & 0 \\ 0 & 0 & 0 \\ 0 & 0 & 0 \end{bmatrix} - \sin\theta_A^B\begin{bmatrix} 0 & 0 & 0 \\ 0 & 0 & -1 \\ 0 & 1 & 0 \end{bmatrix}$$

$$= \begin{bmatrix} 1 & 0 & 0 \\ 0 & \cos\theta_A^B & \sin\theta_A^B \\ 0 & -\sin\theta_A^B & \cos\theta_A^B \end{bmatrix}$$

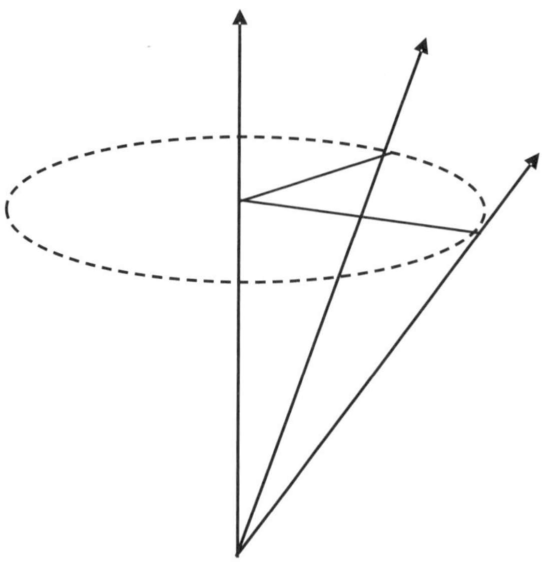

Figure 2.2 Euler rotation.

As a useful identity, we have

$$[\vec{e}\times][\vec{f}\times] = -(\vec{e}\bullet\vec{f})I + \vec{f}\vec{e}^T \tag{2.15}$$

Using the above, it can be shown that

$$\begin{aligned} C_A^B &= R_{\vec{e}}(\theta_A^B) = (\cos\theta_A^B)I + (1-\cos\theta_A^B)\vec{e}\vec{e}^T - \sin\theta_A^B[\vec{e}\times] \\ &= I - \sin\theta_A^B[\vec{e}\times] + (1-\cos\theta_A^B)[\vec{e}\times]^2 \end{aligned} \tag{2.16}$$

An extremely useful equation is the small angle approximation of the above equations, which, as will be found later, are frequently used when modeling the errors in navigation equations.

In the equations above, for small angles, we can use

$$\cos\theta_A^B \approx 1, \ \sin\theta_A^B \approx \theta_A^B$$

and we have

$$C_A^B = R_{\vec{e}}(\theta_A^B) = I - \theta_A^B[\vec{e}\times] = I - [\vec{\theta}_A^B\times] \tag{2.17}$$

It is important to realize that the rotation sequence is no longer an issue for small angles.

As an interesting application of small angle rotation formulae, we can show that the DCM can be written as a matrix exponential:

$$C_A^B = R_{\vec{e}}(\theta_A^B) = R_{\vec{e}}\left(\frac{\theta_A^B}{N}\right)R_{\vec{e}}\left(\frac{\theta_A^B}{N}\right)\ldots R_{\vec{e}}\left(\frac{\theta_A^B}{N}\right) = \left(R_{\vec{e}}\left(\frac{\theta_A^B}{N}\right)\right)^N = \left(I - \frac{1}{N}[\vec{\theta}_A^B\times]\right)^N$$

Let $N \rightarrow \infty$, we have

$$\begin{aligned} C_A^B &= R_{\vec{e}}(\theta_A^B) = \underset{N\rightarrow\infty}{\text{Limit}}\left(I - \frac{1}{N}[\vec{\theta}_A^B\times]\right)^N = \exp(-[\vec{\theta}_A^B\times]) \\ &= I - \sin\theta_A^B[\vec{e}\times] + (1-\cos\theta_A^B)[\vec{e}\times]^2 \end{aligned} \tag{2.18}$$

Euler's eigen-axis rotation theorem essentially states that the most general attitude change of a rigid body is a rotation about some axis. In other words, every DCM (between orthogonal frames) can be expressed as

$$C_A^B = R_{\vec{e}}(\theta_A^B) = (\cos\theta_A^B)I + (1-\cos\theta_A^B)\vec{e}\vec{e}^T - \sin\theta_A^B[\vec{e}\times]$$

2.3.6 Quaternions

As discussed before, quaternions have the advantages of only using four parameters to describe attitude without having singularities. Quaternion is defined based on the Euler eigen-axis rotation theorem.

$$q \equiv \begin{bmatrix} \vec{q} \\ q_4 \end{bmatrix}, \ \vec{q} = \sin\frac{\theta}{2}\vec{e}, \ q_4 = \cos\frac{\theta}{2} \tag{2.19}$$

where $\vec{e}$ is the unit vector of the axis of the rotation and θ is the angle of the rotation.

The following are some immediate facts related to quaternion:

1. The product of quaternion is defined to be

$$p \otimes q = \begin{bmatrix} q_4\vec{p} + p_4\vec{q} - \vec{p} \times \vec{q} \\ p_4 q_4 - \vec{p} \bullet \vec{q} \end{bmatrix} \neq q \otimes p \tag{2.20}$$

2. Quaternion is defined to be normalized

$$q^T q = 1 \tag{2.21}$$

3. Unit quaternion

$$u_q = \begin{bmatrix} 0 \\ 0 \\ 0 \\ 1 \end{bmatrix}, \ q \otimes u_q = q, \ u_q \otimes q = q \tag{2.22}$$

4. Inverse quaternion

$$q^{-1} = \begin{bmatrix} -\vec{q} \\ q_4 \end{bmatrix}, \ q \otimes q^{-1} = u_q, \ q^{-1} \otimes q = u_q \tag{2.23}$$

For each rotation, (at least) two quaternions can be defined. That is why quaternion convention has to be defined when a quaternion is used to represent attitude. We introduce the following two symbols: both are quaternions that represent the coordinate transformation (as well as a rotation) from frame A to frame B.

$$q_{B/A} \equiv \begin{bmatrix} \vec{q}_{B/A} \\ q_4 \end{bmatrix}, \ \vec{q}_{B/A} = \sin\frac{\theta_A^B}{2}\vec{e}, \ q_4 = \cos\frac{\theta_A^B}{2} \tag{2.24}$$

$$q_A^B \equiv \begin{bmatrix} \vec{q}_A^B \\ q_4 \end{bmatrix}, \ \vec{q}_A^B = -\sin\frac{\theta_A^B}{2}\vec{e}, \ q_4 = \cos\frac{\theta_A^B}{2} \tag{2.25}$$

It is not hard to realize that they above two quaternions are inverse of one another

$$\left(q_{B/A}\right)^{-1} = q_A^B$$

The most important difference between the two conventions is the equation for the chain rule (i.e., how transformations can be combined into one):

$$q_{B/A} \otimes q_{C/B} = q_{C/A} \tag{2.26}$$

$$q_B^C \otimes q_A^B = q_A^C \tag{2.27}$$

One can observe that, for the first convention, the chain rule is the opposite of that of DCM, and for the second convention, the chain rule is the same as that of DCM. For this reason, we will be using the second convention most of the time, but it is important to realize that the industry, the educational institutions, and the government agencies are all over the place in terms of which convention to use and it is important to know both.

As mentioned before, quaternion is defined to be normalized:

$$q^T q = 1$$

If for any reason (such as numerical errors), the above equation is not satisfied, it can be normalized using the following method:

$$q \Leftarrow \frac{q}{\sqrt{q^T q}} \tag{2.28}$$

Although not necessary, it is often convenient to keep the quaternion canonical, that is,

$$q_4 \geq 0$$

A quaternion can be canonicalized using the following algorithm:

$$\begin{aligned} &\text{if } (q_4 < 0)\{ \\ &q \Leftarrow -q \\ &\} \end{aligned} \tag{2.29}$$

The quaternions defined above are related to the DCM by the following equations:

$$\begin{aligned} C_A^B &= \left(q_4^2 - \left\|\vec{q}_{B/A}\right\|^2\right)I + 2\vec{q}_{B/A}\left(\vec{q}_{B/A}\right)^T - 2q_4\left[\vec{q}_{B/A}\times\right] \\ C_A^B &= \left(q_4^2 - \left\|\vec{q}_A^B\right\|^2\right)I + 2\vec{q}_A^B\left(\vec{q}_A^B\right)^T + 2q_4\left[\vec{q}_A^B\times\right] \end{aligned} \tag{2.30}$$

which can be easily derived from

$$C_A^B = R_{\vec{e}}\left(\theta_A^B\right) = \left(\cos\theta_A^B\right)I + \left(1 - \cos\theta_A^B\right)\vec{e}\vec{e}^T - \sin\theta_A^B[\vec{e}\times]$$

by observing that

$$\begin{aligned} \sin\theta_A^B &= 2\sin\left(\frac{\theta_A^B}{2}\right)\cos\left(\frac{\theta_A^B}{2}\right) \\ \cos\theta_A^B &= \cos^2\left(\frac{\theta_A^B}{2}\right) - \sin^2\left(\frac{\theta_A^B}{2}\right) = 1 - 2\sin^2\left(\frac{\theta_A^B}{2}\right) \end{aligned}$$

2.4 ECI Frame Revisited

The ECI frame uses the Earth axis of rotation as its z-axis. The Earth axis of rotation is not fixed; it has both precession and nutation motions. As a result, we use two types of the true inertial frames:

1. The fixed epoch ECI frame. The ECI frame at a particular time moment. For example, J2000 ECI is the most frequently used ECI frame.
2. The true of date/mean of date ECI. The average ECI frame for a particular day.

2.4.1 Precession of Earth Axis of Rotation

Neither the plane of the Earth's orbit, the ecliptic, or the plane of the Earth's equator are fixed with respect to distant objects. The dominant motion is the precession of the Earth's polar axis around the ecliptic pole, mainly due to torques on the earth

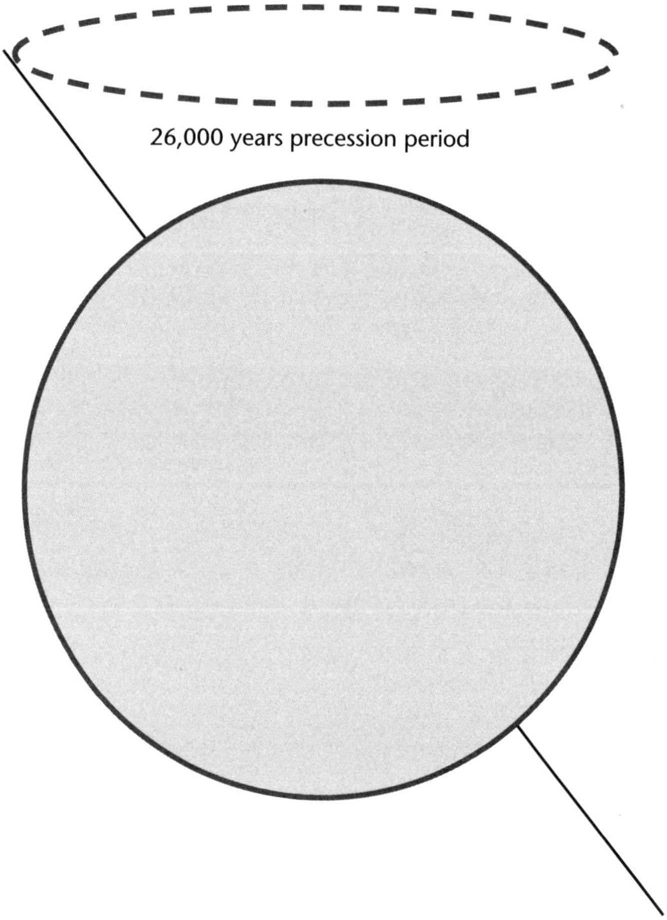

Figure 2.3 Precession of Earth axis of rotation.

caused by the Moon and the Sun. The earth's axis sweeps out a cone of 23.5° half-angle in 26,000 years, about 20 arcsec/year. Figure 2.3 illustrates the precession of the Earth axis of rotation.

The approximate correction of star position measured in ECI frame can be given by

$$\begin{aligned} RA &= RA(2000) + \left(3.075 + 1.336\sin(RA)\tan(Dec)\right)y \\ Dec &= Dec(2000) + 20.04\cos(RA)y \end{aligned} \tag{2.31}$$

where y is the time from January 1, 2000, in fractional years, and the offsets in RA and Dec are in arcseconds.

2.4.2 Nutation of Earth Rotational Axis

In the standard nutation theory, there are 106 nonharmonically related sine and cosine components, mainly due to second-order torque effects from the Sun and the Moon. In addition, there are 85 planetary correction terms. Among these terms, there are four dominant periods of nutation:

- 18.6 years (precession period of the lunar orbit);
- 182.6 days (half a year);
- 13.7 days (half a month);
- 9.3 years (rotation period of the Moon's perigee).

The dominant term has a magnitude around 18 arcsec.

The following approximation for nutation is good to about an arcsecond:

$$\begin{aligned} delta\ RA &= \left(0.9175 + 0.3978\ \sin(RA)\tan(Dec)\right)dL - \cos(RA)\tan(Dec)dE \\ delta\ Dec &= 0.3978\cos(RA)dL + \sin(RA)dE \end{aligned} \tag{2.32}$$

where *delta RA* and *delta Dec* are added to mean coordinates to get apparent coordinates. The nutation in longitude, dL, and the nutation in obliquity of the ecliptic, dE, are computed from the two largest terms in the general theory with

$$\begin{aligned} dL &= -17.3\ \sin(125.0 - 0.05295d) - 1.4\sin(200.0 + 1.97129d) \\ dE &= 9.4\ \cos(125.0 - 0.05295d) + 0.7\cos(200.0 + 1.97129d) \end{aligned} \tag{2.33}$$

where d = Julian Date – 2,451,545.0, the sine and cosine arguments are in degrees, and dL and dE are in arcseconds.

2.4.3 Standard J2000 ECI to True of Date ECI Transformation

The algorithm is described as

1. Compute the Terrestrail Dynamic Time (TDT) relative to GPS time:

$$TDT = t + 51.184 \text{ (time in second)} \tag{2.34}$$

2. Compute the Julian ephemeris date (JED) in days and in centuries (100 years):

$$\begin{aligned} JED &= TDT \text{ expressed in days} \\ g &= \frac{\pi}{180}\left(357.528 + 35999.05\left(\frac{JED - 2451545}{36525}\right)\right) \\ JBD &= JED + \frac{0.001658\sin(g + 0.0167\sin g)}{86400} \\ d &= (JBD - 2451545) \text{ (days)}, \quad T = \frac{d}{36525} \text{ (100 Years)} \end{aligned} \tag{2.35}$$

3. Compute the precession-related variables:

$$\begin{aligned} \zeta &= \frac{\pi}{(3600)(180)}\left(2306.2181T + 0.30188T^2 + 0.017998T^3\right) \\ z &= \frac{\pi}{(3600)(180)}\left(2306.2181T + 1.09468T^2 + 0.018203T^3\right) \\ \theta &= \frac{\pi}{(3600)(180)}\left(2004.3109T + 0.42665T^2 - 0.041833T^3\right) \end{aligned} \tag{2.36}$$

The dominant term is on the order of 2,000 arcsec per 100 years or 20 arsec per year.

4. The precession rotation matrix is then calculated as

$$D = R_z\left(-\frac{\pi}{2} - z\right)R_y(\theta)R_z\left(\frac{\pi}{2} + \zeta\right) \tag{2.37}$$

5. Compute the mean obliquity for the date, the nutation of the obliquity, and the longitudinal nuation. In the standard theory, there are 106 different frequencies. The equations below capture the two dominant frequencies:

$$\bar{\varepsilon} = \frac{\pi}{180}\left(23.439291111 - 0.013004167T - 1.63889 \times 10^{-7}T^2 + 5.03611 \times 10^{-7}T^3\right)$$

The nutation of obliquity is approximated by

$$\Delta\varepsilon = \frac{\pi}{180}\left(0.0026\cos\left(\frac{\pi}{180}(125 - 0.05295d)\right) + 0.0002\cos\left(\frac{\pi}{180}(200.9 + 1.97129d)\right)\right)$$

The longitudinal nutation is approximated by

$$\Delta\psi = \frac{\pi}{180}\left(-0.0048\sin\left(\frac{\pi}{180}(125 - 0.05295d)\right) - 0.0004\sin\left(\frac{\pi}{180}(200.9 + 1.97129d)\right)\right) \tag{2.38}$$

The first term has a period of 360/0.05295 = 6,978.866 days = 18.6143 years. The second term has a period of 360/1.97129 = 182.62 days.

6. The nutation rotation matrix is then calculated as

$$C = R_x\left(-(\bar{\varepsilon} + \Delta\varepsilon)\right)R_z(-\Delta\psi)R_x(\bar{\varepsilon}) \tag{2.39}$$

7. Finally, the transformation from the J2000 ECI to the true-of-date ECI is given by

$$C_{\mathrm{ECI}_{\mathrm{J2000}}}^{\mathrm{ECI}_{\mathrm{tod}}} = (C)(D) \tag{2.40}$$

where D and C are the precession and nutation matrices calculated earlier.

2.5 Earth Model, Earth Gravity Field Model, and Earth Magnetic Field Model

2.5.1 Earth Model

The World Geodetic System WGS84 comprises of a reference ellipsoid, a standard coordinate system, altitude data, and a geoid [8–27].

2.5.1.1 Basic Parameters of WGS84

The coordinate origin of WGS84 is intended to be located at the Earth's center of mass; the uncertainty is believed to be less than 2 cm.

The WGS84 meridian of zero longitude is the IERS Reference Meridian 5.3 arcseconds or 102m (335 ft) East of the Greenwich meridian at the latitude of the Royal Observatory.

The WGS84 datum surface is an oblate spheroid with equatorial radius a = 6,378,137m at the equator and flattening f = 1/298.257223563.

The refined value of the WGS84 gravitational constant (the mass of the Earth's atmosphere included) is $GM = 3{,}986{,}004.418 \times 10^8$ m^3/s^2.

The angular velocity of the Earth is defined to be $\omega = 72.92115 \times 10^{-6}$ rad/s.

2.5.1.2 Derived Parameters of WGS84

The polar semi-minor axis b equals $a(1 - f)$ = 6,356,752.3142m, and the first eccentricity squared, $e^2 = 6.69437999014 \times 10^{-3}$.

2.5.2 Earth Gravitational Model and World Magnetic Model

WGS84 uses the Earth Gravitational Model 2008 (EGM2008). WGS84 was using the World Magnetic Model 2015v2. The new version of WMM 2015 became necessary due to extraordinarily large and erratic movements of the north magnetic pole. WGS84 has recently switched to WMM2020, which was released in December 2019.

2.5.3 Conversion of the Geodetic Position to the ECEF Position

The following is the standard equation used to convert geodetic position (LLA: Latitude, Longitude, Altitude) to the ECEF position

$$
{}^{E}r_{E}^{B} = \begin{bmatrix} x \\ y \\ z \end{bmatrix} = \text{geodetic2ecef}(Lat, Lon, h) = \begin{bmatrix} \left(R_N(lat) + h\right)\cos(lat)\cos(lon) \\ \left(R_N(lat) + h\right)\cos(lat)\sin(lon) \\ \left(\left(\dfrac{b}{a}\right)^2 R_N(lat) + h\right)\sin(lat) \end{bmatrix}
$$

where a is the semimajor axis of the earth ellipsoidal model (2.41)

$$
R_N(lat) = \frac{a^2}{\sqrt{a^2\cos^2(lat) + b^2\sin^2(lat)}} = \frac{a}{\sqrt{1 + e^2\sin^2(lat)}}
$$

$$
e^2 = 1 - \left(\frac{b}{a}\right)^2
$$

2.5.4 Conversion of the ECEF to the Geodetic Position

First of all, the longitude can be easily calculated as

$$
lon = \text{atan2}(y, x) \tag{2.42}
$$

which is the same for geodetic and geocentric coordinate frames.

We can then calculate the radius on the x-y plane as

$$
r_h = \sqrt{x^2 + y^2} \tag{2.43}
$$

The geocentric latitude is easy to determine:

$$
lat_c = \text{atan2}\left(r_h, z\right) \tag{2.44}
$$

The geodetic latitude can be calculated through an iteration using the geocentric latitude as the initial value.

The algorithm is therefore described as:

Iterative algorithm

Initialization:

$$
\begin{aligned} k &= 0 \\ lat(k) &= lat_c \end{aligned} \tag{2.45}
$$

Iteration:

$$R_N(k) = R_N\left(lat(k)\right) = \frac{a}{\sqrt{1 + e^2 \sin^2\left(lat(k)\right)}}$$

$$h(k) = \frac{r_h}{\cos\left(lat(k)\right)} - R_N(k) \tag{2.46}$$

$$lat(k+1) = \operatorname{atan}\left(\frac{z}{r_h}\left(1 - e^2 \frac{R_N(k)}{R_N(k) + h(k)}\right)^{-1}\right)$$

Typically, we only have to iterate 4 or 5 times, and this iteration process will give us both the geodetic latitude *lat* and the geodetic altitude *h*.

2.5.5 Exact Equation-Based Algorithm for the ECEF to Geodetic Conversion

There are also closed-form exact equations for the ECEF to geodetic conversion algorithms available. However, it is important to point out that, for all practical purposes, the above iterative algorithm is computationally cheaper and numerically more accurate than most of the exact equation-based algorithms.

The following closed-form equations were published in 2017 to 2019 and represent probably the best results for the exact equation algorithms. The algorithm was developed by Osen [9] and it is an enhanced version of the algorithm developed by Zhu [10, 11]. The algorithm is listed below.

Part 1: Basic inexpensive calculations before feasibility can be determined

$$\begin{aligned}
w^2 &= x^2 + y^2 \\
l &= \frac{e^2}{2} \\
m &= \frac{w^2}{a^2} \\
n &= \frac{z^2(1 - e^2)}{a^2} \\
p &= \frac{\left(m + n - 4l^2\right)}{6} \\
G &= mnl^2 \\
H &= 2p^3 + G
\end{aligned} \tag{2.47}$$

Part 2: Feasibility test

$$\begin{aligned}
&\text{if } \left(H < H_{\min}\right) \text{ then abort (i.e., problem is not numerically stable)} \\
&H_{\min} = 2.25010182030430273673\text{e-}14
\end{aligned} \tag{2.48}$$

Part 3: More expensive calculations that require square and cubic roots

$$
\begin{aligned}
C &= \sqrt[3]{(\frac{H + G + 2\sqrt{HG})}{2}} \\
i &= \frac{-(2l^2 + m + n)}{2} \\
P &= p^2 \\
\beta &= i / 3 - C - P / C \\
k &= l^2\left(l^2 - m - n\right) \\
t &= \sqrt{\sqrt{\beta^2 - k} - \frac{(\beta + i)}{2}} - sign(m - n)\sqrt{\frac{|\beta - i|}{2}} \\
w &= \sqrt{w^2}
\end{aligned}
\tag{2.49}
$$

Part 4: Some further inexpensive calculations

$$
\begin{aligned}
F &= t^4 + 2it^2 + 2l(m - n)t + k \\
\frac{dF}{dt} &= 4t^3 + 4it + 2l(m - n) \\
\Delta t &= \frac{-F}{dF/dt} \\
u &= t + \Delta t + l \\
v &= t + \Delta t - l \\
\Delta w &= w(1 - 1/u) \\
\Delta z &= z\left(1 - (1 - e^2)/v\right)
\end{aligned}
\tag{2.50}
$$

Part 5: Final answer

$$
\begin{aligned}
lat &= \operatorname{atan}(zu, wv) \\
h &= sign(u - 1)\sqrt{(\Delta w)^2 + (\Delta z)^2}
\end{aligned}
\tag{2.51}
$$

2.5.6 Earth Gravity Field Models

Spherical harmonics are used to model the gravity and magnetic field of the Earth (and other planets), and there has been a significant amount of work to develop and maintain accurate models of Earth gravity field and Earth magnetic field [12–27].

Specifically, let V_g be the scalar gravitational potential that is a function of the position in the ECEF frame:

$$
V_g = V_g\left({}^E R_E^B\right) \tag{2.52}
$$

The gravitational acceleration (or specific force) is given by

$$ {}^{E}g = \frac{\partial V_g\left({}^{E}R_E^B\right)}{\partial\, {}^{E}R_E^B} = \nabla V_g\left({}^{E}R_E^B\right) \tag{2.53} $$

which is the gradient of the gravitational potential function.

The gravitational potential function is modeled by spherical harmonics as

$$ \begin{aligned} V_g\left({}^{E}R_E^B\right) &= -\frac{\mu}{r}\left(1+\sum_{n=2}^{\infty}\left(\frac{a}{r}\right)^n \sum_{m=0}^{n} P_{n,m}(\sin\lambda)\left(C_{n,m}\cos m\phi + S_{n,m}\sin m\phi\right)\right) \\ &= -\left(\frac{\mu}{r}+\sum_{n=2}^{\infty}\sum_{m=0}^{n} V_{n,m}\right) \end{aligned} \tag{2.54} $$

where the ECEF position vector is converted to the geocentric (note that we are not using geodetic here) spherical coordinates in terms of distance from the center of the Earth r, geocentric latitude λ, and longitude ϕ, which satisfy the following equation

$$ {}^{E}R_E^B = \begin{bmatrix} r\cos(\lambda)\cos(\phi) \\ r\cos(\lambda)\sin(\phi) \\ r\sin(\lambda) \end{bmatrix} \tag{2.55} $$

a is the semi-major axis of the Earth, which is used to normalize the coefficients to unitless, $P_{n,0}$ are the Legendre polynomials, and $P_{n,m}$ for $1 \le m \le n$ are the associated Legendre functions that can be recursively derived from $P_{n,0}$.

The gravity potential function can be written in the following form to be more consistent with traditional naming conventions and terms, especially the J-terms in the traditional gravity model:

$$ V_g\left({}^{E}R_E^B\right) = -\frac{\mu}{r}\left(1+\sum_{n=2}^{N_z}\bar{J}_n\left(\frac{a}{r}\right)^n P_{n,0}(\sin\lambda) + \sum_{n=2}^{N_t}\sum_{m=1}^{n}\left(\frac{a}{r}\right)^n P_{n,m}(\sin\lambda)\left(C_{n,m}\cos m\phi + S_{n,m}\sin m\phi\right)\right) \tag{2.56} $$

Clearly,

$$ \bar{J}_n = C_{n,0} \tag{2.57} $$

The first summation of terms in the above are independent of the longitude and are called the zonal terms. The second summation of terms is called the tesseral terms. The terms for $n = 1$ are missing because $n = 1$ with $m = 0$ and $m = 1$ terms represent the dipole term in the multiple-pole expansion. Gravity simply does not exhibit any dipole character and that is why these terms are zero.

The Legendre polynomials are defined as

$$ \begin{aligned} P_{0,0}(x) &= 1 \\ P_{n,0}(x) &= \frac{1}{2^n n!}\frac{d^n}{dx^n}\left(x^2-1\right)^n, \quad n \ge 1 \end{aligned} \tag{2.58} $$

The first few of them are

$$
\begin{aligned}
&P_{0,0}(x) = 1, \;\; P_{1,0}(x) = x \\
&P_{2,0}(x) = \frac{1}{2}\left(3x^2 - 1\right), \;\; P_{3,0}(x) = \frac{1}{2}\left(5x^3 - 3x\right) \\
&P_{4,0}(x) = \frac{1}{8}\left(35x^4 - 30x^3 + 3\right) \\
&P_{5,0}(x) = \frac{1}{8}\left(63x^4 - 70x^3 + 15\right)
\end{aligned} \tag{2.59}
$$

The associated Legendre functions can be derived from the Legendre polynomials using the following equation

$$
P_{n,m}(x) = \left(x^2 - 1\right)^{\frac{m}{2}} \frac{d^m P_{n,0}(x)}{dx^m}, \;\; 1 \le m \le n \tag{2.60}
$$

A useful equation is

$$
P_{n,m}(\sin\lambda) = (\cos\lambda)^m \left.\frac{d^m P_{n,0}(x)}{dx^m}\right|_{x=\sin\lambda}, \;\; 1 \le m \le n \tag{2.61}
$$

It is worthwhile to mention that, in some literature, the associated Legendre functions are replaced by normalized Legendre functions, which will lead to different definitions of the coefficients:

$$
\bar{P}_{n,m}(x) = \begin{cases} \sqrt{2\dfrac{(n-m)!}{(n+m)!}(2n+1)}P_{n,m}(x) & (m = 1,\ldots,n) \\ \sqrt{2n+1}P_{n,m}(x) & (m = 0) \end{cases} \tag{2.62}
$$

Of special interest is the J2 model when we ignore all the tesseral terms and only keep the first dominant term beyond the point-mass gravity model:

$$
\begin{aligned}
V_g\left({}^E R_E^B\right) &= -\frac{\mu}{r}(1 + \bar{J}_2(\frac{a}{r})^n P_{2,0}(\sin\lambda)) \\
&= -\frac{\mu}{r}\left(1 + \bar{J}_2\left(\frac{a}{r}\right)^2 \frac{1}{2}(3(\sin\lambda)^2 - 1)\right) \\
&= -\frac{\mu}{r} - \bar{J}_2 \frac{\mu a^2}{r^3}\frac{1}{2}\left(3(\sin\lambda)^2 - 1)\right)
\end{aligned} \tag{2.63}
$$

Let

$$
J_2 = \left(\mu a^2\right)\bar{J}_2 \tag{2.64}
$$

and we have

$$
V_g\left({}^E R_E^B\right) = -\frac{\mu}{r} - J_2\frac{1}{r^3}\frac{1}{2}\left(3(\sin\lambda)^2 - 1)\right) = -\frac{\mu}{r} - J_2\frac{1}{r^5}\frac{1}{2}\left(3z^2 - r^2\right) \tag{2.65}
$$

and the gradient in ECEF (i.e., the gravity acceleration vector) can be written as

$$g = \nabla V_g = -\frac{\mu}{r^3}\begin{bmatrix} x \\ y \\ z \end{bmatrix} + J_2 \frac{1}{r^7}\begin{bmatrix} x\left(6z^2 - \frac{3}{2}\left(x^2 + y^2\right)\right) \\ y\left(6z^2 - \frac{3}{2}\left(x^2 + y^2\right)\right) \\ x\left(3z^2 - \frac{9}{2}\left(x^2 + y^2\right)\right) \end{bmatrix} \tag{2.66}$$

In the JGM-3 model,

$$J_2 = \bar{J}_2 \mu a^2 = \left(0.1082635854 \times 10^{-2}\right)(398600.4415)(6378.1363)^2 \ (\text{km})^5/\text{s}^2$$

It is important to note that there have been many efficient algorithms developed in the past and publicly available to allow us to calculate:

1. The terms in the geo-potential function;
2. The gradient of the geo-potential function (i.e., the gravitational acceleration);
3. The gravity anomaly;
4. The gravity gradient.

The readers are referred to [27] for details.

In particular, the MATLAB function `gravitysphericalharmonic` implements the spherical harmonics model.

The spherical harmonics model is used to fit measurement data from different sources, which leads to models such as:

1. Joint Earth Gravity Models (JGM-1, JGM-2, JGM-3) developed by Goddard Space Flight Center in cooperation with other institutions;
2. The EGM96 with $Nz = Nt = 360$ with 130,317 coefficients;
3. The EGM2008 model that represents the state of art of comprehensive Earth gravity model development.

It is important to point out that, depending on the applications, it is often not necessary to use a very high-order gravity model due to the fact that other uncertainties can easily exceed the accuracy provided by the higher-order terms. For example, for spacecraft orbit determination, the uncertainty of aerodynamics force model and solar radiation model will often make it unnecessary to go over anything higher than $n = m = 36$.

Among the models mentioned above, the official Earth Gravitational Model EGM2008 is complete to spherical harmonic degree and order 2,159 ($n = m = 2{,}159$) and contains additional coefficients extending to degree 2,190 and order 2,159.

EGM2008 leverages the following important efforts and results in humans' capability of characterizing the Earth gravity field and represents a factor of 6 improvements in resolution, and factors of 3 to 6 in accuracy over EGM96:

- Data collected by the CHAMP (CHAllenging Minisatellite Payload) mission launched in July 2000;

- Data collected by the GRACE (Gravity Recovery and Climate Experiment) mission launched in March 2002;
- Data collected by the GOCE (Gravity Field and Steady-State Ocean Circulation Explorer) mission launched in March 2009;
- The most comprehensive compilation of a global 5 arc-minute equiangular grid of area-mean free-air gravity anomalies.

2.5.7 Earth Magnetic Field Models

Very similar to gravity model, spherical harmonics potential functions are used to parameterize Earth magnetic field models and the field vector can be computed as gradient of the potential function.

Let V_m be the scalar earth magnetic field potential that is a function of the position in ECEF frame:

$$V_m = V_m\left({}^{E}R_{E}^{B}\right) \tag{2.67}$$

The Earth magnetic field B is given by is the gradient of the magnetic potential function:

$${}^{E}B = -\frac{\partial V_m\left({}^{E}R_{E}^{B}\right)}{\partial\, {}^{E}R_{E}^{B}} = -\nabla V_m\left({}^{E}R_{E}^{B}\right) \tag{2.68}$$

which is modeled by spherical harmonics as

$$V_m\left({}^{E}R_{E}^{B}\right) = a\left(\sum_{n=1}^{\infty}\left(\frac{a}{r}\right)^{n+1}\sum_{m=0}^{n}P_{n,m}(\sin\lambda)\left(g_{n,m}\cos m\phi + h_{n,m}\sin m\phi\right)\right) \tag{2.69}$$

where the ECEF position vector is converted to the geocentric spherical coordinates in terms of distance from the center of the earth r, geocentric latitude λ, and longitude ϕ, which satisfy the following equation

$${}^{E}R_{E}^{B} = \begin{bmatrix} r\cos(\lambda)\cos(\phi) \\ r\cos(\lambda)\sin(\phi) \\ r\sin(\lambda) \end{bmatrix}$$

a is the semi-major axis of the Earth, which is used to normalize the coefficients to unitless, $P_{n,0}$ are the Legendre polynomials, and $P_{n,m}$ for $1 \le m \le n$ are the associated Legendre functions, which are identical to the ones used for the gravity model.

Note that, in some literature, the associated Legendre functions are replaced by the Schmidt functions P_n^m (versus $P_{n,m}$), which is slightly different from the normalized Legendre functions used for the gravity model. These differences are nonessential but do lead to different definitions of the coefficients. The Schmidt functions are defined to be

$$P_{n,m} = \begin{cases} \sqrt{2\frac{(n-m)!}{(n+m)!}}P_{n,m}(x) & (m = 1,\ldots,n) \\ P_{n,m}(x) & (m = 0) \end{cases} \tag{2.70}$$

The International Geomagnetic Reference Field (IGRF) and the World Magnetic Model (WMM) are the most commonly used models for Earth magnetic field models. Since the Earth magnetic field are time varying, these models are updated every 5 years. WMM in general is more useful for navigation purposes for the next few years.

2.6 Kalman Filters and Variations

The Kalman filter and its various variations [28–33] have been some of the most important tools used to perform integration and fusion of sensors and information in general. They are naturally one of the most important tools for All Source PNT.

2.6.1 Kalman Filter

Given a linear dynamical system with deterministic input, random input, and random measurement noise, modeled by

$$\begin{aligned} x(k) &= \Phi(k)x(k-1) + \Gamma(k)u(k) + w(k) \\ z(k) &= H(k)x(k) + v(k) \\ x(0) &= x_0 \end{aligned} \tag{2.71}$$

where

$x(k)$ is a n-dimensional state vector

$u(k)$ is a m-dimensional input vector (known, deterministic)

$z(k)$ is a p-dimensional output vector (known, measured)

$\Phi(k)$, $\Gamma(k)$, $H(k)$ are appropriately dimensioned system matrices (known)

$w(k)$, $v(k)$ are zero-mean, white noise with (known) covariance matrices $Q(k)$, $R(k)$

x_0 is the initial value of the state is uncorrelated with $w(k)$ and $v(k)$ and has the covariance matrix P_0

Definitions of symbols:

$\hat{x}(k+1 \mid k)$: estimation of $x(k+1)$ using measurements up to time k

$\hat{x}(k \mid k)$: estimation of $x(k)$ using measurements up to time k

$\tilde{x}(k \mid k)$: estimation error of $\hat{x}(k \mid k)$ using measurements up to time k,i.e.

$\tilde{x}(k \mid k) = \hat{x}(k \mid k) - x(k)$ or $\hat{x}(k \mid k) = x(k) + \tilde{x}(k \mid k)$

$\tilde{x}(k+1 \mid k)$: estimation error of $\hat{x}(k+1 \mid k)$ using measurements up to time k, i.e.,

$\tilde{x}(k+1 \mid k) = \hat{x}(k+1 \mid k) - x(k+1)$ or $\hat{x}(k+1 \mid k) = x(k+1) + \tilde{x}(k+1 \mid k)$

$P(k+1 \mid k)$: covariance matrix of $\tilde{x}(k+1 \mid k)$

$P(k \mid k)$: covariance matrix of $\tilde{x}(k \mid k)$

Without assuming Gaussian distribution for all the random process/variables, the recursive estimator as constructed using the following equations is an optimal minimum variance estimator in the class of linear estimators; this filter is the Kalman filter. With Gaussian assumption, the estimator is optimal in all classes of estimators.

1. State propagation and state covariance propagation

$$\hat{x}(k \mid k-1) = \Phi(k-1)\hat{x}(k-1 \mid k-1) + \Gamma(k)u(k) \tag{2.72}$$

$$P(k \mid k-1) = \Phi(k-1)P(k-1 \mid k-1)\Phi^T(k-1) + Q(k) \tag{2.73}$$

2. Prediction of measurement

$$\hat{z}(k \mid k-1) = H(k)\hat{x}(k \mid k-1) \tag{2.74}$$

3. Innovation/residual computation

$$\mu = \left(z(k) - \hat{z}(k \mid k-1)\right) \tag{2.75}$$

   (Note that, in some literature, residual and innovation are given different meanings. In this book, we use the convention that the Kalman filter residual and Kalman filter innovation are to mean exactly the same thing.)

4. Gain computation

$$K(k) = P(k \mid k-1)H^T(k)\left(H(k)P(k \mid k-1)H^T(k) + R(k)\right)^{-1} \tag{2.76}$$

5. State and covariance update

$$\hat{x}(k \mid k) = \hat{x}(k \mid k-1) + K\mu(k) \tag{2.77}$$

$$P(k \mid k) = \left(I - K(k)H(k)\right)P(k \mid k-1)\left(I - K(k)H(k)\right)^T + K(k)R(k)K^T(k) \tag{2.78}$$

or

$$P(k \mid k) = \left(I - K(k)H(k)\right)P(k \mid k-1) \tag{2.79}$$

Figure 2.4 illustrates the architecture of the Kalman filter.

2.6.2 Extended Kalman Filter (EKF)

Extending the Kalman filters to nonlinear (yet smooth, differentiable) systems, we get the EKF.

Given a nonlinear discrete system:

$$\begin{aligned} x(k) &= f\left(x(k-1), u(k), k\right) + w(k), \\ z(k) &= g\left(x(k), u(k), k\right) + v(k) \end{aligned} \tag{2.80}$$

where the variables are defined the same as the linear system mentioned before. The only difference is that we have nonlinear difference equation for the state and nonlinear algebraic equation for the measurement equation.

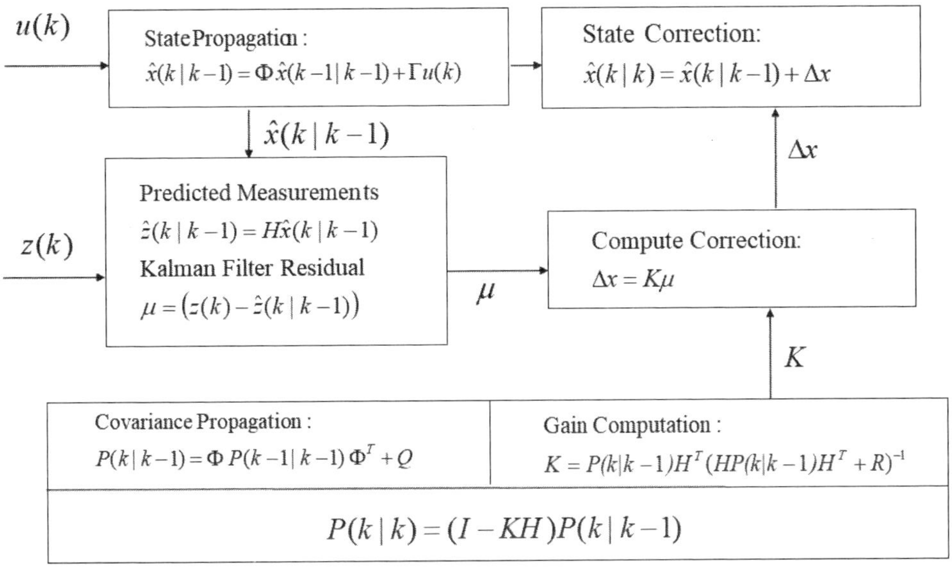

Figure 2.4 Architecture of Kalman filter.

The minimum variance recursive estimator for the above nonlinear system can be described by the following set of equations.

1. State propagation and state covariance propagation

$$\hat{x}(k \mid k-1) = f\left(\hat{x}(k-1 \mid k-1), u(k-1), k-1\right) \tag{2.81}$$

$$P(k \mid k-1) = \Phi(k-1)P(k-1 \mid k-1)\Phi^T(k-1) + Q(k) \tag{2.82}$$

where

$$\Phi(k-1) \triangleq \frac{\partial f}{\partial x}_{|x=\hat{x}(k-1|k-1)} \triangleq f_{x|x=\hat{x}(k-1|k-1)}$$

2. Prediction of measurement

$$\hat{z}(k \mid k-1) = g\left(\hat{x}(k \mid k-1), u(k-1), k\right) \tag{2.83}$$

3. Residual computation

$$\mu = \left(z(k) - \hat{z}(k \mid k-1)\right) \tag{2.84}$$

4. Gain computation

$$K(k) = P(k \mid k-1)H^T(k)\left(H(k)P(k \mid k-1)H^T(k) + R(k)\right)^{-1} \tag{2.85}$$

where

$$H(k) \triangleq \frac{\partial g}{\partial x}_{|x=\hat{x}(k|k-1)} \triangleq g_{x|x=\hat{x}(k|k-1)}$$

5. State and covariance update

$$\hat{x}(k \mid k) = \hat{x}(k \mid k-1) + K\mu(k) \tag{2.86}$$

$$P(k \mid k) = \left(I - K(k)H(k)\right)P(k \mid k-1)\left(I - K(k)H(k)\right)^T + K(k)R(k)K^T(k) \tag{2.87}$$

or

$$P(k \mid k) = \left(I - K(k)H(k)\right)P(k \mid k-1) \tag{2.88}$$

Figure 2.5 illustrates the architecture of the extended Kalman filter.

2.6.3 Generic Propagation/Prediction/Correction Architecture and UKF/Particle Filters

The KF and EKF described in earlier sections have the following general architectural features:

1. We use a known state propagation model to propagate the estimated state.
2. We use a known measurement model to predict the measurement using the best estimated state.
3. We calculate the residual, which is the difference between the measurement and the predicted measurement.
4. We multiply the residual by a gain matrix to generate the correction, which is applied to correct the state estimate.

The above scheme is actually general and depending on how the gains are designed or calculated, we arrive at different type of estimators.

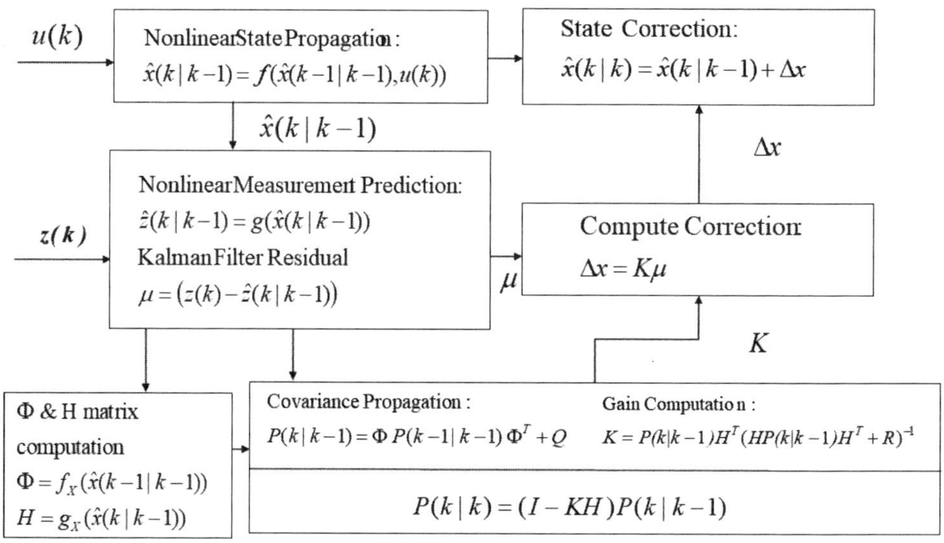

Figure 2.5 Architecture of extended Kalman filter.

For example, the Kalman filter is designed to optimize the H_2 norm of the estimation error. So the Kalman filter is a kind of optimal H_2 estimator.

Similarly, we could design the gain matrix such that it optimizes the H_∞ norm of the estimation error and design an optimal H_∞ estimator.

In many cases, especially for linear time-invariant systems, we could choose a constant gain matrix designed to ensure that the estimator is stable and has reasonable performance. In this case, we would have designed a fixed-gain estimator. One special case of fixed gain estimator is the steady-state Kalman filter.

Of special interests are the unscented Kalman filter (UKF)/sigma-point Kalman filter as well as particle Kalman filter. In both cases, we will use similar equations as the EKF to calculate the gain matrix and update the covariance matrix after the gains are applied. What is different is that instead of using analytical equations to propagate the covariance matrix and to calculate the cross-correlation matrix between the state and the measurement, we will do that statistically.

In the case of UKF, this is achieved by passing a number of strategically selected sigma points through the propagation equation and through the measurement equation so that we can use the results to statistically calculate the covariance matrix and the cross-covariance matrix. In the case of a particle filter, such points are randomly chosen.

Figure 2.6 illustrates the generic architecture. Figure 2.7 illustrates the UKF and particle filter architecture.

The general UKF/particle filter algorithms are described here:

1. Nonlinear state propagation and state covariance propagation:
   a. Given $\hat{x}(k-1 \mid k-1)$ and $P(k-1 \mid k-1)$, form the sigma points or particles $\hat{x}_j(k-1 \mid k-1)$.
   b. Propagate the samples (i.e., the sigma points or particles):

$$\hat{x}_j(k \mid k-1) = f\left(\hat{x}_j(k-1 \mid k-1), u(k)\right) \tag{2.89}$$

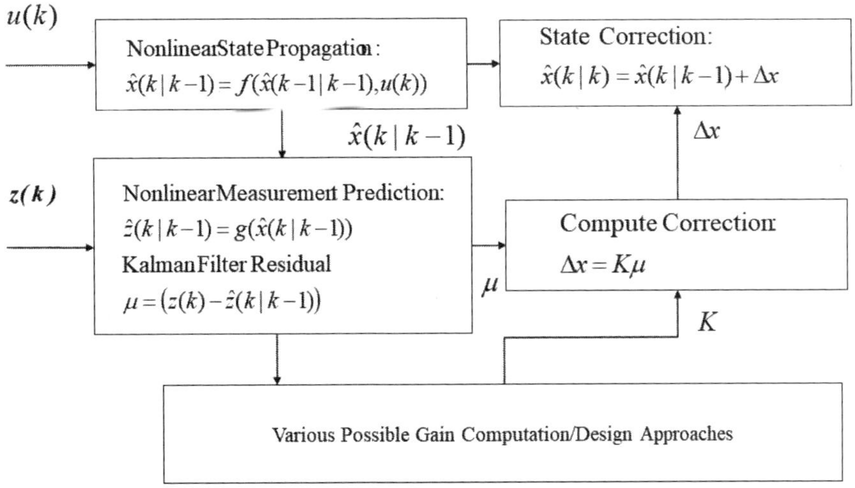

Figure 2.6 Generic architecture of the propagation and correction-based estimator.

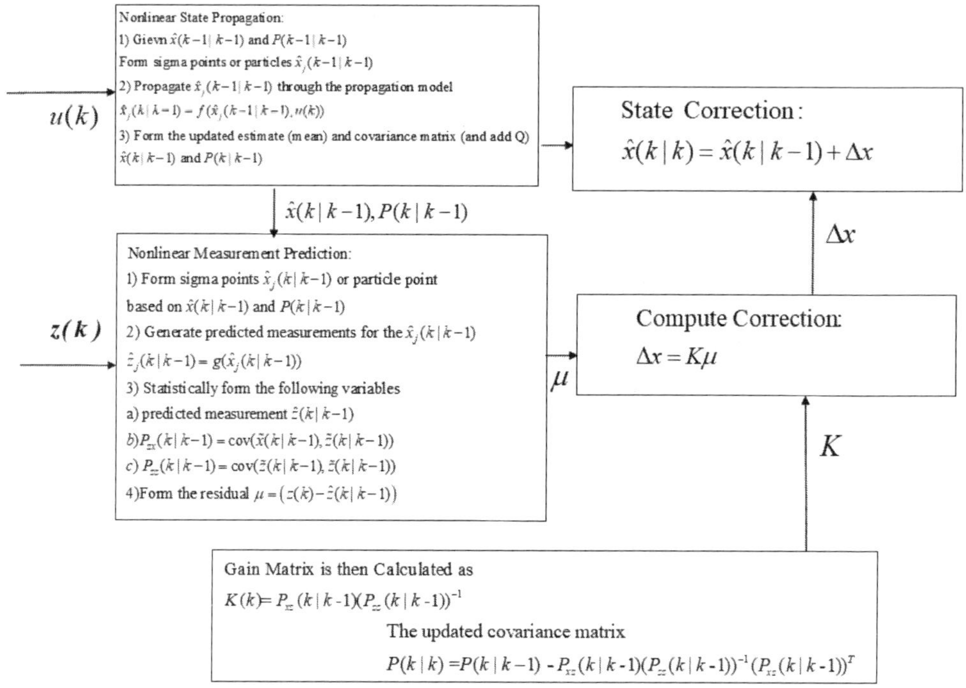

Figure 2.7 UKF/particle filter architecture.

c. Using the above samples, statistically form the updated estimate (mean) and the covariance matrix:

$$\hat{x}(k \mid k-1) \text{ and } P(k \mid k-1)$$

2. Prediction of measurement and its covariance:
   a. Form sigma points/particles $\hat{x}_j(k \mid k-1)$ based the predicted state $\hat{x}(k \mid k-1)$ and covariance matrix $P(k \mid k-1)$.
   b. Generate predicted measurements for $\hat{x}_j(k \mid k-1)$ by

$$\hat{z}_j(k \mid k-1) = g\left(\hat{x}_j(k \mid k-1)\right) \tag{2.90}$$

   c. Statistically form the following variables:
   $\hat{z}(k \mid k-1)$: predicted measurement
   $P_{zx}(k \mid k-1) = \mathrm{cov}\left(\tilde{x}(k \mid k-1), \tilde{z}(k \mid k-1)\right)$: cross-covariance matrix
   $P_{zz}(k \mid k-1) = \mathrm{cov}\left(\tilde{z}(k \mid k-1), \tilde{z}(k \mid k-1)\right)$: covariance matrix
3. Form residual:

$$\mu = \left(z(k) - \hat{z}(k \mid k-1)\right)$$

4. Gain computation. The gain matrix is calculated as

$$K(k) = P_{xz}(k \mid k-1)\left(P_{zz}(k \mid k-1)\right)^{-1} \tag{2.91}$$

5. State and covariance update:

$$\hat{x}(k \mid k) = \hat{x}(k \mid k-1) + K\mu(k) \tag{2.92}$$

$$P(k \mid k) = P(k \mid k-1) - P_{xz}(k \mid k-1)\left(P_{zz}(k \mid k-1)\right)^{-1}\left(P_{xz}(k \mid k-1)\right)^{T} \tag{2.93}$$

Note that we have described UKF/sigma point/particle Kalman filters in a general way. Among these methods, UKF, in particular, uses strategically selected sigma points to minimize the amount of computations that have to be carried out in order to statistically calculate the covariance matrices and the estimates. Further details of UKF are provided here:

1. UKF state and covariance propagation:
   a. Define the augmented state for propagation:

$$x^{a}(k) = \begin{bmatrix} x(k) \\ w(k) \end{bmatrix} \tag{2.94}$$

   b. The estimated value at $k-1$ is given by

$$x^{a}(k-1 \mid k-1) = \begin{bmatrix} \hat{x}(k-1 \mid k-1) \\ 0 \end{bmatrix} \tag{2.95}$$

   c. The corresponding error covariance matrix will be given by

$$P^{a}(k-1 \mid k-1) = \begin{bmatrix} P(k-1 \mid k-1) & 0 \\ 0 & Q(k) \end{bmatrix} \tag{2.96}$$

   d. Assuming that the size of the augmented state vector is L, UKF defines the following strategically selected $2L+1$ sigma points:

$$\begin{aligned} \chi^{a0}(k-1 \mid k-1) &= \hat{x}^{a}(k-1 \mid k-1) \\ \chi^{ai}(k-1 \mid k-1) &= \hat{x}^{a}(k-1 \mid k-1) + \left(\sqrt{(L+\lambda)P^{a}(k-1 \mid k-1)}\right)_{i}, \\ &\quad (i = 1,2,\cdots,L) \\ \chi^{ai}(k-1 \mid k-1) &= \hat{x}^{a}(k-1 \mid k-1) - \left(\sqrt{(L+\lambda)P^{a}(k-1 \mid k-1)}\right)_{i}, \\ &\quad (i = L+1, L+2,\cdots,2L) \end{aligned} \tag{2.97}$$

   where $\sqrt{(L+\lambda)P^{a}(k-1 \mid k-1)}$ is the ith column of the matrix square root of the matrix $(L+\lambda)P^{a}(k-1 \mid k-1)$.

   (Note that, given a semi-positive definite matrix P, the square root of P, denoted as S, can be calculated as a Cholesky factorization of P, which satisfies $P = SS^{T}$.)

   In the above,

$$\chi^{ai}(k-1 \mid k-1) = \begin{bmatrix} \chi^{i}(k-1 \mid k-1) \\ w^{i}(k-1) \end{bmatrix} \tag{2.98}$$

e. The $2L + 1$ sigma points are then propagated through the nonlinear propagation function:

$$\chi^i(k \mid k-1) - f\left(\chi^{ai}(k-1 \mid k-1)\right), \quad i = 0,\cdots,2L \tag{2.99}$$

Note that the input is augmented sigma points but the output is unaugmented.

f. The propagated state estimate and covariance matrix can then be calculated using the output sigma points (unaugmented) as

$$\begin{aligned}\hat{x}(k \mid k-1) &= \sum_{i=0}^{2L} W_s^i \chi^i(k \mid k-1) \\ P(k \mid k-1) &= \sum_{i=0}^{2L} W_c^i \left(\chi^i(k \mid k-1) - \hat{x}(k \mid k-1)\right)\left(\chi^i(k \mid k-1) - \hat{x}(k \mid k-1)\right)^T\end{aligned} \tag{2.100}$$

where

$$\begin{aligned}W_s^0 &= \frac{\lambda}{(L+\lambda)}, \quad W_c^0 = \frac{\lambda}{(L+\lambda)} + \left(1 - \alpha^2 + \beta\right) \\ W_s^i &= W_c^i = \frac{1}{2(L+\lambda)}, \quad \lambda = \alpha^2(L+\kappa) - L\end{aligned} \tag{2.101}$$

The typical values for tunable parameters are $\alpha = 1e-3$, $\beta = 2$, $\kappa = 0$.

2. State and covariance update

a. Starting from $\hat{x}(k \mid k-1)$ and $P(k \mid k-1)$, we will form a new augmented state and covariance matrix for the purposes of state and covariance update:

$$x^a(k) = \begin{bmatrix} x(k) \\ v(k) \end{bmatrix} \tag{2.102}$$

b. We have

$$\hat{x}^a(k \mid k-1) = \begin{bmatrix} \hat{x}(k \mid k-1) \\ 0 \end{bmatrix} \tag{2.103}$$

and the corresponding covariance matrix will be given by

$$P^a(k \mid k-1) = \begin{bmatrix} P(k \mid k-1) & 0 \\ 0 & R(k) \end{bmatrix} \tag{2.104}$$

We will still use L as the size of the augmented state. However, be aware that, at this point, the augmented portion is the measurement noise, not the process noise, the actual value of L will likely to be different from the propagation step.

c. At this moment, we will choose a new set of sigma points for the purpose of measurement prediction and the covariance matrices associated with it:

$$\chi^{a0}(k \mid k-1) = \hat{x}^a(k \mid k-1)$$
$$\chi^{ai}(k \mid k-1) = \hat{x}^a(k \mid k-1) + \left(\sqrt{(L+\lambda)P^a(k \mid k-1)}\right)_i, \quad (i = 1,2,\cdots,L) \tag{2.105}$$
$$\chi^{ai}(k \mid k-1) = \hat{x}^a(k \mid k-1) - \left(\sqrt{(L+\lambda)P^a(k \mid k-1)}\right)_i, \quad (i = L+1, L+2,\cdots,2L)$$

d. Again note that

$$\chi^{ai}(k \mid k-1) = \begin{bmatrix} \chi^i(k \mid k-1) \\ v^i(k) \end{bmatrix} \tag{2.106}$$

e. We will then pass the sigma points through the measurement equation:

$$\gamma^i(k \mid k-1) = g\left(\chi^{ai}(k \mid k-1)\right), \quad i = 0,\cdots,2L \tag{2.107}$$

(Again, note that the input is the augmented sigma points, the output is unaugmented sigma points.)

f. The predicted measurement and its covariance matrix are then computed as

$$\hat{z}(k \mid k-1) = \sum_{i=0}^{2L} W_s^i \gamma^i(k \mid k-1)$$
$$P_{zz}(k \mid k-1) = \sum_{i=0}^{2L} W_c^i \left(\gamma^i(k \mid k-1) - \hat{z}(k \mid k-1)\right)\left(\gamma^i(k \mid k-1) - \hat{z}(k \mid k-1)\right)^T \tag{2.108}$$

g. The state/measurement cross-correlation matrix is given by

$$P_{xz}(k \mid k-1) = \sum_{i=0}^{2L} W_c^i \left(\chi^i(k \mid k-1) - \hat{x}(k \mid k-1)\right)\left(\gamma^i(k \mid k-1) - \hat{z}(k \mid k-1)\right)^T \tag{2.109}$$

h. The gain matrix is then computed as

$$K(k) = P_{xz}(k \mid k-1)\left(P_{zz}(k \mid k-1)\right)^{-1} \tag{2.110}$$

i. Finally, the standard Kalman filter equations are used to perform the state and covariance update:

$$\hat{x}(k \mid k) = \hat{x}(k \mid k-1) + K(k)\left(z(k) - \hat{z}(k \mid k-1)\right)$$
$$P(k \mid k) = P(k \mid k-1) - K(k)P_{zz}(k \mid k-1)K^T(k) \tag{2.111}$$

2.6.4 Practical Issues in the Application of the Kalman Filters

2.6.4.1 Time-Matching

One of the most important variables in the Kalman filter is the Kalman filter residual. This is the place the predicted measurements meet the measurements.

The residual is the difference between the measurement (representing the measured solution) and the predicted measurement (representing the propagated solution). It is the driver of all the corrections generated by the Kalman filter.

It is critically important that the residual is computed correctly. One of the most important issues for this correctness is that both the predicted measurement and the measurement need to be valid for the same time moment. Significant errors can be generated if the measurements change fast and are not time-matched with the predicted measurements.

A general technique to ensure the time-matching is described in [28]. The key ideas are outlined here:

1. The propagation sensor data is time-tagged.
2. The measurement sensor data is time-tagged.
3. Clock reset/synchronization schemes are used to ensure that the clock of the computer (which receives the sensor data), the propagation sensor clock, and measurement sensor clock are all synchronized with respect to the same time reference.
4. The propagated state and the corresponding time tag (typically available at high rate) are saved into a circular buffer.
5. For each time-tagged measurement, a time-matched state is retrieved from the time-matching buffer. Interpolation schemes are used to match the time precisely.
6. The time-matched state is then used in the measurement model to compute the predicted measurement.
7. Finally, the time-matched residual is computed.

2.6.4.2 Nonwhite Noise

In practical systems, it is very rare that we have real white noises. The essence of being white is that the noise is almost uncorrelated over the time period of interest.

For process noise, being white means that the random noise is uncorrelated over the time interval during which the covariance propagation is performed. For measurement noise, it means that the random noise is uncorrelated between updates. With this practical understanding, it will not be hard to deal with colored noises in practice.

Solutions for Nonwhite Process Noise

Two approaches are frequently used in practical designs.

- *Approach 1:* We will model the noise using a linear dynamic model such as first-order Markov model. When this approach is taken, additional states are added to the Kalman filter to model the correlated behavior of the noise.
- *Approach 2:* We will perform the covariance propagation over a longer interval. In this case, we need to ensure that the state transition matrix and the Q-matrix are properly formulated to account for the integrated impact of the process noise over the propagation interval.

Solutions for Non-White Measurement Noise

There are three frequently used approaches to deal with nonwhite measurement noise.

- *Approach 1:* We will model the noise using linear dynamic models such as the first-order Markov model. When this approach is taken, additional states are added to the Kalman filter to model the correlated behavior of the noise.
- *Approach 2:* We will reduce the frequency at which the updates are performed so that the errors are uncorrelated between updates. Fundamentally, when the measurement noise is correlated, the updates performed at a high rate do not introduce new information into the system as it should. Without recognizing this fact, the Kalman filter will become overly optimistic by lowering the covariance matrix each time an update is performed. Reducing the update frequency brings the system back to reality.
- *Approach 3:* We will use an artificially larger R-matrix to remove the over-optimism introduced by the white noise assumption, achieving similar effects as in approach 2.

2.6.4.3 Integrity Monitoring

Since Kalman filter residual is so important, it is often a good idea to perform integrity monitoring on the residual. For example, the residuals can be compared against the 3 (or higher) sigma covariance bound to determine if the residual should be rejected for Kalman filter correction computation.

In addition, if rejections happen consistently over some time, it is an indication of a failure or fault condition in the system. By properly choosing the consistency threshold, useful fault detection algorithms can be designed using this approach.

2.6.4.4 Kalman Filter Divergence

Kalman filter divergence happens when Kalman filter no longer provides a good estimate of the state. Kalman filter divergence can be detected by checking the Kalman filter residuals. If the residual persistently exceeds what is predicted by the covariance, divergence is happening.

The following are most frequently seen reasons for divergence:

1. Error models used by the Kalman filter do not properly reflect reality.
2. Computational errors are ignored.
3. Linearization errors are not properly accounted for.
4. Nonwhite noise is modeled as white noise, leading to over-optimistic covariance and very small gains and virtually no corrections.

The basic rule to avoid divergence is to not lie to a Kalman filter; what this means is that we need to properly model all the different error sources for a Kalman filter design task. When a particular error source is not modeled in the system, it is very important to have an assessment of the unmodeled errors.

For example, the computational errors and linearization errors are frequently ignored by inexperienced designers. What is critically important is to ensure that the computational errors and linearization errors are significantly smaller than the process noise for the state transition model and significantly smaller than the measurement noise for the measurement model.

If the results of the assessment mentioned above are such that the unmodeled errors are not ignorable, the following are a number of remedies that could be used to solve the problem:

1. If computational errors are relatively too large to ignore, we will want to use higher precision computations and higher-precision algorithms. This problem happens frequently when the sensors used are of high precision and it is highly likely that computational errors can overwhelm the sensor errors, leading to a waste of expensive instruments.
2. If linearization and other modeling errors are larger than modeled, we will first want to try to use a higher-precision linearized model. If that is not feasible, it may be time to try UKF or particle filters.
3. There is always another easy option we can use. That is to increase the Q-matrix to account for the unmodeled propagation model errors and to increase the R-matrix to account for the unmodeled measurement model errors. This is done at the cost of degraded performance and therefore should only be used when everything else fails.

References

[1] Wertz, J. R., *Spacecraft Attitude Determination and Control*, Boston, MA: D. Reidel Publishing Company, 1986.

[2] Sidi, M. I., *Spacecraft Dynamics and Control*, Cambridge, U.K.: Cambridge University Press, 1997.

[3] Shuster, M. D., "A Survey of Attitude Representations," *The Astronautical Science*, Vol. 41, 1993, pp. 439–517.

[4] Tamer, M., et al., "Comparative Study of Spacecraft Attitude Determination and Estimation Algorithms (A Cost-Benefit Approach)," *Aerospace Science and Technology*, Vol. 26, No. 1, April-May 2013, pp. 211–215.

[5] Mohd, Z. H., et al., "Review on Attitude Estimation Algorithm of Attitude Determination System," *ARPN Journal of Engineering and Applied Sciences*, Vol. 11, 2016, pp. 4455–4460.

[6] Savage, P. G., "Strapdown Inertial Navigation Integration Algorithm Design Part 1: Attitude Algorithms," *Journal of Guidance, Control, and Dynamics*, Vol. 21, No. 1, 1998, pp. 19–28.

[7] Savage, P. G., "Strapdown Inertial Navigation Integration Algorithm Design Part 2: Velocity and Position Algorithms," *Journal of Guidance, Control, and Dynamics*, Vol. 21, No. 2, March–April 1998.

[8] Pisacane, V. L., (ed.), *Fundamentals of Space Systems*, 2nd ed., Oxford, U.K.: Oxford University Press, 2005.

[9] Osen, K., *Accurate Conversion of Earth-Fixed Earth-Centered Coordinates to Geodetic Coordinates*, Research Report, Norwegian University of Science and Technology, 2017, hal-01704943v2, https://hal.archives-ouvertes.fr/hal-01704943v2.

[10] Zhu, J., "Exact Conversion of Earth-Centered Earth-Fixed Coordinates to Geodetic Coordinates" *Journal of Guidance, Control, and Dynamics*, Vol. 16, No. 2, March-April 1993.

[11] Zhu, J., "Conversion of Earth-Centered Earth-Fixed Coordinates to Geodetic Coordinates," *IEEE Transactions on Aerospace and Electronic Systems*, Vol. 30, July 1994.

[12] Pavlis, N. K., et al., "The Development and Evaluation of the Earth Gravitational Model 2008 (EGM2008)," *Journal of Geophysical Research: Solid Earth*, Vol. 117, No. B4, April 2012.

[13] Roithmayr, C. M., "Contributions of Spherical Harmonics to Magnetic and Gravitational Fields," NASA/TM–2004–213007, NASA Technical Report, March 2004.

[14] https://earth-info.nga.mil/GandG/wgs84/gravitymod/egm2008/egm08_wgs84.html.

[15] https://www.esri.com/news/arcuser/0703/geoid1of3.html.

[16] "WGS 84 Earth Gravitational Model," earth-info.nga.mil.

[17] "WGS 84, N=M=180 Earth Gravitational Model," earth-info.nga.mil.

[18] Lemoine, F. G., et al., "The Development of the Joint NASA GSFC and the National Imagery and Mapping Agency (NIMA) Geopotential Model EGM96," NASA/TP-1998-206861, July 1998.

[19] Gottlieb, R. G., "Fast Gravity, Gravity Partials, Normalized Gravity, Gravity Gradient Torque and Magnetic Field: Derivation, Code and Data," *Technical Report NASA Contractor Report 188243*, NASA Lyndon B. Johnson Space Center, Houston, TX, February 1993.

[20] "Department of Defense World Geodetic System 1984, Its Definition and Relationship with Local Geodetic Systems", National Imagery And Mapping Agency Technical Report, NIMA TR8350.2, 1984.

[21] Kenyon S., et al., "Towards the Next Earth Gravitational Model," *Society of Exploration Geophysicists 77th Annual Meeting*, San Antonio, TX, September 23–28, 2007.

[22] Pavlis, N. K., et al., "An Earth Gravitational Model to Degree 2160: EGM2008," *2008 General Assembly of the European Geosciences Union*, Vienna, Austria, April 13–18, 2008.

[23] Grueber, T., and A. Köhl, "Validation of the EGM2008 Gravity Field with GPS-Leveling and Oceanographic Analyses," *IAG International Symposium on Gravity, Geoid & Earth Observation 2008*, Chania, Greece, June 23–27, 2008.

[24] Förste, C., et al., "A Mean Global Gravity Field Model from the Combination of Satellite Mission and Altimetry/Gravimetry Surface Data—EIGEN-GL04C," European Geosciences Union, General Assembly 2006, Vienna, Austria, 02 - 07 April 2006.

[25] Colombo, O. L., "Numerical Methods for Harmonic Analysis on the Sphere," Reports of the Department of Geodetic Science, Report No. 310, The Ohio State University, Columbus, OH., March 1981.

[26] Colombo, O. L., "The Global Mapping of Gravity with Two Satellites," *Netherlands Geodetic Commission*, Vol. 7, No. 3, Delft, The Netherlands, 1984.

[27] Jones, B. A., "Efficient Models for the Evaluation and Estimation of the Gravity Field," Doctoral dissertation, University of Colorado, Boulder, CO, 2010.

[28] Li, R., and Y. -W. Wu, "Spacecraft Attitude Determination System and Method," U.S. Patent 6285927 (Expired), 1999.

[29] Gelb, M., *Applied Optimal Estimation*, Cambridge, MA: MIT Press, 1974.

[30] Brown, R. G., and P. Hwang, *Introduction to Random Signals and Applied Kalman Filtering*, New York: John Wiley & Sons, 1996.

[31] Li, R., and Y. -W. Wu, "Kalman Filtering & Applications," *UCLA Lecture Notes*, 2005–2019.

[32] Julier, S., J. K. Uhlmann, and H. Durrant-Whyte, "A New Approach for Filtering Nonlinear System," *Proceedings of the American Control Conference*, 1995, pp. 1628–1632.

[33] Julier, S. J., and J. K. Uhlmann, *A General Method for Approximating Nonlinear Transformation of Probability Distributions*, Technical Report RRG, Department of Engineering Science, University of Oxford, November 1996.

CHAPTER 3

Various Propagation-Based Solutions

3.1 Propagation Model of Constant and Almost Constant Parameters

3.1.1 Random Walk and First-Order Markov Process

In PNT systems, it is often necessary to model parameters of various sensors and systems that are to be used as part of the All Source PNT system. Typical examples are gyro and accelerometer biases, scale factors, misalignments, G-sensitivity coefficients, optical sensor alignment errors, optical sensor effective focal length errors, clock bias, and optical sensor distortion parameters.

Modeling these parameters allows us to be able to include them as part of the Kalman filter states and to be able to possibly estimate or calibrate these parameters [1–5].

Given an unknown parameter p that is absolutely a constant, it can be modeled by the following differential equation in the continuous time domain and the difference equation in the discrete time domain.

Continuous model:

$$\frac{d}{dt}p(t) = 0 \tag{3.1}$$

$$E\left(p^2\left(t_0\right)\right) = \sigma^2\left(p\left(t_0\right)\right) \tag{3.2}$$

Discrete model

$$p(k+1) = p(k) \tag{3.3}$$

$$E\left(p^2\left(k_0\right)\right) = \sigma^2\left(p\left(k_0\right)\right) \tag{3.4}$$

Even though p is a constant, because we do not have full knowledge of its value, it is being modeled as a random process and its initial value is modeled as a random variable. By using an estimator, we can estimate the value of p by using measurements related to p to achieve reduced variance on its uncertainty over time.

When a parameter p is known to be changing slowly, for example, a gyro bias that varies slowly with temepratura, we will typically model the parameter either as a random walk process or a first-order Markov process.

The equations for the random walk model are listed next.

Continuous model:

$$\frac{d}{dt}p(t) = w_{prw}(t) \tag{3.5}$$

where $w_{prw}(t)$ is a white noise process that generates the random walk.

$$\begin{aligned} E\left(w_{prw}(t)w_{prw}(s)\right) &= \sigma_{prw}^2\delta(t-s) \\ E\left(p^2\left(t_0\right)\right) &= \sigma^2\left(p\left(t_0\right)\right) \end{aligned} \tag{3.6}$$

Discrete model:

$$p(k+1) = p(k) + w_{dprw}(k) \tag{3.7}$$

where $w_{dprw}(k)$ is a white noise sequence that generates the random walk.

$$\begin{aligned} E\left(w_{dprw}(k)w_{dprw}(j)\right) &= \left(\sigma_{dprw}^2\right)\delta_{k,j} = \left(\sigma_{prw}^2\delta t\right)\delta_{k,j} \\ E\left(p^2\left(k_0\right)\right) &= \sigma^2\left(p\left(k_0\right)\right) \end{aligned} \tag{3.8}$$

where δt is the sampling period.

The continuous model and the discrete model can be related by the following equations

$$\begin{aligned} w_{dprw}(k) &= \int_{(k-1)\delta t}^{k\delta t} w_{prw}(t)dt \\ \sigma_{dprw} &= \sigma_{dprw}\sqrt{\delta t} \end{aligned} \tag{3.9}$$

The random walk process is simple and easy to understand. One thing to note about the random walk process model is that the variance of the parameter can grow unbounded over time. It is easy to show that

$$\begin{aligned} &P(t) = E\left(p^2(t)\right) = \sigma_{prw}^2\left(t-t_0\right) \\ &\text{or} \\ &\sigma\left(p(t)\right) = \sqrt{P(t)} = \sigma_{prw}\sqrt{t-t_0} \end{aligned} \tag{3.10}$$

In many cases, a parameter just varies randomly but stays bounded with some correlation time. In these cases, the first-order Markov process is more approprite since the parameter is modeled as the ouput of a first-order, stable, lowpass filter with bounded variance.

The following are the equations for the first-order Markov process model.

Continous model:

$$\frac{d}{dt}p(t) = -\frac{1}{T}p(t) + w(t) \tag{3.11}$$

where $w(t)$ is the white noise that generates the first-order Markov process. T is the correlation time. We have

$$\begin{aligned} E\big(w(t)w(s)\big) &= \sigma_c^2\delta(t-s) \\ E\big(p^2(t_0)\big) &= \sigma^2\big(p(t_0)\big) \end{aligned} \tag{3.12}$$

Discrete model:

$$p(k+1) = e^{-(\delta t/T)}p(k) + w_d(k) \tag{3.13}$$

where $w_d(k)$ is the white noise sequence that generates the first-order Markov process. T is the correlation time. We have

$$\begin{aligned} E\big(w_d(k)w_d(j)\big) &= \big(\sigma_d^2\big)\delta_{k,j} = \big(\sigma_c^2\delta t\big)\delta_{k,j} \\ E\big(p^2(k_0)\big) &= \sigma^2\big(p(k_0)\big) \end{aligned} \tag{3.14}$$

3.1.2 Differential and Algebraic Lyapunov Equations

In order to predict the covariance matrix of a general first-order Markov process in a vector form, let us study the random process $x(t)$ modeled by a vector differential equation:

$$\dot{x}(t) = A(t)x(t) + G(t)w(t) \tag{3.15}$$

where $w(t)$ is a white noise process with

$$\begin{aligned} E\big(w(t)\big) &= 0 \\ E\big(w(t)w^T(s)\big) &= Q_w\delta(t-s) \\ E\big(w(t)x^T(t_0)\big) &= 0 \end{aligned} \tag{3.16}$$

Let the covariance matrix be

$$P(t) = E\big(x(t)x(t)^T\big)$$

We have

$$\begin{aligned} \frac{d}{dt}P(t) &= \frac{d}{dt}E\big(x(t)x(t)^T\big) = E\big(\dot{x}(t)x(t)^T\big) + E\big(x(t)\dot{x}(t)^T\big) \\ &= E\big((Ax + Gw)x^T\big) + E\big(x\big(x^TA^T + w^TG^T\big)\big) \\ &= AE\big(xx^T\big) + E\big(xx^T\big)A^T + GE\big(wx^T\big) + E\big(xw^T\big)G^T \end{aligned}$$

Note that

$$x(t) = \Phi(t,t_0)x(t_0) + \int_{t_0}^{t} \Phi(t,s)G(s)w(s)ds$$

$$E(x(t)w(t)^T) = E(\Phi(t,t_0)x(t_0)w^T) + E\left(\int_{t_0}^{t} \Phi(t,s)G(s)w(s)w^T(t)ds\right)$$

$$= 0 + \int_{t_0}^{t} \Phi(t,s)G(s)E\{w(s)w^T(\tau)\}ds = \int_{t_0}^{t} \Phi(t,s)G(s)Q\delta(s-t)ds$$

$$= \frac{1}{2}\Phi(t,t)G(t)Q = \frac{1}{2}G(t)Q$$

where 1/2 is due to the fact that t happens to be the integration boundary point.

Similarly, we have

$$E(w(t)x(t)^T) = \frac{1}{2}QG^T(t)$$

Consequently, we have

$$\dot{P}(t) = A(t)P(t) + P(t)A^T(t) + G(t)Q(t)G^T(t) \tag{3.17}$$

The above equation is called the differential Lyapunov equation, and we can use this equation to predict the covariance matrix of a random process that is modeled by a linear differential equation.

For stable, linear time-invariant systems, the solution of the differential Lyapunov equation will converge to a steady-state value.

At the steady state, we have $\dot{P} = 0$; consequently, we have the following algebraic Lyapunov equation:

$$AP + PA^T + GQG^T = 0 \tag{3.18}$$

which is satisfied by the steady-state covariance matrix.

Similarly, for linear discrete system, we can model the process as

$$x(k+1) = \Phi(k+1,k)x(k) + G(k)w(k) \tag{3.19}$$

where $w(k)$ is a discrete white noise sequence with

$$\begin{aligned} E(w(k)) &= 0 \\ E(w(k)w^T(j)) &= Q(k)\delta_{k,j} \end{aligned} \tag{3.20}$$

It can be easily shown that that the covariance matrix satisfies the following discrete Lyapunov equation:

$$P(k+1) = \Phi(k+1,k)P(k)\Phi^T(k+1,k) + G(k)Q(k)G^T(k) \tag{3.21}$$

For a stable, linear time-invariant discrete system, the covariance matrix will converge to a steady-state value and the steady-state covariance matrix satisfies the following discrete Lyapunov equation:

$$P = \Phi P \Phi^T + GQG^T \tag{3.22}$$

The first-order Markov process that we used is modeled by a linear time-invariant system and it has a bounded variance value that can be found by solving the algebraic Lyapunov equation.

These equations will be very useful in the following sections.

3.2 Clock and Clock Propagation and Error Models

3.2.1 Overview

Precise timing is critically important for many applications. Typically, a precision time reference consists of an oscillator (XO) aided by an update or correction mechanism. The correction mechanism can use another time reference made available through direct updates or through, for example, RF or optical-based time transfer mechanisms.

There are many different types of XOs available and they are being used in various types of systems. Table 3.1 provides a reference point before the chip scale atomic clock (CSAS) becomes reality.

The SWAP (size, weight, and power) has changed significantly because of CSAC technology. Figure 3.1 is an example of CSAC by Teledyne Scientific & Imaging [6].

3.2.2 Time and Frequency Errors, Allan Variance, and Power Spectral Density (PSD)

Given an oscillator that outputs a sine wave

$$V(t) = A\sin\left(2\pi f_0 t + \phi(t)\right) \tag{3.23}$$

where f_0 is the nominal frequency with a unit of hertz, and $\phi(t)$ is the phase error, deviation, or variation as the result of clock instability. The unit of $\phi(t)$ is radian.

Table 3.1 Comparison of Various Types of Oscillators

	Crystal Oscillator (CXO)	*Temperature-Compensated CXO (TCXO)*	*Microcomputer-Compensated CXO (MCXO)*	*Oven Controlled CXO (OCXO)*	*Rubidium Atomic Frequency Standard (RbXO)*	*Cesium Atomic Standard (CsXO)*
Typical performance	100 to 10 PPM	1 PPM	0.1–0.01 PPM	0.5–0.01 PPM	1e-3 PPM	1e-5–1e-6 PPM
Aging (1/10 year)	10–20 PPM	2–5 PPM	1–3 PPM	1–0.01 PPM	5e-3 to 0.5e-3 PPM	1-e5–1e-6 PPM
Radiation	2.00E-12	2.00E-12	2.00E-12	2.00E-12	2.00E-13	2.00E-14
Power (W)	2.00E-05	1.00E-04	2.00E-04	1–3	6–12	25–40
Weight (gram)	20	50	100	200–500	1,500–2,500	10,000–20,000

TCSAC Key Features Include:
< 3E-10 / month Aging Rate
± 5.0E-11 accuracy at shipment
< -56 dBc /Hz at 1 Hz
< 5E-10 Df/f over the full operating temperature range
< 1E-11 @1000s Short Term Stability (typical ADEV)
-10oC to +70oC Operating Temperature
1PPS input for synchronization
RS-232 interface for monitoring and control

Figure 3.1 Example of a CSAC. Authorized use, credit: Teledyne Scientific & Imaging

A number of variables can be defined and is used in various literature.
The variable

$$x(t) = \frac{1}{2\pi f_0}\phi(t) \tag{3.24}$$

has the unit of time (seconds). $x(t)$ is the time phase in seconds or it is simply the time output of the clock.

The instantaneous frequency departure from the nominal frequency in fractions of nominal frequency or clock rate error can be given by

$$y(t) = \frac{1}{2\pi f_0}\dot{\phi}(t) = \dot{x}(t) \tag{3.25}$$

$y(t)$ has no unit or we can use second/second. $y(t)$ is the time rate of the clock. The angle $\phi(t)$, the time phase $x(t)$, and the fractional frequency/time rate $y(t)$ are treated as random processes and their corresponding PSD functions satisfy the following equations:

$$\begin{aligned} S_y(f) &= \frac{1}{\left(2\pi f_0\right)^2} S_{\dot{\phi}}(f) = \frac{(2\pi f)^2}{\left(2\pi f_0\right)^2} S_\phi(f) = \left(\frac{f}{f_0}\right)^2 S_\phi(f) \\ S_x(f) &= \frac{1}{\left(2\pi f_0\right)^2} S_\phi(f) \\ S_y(f) &= \left(\frac{f}{f_0}\right)^2 S_\phi(f) = (2\pi f)^2 S_x(f) \end{aligned} \tag{3.26}$$

The PSD provides information about the contents of the noise at different frequencies.

The Allan variance is defined to be

$$\sigma_y^2(\tau) = \frac{1}{2}E\left(\left(\bar{y}_{k+1} - \bar{y}_k\right)^2\right) = \frac{1}{2}E\left(\left(x_{k+2} - 2x_{k+1} + x_k\right)^2\right) \tag{3.27}$$

where $\bar{y}_k$ is the average value of $y(t)$, that is, the average clock rate, over a period of τ:

$$\bar{y}_k = \frac{1}{\tau}\int_{t_k}^{t_k+\tau} y(t)\,dt = \frac{x_{k+1} - x_k}{\tau} \tag{3.28}$$

The Allan variance, as a function of averaging time τ, provides information about how the clock rate or in general, how the variable $y(t)$ is changing over the averaging period. Shorter averaging time captures the higher-frequency contents, and longer averaging time captures lower-frequency contents.

As we can intuitively imagine, the Allan variance and the PSD provide mathematically equivalent information and they can be derived from one another using the following equation:

$$\sigma_y^2(\tau) = 2\int_0^\infty S_y(f)\frac{\sin^4(\pi f\tau)}{(\pi f\tau)^2}\,df \tag{3.29}$$

(Note that if we use single-sided PSD, the factor is 2; if we use two-sided PSD, the factor will be 4; here we are using single-sided PSD.)

For most applications, the PSD can be approximated by the following power series expansion:

$$S_y(f) = \frac{h_{-2}}{f^2} + \frac{h_{-1}}{f} + h_0 + h_1 f + h_2 f^2 \tag{3.30}$$

where

h_{-2}/f^2 is the PSD of a random process called frequency/rate random walk.

h_{-1}/f is the PSD of a random process called frequency flicker noise.

h_0 is the PSD of a random process called frequency white noise or phase random walk.

$h_1 f$ is the PSD of a random process called phase flicker noise.

$h_2 f^2$ is the PSD of a random process called phase white noise.

As we know, a colored noise can be generated by a linear finite-dimensional system, if its PSD can be factored into the inner product of rational transfer functions.

The linear system that can be used to generate the frequency random walk will be an integrator. The linear system to generate frequency flicker noise unfortunately will not be a finite dimensional system and so is true for the phase flicker noise. The phase white noise can be generated using a differentiator.

When the PSD is plotted in a log-log scale, each one of the terms shows different slope in the PSD chart. Similarly, the Allan variance has an approximate power law expansion with respect to the average time τ. We have the following equation

$$\sigma_y^2(\tau) = \frac{(2\pi)^2}{6}h_{-2}\tau + (2\ln 2)h_{-1} + 0.5\frac{h_0}{\tau} + \frac{1.038 + 3\ln\left(2\pi f_h\tau\right)}{(2\pi)^2}\frac{h_1}{\tau^2} + \frac{3f_h}{(2\pi)^2}\frac{h_2}{\tau^2} \tag{3.31}$$

where f_h is the high-frequency cutoff beyond which the PSD is considered as 0.

Note that it will be difficult to distinguish between phase flicker and phase white noise in an Allan variance plot since they both have −2 power term of τ.

In order to use a Kalman filter, we will typically need to use realizable models to bound the PSD. For clocks, we will use the following realizable models:

3.2.2.1 Phase White Noise

This will be a white noise on the clock phase. For the fractional frequency, it can be modeled as a white noise passing through a differentiator. Consequently, the PSD can be written as

$$psd = \left(\omega^2\right)\sigma_{pwn}^2 = (2\pi)^2\sigma_{pwn}^2 f^2 = h_2 f^2 \tag{3.32}$$

and we have

$$h_2 = (2\pi)^2\sigma_{pwn}^2 \tag{3.33}$$

The corresponding Allan variance is then

$$\sigma_y^2(\tau) = \frac{3f_h}{(2\pi)^2}\frac{h_2}{\tau^2} = \left(3f_h\right)\frac{\sigma_{pwn}^2}{\tau^2} \tag{3.34}$$

3.2.2.2 Phase Random Walk

This will be modeled as white noise on the fractional frequency (or, equivalently, clock rate), which will be a random walk on the phase. Consequently, the PSD can be written as

$$psd = \sigma_{prw}^2 = h_0 \tag{3.35}$$

The corresponding Allan variance is then given by

$$\sigma_y^2(\tau) = 0.5\frac{h_0}{\tau} = \frac{\sigma_{PRW}^2}{2}\tau^{-1} \tag{3.36}$$

3.2.2.3 Frequency/Rate Random Walk

This will be modeled as the integration of white noise on the fractional frequency, consequently the PSD can be written as

$$psd = \frac{\sigma_{rrw}^2}{\omega^2} = \frac{\dfrac{\sigma_{rrw}^2}{(2\pi)^2}}{f^2} = \frac{h_{-2}}{f^2} \tag{3.37}$$

as the result

$$h_{-2} = \frac{\sigma_{rwn}^2}{(2\pi)^2} \tag{3.38}$$

The corresponding Allan variance is then given by

$$\sigma_y^2(\tau) = \frac{\sigma_{rrw}^2}{6}\tau \tag{3.39}$$

3.2.3 Rational Model for Clock Propagation

Using phase white noise, phase random walk, and rate random walk model, we can write the following clock propagation error model:

$$\frac{dx}{dt} = 1 + b + w_{PRW} \tag{3.40}$$

$$\frac{db}{dt} = w_{RRW} \tag{3.41}$$

$$mT = x + w_{pwn} \tag{3.42}$$

where $x(t)$ is the time phase in time unit.

$b(t)$ is the raw clock frequency or rate error. $b(t)$ is normally a slow-changing, almost constant variable, which is modeled as a random walk process.

w_{PRW} is the white noise process that integrates into the phase random walk error and

$$E\left(w_{PRW}\right) = 0 \tag{3.43}$$

$$E\left(w_{PRW}(t)w_{PRW}(s)\right) = \sigma_{prw}^2\delta(t - s) \tag{3.44}$$

w_{RRW} is the white noise process that integrates into the rate random walk, that is, the random walk in the clock rate bias $b(t)$ and

$$E\left(w_{RRW}\right) = 0 \tag{3.45}$$

$$E\left(w_{RRW}(t)w_{RRW}(s)\right) = \sigma_{rrw}^2\delta(t - s) \tag{3.46}$$

w_{PWN} is the white noise process that adds to the phase output and

$$E\left(w_{PWN}\right) = 0 \tag{3.47}$$

$$E\left(w_{PWN}(t)w_{PWN}(s)\right) = \sigma_{pwn}^2\delta(t - s) \tag{3.48}$$

If we just care about the clock error, we can use

$$\delta x(t) = x(t) - t \tag{3.49}$$

and we have

$$\begin{aligned} \frac{d}{dt}\begin{bmatrix} \delta x \\ b \end{bmatrix} &= \begin{bmatrix} 1 & 0 \\ 0 & 0 \end{bmatrix}\begin{bmatrix} \delta x \\ b \end{bmatrix} + \begin{bmatrix} w_{PRW} \\ w_{RRW} \end{bmatrix} \\ \delta T &= \delta x + w_{PRW} \end{aligned} \tag{3.50}$$

Using the differential Lyapunov equation, we can write the differential equation for the time phase error variance due to the phase random walk

$$\dot{p}_{prw} = 0 \bullet p_{prw} + \sigma_{PRW}^2 \tag{3.51}$$

which leads to

$$p_{prw} = \sigma^2{}_{PRW}t \text{ or } \sigma_{prw} = \sigma_{PRW}\sqrt{t} \tag{3.52}$$

In other words, the time error standard deviation grows with square root of the time.

For the rate random walk, we have the following differential Lyapunov equation:

$$\begin{bmatrix} \dot{p}_{11} & \dot{p}_{12} \\ \dot{p}_{12} & \dot{p}_{22} \end{bmatrix} = \begin{bmatrix} 0 & 1 \\ 0 & 0 \end{bmatrix} \begin{bmatrix} p_{11} & p_{12} \\ p_{12} & p_{22} \end{bmatrix} + \begin{bmatrix} p_{11} & p_{12} \\ p_{12} & p_{22} \end{bmatrix} \begin{bmatrix} 0 & 0 \\ 1 & 0 \end{bmatrix} + \begin{bmatrix} 0 & 0 \\ 0 & \sigma_{RRW}^2 \end{bmatrix} = \begin{bmatrix} 2p_{12} & p_{22} \\ p_{22} & \sigma_{RRW}^2 \end{bmatrix} \tag{3.53}$$

which can be simplified as

$$\begin{aligned} \dot{p}_{22} &= \sigma_{RRW}^2 \\ p_{22} &= \sigma_{RRW}^2 t \end{aligned} \tag{3.54}$$

and

$$\begin{aligned} \dot{p}_{12} &= p_{22} = \sigma_{RRW}^2 t \\ p_{12} &= \frac{1}{2}\sigma_{RRW}^2 t^2 \end{aligned} \tag{3.55}$$

and

$$\begin{aligned} \dot{p}_{11} &= 2p_{12} = \sigma_{RRW}^2 t^2 \\ p_{11} &= \frac{1}{3}\sigma_{RRW}^2 t^3 \end{aligned} \tag{3.56}$$

In conclusion, we have

$$\begin{aligned} \sigma_b(t) &= \sqrt{p_{22}} = \sigma_{RRW}\sqrt{t} \\ \sigma_{\delta x}(t) &= \sqrt{p_{11}} = \frac{1}{\sqrt{3}}\sigma_{RRW} t^{\frac{3}{2}} \\ \operatorname{cov}(\delta x, b) &= p_{12} = \frac{1}{2}\sigma_{RRW}^2 t^2 \end{aligned} \tag{3.57}$$

The above equations give us a clear sense how each type of the errors propagate forward with time. Specifically, the phase random walk-induced errors grow with $\sqrt{t}$ and the rate random walk-induced errors grow with $t^{3/2}$.

3.3 Inertial Measurement Unit (IMU) Gyro and Accelerometer Error Parameter Models

The three-axis gyros in an IMU measure the rotation rate vector of the IMU body frame (B) with respect to the inertial frame (I): ${}^{B}\omega_{I}^{B}$. The measured rate is denoted by $m^{B}\omega_{I}^{B}$ and the error in the measurement is given by

$$\delta^{B}\omega_{I}^{B} = m^{B}\omega_{I}^{B} - {}^{B}\omega_{I}^{B} = \delta S_{g}\,{}^{B}\omega_{I}^{B} + b_{g} + w_{arw} \tag{3.58}$$

where the 3 × 3 matrix δS_g models the scale factor errors (the diagonal elements) and the misalignments (off-diagonal elements) and the 3 × 1 vector b_g models the gyro bias errors.

Both the scale factor error/misalignment matrix and the bias vector are almost constant parameters that can be modeled using a random walk or the first-order Markov process mentioned earlier in this chapter.

w_{arw} is a 3 × 1 white noise vector that generates the angle random walk and

$$E\left(w_{arw}\right) = 0$$

$$E\left(w_{arw}(t)^{T} w_{arw}(s)\right) = \begin{bmatrix} \left(\sigma_{ARW}(1)\right)^{2} & 0 & 0 \\ 0 & \left(\sigma_{ARW}(2)\right)^{2} & 0 \\ 0 & & \left(\sigma_{ARW}(3)\right)^{2} \end{bmatrix} \delta(t-s) \tag{3.59}$$

Similarly, we can model the accelerometer errors.

The three-axis accelerometers in a IMU measure specific force (acceleration – gravity) of the IMU body frame (B) with respect to the inertial frame (I): ${}^{B}a_{I}^{B}$. The measured specific force is denoted by $m^{B}a_{I}^{B}$ and the error in the measurement is given by

$$\delta^{B}a_{I}^{B} = m^{B}a_{I}^{B} - {}^{B}a_{I}^{B} = \delta S_{a}a_{I}^{B} + b_{a} + w_{vrw} \tag{3.60}$$

where the 3 × 3 matrix δS_a models the scale factor errors (the diagonal elements) and the misalignments (off-diagonal elements) and the 3 × 1 vector b_a models the accelerometer bias errors.

Both the scale factor error/misalignment matrix and the bias vector are almost constant parameters, which can be modeled using a random walk or the first-order Markov process mentioned earlier in this chapter.

w_{vrw} is a 3 × 1 white noise vector that generates the velocity random walk and

$$E\left(w_{vrw}\right) = 0$$

$$E\left(w_{vrw}(t)^{T} w_{arw}(s)\right) = \begin{bmatrix} \left(\sigma_{VRW}(1)\right)^{2} & 0 & 0 \\ 0 & \left(\sigma_{VRW}(2)\right)^{2} & 0 \\ 0 & & \left(\sigma_{VRW}(3)\right)^{2} \end{bmatrix} \delta(t-s) \tag{3.61}$$

If we use the x-axis accelerometer as a reference axis and the plane spanned by the x-axis accelerometer and the y-axis accelerometer as the plane for the IMU body x-y plane, then the x-axis will not have any misalignment errors, and the y-axis will not have any errors normal to the plane spanned by x and y. Consequently, the scale factor error and misalignment matrix will carry the following form:

$$\delta S_a = \begin{bmatrix} s_{a11} & 0 & 0 \\ s_{a21} & s_{a22} & 0 \\ s_{a31} & s_{a32} & s_{a33} \end{bmatrix} \tag{3.62}$$

which contains three scale factor errors and only three misalignment errors.

This particular way of choosing the IMU frame is illustrated in Figure 3.2.

3.4 Attitude Propagation and Error Models

3.4.1 Angular Velocity and Attitude Kinematics

3.4.1.1 Angular Velocity

We start by defining the angular velocity using the Coriolis equation.

Given a vector $\vec{R}$ constant in the body frame B, the linear velocity of vector $\vec{R}$ with respect to the frame R is given by $(d\vec{R}/dt)_R$ where $(\ )_R$ denotes that the differentiation is done with respect to frame R.

The angular velocity is then implicitly defined by

$$\vec{V} = \left(\frac{d\vec{R}}{dt}\right)_R = \vec{\omega}_R^B \times \vec{R} \tag{3.63}$$

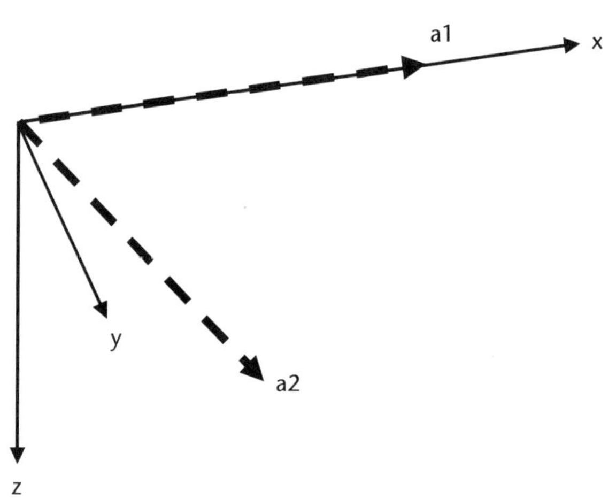

Figure 3.2 Frame defined using accelerometers.

Without a loss of generality, the angular velocity can be defined using a unit vector in the body frame

$$\left(\frac{d\vec{U}}{dt}\right)_R = \vec{\omega}_R^B \times \vec{U} \tag{3.64}$$

3.4.1.2 Attitude Kinematics Equation in DCM

Since $C_B^R = [{}^R u_{1B} \ {}^R u_{2B} \ {}^R u_{3B}]$ and ${}^R u_{kB}$ $(k = 1, 2, 3)$ are unit vectors in the body frame viewed in the reference frame, we have

$$\frac{dC_B^R}{dt} = \left[\frac{d\,{}^R u_{1B}}{dt} \ \frac{d\,{}^R u_{2B}}{dt} \ \frac{d\,{}^R u_{3B}}{dt}\right] = \left[{}^R\omega_R^B \times {}^R u_{1B} \ \ {}^R\omega_R^B \times {}^R u_{2B} \ \ {}^R\omega_R^B \times {}^R u_{3B}\right] \tag{3.65}$$

or

$$\frac{dC_B^R}{dt} = \left[{}^R\omega_R^B \times\right]\left[{}^R u_{1B} \ \ {}^R u_{2B} \ \ {}^R u_{3B}\right] = \left[{}^R\omega_R^B \times\right] C_B^R \tag{3.66}$$

and we have just derived a very important equation for the kinematics of the DCM:

$$\frac{dC_B^R}{dt} = \left[{}^R\omega_R^B \times\right] C_B^R \tag{3.67}$$

The above provides a differential equation that allows us to determine how the attitude (as represented by the DCM) propagates given the angular velocity expressed in the reference frame.

Alternatively, we can also use the angular velocity expressed in the body frame. Note that ${}^R\omega_R^B = C_B^R \, {}^B\omega_R^B$ and

$$\left[{}^R\omega_R^B \times\right] = \left[C_B^R \, {}^B\omega_R^B \times\right] = C_B^R\left[{}^B\omega_R^B \times\right] C_R^B \tag{3.68}$$

We have

$$\dot{C}_B^R = \left[{}^R\omega_R^B \times\right] C_B^R = C_B^R\left[{}^B\omega_R^B \times\right] C_R^B C_B^R = C_B^R\left[{}^B\omega_R^B \times\right] \tag{3.69}$$

In conclusion, we have

$$\dot{C}_B^R = \left[{}^R\omega_R^B \times\right] C_B^R = C_B^R\left[{}^B\omega_R^B \times\right] \tag{3.70}$$

It is also possible to use the rotation rate expressed in the body frame for the rotation rate of the body with respect to the inertial frame, and the rate expressed in the reference frame for the rotation rate of the reference frame with respect to the inertial frame.

Using $C_B^R = C_I^R C_B^I$, we can show that

$$\dot{C}_B^R = \dot{C}_I^R C_B^I + C_I^R \dot{C}_B^I = \left[{}^R\omega_R^I \times\right] C_I^R C_B^I + C_I^R C_B^I \left[{}^B\omega_I^B \times\right] = -\left[{}^R\omega_I^R \times\right] C_B^R + C_B^R\left[{}^B\omega_I^B \times\right] \tag{3.71}$$

So we have

$$\dot{C}_B^R = C_B^R\left[{}^B\omega_I^B\times\right] - \left[{}^R\omega_I^R\times\right]C_B^R \tag{3.72}$$

3.4.1.3 Attitude Kinematics Equation in Quaternion Form

Similarly, it is not hard to derive the following differential equations in the quaternion form

$$\dot{q}_B^R = \frac{1}{2}\,{}^R\omega_R^B \otimes q_B^R = \frac{1}{2}q_B^R \otimes {}^B\omega_R^B \tag{3.73}$$

$$\dot{q}_B^R = \frac{1}{2}\left(q_B^R \otimes {}^B\omega_I^B - {}^R\omega_I^R \otimes q_B^R\right) \tag{3.74}$$

where

$$v \otimes q \equiv \begin{bmatrix} v \\ 0 \end{bmatrix} \otimes q, \quad q \otimes v \equiv q \otimes \begin{bmatrix} v \\ 0 \end{bmatrix}$$

3.4.1.4 Attitude Kinematics in the Discrete Form

By simply using the chain rules, we have the following DCM equations:

$$C_R^{B(k+1)} = C_{B(k)}^{B(k+1)}C_R^{B(k)} \tag{3.75}$$

$$C_{R(k+1)}^{B} = C_{R(k)}^{B}C_{R(k+1)}^{R(k)} \tag{3.76}$$

$$C_{R(k+1)}^{B(k+1)} = C_{B(k)}^{B(k+1)}C_{R(k)}^{B(k)}C_{R(k+1)}^{R(k)} \tag{3.77}$$

Similarly, we have the quaternion equations

$$q_R^{B(k+1)} = q_{B(k)}^{B(k+1)}q_R^{B(k)} \tag{3.78}$$

$$q_{R(k+1)}^{B} = q_{R(k)}^{B}q_{R(k+1)}^{R(k)} \tag{3.79}$$

$$q_{R(k+1)}^{B(k+1)} = q_{B(k)}^{B(k+1)}q_{R(k)}^{B(k)}q_{R(k+1)}^{R(k)} \tag{3.80}$$

The delta DCM and the delta quaternion can be calculated using the following equations.

Assuming that the body frame is moved by $\delta\,{}^B\vec{\theta}_B(k)$ and the reference frame is moved by $\delta\,{}^R\vec{\theta}_R(k)$, we then have

$$\begin{aligned} C_{B(k)}^{B(k+1)} &= I - \sin\delta\,{}^B\theta_B\left[\left(\frac{\delta\,{}^B\vec{\theta}_B(k)}{\delta\,{}^B\theta_B(k)}\right)\times\right] + \left(1-\cos\delta\,{}^B\theta_B\right)\left[\left(\frac{\delta\,{}^B\vec{\theta}_B(k)}{\delta\,{}^B\theta_B(k)}\right)\times\right]^2 \\ &= I - \frac{\sin\delta\,{}^B\theta_B(k)}{\delta\,{}^B\theta_B(k)}\left[\delta\,{}^B\vec{\theta}_B(k)\times\right] + \frac{\left(1-\cos\delta\,{}^B\theta_B(k)\right)}{\left(\delta\,{}^B\theta_B(k)\right)^2}\left[\delta\,{}^B\vec{\theta}_B(k)\times\right]^2 \end{aligned} \tag{3.81}$$

and

$$C_{R(k)}^{R(k+1)} = I - \frac{\sin\delta^R\theta_R(k)}{\delta^R\theta_R(k)}\left[\delta^R\vec{\theta}_R(k)\times\right] + \frac{\left(1-\cos\delta^R\theta_R(k)\right)}{\left(\delta^R\theta_R(k)\right)^2}\left[\delta^R\vec{\theta}_R(k)\times\right]^2$$

$$C_{R(k+1)}^{R(k)} = I + \frac{\sin\delta^R\theta_R(k)}{\delta^R\theta_R(k)}\left[\delta^R\vec{\theta}_R(k)\times\right] + \frac{\left(1-\cos\delta^R\theta_R(k)\right)}{\left(\delta^R\theta_R(k)\right)^2}\left[\delta^R\vec{\theta}_R(k)\times\right]^2 \tag{3.82}$$

The quaternion versions of these equations are listed here:

$$q_{B(k)}^{B(k+1)} = \begin{bmatrix} \left(\dfrac{\delta^B\vec{\theta}_B(k)}{\delta^B\theta_B(k)}\right)\sin\dfrac{\delta^B\theta_B}{2} \\ \cos\dfrac{\delta^B\theta_B}{2} \end{bmatrix} \tag{3.83}$$

$$q_{R(k)}^{R(k+1)} = \begin{bmatrix} \left(\dfrac{\delta^R\vec{\theta}_R(k)}{\delta^R\theta_R(k)}\right)\sin\dfrac{\delta^R\theta_R}{2} \\ \cos\dfrac{\delta^R\theta_R}{2} \end{bmatrix}$$

$$q_{R(k+1)}^{R(k)} = \begin{bmatrix} -\left(\dfrac{\delta^R\vec{\theta}_R(k)}{\delta^R\theta_R(k)}\right)\sin\dfrac{\delta^R\theta_R}{2} \\ \cos\dfrac{\delta^R\theta_R}{2} \end{bmatrix} \tag{3.84}$$

3.4.2 Attitude Propagation/Numerical Integration Algorithms

We will describe three types of attitude propagation algorithms:

- *Type 1:* The algorithm will be running at the rate that the gyro data is available and it will assume that the axis of rotation remains unchanged during the gyro sampling interval.
- *Type 2:* With the same assumption as the type 1 algorithm, the gyro delta angle will be first integrated using coning correction algorithm and then propagated using the type 1 algorithm at lower rate.
- *Type 3:* This algorithm assumes that the instantaneous axis of rotation may be changing and attempts to achieve a better performance.

Without a loss of generality, we will focus on quaternion-based equations. DCM-based equations can be similarly derived. For all practical purposes, quaternion algorithms will be used in real implementations due to much less computational burden.

3.4.2.1 Type 1 Algorithm

The type 1 algorithm uses the chain rules for quaternions and DCMs. In the following, we will list a few important versions of this algorithm.

If the reference frame is the inertial frame, for example, ECI or any other inertial frame, the attitude propagation algorithm bears the simplest form:

$$q_I^{B(k+1)} = \begin{bmatrix} \left(\dfrac{\delta^B \vec{\theta}_B(k)}{\delta^B \theta_B(k)} \right) \sin \dfrac{\delta^B \theta_B}{2} \\ \cos \dfrac{\delta^B \theta_B}{2} \end{bmatrix} \otimes q_I^{B(k)} \tag{3.85}$$

For spacecraft applications, it is often useful to use the orbit frame as the reference frame, that is, $R = O$ (orbit frame), and the orbit frame is rotating around the negative orbit-normal axis ($-y$-axis) with respect to the inertial frame; consequently, we have

$$\delta^O \vec{\theta}_O(k) = \begin{bmatrix} 0 \\ -\Omega \delta t \\ 0 \end{bmatrix} \tag{3.86}$$

where Ω is the orbit rate.

The attitude propagation algorithm will then be given by

$$q_O^{B(k+1)} = \begin{bmatrix} \left(\dfrac{\delta^B \vec{\theta}_B(k)}{\delta^B \theta_B(k)} \right) \sin \dfrac{\delta^B \theta_B}{2} \\ \cos \dfrac{\delta^B \theta_B}{2} \end{bmatrix} \otimes q_O^{B(k)} \otimes \begin{bmatrix} 0 \\ -\sin 0.5\Omega \delta t \\ 0 \\ \cos 0.5\Omega \delta t \end{bmatrix} \tag{3.87}$$

For Earth navigation, it is often convenient to use ECEF frame as the reference frame, that is, R = ECEF and since ECEF frame is rotating around its z-axis at the Earth rotation rate, we have

$$\delta^E \vec{\theta}_E(k) = \begin{bmatrix} 0 \\ 0 \\ \Omega_e \delta t \end{bmatrix} \tag{3.88}$$

where Ω_e is the Earth rotation rate and the attitude propagation algorithm will then be given by

$$q_E^{B(k+1)} = \begin{bmatrix} \left(\dfrac{\delta^B \vec{\theta}_B(k)}{\delta^B \theta_B(k)} \right) \sin \dfrac{\delta^B \theta_B}{2} \\ \cos \dfrac{\delta^B \theta_B}{2} \end{bmatrix} \otimes q_E^{B(k)} \otimes \begin{bmatrix} 0 \\ 0 \\ \sin 0.5\Omega_e \delta t \\ \cos 0.5\Omega_e \delta t \end{bmatrix} \tag{3.89}$$

Another frequently used reference frame is the local level frame, often called the navigation frame (i.e., $R = N$). The navigation frame rotates with respect to the inertial frame at the summation of two rotation rates: the craft rate, the rotation rate of the local level frame with respect to Earth due to the velocity respect to the Earth; and the Earth rotation rate with respect to the inertial frame. In other words, we have

$$ {}^{N}\omega_{I}^{N} = {}^{N}\omega_{E}^{N} + {}^{N}\omega_{I}^{E} \tag{3.90} $$

For most applications, ${}^{N}\omega_{I}^{N}$ changes much more slowly than the body rotation rate and can therefore be safely treated as constant during the attitude propagation interval; consequently, we have

$$ \delta^{N}\vec{\theta}_{N}(k) = {}^{N}\omega_{I}^{N}\delta t \tag{3.91} $$

The attitude propagation algorithm will then be given by

$$ q_{E}^{B(k+1)} = \begin{bmatrix} \left(\dfrac{\delta^{B}\vec{\theta}_{B}(k)}{\delta^{B}\theta_{B}(k)}\right)\sin\dfrac{\delta^{B}\theta_{B}}{2} \\ \cos\dfrac{\delta^{B}\theta_{B}}{2} \end{bmatrix} \otimes q_{E}^{B(k)} \otimes \begin{bmatrix} \dfrac{{}^{N}\omega_{I}^{N}}{\left\|{}^{N}\omega_{I}^{N}\right\|}\sin\dfrac{\left\|{}^{N}\omega_{I}^{N}\right\|\delta t}{2} \\ \cos\dfrac{\left\|{}^{N}\omega_{I}^{N}\right\|\delta t}{2} \end{bmatrix} \tag{3.92} $$

3.4.2.2 Type 2 Algorithm

As described before, the type 2 algorithm uses the identical equations as type 1 algorithm at a lower rate using delta angles that are the result of coning correction integration algorithm.

The foundation of the approach is that the instantaneous angle of rotation satisfy the following equation

$$ \dot{\vec{\theta}}(t) = {}^{B}\omega_{I}^{B}(t) + \frac{1}{2}\vec{\theta}(t) \times {}^{B}\omega_{I}^{B}(t) + \frac{1}{\theta^{2}}\left(1 - \frac{\theta\sin\theta}{2(1-\cos\theta)}\right)\vec{\theta}(t) \times \left(\vec{\theta}(t) \times {}^{B}\omega_{I}^{B}(t)\right) \tag{3.93} $$

One thing important to observe about the above equation is that when the axis of rotation does not change, then $\vec{\theta}(t)$ and ${}^{B}\omega_{I}^{B}(t)$ are in the same direction, and $\vec{\theta}(t) \times {}^{B}\omega_{I}^{B}(t) = 0$, and we have

$$ \dot{\vec{\theta}}(t) = {}^{B}\omega_{I}^{B}(t) $$

which matches with our intuitive expectation that that there is no coning correction needed if the axis of rotation remains constant.

Typically, the third term in the exact equation above is very small compared with the other terms and is consequently ignored, and the following is the standard equation used by almost everyone for coning correction:

$$ \dot{\vec{\theta}}(t) = {}^{B}\omega_{I}^{B}(t) + \frac{1}{2}\vec{\theta}(t) \times {}^{B}\omega_{I}^{B}(t) \tag{3.94} $$

Figure 3.3 illustrates the difference between the type 2 algorithms and type 1 algorithm.

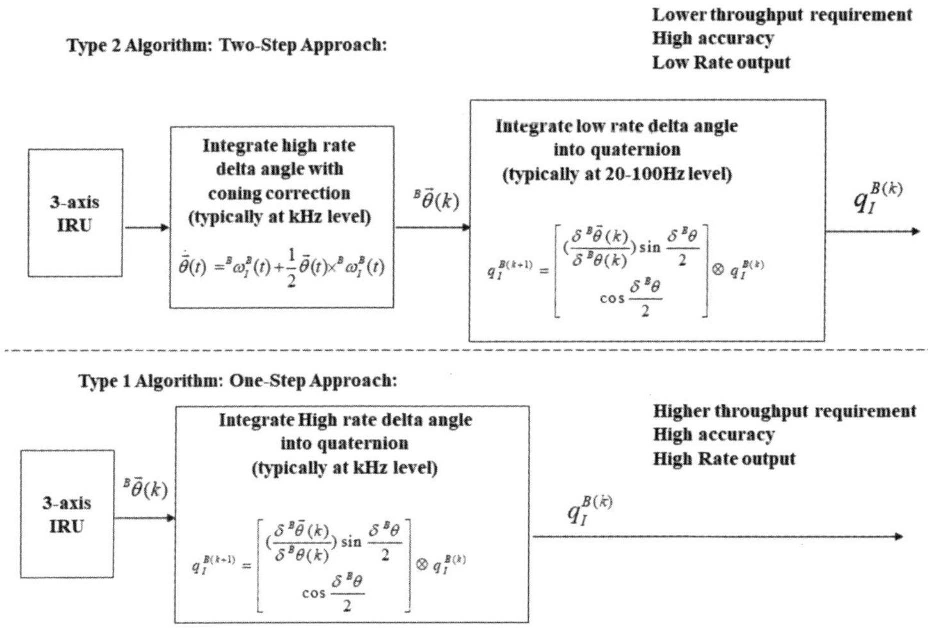

Figure 3.3 The type 1 versus the type 2 attitude propagation algorithms.

3.4.2.3 Type 3 Algorithm

The type 3 algorithm does not assume that the axis of rotation even stay fixed within the interval of the gyro delta angle output and attempts to derive useful information from consecutive samples. The type 3 algorithm typically can achieve higher accuracy and is often the algorithm of choice.

We start with the following differential equation:

$$\dot{q}_B^R = \frac{1}{2} q_B^R \otimes {}^B\omega_R^B = \frac{1}{2} q_B^R \, {}^B\omega_R^B$$
$$\text{where} \quad q \otimes v \equiv q \otimes \begin{bmatrix} v \\ 0 \end{bmatrix} \equiv q \begin{bmatrix} v \\ 0 \end{bmatrix} \tag{3.95}$$

Our purpose is to develop a numerical integration method for the above differential equation. We achieve fairly good performance if we use RK4 (fourth-order Runge-Kutta algorithm). In the following, we will derive an example algorithm that has been proven to have a good performance.

To further simplify the symbols, we will be simply using

$$q = q_B^R, \quad W = \begin{bmatrix} \omega_1 \\ \omega_2 \\ \omega_3 \\ 0 \end{bmatrix} = \begin{bmatrix} {}^B\omega_R^B \\ 0 \end{bmatrix} \quad \text{and} \quad \dot{q} = \frac{1}{2} q W \tag{3.96}$$

Similar to the derivation of other numerical integration method, we will use the Taylor expansion

$$q_p = q(t - \Delta t) = q - \frac{dq}{dt}\Delta t + \frac{1}{2}\frac{d^2q}{dt^2}(\Delta t)^2 - \frac{1}{6}\frac{d^3q}{dt^3}(\Delta t)^3 + \frac{1}{24}\frac{d^4q}{dt^4}(\Delta t)^4 + \cdots \quad (3.97)$$

and

$$\begin{aligned}
\frac{dq}{dt} &= q\left(\frac{1}{2}W\right)\\
\frac{d^2q}{dt^2} &= q\left(\frac{1}{4}W^2 + \frac{1}{2}\dot{W}\right)\\
\frac{d^3q}{dt^3} &= q\left[\frac{1}{8}W^3 + \frac{1}{4}W\dot{W} + \frac{1}{4}\left(\dot{W}W + W\dot{W}\right) + \frac{1}{2}\ddot{W}\right]\\
\frac{d^4q}{dt^4} &= q\left[\frac{1}{16}W^4 + \frac{1}{8}\left(W^2\dot{W} + \dot{W}W^2\right) + \frac{1}{4}W\left(W\dot{W} + \dot{W}W\right) + \frac{3}{4}\dot{W}^2 + \frac{1}{4}\left(W\ddot{W} + \ddot{W}W\right) + \frac{1}{2}W\ddot{W} + \frac{1}{2}\dddot{W}\right]
\end{aligned} \quad (3.98)$$

The following symbols will be used to simplify the equations:

$$\begin{aligned}
\frac{1}{2}\Delta t\omega_i(t) &= \delta_i\\
\frac{1}{2}\Delta t\omega_i(t - \Delta t) &= \delta_i^p\\
\delta_0^2 &= \delta_1^2 + \delta_2^2 + \delta_3^2
\end{aligned}$$

The following equations are useful for the third-order and fourth-order equations

$$\frac{1}{2}W\Delta t = \begin{pmatrix} \delta_1 \\ \delta_2 \\ \delta_3 \\ 0 \end{pmatrix}$$

$$\frac{1}{2}\dot{W}\Delta t = \begin{pmatrix} \dot{\delta}_1 \\ \dot{\delta}_2 \\ \dot{\delta}_3 \\ 0 \end{pmatrix}$$

$$\frac{1}{2}\ddot{W}\Delta t = \begin{pmatrix} \ddot{\delta}_1 \\ \ddot{\delta}_2 \\ \ddot{\delta}_3 \\ 0 \end{pmatrix}$$

$$\frac{1}{2}\dddot{W}\Delta t = \begin{pmatrix} \dddot{\delta}_1 \\ \dddot{\delta}_2 \\ \dddot{\delta}_3 \\ 0 \end{pmatrix}$$

$$\frac{1}{8}W^3(\Delta t)^3 = -\delta_0^2\begin{pmatrix}\delta_1\\ \delta_2\\ \delta_3\\ 0\end{pmatrix}$$

$$\frac{1}{8}\left(W^2\dot{W} + \dot{W}W^2\right)(\Delta t)^3 = -2\delta_0^2\begin{pmatrix}\dot{\delta}_1\\ \dot{\delta}_2\\ \dot{\delta}_3\\ 0\end{pmatrix}$$

$$\frac{1}{4}W\dot{W}(\Delta t)^2 = \begin{pmatrix}\delta_2\dot{\delta}_3 - \delta_3\dot{\delta}_2\\ \delta_3\dot{\delta}_1 - \delta_1\dot{\delta}_3\\ \delta_1\dot{\delta}_2 - \delta_2\dot{\delta}_1\\ -\left(\delta_1\dot{\delta}_1 + \delta_2\dot{\delta}_2 + \delta_3\dot{\delta}_3\right)\end{pmatrix}$$

$$\frac{1}{4}W\ddot{W}(\Delta t)^2 = \begin{pmatrix}\delta_2\ddot{\delta}_3 - \delta_3\ddot{\delta}_2\\ \delta_3\ddot{\delta}_1 - \delta_1\ddot{\delta}_3\\ \delta_1\ddot{\delta}_2 - \delta_2\ddot{\delta}_1\\ -\left(\delta_1\ddot{\delta}_1 + \delta_2\ddot{\delta}_2 + \delta_3\ddot{\delta}_3\right)\end{pmatrix}$$

$$\frac{1}{8}W\left(W\dot{W} + \dot{W}W\right)(\Delta t)^3 = -2\left(\delta_1\dot{\delta}_1 + \delta_2\dot{\delta}_2 + \delta_3\dot{\delta}_3\right)\begin{pmatrix}\delta_1\\ \delta_2\\ \delta_3\\ 0\end{pmatrix}$$

$$\frac{1}{4}W^2(\Delta t)^2 = -\delta_0^2 q_u$$

$$\frac{1}{4}\left(\dot{W}W + W\dot{W}\right)(\Delta t)^2 = -2\left(\delta_1\dot{\delta}_1 + \delta_2\dot{\delta}_2 + \delta_3\dot{\delta}_3\right)q_u$$

$$\frac{1}{16}W^4(\Delta t)^4 = \delta_0^4 q_u \dot{W}^2(\Delta t)^2 = -\left(\dot{\delta}_1^2 + \dot{\delta}_2^2 + \dot{\delta}_3^2\right)q_u$$

$$\frac{1}{4}\dot{W}^2(\Delta t)^2 = -\left(\dot{\delta}_1^2 + \dot{\delta}_2^2 + \dot{\delta}_3^2\right)q_u$$

$$\frac{1}{4}\left(W\ddot{W} + \ddot{W}W\right)(\Delta t)^2 = -2\left(\delta_1\ddot{\delta}_1 + \delta_2\ddot{\delta}_2 + \delta_3\ddot{\delta}_3\right)q_u$$

For attitude propagation, we have

$$q_p = q(t - \Delta t) = q(t)\Phi(t, t - \Delta t) \equiv q(t)\begin{pmatrix} \Phi_1 \\ \Phi_2 \\ \Phi_3 \\ \Phi_4 \end{pmatrix} \tag{3.99}$$

or

$$q(t) = q(t - \Delta t)\begin{pmatrix} -\Phi_1 \\ -\Phi_2 \\ -\Phi_3 \\ \Phi_4 \end{pmatrix} \tag{3.100}$$

We will use the following equations to calculate the rate and the acceleration of the delta angle:

$$\begin{aligned} \dot{\delta}_i \Delta t &= \delta_i - \delta_i^{\,p} \\ \ddot{\delta}_i (\Delta t)^2 &= \delta_i - 2\delta_i^{\,p} + \delta_i^{\,pp} \\ \dddot{\delta}_i (\Delta t)^3 &= \delta_i - 3\delta_i^{\,p} + 3\delta_i^{\,pp} - \delta_i^{\,ppp} \end{aligned}$$

For the third-order algorithm, we have

$$\Phi(t, t' - \Delta t) = \begin{pmatrix} \left(-\frac{2}{3} + \frac{1}{6}\delta_0^2\right)\delta_1 + \frac{1}{6}\left(\delta_2\delta_3^p - \delta_3\delta_2^p\right) - \frac{1}{6}\left(\delta_1^p + \delta_1^{pp}\right) \\ \left(-\frac{2}{3} + \frac{1}{6}\delta_0^2\right)\delta_2 + \frac{1}{6}\left(\delta_3\delta_1^p - \delta_1\delta_3^p\right) - \frac{1}{6}\left(\delta_2^p + \delta_2^{pp}\right) \\ \left(-\frac{2}{3} + \frac{1}{6}\delta_0^2\right)\delta_3 + \frac{1}{6}\left(\delta_1\delta_2^p - \delta_2\delta_1^p\right) - \frac{1}{6}\left(\delta_3^p + \delta_3^{pp}\right) \\ 1 - \frac{1}{2}\left(\delta_1\delta_1^p + \delta_2\delta_2^p + \delta_3\delta_3^p\right) \end{pmatrix} \tag{3.101}$$

$$\begin{pmatrix} -\Phi_1 \\ -\Phi_2 \\ -\Phi_3 \\ \Phi_4 \end{pmatrix} = \begin{pmatrix} \left(\frac{2}{3} - \frac{1}{6}\delta_0^2\right)\delta_1 \quad \frac{1}{6}\left(\delta_2\delta_3^p - \delta_3\delta_2^p\right) + \frac{1}{6}\left(\delta_1^p + \delta_1^{pp}\right) \\ \left(\frac{2}{3} - \frac{1}{6}\delta_0^2\right)\delta_2 - \frac{1}{6}\left(\delta_3\delta_1^p - \delta_1\delta_3^p\right) + \frac{1}{6}\left(\delta_2^p + \delta_2^{pp}\right) \\ \left(\frac{2}{3} - \frac{1}{6}\delta_0^2\right)\delta_3 - \frac{1}{6}\left(\delta_1\delta_2^p - \delta_2\delta_1^p\right) + \frac{1}{6}\left(\delta_3^p + \delta_3^{pp}\right) \\ 1 - \frac{1}{2}\left(\delta_1\delta_1^p + \delta_2\delta_2^p + \delta_3\delta_3^p\right) \end{pmatrix} \tag{3.102}$$

We can use the following simplifications:

$$\frac{1}{6}\left(\delta_i^p + \delta_i^{pp}\right) = \frac{1}{3}\delta_i$$
$$1 - \frac{1}{2}\left(\delta_1\delta_1^p + \delta_2\delta_2^p + \delta_3\delta_3^p\right) = 1 - \delta_0^2$$

and we end up with

$$\begin{pmatrix} -\Phi_1 \\ -\Phi_2 \\ -\Phi_3 \\ \Phi_4 \end{pmatrix} = \begin{pmatrix} \left(1 - \frac{1}{6}\delta_0^2\right)\delta_1 - \frac{1}{6}\left(\delta_2\delta_3^p - \delta_3\delta_2^p\right) \\ \left(1 - \frac{1}{6}\delta_0^2\right)\delta_2 - \frac{1}{6}\left(\delta_3\delta_1^p - \delta_1\delta_3^p\right) \\ \left(1 - \frac{1}{6}\delta_0^2\right)\delta_3 - \frac{1}{6}\left(\delta_1\delta_2^p - \delta_2\delta_1^p\right) \\ 1 - \frac{1}{2}\delta_0^2 \end{pmatrix} \tag{3.103}$$

We can also derive the fourth-order algorithm:

$$\Phi(t,t-\Delta t) =$$
$$\begin{pmatrix} \left[-\frac{15}{24} - \frac{1}{12}\delta_0^2 + \frac{1}{6}\left(\delta_1\delta_1^p + \delta_2\delta_2^p + \delta_3\delta_3^p\right)\right]\delta_1 - \left(\frac{7}{24} - \frac{1}{12}\delta_0^2\right)\delta_1^p - \frac{1}{24}\left(\delta_1^{pp} + \delta_1^{ppp}\right) + \frac{1}{12}\left(\delta_2\delta_3^{pp} - \delta_3\delta_2^{pp}\right) \\ \left[-\frac{15}{24} - \frac{1}{12}\delta_0^2 + \frac{1}{6}\left(\delta_1\delta_1^p + \delta_2\delta_2^p + \delta_3\delta_3^p\right)\right]\delta_2 - \left(\frac{7}{24} - \frac{1}{12}\delta_0^2\right)\delta_2^p - \frac{1}{24}\left(\delta_2^{pp} + \delta_2^{ppp}\right) + \frac{1}{12}\left(\delta_3\delta_1^{pp} - \delta_1\delta_3^{pp}\right) \\ \left[-\frac{15}{24} - \frac{1}{12}\delta_0^2 + \frac{1}{6}\left(\delta_1\delta_1^p + \delta_2\delta_2^p + \delta_3\delta_3^p\right)\right]\delta_3 - \left(\frac{7}{24} - \frac{1}{12}\delta_0^2\right)\delta_3^p - \frac{1}{24}\left(\delta_3^{pp} + \delta_3^{ppp}\right) + \frac{1}{12}\left(\delta_1\delta_2^{pp} - \delta_2\delta_1^{pp}\right) \\ 1 - \frac{7}{24}\delta_0^2 + \frac{1}{24}\delta_0^4 + \frac{1}{12}\left(\delta_1\delta_1^p + \delta_2\delta_2^p + \delta_3\delta_3^p\right) - \frac{1}{8}\left(\delta_0^p\right)^2 - \frac{1}{6}\left(\delta_1\delta_1^{pp} + \delta_2\delta_2^{pp} + \delta_3\delta_3^{pp}\right) \end{pmatrix} \tag{3.104}$$

3.4.3 Euler Angle Attitude Propagation

For analysis purposes, it is sometimes useful to use Euler angle-based equations. For that purpose, we next list the differential equations for the typical 321-yaw-pitch-roll Euler angles

$$\begin{aligned} \dot{\varphi} &= \left(\omega_3 \cos\psi - \omega_1 \sin\psi\right)\sec\theta \\ \dot{\theta} &= \omega_3 \sin\psi + \omega_1 \cos\psi \\ \dot{\psi} &= \omega_2 - \left(\omega_3 \cos\psi - \omega_1 \sin\psi\right)\tan\theta \end{aligned} \tag{3.105}$$

and

$$\begin{aligned} \omega_1 &= \dot{\theta}\cos\psi - \dot{\varphi}\cos\theta\sin\psi \\ \omega_2 &= \dot{\psi} + \dot{\varphi}\sin\theta \\ \omega_3 &= \dot{\theta}\sin\psi + \dot{\varphi}\cos\theta\cos\psi \end{aligned} \tag{3.106}$$

3.4.4 Attitude Propagation Error Models

The attitude propagation error models will at least serve the following purposes:

1. Allow us to analyze how the different sensor parameters will impact the attitude propagation error.
2. Allow us to build linearized equations that will be useful for Kalman filter development.

The attitude is about the relationship between two coordinate frames. Without loss of generality, we can call them body and reference frames. It is important to realize that the attitude error can be expressed in either the body frame or the reference frame. Although the two expressions are mathematically equivalent, they do have very different implications in terms of practical applications and convenience.

3.4.4.1 Two Types of Attitude Errors

Given a DCM C_B^R and its knowledge or estimate $\hat{C}_B^R$, the difference between the two DCMs can be viewed entirely as an error of the knowledge of the reference frame; in other words, we have

$$\hat{C}_B^R = C_R^{\hat{R}} C_B^R = \left(I - \left[\theta_R^{\hat{R}} \times\right]\right) C_B^R = \left(I - \left[\delta^R \theta_B^R \times\right]\right) C_B^R \tag{3.107}$$

where $\delta^R \theta_B^R = \theta_R^{\hat{R}}$ is the error of the attitude (R with respect to B) expressed in the R frame.

$\delta^R \theta_B^R$ is a better symbol than $\theta_R^{\hat{R}}$ as it provides the context explicitly.

Similarly, we can also view the difference between the two DCMs as an error of the knowledge of the body frame:

$$\hat{C}_B^R = C_B^R C_{\hat{B}}^B = C_B^R\left(I - \left[\theta_{\hat{B}}^B \times\right]\right) = C_B^R\left(I + \left[\theta_B^{\hat{B}} \times\right]\right) = C_B^R\left(I + \left[\delta^B \theta_R^B \times\right]\right) \tag{3.108}$$

where $\delta^B \theta_R^B = \theta_B^{\hat{B}} = -\delta^B \theta_B^R$ is the error of the attitude (B with respect to R) expressed in the B frame.

Clearly, for the same attitude error described in two different ways, we should have

$$C_B^R\left(I + \left[\theta_B^{\hat{B}} \times\right]\right) = \left(I - \left[\theta_R^{\hat{R}} \times\right]\right) C_B^R$$

Let us do some simple algebra:

$$\begin{aligned} C_B^R\left(I + \left[[\theta_B^{\hat{B}} \times\right]\right) &= C_B^R I + C_B^R\left[\theta_B^{\hat{B}} \times\right] = C_B^R I + \left[\left(C_B^R \theta_B^{\hat{B}}\right) \times\right] C_B^R \\ &= \left(I + \left[\left(C_B^R \theta_B^{\hat{B}}\right) \times\right]\right) C_B^R = \left(I - \left[\theta_R^{\hat{R}} \times\right]\right) C_B^R \end{aligned}$$

In order for the above to be true, we must have:

$$\theta_R^{\hat{R}} = -C_B^R \theta_B^{\hat{B}} \quad \text{or} \quad \theta_B^{\hat{B}} = -C_R^B \theta_R^{\hat{R}} \tag{3.109}$$

The minus signs become a lot more understandable if we use the more explicit symbols:

$$\delta^R\theta_B^R = -C_B^R\delta^B\theta_R^B - C_B^R\delta^B\theta_B^R \tag{3.110}$$

or

$$\delta^B\theta_R^B = \theta_B^{\hat{B}} = -C_R^B\delta^R\theta_B^R = C_R^B\delta^R\theta_R^B \tag{3.111}$$

3.4.4.2 Attitude Error Kinematics

Let us first look at the kinematic equation for attitude error represented in the reference frame.

Given that we have three previously derived equations, the definition of attitude error in the reference frame:

$$C_B^{\hat{R}} = \left(I - \left[\delta^R\theta_I^R\times\right]\right)C_B^R \tag{3.112}$$

The kinematic equation of the truth attitude in DCM form:

$$\dot{C}_B^R = C_B^R\left[{}^B\omega_R^B\times\right] \tag{3.113}$$

The kinematic equation of the estimated attitude in DCM form:

$$\dot{\hat{C}}_B^R = \hat{C}_B^R\left[{}^B\hat{\omega}_R^B\times\right] \tag{3.114}$$

On one side, we have

$$\dot{\hat{C}}_B^R = \hat{C}_B^R\left[{}^B\omega_R^B\times\right] = \left(I - \left[\delta^R\theta_I^R\times\right]\right)C_B^R\left[{}^B\hat{\omega}_R^B\times\right] = \left(I - \left[\delta^R\theta_I^R\times\right]\right)\left[C_B^R\,{}^B\hat{\omega}_R^B\times\right]C_B^R$$

On the other side,

$$\begin{aligned}
\dot{\hat{C}}_B^R &= \frac{d}{dt}\left(I - \left[\delta^R\theta_I^R\times\right]\right)C_B^R = -\left[\delta^R\dot{\theta}_I^R\times\right]C_B^R + \left(I - \left[\delta^R\theta_I^R\times\right]\right)\dot{C}_B^R \\
&= -\left[\delta^R\dot{\theta}_I^R\times\right]C_B^R + \left(I - \left[\delta^R\dot{\theta}_I^R\times\right]\right)C_B^R\left[{}^B\omega_R^B\times\right] \\
&= -\left[\delta^R\dot{\theta}_I^R\times\right]C_B^R + \left(I - \left[\delta^R\theta_I^R\times\right]\right)\left[C_B^R\,{}^B\omega_R^B\times\right]C_B^R
\end{aligned}$$

Equating both sides, we have

$$-\left[\delta^R\dot{\theta}_I^R\times\right] = \left(I - \left[\delta^R\theta_I^R\times\right]\right)\left[C_B^R\left({}^B\hat{\omega}_R^B - {}^B\omega_R^B\right)\times\right] = \left(I - \left[\delta^R\theta_I^R\times\right]\right)\left[C_B^R\left(\delta^B\omega_R^B\right)\times\right]$$

Ignoring the second-order terms, we have

$$\left[\delta^R\dot{\theta}_I^R\times\right] = -\left[C_B^R\left(\delta^B\omega_R^B\right)\times\right]$$

Finally, we have the following attitude error kinematic equation for attitude error represented in the reference frame

$$\delta^R\dot{\theta}_B^R = -C_B^R\delta^B\omega_R^B \tag{3.115}$$

where $\delta^B\omega_R^B$ is the rate error in the body frame.

The minus sign is easy to understand: a positive error in the B with respect to R rate will cause an attitude error in the B with respect to R attitude and a negative error in the R with respect to B attitude.

We can also derive the kinematic equation for the attitude error represented in the body frame.

Starting from equation (3.111), we can write

$$\delta^B\dot{\theta}_R^B = -\dot{C}_R^B\delta^R\theta_B^R - C_R^B\delta^R\dot{\theta}_B^R$$

Using

$$\dot{C}_R^B = -\left[{}^B\omega_R^B\times\right]C_R^B$$

and

$$\delta^R\dot{\theta}_B^R = -C_B^R\delta^B\omega_R^B$$

We have

$$\begin{aligned}\delta^B\dot{\theta}_R^B &= -\dot{C}_R^B\delta^R\theta_B^R - C_R^B\delta^R\dot{\theta}_B^R = \left[{}^B\omega_R^B\times\right]C_R^B\delta^R\theta_B^R + C_R^BC_B^R\delta^B\omega_R^B \\ &= -\left[{}^B\omega_R^B\times\right]\delta^B\theta_R^B + \delta^B\omega_R^B\end{aligned}$$

Finally, the kinematic equation for the body frame attitude errors can be written as

$$\delta^B\dot{\theta}_R^B = -\left[{}^B\omega_R^B\times\right]\delta^B\theta_R^B + \delta^B\omega_R^B \tag{3.116}$$

It is important to realize that, even if the IMU were calibrated perfectly, the attitude error in the body frame would still change because of the rotation of the body frame itself. However, the attitude error in the reference frame stays constant when there are no contributions from the gyros. This is exactly why we most frequently want to use attitude error in the reference frame for the Kalman filter designs: when an attitude error is estimated, it tends to stay constant or change slowly for some time and, consequently, the application of the correction does not have to be done with minimum time delays. However, attitude error in the body frame rotates with the body frame and the application of the corrections in the body frame must be done in a timely manner.

It is often necessary to propagate the attitude when both the body frame and the reference frame are moving in the inertial frame and we showed earlier that we can use

$$\dot{C}_B^R = C_B^R\left[{}^B\omega_I^B\times\right]-\left[{}^R\omega_I^R\times\right]C_B^R$$

or, in quaternion form:

$$\dot{q}_B^R = \frac{1}{2}\left(q_B^R \otimes {}^B\omega_I^B - {}^R\omega_I^R \otimes q_B^R\right)$$

An easy way to derive the error equation in this case is to look at the frame R attitude with respect to the inertial frame and the frame B attitude with respect to the inertial frame separately but both represented in the reference frame:

$$\begin{aligned}
\delta\,{}^R\dot{\theta}_I^R &= -\left[{}^R\omega_I^R\times\right]\delta\,{}^R\theta_I^R + \delta\,{}^R\omega_I^R \\
\delta\,{}^R\dot{\theta}_I^B &= -\left[{}^R\omega_I^R\times\right]\delta\,{}^R\theta_I^B + C_B^R\delta\,{}^B\omega_I^B \\
\delta\,{}^R\theta_B^R &= \delta\,{}^R\theta_I^R - \delta\,{}^R\theta_I^B
\end{aligned}$$

and we end up with the following very important equation:

$$\delta\,{}^R\dot{\theta}_B^R = -\left[{}^R\omega_I^R\times\right]\delta\,{}^R\theta_B^R - C_B^R\delta\,{}^R\omega_I^B + \delta\,{}^R\omega_I^R \tag{3.117}$$

3.4.4.3 Interesting Conclusions That Can Be Drawn from the Attitude Error Kinematic Equations

There are many interesting conclusions from which we can draw from the attitude error kinematic equations.

Bounding Three-Axis Attitude Error by a Single-Axis LOS Sensor

Assuming that we use the gyro output to integrate into attitude for a spacecraft traveling in a circular orbit, interestingly enough, we only need a single LOS sensor (such as an Earth sensor) to be able to stop the drifting of attitude error in all three axes.

The LOS sensor is in the body frame; at any moment, it can only measure the instantaneous roll and pitch attitude errors. However, because the attitude error in the body frame rotates with the body frame, a quarter-orbit later, the yaw error that was not observable has become a roll error and can consequently be observed using the single LOS sensor. As the result, as the spacecraft travels through the orbit, all three axes of the attitude errors become observable through the measurements of a single LOS sensor. This effect is the roll/yaw coupling in the spacecraft attitude determination world and is also fundamentally related to gyro-compassing. Figure 3.4 illustrates this point.

Contribution of IMU Error Parameters

As we discussed before, the following equation is a typical model for the measured rate of a three-axis gyro triad:

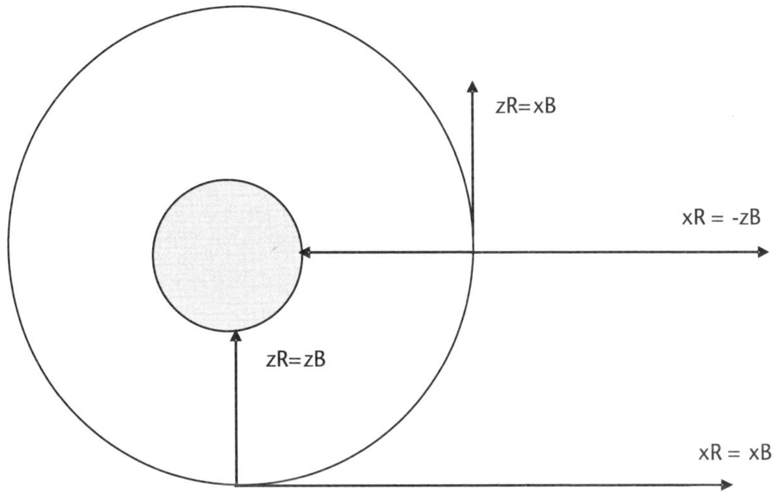

Figure 3.4 Roll/yaw coupling.

$$\omega_m = (I + S)\omega + b + w_{ARW}$$

The rate error is then

$$\delta\omega = \omega_m - \omega = (I + S)\omega + b + w_{ARW} - \omega = S\omega + b + w_{ARW}$$

Let us first look at the simplest case when there is no motion at all. In this case,

$$\delta\omega = S\omega + b + w_{ARW} = S0 + b + w_{ARW} = b + w_{ARW}$$

We then have

$$\dot{\theta}_B^{\hat{B}} = -\left[{}^B\omega_R^B \times\right]\theta_B^{\hat{B}} + \delta\,{}^B\omega_R^B = b + w_{ARW}$$

We can easily integrate the above equation and arrive at the following results:

$$\theta_B^{\hat{B}} \doteq \theta_B^{\hat{B}}(0) + bt + \theta_B^{\hat{B}}(t)_{ARW}$$

where

$$\sigma\left(\theta_B^{\hat{B}}(t)_{ARW}\right) = \sigma_{ARW}\sqrt{t}$$

So we have following two simple conclusions. When there is no motion:

1. Gyro bias introduces linearly growing attitude errors over time.
2. The angle random walk (ARW) introduces a random error with its uncertainty growing with square root of time.

Now let us look at the scenario that the body frame moves around a fixed axis in the body frame and at a constant rate.

The truth attitude DCM can be written as

$$
\begin{aligned}
C_B^R(t) &= I - \sin\theta_A^B[\vec{e}\times] + \left(1 - \cos\theta_A^B\right)[\vec{e}\times]^2 \\
\theta_B^R &= -\omega t \\
C_B^R(t) &= I + \sin\omega t[\dot{e}\times] + (1 - \cos\omega t)[\vec{e}\times]^2
\end{aligned}
$$

We then have

$$
\begin{aligned}
\dot{\theta}_R^{\hat{R}} &= -C_B^R\delta\,{}^B\omega_R^B = -\left(I + \sin\omega t[\vec{e}\times] + (1 - \cos\omega t)[\vec{e}\times]^2\right)\delta\,{}^B\omega_R^B \\
\theta_R^{\hat{R}}(t) &= \theta_R^{\hat{R}}(0) - \left(tI - (\cos\omega t - 1)[\vec{e}\times] + (t - \sin\omega t)[\vec{e}\times]^2\right)\delta\,{}^B\omega_R^B \\
&= \theta_R^{\hat{R}}(0) - \left([\vec{e}\times] + t\left(I + [\vec{e}\times]^2\right) - \cos\omega t[\vec{e}\times] - \sin\omega t)[\vec{e}\times]^2\right)\delta\,{}^B\omega_R^B \\
&= \theta_R^{\hat{R}}(0) - \left([\vec{e}\times] + t\left(\vec{e}\vec{e}^T\right) - \cos\omega t[\vec{e}\times] - \sin\omega t)[\vec{e}\times]^2\right)\delta\,{}^B\omega_R^B
\end{aligned}
$$

Now we can analyze the details using the following equations

$$
\begin{aligned}
\theta_R^{\hat{R}}(t) &= \theta_R^{\hat{R}}(0) - \left([\vec{e}\times] + t\left(\vec{e}\vec{e}^T\right) - \cos\omega t[\vec{e}\times] - \sin\omega t)[\vec{e}\times]^2\right)\delta\,{}^B\omega_R^B \\
\delta\,{}^B\omega_R^B &= S\omega\vec{e} + b
\end{aligned}
$$

The following conclusions are most important:

1. There is a constant term due to the rate error.
2. There is a linearly growing term due to the rate error.
3. There is a sinusoidal term at the frequency of the constant rotation rate.

We can greatly simplify the analysis if we change the body frame such that one of the coordinate axes is the axis of rotation. Without loss of generality, let us assume we already did that and the axis of rotation is the z-axis. We have the following greatly simplified equations:

$$
C_B^R = \begin{bmatrix} \cos\omega t & \sin\omega t & 0 \\ -\sin\omega t & \cos\omega t & 0 \\ 0 & 0 & 1 \end{bmatrix}, \quad {}^B\omega_R^B = \begin{bmatrix} 0 \\ 0 \\ \omega \end{bmatrix}
$$

and

$$
\dot{\theta}_R^{\hat{R}} = -\begin{bmatrix} \cos\omega t & \sin\omega t & 0 \\ -\sin\omega t & \cos\omega t & 0 \\ 0 & 0 & 1 \end{bmatrix}\delta\,{}^B\omega_R^B
$$

Consequently,

$$
\theta_R^{\hat{R}}(t) = \begin{bmatrix} -\dfrac{1}{\omega}\sin\omega t & -\dfrac{1}{\omega}(\cos\omega t - 1) & 0 \\ \dfrac{1}{\omega}(\cos\omega t - 1) & -\dfrac{1}{\omega}\sin\omega t & 0 \\ 0 & 0 & -t \end{bmatrix}\begin{bmatrix} \omega S_{13} + b_1 \\ \omega S_{23} + b_2 \\ S_{33}\omega + b_3 \end{bmatrix}
$$

and we have

$$
\begin{aligned}
\Delta\theta_{1R}^{\hat{R}}(t) &= -\sin\omega t\left(S_{13} + b_1/\omega\right) - (\cos\omega t - 1)\left(S_{23} + b_2/\omega\right) \\
\Delta\theta_{2R}^{\hat{R}}(t) &= (\cos\omega t - 1)\left(S_{13} + b_1/\omega\right) - \sin\omega t\left(S_{23} + b_2/\omega\right) \\
\Delta\theta_{3R}^{\hat{R}}(t) &= -t\left(S_{33}\omega + b_3\right)
\end{aligned}
\tag{3.118}
$$

We can draw the following important conclusions (only true for the single-axis motion):

1. The contributions of the gyro biases along the nonspinning axes become bounded due to spinning. The errors are proportional to the ratios of the biases and the spinning rate (gyro-compassing effect).
2. The contributions of the misalignments during a fixed single-axis rotation are bounded. They are either zero mean or nonzero mean. The magnitudes of the errors are exactly the misalignments themselves.
3. Around the spinning axis, the bias contribution grows with time and is unbounded. The scale factor error contribution is proportional to the integrated rate (or total angle slewed) and is also unbounded.

Some of the above serve as the foundation of the modulation strategy for gimbaled platform based navigation systems. By moving the platform with proper strategy, we can convert unbounded linear drifting errors into bounded nondrifting errors.

Another interesting phenomenon is the accumulation of the misalignment errors, in particular, nonorthogonality errors, in coning motions.

Let us assume that the body frame goes through a sequence of three slews.

- Slew 1: 180° rotation around the z-axis of the body and the reference frames. We have

$$
\delta^R\vec{\theta}(t) = -\frac{1}{\omega}\begin{bmatrix} \sin\omega t & -(1-\cos\omega t) & 0 \\ (1-\cos\omega t) & \sin\omega t & 0 \\ 0 & 0 & \omega t \end{bmatrix}\begin{bmatrix} S_{13}\omega + b_1 \\ S_{23}\omega + b_2 \\ S_{33}\omega + b_3 \end{bmatrix}
$$

$$
\delta^R\theta(t) = \begin{bmatrix} -\sin\omega t\left(S_{13} + b_1/\omega\right) - (\cos\omega t - 1)\cdot\left(S_{23} + b_2/\omega\right) \\ -\sin\omega t\left(S_{23} + b_2/\omega\right) - (1 - \cos\omega t)\cdot\left(S_{13} + b_1/\omega\right) \\ -t\left(S_{33} + b_3/\omega\right) \end{bmatrix}
$$

Setting the scale factors and biases to 0, and$\omega t = \pi$, we have

$$
\delta^R\theta_1 = 2S_{23},\quad \delta^R\theta_2 = -2S_{13},\quad \delta^R\theta_2 = 0
$$

Figure 3.5 illustrates the rotation.

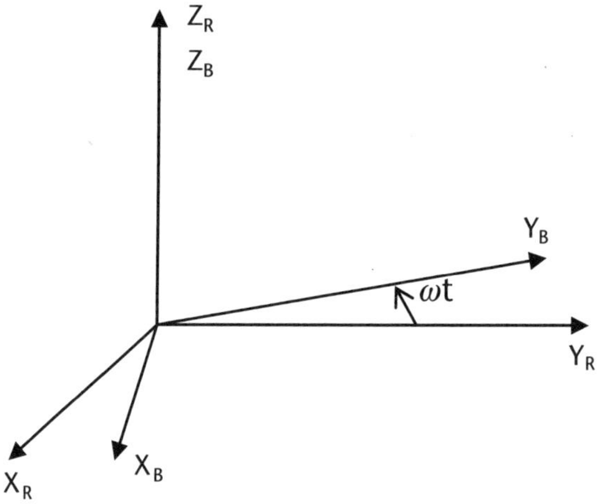

Figure 3.5 Rotation around the *z*-axis.

- Slew 2: Rotation around the x-axis of the new body frame for another 180°. At the end of the last slew, we arrived at a new body frame B_1 and we have the following DCM:

$$C_{B_1}^R = C_B^R(\omega t = \pi) = \begin{bmatrix} -1 & 0 & 0 \\ 0 & -1 & 0 \\ 0 & 0 & 1 \end{bmatrix}$$

Starting from this new frame, we rotate around the x-axis of the new frame from B_1 to B_2

$$\delta^{B_1}\theta(t) = \frac{1}{\omega}\begin{bmatrix} -\omega t & 0 & 0 \\ 0 & -\sin\omega t & (1-\cos\omega t) \\ 0 & -(1-\cos\omega t) & -\sin\omega t \end{bmatrix}\begin{bmatrix} S_{11}\omega + b_1 \\ S_{21}\omega + b_2 \\ S_{31}\omega + b_3 \end{bmatrix}$$

We will need to transform the error result back to the reference frame and we have:

$$\delta^R\theta(t) = C_{B_1}^R \cdot \left(\delta^{B_1}\theta(t)\right) = \begin{bmatrix} -1 & 0 & 0 \\ 0 & -1 & 0 \\ 0 & 0 & 1 \end{bmatrix}\begin{bmatrix} -t\cdot\left(S_{11} + b_1/\omega\right) \\ -\sin\omega t\cdot\left(S_{21} + b_2/\omega\right) - (\cos\omega t - 1)\cdot\left(S_{31} + b_3/\omega\right) \\ -\sin\omega t\cdot\left(S_{31} + b_3/\omega\right) + (\cos\omega t - 1)\cdot\left(S_{21} + b_2/\omega\right) \end{bmatrix}$$

Setting the scale factors and biases to 0, and $\omega t = \pi$, we have

$$\delta^R\theta_1 = 0; \quad \delta^R\theta_2 = -2S_{31}; \quad \delta^R\theta_3 = -2S_{21}$$

- Slew 3: Rotation around the y-axis of the new body frame for another 180°. At the end of the last slew, we end up with frame B_2 and we have

$$C_{B_2}^R = C_{B_1}^R C_{B_2}^{B_1} = \begin{bmatrix} -1 & 0 & 0 \\ 0 & -1 & 0 \\ 0 & 0 & 1 \end{bmatrix} \begin{bmatrix} 1 & 0 & 0 \\ 0 & -1 & 0 \\ 0 & 0 & -1 \end{bmatrix} = \begin{bmatrix} -1 & 0 & 0 \\ 0 & 1 & 0 \\ 0 & 0 & -1 \end{bmatrix}$$

Starting from this new frame, we will rotate around the y-axis for 180° and we have

$$\delta^{B_2}\theta(t) = \frac{1}{\omega}\begin{bmatrix} \sin\omega t & 0 & -(1-\cos\omega t) \\ 0 & \omega t & 0 \\ (1-\cos\omega t) & 0 & \sin\omega t \end{bmatrix} \cdot \begin{bmatrix} S_{12}\omega + b_1 \\ S_{22}\omega + b_2 \\ S_{32}\omega + b_3 \end{bmatrix}$$

and we will want to transform the error back to the original reference frame:

$$\delta^R\theta(t) = C_{B_2}^R \cdot \left(\delta^{B_2}\theta(t)\right) = \begin{bmatrix} -1 & 0 & 0 \\ 0 & 1 & 0 \\ 0 & 0 & -1 \end{bmatrix} \begin{bmatrix} \sin\omega t \cdot \left(S_{12} + b_1/\omega\right) - (1-\cos\omega t) \cdot \left(S_{32} + b_3/\omega\right) \\ -t \cdot \left(S_{22} + b_2/\omega\right) \\ \sin\omega t \cdot \left(S_{32} + b_3/\omega\right) + (1-\cos\omega t) \cdot \left(S_{12} + b_1/\omega\right) \end{bmatrix}$$

Setting the scale factors and biases to 0 and $\omega t = \pi$, we have

$$\delta^R\theta_1 = 2S_{32}; \quad \delta^R\theta_2 = 0; \quad \delta^R\theta_3 = -2S_{12}$$

After 180° rotation around the z-axis, 180° rotation around the x-axis, and 180° rotation around the y-axis, we return to the original attitude.

Since the attitude knowledge errors have been transformed to the same reference frame, the total error will be simply the summation of the errors resulted from each of the three rotations:

$$\Delta^R\theta_1 = 2\left(S_{23} + S_{32}\right), \quad \Delta^R\theta_2 = -2\left(S_{13} + S_{31}\right), \quad \Delta^R\theta_3 = -2\left(S_{12} + S_{21}\right)$$

If we keep doing the sequence of the three rotations for N times, the error will accumulate to

$$\Delta^R\theta_1 = 2N\left(S_{23} + S_{32}\right), \quad \Delta^R\theta_2 = -2N\left(S_{13} + S_{31}\right), \quad \delta^R\theta_3 = -2N\left(S_{12} + S_{21}\right)$$

It is important to know that if the misalignment matrix is skew-symmetric, that is,

$$\begin{aligned} S_{23} &= -S_{23} \text{ then } S_{23} + S_{32} = 0 \\ S_{13} &= -S_{31} \text{ then } S_{13} + S_{31} = 0 \\ S_{12} &= -S_{21} \text{ then } S_{12} + S_{21} = 0 \end{aligned}$$

then the errors after the coning motion will be 0.

It turns out that when the misalignment error represents a rotation of the coordinate frame, the matrix is indeed skew-symmetric.

In other words, only the errors due to nonorthogonality misalignment will accumulate through coning motions. That is why in later sections we will want to decompose the misalignment errors into rotational terms and nonorthogonality terms.

3.5 Body Frame Speed and Attitude/Heading Sensor-Based Navigation

3.5.1 Introduction

As another example of a propagation-based solution, we will describe the vehicle navigation solution based on sensors that can measure velocity in the vehicle body frame combined with a separate attitude/heading reference system (AHRS).

In real systems, the body frame speed can be provided by odometers for ground vehicles or by flow meters for marine or submarine vehicles and by air speed sensors or Doppler radar sensors for aircraft.

The AHRS can be provided by low-cost gyros slaved to gravity vector as the vertical reference and a magnetometer for the heading reference, which itself is an integrated system that uses the combination of propagated solution and measured solution.

In this section, we will make simple assumptions on both the speed sensor and the AHRS for the purpose of simplicity and the possible more comprehensive treatment of this topic will become obvious and easy after we cover the topic of inertial navigation (Section 3.6) and the topic of the integration of sensors using Kalman filters.

3.5.2 Navigation Equations

Navigation equations can be written in many different frames to achieve different types of convenience. We will use both the ECEF frame (E) and the local level navigation frame (N).

Let's start with ECEF frame E. Let C_B^E be the body to the reference frame DCM that represents the attitude. We will assume that the velocity is along the vehicle body x-axis. In this ideal case, the velocity vector in the body frame can be given by

$$ {}^B v_E^B = \begin{bmatrix} v \\ 0 \\ 0 \end{bmatrix} $$

Consequently, the velocity in the reference frame is given by

$$ {}^E v_E^B = C_B^E \, {}^B v_E^B = C_B^E \begin{bmatrix} v \\ 0 \\ 0 \end{bmatrix} = C_B^E(:,1)v $$

The position in the reference frame can then be computed as the integration of the velocity:

$$ \frac{d}{dt} {}^E r_E^B = {}^E v_E^B = C_B^E(:,1)v \tag{3.119} $$

The equations for the local-level navigation frame for the velocity will be the same. Since the local-level frame moves with the vehicle, the position has to use separate equations.

Take NED local-level frame, for example. As mentioned before, we have the same velocity equations:

$$ {}^{N}v_E^B = C_B^{N\,B}v_E^B = C_B^N \begin{bmatrix} v \\ 0 \\ 0 \end{bmatrix} = C_B^N(:,1)v $$

The latitude, longitude, and altitude can be calculated by integrating the following equations

$$ \begin{aligned} \frac{d}{dt}L &= \frac{{}^{N}v_E^B(1)}{R_N(L)+h} \\ \frac{d}{dt}\lambda &= \frac{{}^{N}v_E^B(2)}{\left(R_E(L)+h\right)\cos(L)} \\ \frac{d}{dt}h &= -{}^{N}v_E^B(3) \end{aligned} \tag{3.120} $$

where

$R_N(L)$ is the meridian radius of curvature and is given by

$$ R_N(L) = \frac{R_0\left(1-e^2\right)}{\left(1-e^2\sin^2 L\right)^{3/2}} \tag{3.121} $$

$R_E(L)$ is the transverse radius of curvature and is given by

$$ R_E(L) = \frac{R_0}{\left(1-e^2\sin^2 L\right)^{1/2}} \tag{3.122} $$

3.5.3 Error Models

We start with the velocity error model.

The truth satisfies the following equation

$$ {}^{E}v_E^B = C_B^{EB}v_E^B = C_B^E \begin{bmatrix} v \\ 0 \\ 0 \end{bmatrix} = C_B^E(:,1)v $$

The estimate satisfies the following equation

$$ {}^{E}\hat{v}_E^B = \hat{C}_E^B \begin{bmatrix} \hat{v} \\ 0 \\ 0 \end{bmatrix} $$

The difference

$$\delta^E v_E^B = {}^E\hat{v}_E^B - {}^E v_E^B = \hat{C}_B^E \begin{bmatrix} \hat{v} \\ 0 \\ 0 \end{bmatrix} - C_B^E \begin{bmatrix} v \\ 0 \\ 0 \end{bmatrix} = \hat{C}_B^E \begin{bmatrix} \hat{v} \\ 0 \\ 0 \end{bmatrix} - C_B^E \begin{bmatrix} \hat{v} \\ 0 \\ 0 \end{bmatrix} + C_B^E \begin{bmatrix} \hat{v} \\ 0 \\ 0 \end{bmatrix} - C_B^E \begin{bmatrix} v \\ 0 \\ 0 \end{bmatrix}$$

$$= \left(\hat{C}_B^E - C_B^E\right) \begin{bmatrix} \hat{v} \\ 0 \\ 0 \end{bmatrix} + C_B^E \begin{bmatrix} \hat{v} - v \\ 0 \\ 0 \end{bmatrix}$$

Since

$$\hat{C}_B^E - C_B^E = C_B^E[\delta^B\theta_E^B\times] = -[\delta^E\theta_B^E\times]C_B^E$$

$$\delta^E v_E^B = \left(\hat{C}_B^E - C_B^E\right) \begin{bmatrix} \hat{v} \\ 0 \\ 0 \end{bmatrix} + C_B^E \begin{bmatrix} \hat{v} - v \\ 0 \\ 0 \end{bmatrix} = -\left[\delta^E\theta_B^E\times\right]C_B^E \begin{bmatrix} \hat{v} \\ 0 \\ 0 \end{bmatrix} + C_B^E(1,:)\delta v$$

$$= -\left[\delta^E\theta_B^E\times\right]C_B^E(1,:)\hat{v} + C_B^E(1,:)\delta v = \hat{v}\left[\left(C_B^E(1,:)\right)\times\right]\delta^E\theta_B^E + \left(C_B^E(1,:)\right)\delta v$$

In summary, we have

$$\delta^E v_E^B = \hat{v}\left[\left(C_B^E(1,:)\right)\times\right]\delta^E\theta_B^E + \left(C_B^E(1,:)\right)\delta v \tag{3.123}$$

The position error in ECEF frame is simply

$$\frac{d}{dt}\delta^E r_E^B = \delta^E v_E^B \tag{3.124}$$

Similarly, in the NED frame, we have the same equation for the velocity error

$$\delta^N v_E^B = \hat{v}\left[\left(C_B^N(1,:)\right)\times\right]\delta^N\theta_B^N + \left(C_B^N(1,:)\right)\delta v \tag{3.125}$$

The position error equations are simply

$$\begin{aligned} \frac{d}{dt}\delta L &= \frac{\delta^N v_R^B(1)}{R_N(L) + h} \\ \frac{d}{dt}\delta\lambda &= \frac{\delta^N v_R^B(2)}{\left(R_E(L) + h\right)\cos(L)} \\ \frac{d}{dt}\delta h &= -\delta^N v_R^B(3) \end{aligned} \tag{3.126}$$

3.6 Inertial Navigation and Error Models

3.6.1 Introduction

The inertial navigation algorithm is one of the most mature navigation algorithms that is widely discussed in the literature [3, 7–26]. Instead of repeating the same derivations and discussions that were done many times elsewhere, we will simply write the conclusions using the symbols and conventions of this book.

3.6.2 ECI Frame Navigation Equations and Error Models

The inertial navigation equation in the ECI frame is the simplest, although not the most convenient in terms of application and analysis.

The attitude propagation equation in the quaternion form is simply

$$\dot{q}_B^I = \frac{1}{2} q_B^I \otimes {}^B\omega_B^I$$

The DCM can be derived from the quaternion as

$$C_B^I = q2DCM\left(q_B^I\right)$$

We can use the various numerical integration methods in Section 3.4 in the actual implementation of the above equations in practical systems.

As discussed in Section 3.4, the attitude error can be modeled by the following equation:

$$\delta^I\dot{\theta}_B^I = -C_B^I \delta^B\omega_I^B$$

where the attitude error angle vector is related to the DCM by the following equation:

$$\hat{C}_B^I - C_B^I = C_B^I\left[\delta^B\theta_I^B\times\right] = -\left[\delta^I\theta_B^I\times\right]C_B^I$$

The velocity equation is simply

$$\frac{d}{dt}{}^Iv_I^B = C_B^I\,{}^Bf_I^B + {}^Ig$$

where ${}^Bf_I^B$ is the specific force measured by the accelerometers and ${}^Ig$ is the gravitational acceleration (or in short gravity), computed using gravity model, expressed in the inertial frame (I).

The velocity error is due to three error sources:

1. The accelerometer error $\delta^Bf_I^B = m^Bf_I^B - {}^Bf_I^B$ (measurement – truth);
2. The attitude error $\delta C_B^I = \hat{C}_B^I - C_B^I = C_B^I[\delta^B\theta_I^B\times] = -[\delta^I\theta_B^I\times]C_B^I$;
3. The gravity error due to position error

$$\delta^Ig = {}^Ig\left({}^I\hat{r}_I^B\right) - {}^Ig\left({}^Ir_I^B\right) = \left[\frac{\partial^Ig}{\partial^Ir_I^B}\right]\left({}^I\hat{r}_I^B - {}^Ir_I^B\right) = \left[\frac{\partial^Ig}{\partial^Ir_I^B}\right]\delta^Ir_I^B$$

The velocity error can be written as

$$\delta^Iv_I^B = {}^I\hat{v}_I^B - {}^Iv_I^B$$

which can be written as

$$\begin{aligned}\frac{d}{dt}\delta^I v_I^B - \frac{d}{dt}{}^I\hat{v}_I^B - \frac{d}{dt}{}^I v_I^B &= \hat{C}_B^I m^B f_I^B - C_B^I{}^B f_I^B + {}^I\hat{g} - g \\ &= \hat{C}_B^I m^B f_I^B - C_B^I m^B f_I^B + C_B^I m^B f_I^B - C_B^I{}^B f_I^B + {}^I\hat{g} - {}^I g \\ &= \left(\hat{C}_B^I - C_B^I\right) m^B f_I^B + C_B^I \delta^B f_I^B + \delta^I g\end{aligned}$$

By substituting the three error sources into the equation, we have

$$\frac{d}{dt}\delta^I v_I^B = -\left[\delta^I\theta_B^I \times\right]\left(C_B^I m^B f_I^B\right) + C_B^I \delta^B f_I^B + \left[\frac{\partial^I g}{\partial^I r_I^B}\right]\delta^I r_I^B$$

or

$$\frac{d}{dt}\delta^I v_I^B = \left[\left(C_B^I m^B f_I^B\right)\times\right]\delta^I\theta_B^I + C_B^I \delta^B f_I^B + \left[\frac{\partial^I g}{\partial^I r_I^B}\right]\delta^I r_I^B$$

(Note that for the attitude error term the signs are different from other literature. This is due to the difference of the attitude error convention. We use the inertial with respect to the body attitude error in the inertial frame.)

The position equation in the inertial frame is simply

$$\frac{d}{dt}{}^I r_I^B = {}^I v_I^B$$

and the corresponding error equation is simply

$$\frac{d}{dt}\delta^I r_I^B = \delta^I v_I^B$$

To summarize, we have the following navigation equations:

$$\dot{q}_B^I = \frac{1}{2} q_B^I \otimes {}^B\omega_B^I \tag{3.127}$$

$$C_B^I = q2DCM\left(q_B^I\right) \tag{3.128}$$

$$\frac{d}{dt}{}^I v_I^B = C_B^I{}^B f_I^B + {}^I g \tag{3.129}$$

$$\frac{d}{dt}{}^I r_I^B = {}^I v_I^B \tag{3.130}$$

and the following error models:

$$\delta^I\dot{\theta}_B^I = -C_B^I \delta^B \omega_I^B \tag{3.131}$$

$$\frac{d}{dt}\delta^I v_I^B = \left[\left(C_B^I m^B f_I^B\right)\times\right]\delta^I\theta_B^I + C_B^I \delta^B f_I^B + \left[\frac{\partial^I g}{\partial^I r_I^B}\right]\delta^I r_I^B \tag{3.132}$$

$$\frac{d}{dt}\delta^I r_I^B = \delta^I v_I^B \tag{3.133}$$

(Note that there is a sign difference between what is used here and some other literature. The sign difference is caused by the attitude error convention; if we do not have the minus sign in the attitude error equation, then there will be a minus sign in the attitude error contribution to velocity, which leads to the same equation as the others.)

3.6.3 ECEF Frame (or Other Earth Fixed Frame) Navigation Equations and Error Equations

The inertial navigation equations in the ECEF frame are only slightly more complicated than those in ECI frame. In most cases, ECEF frame equations are also fairly convenient and often represent the best compromise between simplicity and convenience.

The attitude propagation equation in quaternion form is simply

$$\dot{q}_B^E = \frac{1}{2}\left(q_B^E \otimes {}^B\omega_I^B - {}^E\omega_I^E \otimes q_B^E\right)$$

We can use the various numerical integration methods described in Section 3.4 in the actual implementations of the above equation in practical systems.

The DCM can be derived from the quaternion as

$$C_B^E = q2DCM\left(q_B^E\right)$$

The corresponding attitude error, as discussed in Section 3.4, satisfies the following equation:

$$\delta^E\dot{\theta}_B^E = -\left[{}^E\omega_I^E \times\right]\delta^E\theta_B^E - C_B^E\delta^B\omega_I^B + \delta^E\omega_I^E$$

It is generally a good assumption that we know the Earth rotation rate in ECEF perfectly. Consequently,

$$\delta^E\omega_I^E = 0$$

and we simply have

$$\delta^E\dot{\theta}_B^E = -\left[{}^E\omega_I^E \times\right]\delta^E\theta_B^E - C_B^E\delta^B\omega_I^B$$

The velocity equation is slightly more complicated than the inertial frame version:

$$\frac{d}{dt}{}^E v_E^B = C_B^E\,{}^B f_I^B + {}^E g - 2\,{}^E\omega_I^E \times {}^E v_E^B - \left[{}^E\omega_I^E \times\right]\left[{}^E\omega_I^E \times\right]{}^E r_E^B$$

where ${}^B f_I^B$ is the ideal specific force measured by the accelerometers and ${}^E g$ is the gravity, computed using the gravity model, expressed in the Earth frame (E).

The centrifugal acceleration is often combined with gravity as

$$ {}^{E}g_c\left({}^{E}r_E^B\right) = {}^{E}g - \left[{}^{E}\omega_I^E \times\right]\left[{}^{E}\omega_I^E \times\right]{}^{E}r_E^B $$

Consequently, the equation can be written as

$$ \frac{d}{dt}{}^{E}v_E^B = C_B^E\,{}^{B}f_I^B + {}^{E}g_c - 2\,{}^{E}\omega_I^E \times {}^{E}v_E^B $$

The position equation in Earth frame is simply

$$ \frac{d}{dt}{}^{E}r_E^B = {}^{E}v_E^B $$

Similar to what we did before, recognizing that ${}^{E}\omega_I^E$, the Earth rate in ECEF frame, is a known constant vector, the error equations for the velocity and position can be written as below:

$$ \frac{d}{dt}\delta\,{}^{E}v_E^B = \left[\left(C_B^E m\,{}^{B}f_I^B\right)\times\right]\delta\,{}^{E}\theta_B^E + C_B^E\delta\,{}^{B}f_I^B + \left[\frac{\partial\,{}^{E}g}{\partial\,{}^{E}r_E^B}\right]\delta\,{}^{E}r_I^B - 2\left[\left({}^{E}\omega_I^E\right)\times\right]\delta\,{}^{E}v_E^B $$

$$ \frac{d}{dt}\delta\,{}^{E}r_E^B = \delta\,{}^{E}v_E^B $$

In summary, we have the following set of equations:

Navigation equations:

$$ \dot{q}_B^E = \frac{1}{2}\left(q_B^E \otimes {}^{B}\omega_I^B - {}^{E}\omega_I^E \otimes q_B^E\right) \tag{3.134} $$

$$ C_B^E = q2DCM\left(q_B^E\right) \tag{3.135} $$

$$ \frac{d}{dt}{}^{E}v_E^B = C_B^E\,{}^{B}f_I^B + {}^{E}g_c - 2\,{}^{E}\omega_I^E \times {}^{E}v_E^B \tag{3.136} $$

$$ \frac{d}{dt}{}^{E}r_E^B = {}^{E}v_E^B \tag{3.137} $$

and the following error models:

$$ \frac{d}{dt}\delta\,{}^{E}\theta_B^E = -\left[{}^{E}\omega_I^E \times\right]\delta\,{}^{E}\theta_B^E - C_B^E\delta\,{}^{B}\omega_I^B + \delta\,{}^{E}\omega_I^E \tag{3.138} $$

$$ \frac{d}{dt}\delta\,{}^{E}v_E^B = \left[\left(C_B^E m\,{}^{B}f_I^B\right)\times\right]\delta\,{}^{E}\theta_B^E + C_B^E\delta\,{}^{B}f_I^B + \left[\frac{\partial\,{}^{E}g}{\partial\,{}^{E}r_E^B}\right]\delta\,{}^{E}r_I^B - 2\left[\left({}^{E}\omega_I^E\right)\times\right]\delta\,{}^{E}v_E^B \tag{3.139} $$

$$ \frac{d}{dt}\delta\,{}^{E}r_E^B = \delta\,{}^{E}v_E^B \tag{3.140} $$

3.6.4 Local-Level Frame Inertial Navigation Equations and Error Equations

The inertial navigation equations in the local-level frame are, in general, more complicated than the ECI and ECEF frame versions. However, equations in the local-level frame, especially the error equations are most intuitive and most revealing (such as Schuler loop).

The local-level frame and the ECEF frame are related by the following transformation matrix:

$$C_E^N = \begin{bmatrix} \cos\alpha & -\sin\alpha & 0 \\ \sin\alpha & \cos\alpha & 0 \\ 0 & 0 & 1 \end{bmatrix} \begin{bmatrix} \cos L & 0 & \sin L \\ 0 & 1 & 0 \\ -\sin L & 0 & \cos L \end{bmatrix} \begin{bmatrix} 1 & 0 & 0 \\ 0 & \cos\lambda & \sin\lambda \\ 0 & -\sin\lambda & \cos\lambda \end{bmatrix} \begin{bmatrix} 0 & 0 & 1 \\ 0 & 1 & 0 \\ -1 & 0 & 0 \end{bmatrix}$$

$$= \begin{bmatrix} -\cos\alpha\sin L\cos\lambda + \sin\alpha\sin\lambda & -\cos\alpha\sin L\sin\lambda - \sin\alpha\cos\lambda & \cos\alpha\cos L \\ -\sin\alpha\sin L\cos\lambda - \cos\alpha\sin\lambda & -\sin\alpha\sin L\sin\lambda + \cos\alpha\cos\lambda & \sin\alpha\cos L \\ -\cos L\cos\lambda & -\cos L\sin\lambda & \sin L \end{bmatrix} \tag{3.141}$$

where α is the wander-azimuth angle introduced to mitigate the singularity at the North Pole and the South Pole. When $\alpha = 0$, the wander-azimuth local level frame degrades to the NED local-level frame and in this case, we have:

$$C_E^N = \begin{bmatrix} -\sin L\cos\lambda & -\sin L\sin\lambda & \cos L \\ -\sin\lambda & \cos\lambda & 0 \\ -\cos L\cos\lambda & -\cos L\sin\lambda & \sin L \end{bmatrix} \tag{3.142}$$

In general, the latitude, longitude, and the wander-azimuth angle can be calculated from the DCM using the following equations:

$$\begin{aligned} L &= \sin^{-1} C_E^N(3,3) \\ \lambda &= \tan^{1} \frac{C_E^N(3,2)}{C_E^N(3,1)} \\ \alpha &= \tan^{-1} \frac{C_E^N(2,3)}{C_E^N(1,3)} \end{aligned} \tag{3.143}$$

Note that the equations for latitude and longitude remain the same even if N is simply the NED frame.

As seen from the above, the DCM C_E^N fundamentally represents the position on the Earth.

The attitude propagation equation in the quaternion form is simply

$$\dot{q}_B^N = \frac{1}{2}\left(q_B^N \otimes {}^B\omega_I^B - {}^N\omega_I^N \otimes q_B^N\right)$$

and we can use the various numerical integration methods described in Section 3.4 in the actual implementations of the above equation in practical systems.

The DCM can be derived from the quaternion as

$$C_B^N = q2DCM\left(q_B^N\right)$$

The corresponding attitude error, as discussed in Section 3.4, satisfies the following equation:

$$\delta^N\dot{\theta}_B^N = -\left[{}^N\omega_I^N \times\right]\delta^N\theta_B^N - C_B^N\delta^B\omega_I^B + \delta^N\omega_I^N$$

where

$$^N\omega_I^N = {}^N\omega_E^N + {}^N\omega_I^E$$
$$\delta^N\omega_I^N = \delta^N\omega_E^N + \delta^N\omega_I^E$$

The velocity equation will become more complicated than the inertial and ECEF frame equations:

$$\frac{d}{dt}{}^Nv_E^B = C_B^N{}^Bf_I^B + C_E^N\left({}^Eg - \left[{}^E\omega_I^E\times\right]\left[{}^E\omega_I^E\times\right]{}^Er_E^B\right) - {}^N\omega_E^N \times {}^Nv_E^B - 2{}^N\omega_I^E \times {}^Nv_E^B$$

or

$$\frac{d}{dt}{}^Nv_E^B = C_B^N{}^Bf_I^B + C_E^N{}^Eg_c - \left(2{}^N\omega_I^E + {}^N\omega_E^N\right){}^Nv_E^B$$

The craft rate or transport rate $^N\omega_E^N$ is calculated from the velocity.

For the wander-azimuth local level frame, we use the degree of the freedom introduced by the wander-azimuth angle in such a way that $^N\omega_E^N(3) = 0$. We have

$$\begin{bmatrix} {}^N\omega_E^N(1) \\ {}^N\omega_E^N(2) \end{bmatrix} = \begin{bmatrix} \left(\frac{1}{R_N+h} - \frac{1}{R_E+h}\right)\sin\alpha\cos\alpha & \frac{\sin^2\alpha}{R_N+h} + \frac{\cos^2\alpha}{R_E+h} \\ -\left(\frac{\cos^2\alpha}{R_N+h} + \frac{\sin^2\alpha}{R_E+h}\right) & -\left(\frac{1}{R_N+h} - \frac{1}{R_E+h}\right)\sin\alpha\cos\alpha \end{bmatrix} \begin{bmatrix} {}^Nv_E^B(1) \\ {}^Nv_E^B(2) \end{bmatrix}$$

$$^N\omega_E^N(3) = 0$$

where the matrix

$$Cur(L,\lambda,h,\alpha) = \begin{bmatrix} \left(\frac{1}{R_N+h} - \frac{1}{R_E+h}\right)\sin\alpha\cos\alpha & \frac{\sin^2\alpha}{R_N+h} + \frac{\cos^2\alpha}{R_E+h} \\ -\left(\frac{\cos^2\alpha}{R_N+h} + \frac{\sin^2\alpha}{R_E+h}\right) & -\left(\frac{1}{R_N+h} - \frac{1}{R_E+h}\right)\sin\alpha\cos\alpha \end{bmatrix}$$

is the curvature matrix $Cur(L, \lambda, h, \alpha)$ as a function of 3-D position and the wander azimuth angle.

As a point of reference, when N is NED without the wander-azimuth angle, the craft rate can be computed using the following equation

$$^N\omega_E^N = \begin{bmatrix} \frac{{}^Nv_E^B(2)}{R_E+h} \\ -\frac{{}^Nv_E^B(1)}{R_N+h} \\ -\frac{{}^Nv_E^B(2)\tan(L)}{R_N+h} \end{bmatrix}$$

Note that ${}^{N}\omega_{E}^{N}(3)$ becomes undefined when L is 90° or −90° in this case.

The position, represented by the DCM C_E^N and the altitude h, can then be calculated by integrating the following differential equations:

$$\frac{d}{dt}C_E^N = C_E^N\left[{}^{N}\omega_E^N \times\right]$$
$$\frac{d}{dt}h = -{}^{N}v_R^B(3)$$

Alternatively, it could be computationally cheaper to directly integrate into latitude, longitude, and altitude using the following equations:

$$\frac{d}{dt}L = \frac{{}^{N}v_R^B(1)\cos\alpha + {}^{N}v_R^B(2)\sin\alpha}{R_N(L)+h}$$
$$\frac{d}{dt}\lambda = \frac{-{}^{N}v_R^B(1)\sin\alpha + {}^{N}v_R^B(2)\cos\alpha}{\left(R_E(L)+h\right)\cos(L)}$$
$$\frac{d}{dt}h = -{}^{N}v_R^B(3)$$

In reality, directly integrating latitude, longitude, and altitude only makes sense when $N = \text{NED}$, that is, when $\alpha = 0$ (otherwise, we will need to integrate the DCM anyway to be able to calculate α) and in this case, we have

$$\frac{d}{dt}L = \frac{{}^{N}v_R^B(1)}{R_N(L)+h} = -{}^{N}\omega_E^N(2)$$
$$\frac{d}{dt}\lambda = \frac{{}^{N}v_R^B(2)}{\left(R_E(L)+h\right)\cos(L)} = \frac{{}^{N}\omega_E^N(1)}{\cos(L)}$$
$$\frac{d}{dt}h = -{}^{N}v_R^B(3)$$

We summarize the above into two algorithms:

- Algorithm 1 (illustrated in Figure 3.6):
  1. Attitude propagation:

$$\dot{q}_B^N = \frac{1}{2}\left(q_B^N \otimes {}^{B}\omega_I^B - {}^{N}\omega_I^N \otimes q_B^N\right) \tag{3.144}$$

  2. Convert the quaternion to the DCM:

$$C_B^N = q2DCM\left(q_B^N\right) \tag{3.145}$$

  3. Integrate the velocity equation:

$$\frac{d}{dt}{}^{N}v_E^B = C_B^N\,{}^{B}f_I^B + C_E^N\left({}^{E}g - \left[{}^{E}\omega_I^E\times\right]\left[{}^{E}\omega_I^E\times\right]{}^{E}r_E^B\right) - \left({}^{N}\omega_E^N + 2\,{}^{N}\omega_I^E\right)\times{}^{N}v_E^B \tag{3.146}$$

4. Calculate the craft rate:

$$\begin{bmatrix} {}^{N}\omega_E^N(1) \\ {}^{N}\omega_E^N(2) \end{bmatrix} = \begin{bmatrix} \left(\dfrac{1}{R_N+h} - \dfrac{1}{R_E+h}\right)\sin\alpha\cos\alpha & \dfrac{\sin^2\alpha}{R_N+h} + \dfrac{\cos^2\alpha}{R_E+h} \\ -\left(\dfrac{\cos^2\alpha}{R_N+h} + \dfrac{\sin^2\alpha}{R_E+h}\right) & -\left(\dfrac{1}{R_N+h} - \dfrac{1}{R_E+h}\right)\sin\alpha\cos\alpha \end{bmatrix} \begin{bmatrix} {}^{N}v_E^B(1) \\ {}^{N}v_E^B(2) \end{bmatrix}$$

$$ {}^{N}\omega_E^N(3) = 0 \tag{3.147}$$

5. Integrate into position:

$$\begin{aligned} \frac{d}{dt} C_E^N &= C_E^N\left[{}^{N}\omega_E^N \times\right] \\ \frac{d}{dt} h &= -{}^{N}v_R^B(3) \end{aligned} \tag{3.148}$$

6. Calculate Lat, Long, and the wander-azimuth angle from C_E^N:

$$\begin{aligned} L &= \sin^{-1} C_E^N(3,3) \\ \lambda &= \tan^{-1}\frac{C_E^N(3,2)}{C_E^N(3,1)} \\ \alpha &= \tan^{-1}\frac{C_E^N(2,3)}{C_E^N(1,3)} \end{aligned} \tag{3.149}$$

7. Transform the Earth rate into the local-level frame:

$$ {}^{N}\omega_I^E = C_E^N\, {}^{E}\omega_I^E \tag{3.150}$$

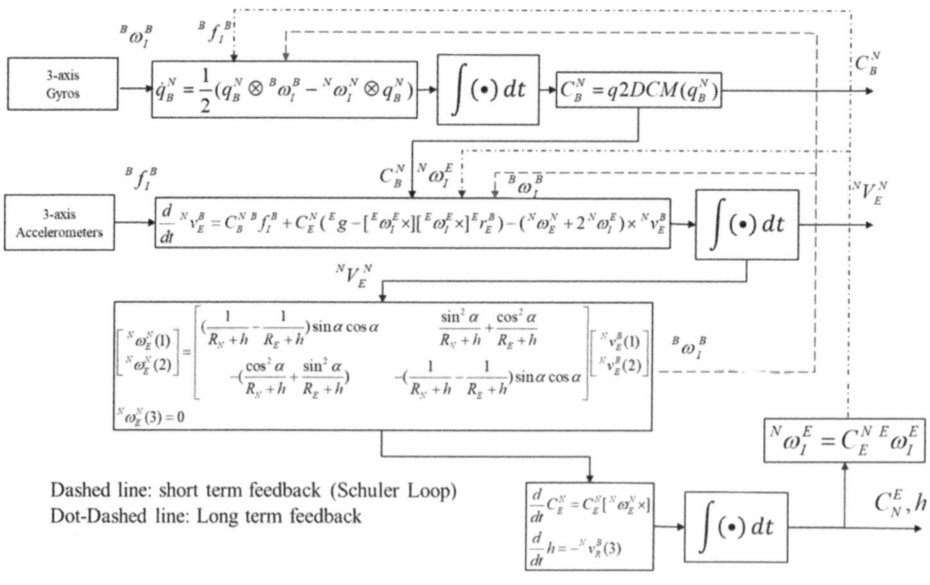

Figure 3.6 Strapdown inertial navigation algorithm in a local-level frame.

- Algorithm 2 (only applicable when N = NED)
  1. Attitude propagation:

$$\dot{q}_B^N = \frac{1}{2}\left(q_B^N \otimes {}^B\omega_I^B - {}^N\omega_I^N \otimes q_B^N\right)$$

  2. Convert the quaternion to the DCM:

$$C_B^N = q2DCM\left(q_B^N\right)$$

  3. Integrate the velocity equation:

$$\frac{d}{dt}{}^N v_E^B = C_B^N\,{}^B f_I^B + C_E^N\left({}^E g - \left[{}^E\omega_I^E \times\right]\left[{}^E\omega_I^E \times\right]{}^E r_E^B\right) - \left({}^N\omega_E^N + 2\,{}^N\omega_I^E\right) \times {}^N v_E^B$$

  4. Calculate the craft rate:

$${}^N\omega_E^N = \begin{bmatrix} \dfrac{{}^N v_E^B(2)}{R_E + h} \\ -\dfrac{{}^N v_E^B(1)}{R_N + h} \\ -\dfrac{{}^N v_E^B(2)\tan(L)}{R_N + h} \end{bmatrix} \quad (3.151)$$

  5. Integrate the velocity into position directly:

$$\begin{aligned} \frac{d}{dt}L &= -{}^N\omega_E^N(2) \\ \frac{d}{dt}\lambda &= \frac{{}^N\omega_E^N(1)}{\cos(L)} \\ \frac{d}{dt}h &= -{}^N v_R^B(3) \end{aligned} \quad (3.152)$$

  6. Calculate C_E^N from Lat, Lon:

$$C_E^N = \begin{bmatrix} \sin L\cos\lambda & -\sin L\sin\lambda & \cos L \\ -\sin\lambda & \cos\lambda & 0 \\ -\cos L\cos\lambda & -\cos L\sin\lambda & \sin L \end{bmatrix}$$

  7. Transform the Earth rate into the local-level frame:

$${}^N\omega_I^E = C_E^N\,{}^E\omega_I^E$$

Although mathematically equivalent, the error equations associated with algorithm 1 take a simpler form and are generally recommended if the algorithm needs to work with a Kalman filter and other sensor or information sources.

As discussed before, the attitude error can be analyzed using the following equation:

$$\delta^N \dot{\theta}_B^N = -\left[{}^N\omega_I^N \times\right]\delta^N\theta_B^N - C_B^N \delta^B \omega_I^B + \delta^N \omega_I^N$$

The velocity error can be analyzed using the following equation:

$$\frac{d}{dt}\delta^N V_E^N = C_B^N \delta^B a_I^B + {}^N a_I^B \times \delta^N \theta_B^N - \left(2\,{}^N\omega_I^E + {}^N\omega_E^N\right) \times \delta^N V_E^N$$
$$-\left(2\delta^N\omega_I^E + \delta^N\omega_E^N\right) \times {}^N V_E^N + \delta^N g$$

Ignoring the second-order effects of the altitude error on the craft rate, the errors in craft rates can be calculated from the velocity error:

$$\begin{bmatrix} \delta^N \omega_E^N(1) \\ \delta^N \omega_E^N(2) \end{bmatrix} = \begin{bmatrix} \left(\dfrac{1}{R_N + h} - \dfrac{1}{R_E + h}\right)\sin\alpha\cos\alpha & \dfrac{\sin^2\alpha}{R_N + h} + \dfrac{\cos^2\alpha}{R_E + h} \\ -\left(\dfrac{\cos^2\alpha}{R_N + h} + \dfrac{\sin^2\alpha}{R_E + h}\right) & -\left(\dfrac{1}{R_N + h} - \dfrac{1}{R_E + h}\right)\sin\alpha\cos\alpha \end{bmatrix} \begin{bmatrix} \delta^N v_E^B(1) \\ \delta^N v_E^B(2) \end{bmatrix}$$

$$\delta^N \omega_E^N(3) = 0$$

The horizontal position error can be described as an angular error $\delta^N\theta_E^N$ such that

$$C_{\hat{N}}^E = C_N^E C_{\hat{N}}^N = C_N^E\left(I + \left[\delta^N\theta_E^N \times\right]\right) = C_N^E\left(I - \left[\delta^N\theta_N^E \times\right]\right)$$

The differential equation for this error is given by

$$\frac{d}{dt}\delta^N\theta_E^N = -{}^N\omega_E^N \times \delta^N\theta_E^N + \delta^N\omega_E^N$$

Let's look at the error of projecting the Earth rate using an incorrect horizontal location.

By definition, the local level to the ECEF DCM knowledge is given by

$$\hat{C}_N^E = C_N^E\left(I + \left[\delta^N\theta_E^N \times\right]\right)$$

which can be rewritten as

$$\hat{C}_E^N = \left(I - \left[\delta^N\theta_E^N \times\right]\right)C_E^N$$

We then can write the Earth rate projection error as

$$\delta^N\omega_I^E = {}^N\hat{\omega}_I^E - {}^N\omega_I^E = \left(I - \left[\delta^N\theta_E^N \times\right]\right)C_E^N\,{}^E\omega_I^E - C_E^N\,{}^E\omega_I^E$$
$$= \left(I - \left[\delta^N\theta_E^N \times\right]\right){}^N\omega_I^E - {}^E\omega_I^E = -\left[\delta^N\theta_E^N \times\right]{}^N\omega_I^E$$

In summary, we have the following collection of equations for the local-level frame navigation error analysis.

The local-level frame error equations are:

1. Attitude error propagation:

$$\delta^N\dot{\theta}_B^N = -\left[{}^N\omega_I^N \times\right]\delta^N\theta_B^N - C_B^N\delta^B\omega_I^B + \delta^N\omega_I^N \tag{3.153}$$

2. Velocity error propagation:

$$\begin{aligned}\frac{d}{dt}\delta^N V_E^N &= C_B^N\delta^B a_I^B + {}^N a_I^B \times \delta^N\theta_B^N - \left(2\,{}^N\omega_I^E + {}^N\omega_E^N\right)\times \delta^N V_E^N \\ &\quad -\left(2\delta^N\omega_I^E + \delta^N\omega_E^N\right)\times {}^N V_E^N + \delta^N g\end{aligned} \tag{3.154}$$

3. Craft rate error:

$$\begin{aligned}\begin{bmatrix}\delta^N\omega_E^N(1)\\ \delta^N\omega_E^N(2)\end{bmatrix} &= \begin{bmatrix}\left(\dfrac{1}{R_N+h}-\dfrac{1}{R_E+h}\right)\sin\alpha\cos\alpha & \dfrac{\sin^2\alpha}{R_N+h}+\dfrac{\cos^2\alpha}{R_E+h}\\ -\left(\dfrac{\cos^2\alpha}{R_N+h}+\dfrac{\sin^2\alpha}{R_E+h}\right) & -\left(\dfrac{1}{R_N+h}-\dfrac{1}{R_E+h}\right)\sin\alpha\cos\alpha\end{bmatrix}\begin{bmatrix}\delta^N v_E^B(1)\\ \delta^N v_E^B(2)\end{bmatrix}\\ \delta^N\omega_E^N(3) &= 0\end{aligned} \tag{3.155}$$

4. Position error propagation:

$$\begin{aligned}\frac{d}{dt}\delta^N\theta_E^N &= -{}^N\omega_E^N \times \delta^N\theta_E^N + \delta^N\omega_E^N\\ \frac{d}{dt}\delta h &= -\delta^N v_R^B(3)\end{aligned} \tag{3.156}$$

5. Earth rate projection error:

$$\delta^N\omega_I^E = -\delta^N\theta_E^N \times {}^N\omega_I^E = {}^N\omega_I^E \times \delta^N\theta_E^N \tag{3.157}$$

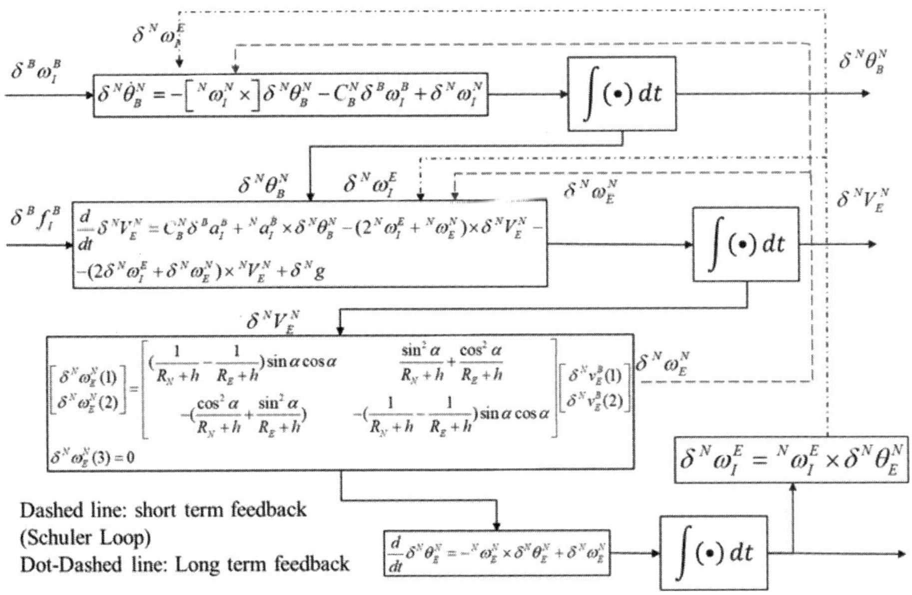

Figure 3.7 Local level frame inertial navigation error model.

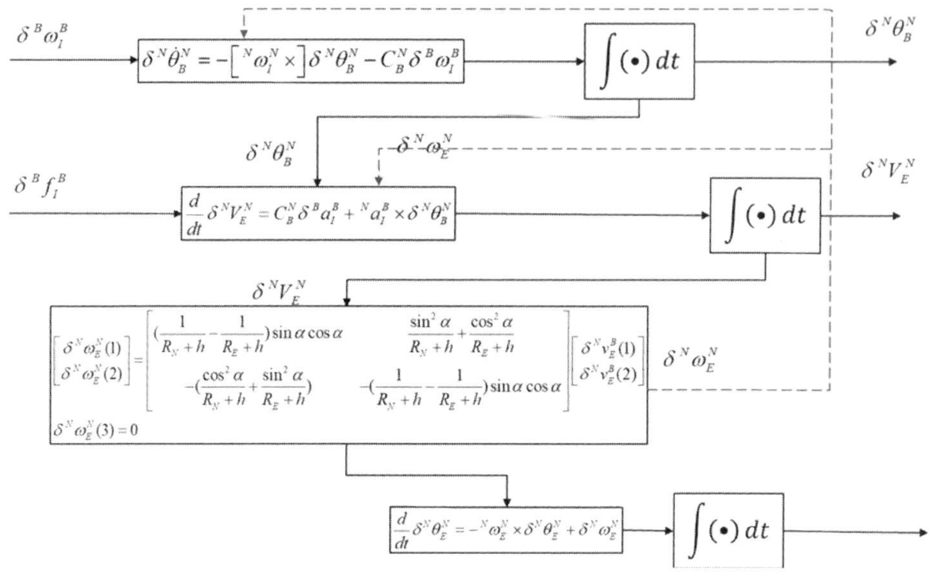

Figure 3.8 Short-term simplified error model.

The above error model can be illustrated by Figure 3.7.

Examining the diagram, we can find two feedback loops:

1. *Short-term feedback loop:* The velocity error generates a craft rate error that is fed back to the platform attitude error (and Coriolis acceleration error to a lesser degree), which then impacts the projection of an accelerometer output to a local-level frame, and, in turn, it generates velocity error. This loop is what is traditionally called the Schuler loop.
2. *Long-term feedback loop:* The position error generates the projection error of the Earth rate, which is fed back to the platform attitude error (and

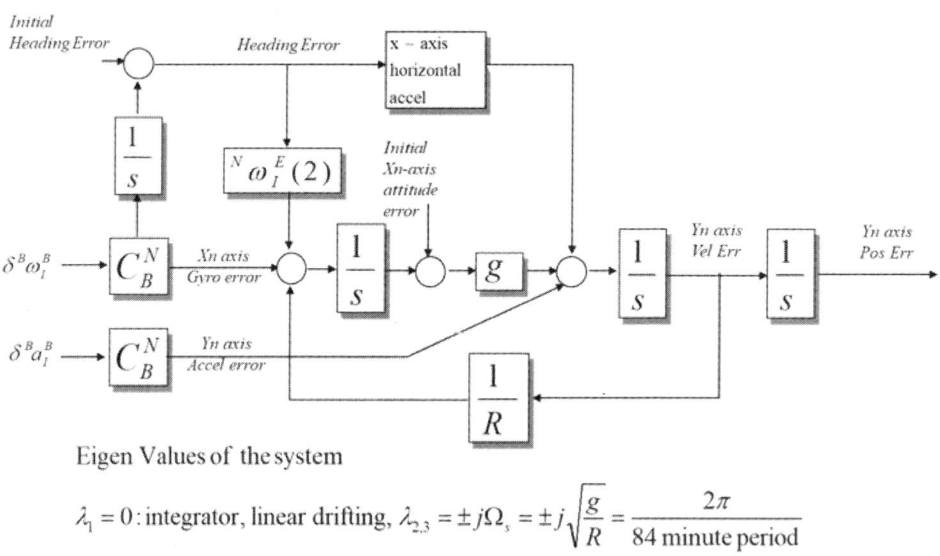

Figure 3.9 Schuler loop.

Coriolis acceleration error to a lesser degree), which then impacts projection of accelerometer output to local-level frame, and, in turn, it generates a velocity error, and the velocity error generates a position error. This loop has a much slower impact as it takes a huge position error to generate a small error in the projection of the Earth rate. As a result, we can safely ignore this loop in most analysis and design work.

Figure 3.8 illustrates the error model with short-term effects only.

By assuming a nominal 0 acceleration, and using the equations in the NED frame, the error equation for any one of the horizontal axes can be simplified to show explicitly the Schuler loop. This is shown in Figure 3.9.

References

[1] Gelb, M., *Applied Optimal Estimation*, Cambridge, MA: MIT Press, 1974.

[2] Brown, R. G., and P. Hwang, *Introduction to Random Signals and Applied Kalman Filtering*, New York: John Wiley & Sons, 1996.

[3] Li, A., and Y. -W. Wu, "Kalman Filtering & Applications," *UCLA Lecture Notes*, 2005–2019.

[4] Julier, S. J., J. K. Uhlmann, and H. Durrant-Whyte, "A New Approach for Filtering Nonlinear System," *Proc. of American Control Conference*, 1995, pp. 1628–1632.

[5] Julier, S. J., and J. K. Uhlmann, *A General Method for Approximating Nonlinear Transformation of Probability Distributions*, Technical Report RRG, Department of Engineering Science, University of Oxford, November 1996.

[6] http://www.teledyne-si.com/products-and-services/scientific-company/csac.

[7] Savage, P. G., "Strapdown Inertial Navigation Integration Algorithm Design Part 1: Attitude Algorithms," *Journal of Guidance, Control, and Dynamics*, Vol. 21, No. 1, 1998, pp. 19–28.

[8] Savage, P. G., "Strapdown Inertial Navigation Integration Algorithm Design Part 2: Velocity and Position Algorithms," *Journal of Guidance, Control, and Dynamics*, Vol. 21, No. 2, March–April 1998.

[9] Titterton, D., and J. Weston, *Strapdown Inertial Navigation Technology*, Second Edition, The American Institute of Aeronautics and Astronautics, 2004.

[10] Groves, P. D., *Principles of GNSS, Inertial, and Multisensor Integrated Navigation Systems*, Second Edition, Norwood, MA: Artech House, 2013.

[11] Roetenberg, D., "Inertial and Magnetic Sensing of Human Motion," Ph.D. thesis, Universiteit Twente, 2006.

[12] Jordan, J. W., "An Accurate Strapdown Direction Cosine Algorithm," NASA TN-D-5384, September 1969.

[13] Bortz, J. E., "A New Mathematical Formulation for Strapdown Inertial Navigation," *IEEE Transactions on Aerospace and Electronic Systems*, Vol. AES-7, No. 1, 1971, pp. 61–66.

[14] Ignagni, M. B., "Optimal Strapdown Attitude Integration Algorithms," *Journal of Guidance, Control, and Dynamics*, Vol. 13, No. 2, 1990, pp. 363–369.

[15] Ignagni, M. B., "Efficient Class of Optimized Coning Compensation Algorithms," *Journal of Guidance, Control, and Dynamics*, Vol. 19, No. 2, 1996, pp. 424–429.

[16] Mark, J. G., and Tazartes, D. A., "On Sculling Algorithms," *3rd St. Petersburg International Conf. on Integrated Navigation Systems*, St. Petersburg, Russia, May 1996.

[17] Miller, R., "A New Strapdown Attitude Algorithm," *Journal of Guidance, Control, and Dynamics*, Vol. 6, No. 4, 1983, pp. 287–291.

[18] Wu, M., et al., "Optimization-Based Alignment for Inertial Navigation Systems: Theory and Algorithm," *Aerosp. Sci. Technol.*, Vol. 15, 2011, pp. 1–17.

[19] Chang, L., J. Li, and S. Chen, "Initial Alignment by Attitude Estimation for Strapdown Inertial Navigation Systems," *IEEE Transactions on Instrumentation and Measurement*, Vol. 64, 2015, pp. 784–794.

[20] Silson, P. M. G., "Coarse Alignment of a Ship's Strapdown Inertial Attitude Reference System Using Velocity Loci," *IEEE Transactions on Instrumentation and Measurement*, Vol. 60, 2011, pp. 1930–1941.

[21] Wu, Y., and X. Pan, "Velocity/Position Integration Formula Part I: Application to In-Flight Coarse Alignment," *IEEE Transactions on Aerospace and Electronic Systems*, Vol. 49, 2012, pp. 1006–1023.

[22] Fang, J., and S. Yang, "Study on Innovation Adaptive EKF for In-Flight Alignment of Airborne POS," *IEEE Transactions on Instrumentation and Measurement*, Vol. 60, 2011, pp. 1378–1388.

[23] El-Sheimy, N., S. Nassar, and A. Noureldin, "Wavelet De-Noising for IMU Alignment," *IEEE Aerospace and Electronic Systems Magazine*, Vol. 19, 2004, pp. 32–39.

[24] Xu, X., et al., "A Kalman Filter for SINS Self-Alignment Based on Vector Observation," *Sensors*, Vol. 17, 2017, p. 264.

[25] Liang, Y., and Y. Jia, "A Nonlinear Quaternion-Based Fault-Tolerant SINS/GNSS Integrated Navigation Method for Autonomous UAVs," *Aerosp. Sci. Technol.*, Vol. 40, 2015, pp. 191–199.

[26] Chang, L., J. Li, and K. Li, "Optimization-Based Alignment for Strapdown Inertial Navigation System: Comparison and Extension," *IEEE Transactions on Aerospace and Electronic Systems*, Vol. 52, 2014, pp. 1697–1713.

CHAPTER 4

Various Measurement-Based Solutions

4.1 Introduction

In Chapter 3, we focused on PNT algorithms and their error analysis when the inputs are integrated or propagated, which, in general, require the integration of differential equations or propagation of difference equations. These algorithms and the information sources, such as a sensor, model, or data, will play the role of providing the propagation solutions when a Kalman filter or similar sensor fusion mechanisms are used to provide the integrated solutions.

In this chapter, we will focus on another type of sensor and information sources that can be used to derive a part or all of the PNT solutions directly or by using algebraic transformations instead of using differential or difference equations. These types of sensors and information sources are often used as the update solution or measurement solution when the different types of information sources are integrated or fused together to achieve better performance.

It is important to note that some sensors and information sources can be used to generate the propagation solutions and can also be used to generate the measurement solutions. For example, we can measure velocity using Doppler radars and we can use the velocity measurements as an update source to bound inertial navigation errors. However, we can also use such velocity to start a propagation solution by integrating the velocity into position.

Similar to Chapter 3, we will focus on the mathematics and algorithms instead of specific sensor technologies. The use of specific sensors in an All Source PNT system will be discussed in later chapters.

4.2 PNT Using Range and Range Rate Measurements

4.2.1 Two-Way Ranging and Time Transfer

Assume that there are radio or optical signal transceivers located at two locations, location 1 and location 2, or node 1 and node 2, and each of the nodes has its own time reference. We can measure the relative range between the two nodes and synchronize the two time references by determining the time offset between the two nodes. This technique is the two-way range measurement and time transfer.

This is achieved by node 1 and node 2 sending to node 2 and node 1 signals (RF, optical, or other types) that bear some type of time signature, such as a pulse or a code/data modulation sequence. The result is that we will have two sets of TOT and TOA measurements. These measurements can form two equations to solve for two unknowns, the range and time offset between the two nodes.

Let the offset between the two time references be given by Δt, that is,

$$^{R2}t = {}^{R1}t + \Delta t$$

where ^{R1}t is the time according to node 1's clock and ^{R2}t is the time according to node 2's clock.

Let $^{R1}t_{11}$ be the time when node 1 sends out a time pulse (or equivalent) according to its own clock, and $^{R2}t_{12}$ be the time when node 2 receives the time pulse (or equivalent) according to node 2's clock. Clearly, we have

$$^{R2}t_{12} = {}^{R1}t_{11} + \Delta t + \frac{r}{c}$$

where r is the range (distance) between the two nodes, and c is the speed of RF or optical or another type of signal traveling through the media.

Similarly, node 2 can send a pulse (or equivalent) to node 1.

Let $^{R2}t_{22}$ be the time when node 2 sends out a time pulse (or equivalent) according to its own clock and $^{R1}t_{21}$ be the time when node 1 receives the time pulse (or equivalent) according to node 1's clock. Similarly, we have

$$^{R1}t_{21} = {}^{R2}t_{22} - \Delta t + \frac{r}{c}$$

We can write the two equations as

$$\begin{aligned} \Delta t + \frac{r}{c} &= \left({}^{R2}t_{12} - {}^{R1}t_{11}\right) \\ \Delta t - \frac{r}{c} &= \left({}^{R2}t_{22} - {}^{R1}t_{21}\right) \end{aligned} \tag{4.1}$$

Consequently, we can solve for the clock difference (to allow time synchronization and time transfer) and the range measurements:

$$\begin{aligned} \Delta t &= \frac{\left({}^{R2}t_{12} - {}^{R1}t_{11}\right) + \left({}^{R2}t_{22} - {}^{R1}t_{21}\right)}{2} \\ r &= c\frac{\left({}^{R2}t_{12} - {}^{R1}t_{11}\right) - \left({}^{R2}t_{22} - {}^{R1}t_{21}\right)}{2} \end{aligned} \tag{4.2}$$

In order for the above equations to provide good results, it is important that the two paths are symmetric. In other words, the path from node 1 to node 2 and the path from node 2 and node 1 need to be the same path. Asymmetry can happen at least in the following two scenarios:

1. When the nodes are moving, and the timing pulses are not sent at the same time. In this case, we will want to send out the pulses from each node with very small time differences. This typically will not become a problem since very high data rate can be achieved over RF or optical links.

2. When the media are changing, such as the troposphere variations over time, and when the timing pulses are not sent at the same time. Again, this typically will not become a problem since very high data rate can be achieved over RF or optical links.

These problems become more significant when one or both of the two nodes are spacecraft. In this case, the nodes are far apart from each other and it takes significant amount of time for the signal to travel to the other node. In addition, the speed of the nodes can be such that the relativistic effects, in addition to the motion during the signal travel time (aberration error), must be considered. In these scenarios, careful modeling of the motion of each of the nodes and the media becomes important. The errors that need to be corrected will then include aberration error due to the motion of the nodes during the signal travel time, relativistic effects, and the aberration by the media.

Without considering the asymmetry mentioned above, and assuming that each of the TOT and TOA measurement errors are independent, we can estimate the standard deviations of the time synchronization error and the range measurement error to be given by

$$\begin{aligned} \sigma(\Delta t) &= \frac{\sqrt{\sigma^2\left({}^{R2}t_{12}\right) + \sigma^2\left({}^{R1}t_{11}\right) + \sigma^2\left({}^{R2}t_{22}\right) + \sigma^2\left({}^{R1}t_{21}\right)}}{2} \\ \sigma(r) &= c\sigma(\Delta t) \end{aligned} \tag{4.3}$$

If the variances of TOA and TOT errors are the same for each node, then the above can be simplified to be

$$\begin{aligned} \sigma(\Delta t) &= \sqrt{\frac{\sigma^2\left({}^{R2}t_{12}\right) + \sigma^2\left({}^{R1}t_{11}\right)}{2}} \\ \sigma(r) &= c\sigma(\Delta t) \end{aligned} \tag{4.4}$$

If all the timing errors have the same standard deviation and are equal to $\sigma(t)$, then the above can be further simplified to be

$$\begin{aligned} \sigma(\Delta t) &= \sigma(t) \\ \sigma(r) &= c\sigma(\Delta t) \end{aligned} \tag{4.5}$$

The important implication of the above equation is that the two-way time transfer does not amplify the timing errors.

When multiple measurements are taken, we can further average out the noise to achieve higher performance.

4.2.2 Simple Range-Based Positioning

A significant number of PNT solutions are based on range or pseudo-range measurements [1–4].

Given n objects with known positions located in a coordinate frame with their positions given as

$$r_k = \begin{bmatrix} x_k \\ y_k \\ z_k \end{bmatrix} \quad (k = 1,2,\cdots n)$$

and a navigator who wants to find its own position:

$$r = \begin{bmatrix} x \\ y \\ z \end{bmatrix}$$

by measuring the ranges to each of the objects:

$$d_k = \|r - r_k\| + v_k = \sqrt{(x - x_k)^2 + (y - y_k)^2 + (z - z_k)^2} + v_k \tag{4.6}$$

where v_k is the range measurement error.

It will be useful to linearize the above equations around an approximate solution:

$${}_0r = \begin{bmatrix} {}_0x \\ {}_0y \\ {}_0z \end{bmatrix}$$

$$d_k = \|r_k - r\| + v_k = \|(r_k - {}_0r) - (r - r_0)\| + v_k \approx \|r_k - {}_0r\| - \frac{(r_k - {}_0r)^T}{\|r_k - {}_0r\|}(r - {}_0r) + v_k$$

which can be written as

$$\Delta d_k({}_0r) = d_k - \|r_k - {}_0r\| = -u_{\text{LOS}}(r_0)_k^T (r - {}_0r) + v_k$$

where

$$u_{\text{LOS}}(r_0)_k = \frac{(r_k - {}_0r)}{\|r_k - {}_0r\|}$$

is the LOS unit vector from ${}_0r$ to r_k.

In summary, for each k, we have the following linearized measurement equation:

$$\Delta d_k({}_0r) = -\left(u_{\text{LOS}}({}_0r)_k\right)^T \Delta r + v_k \tag{4.7}$$

The geometric meaning of the above equation is easily observed: Δr is the vector from ${}_0r$ to r, the range deviation due to Δr is simply the projection of Δr on to the LOS direction.

This equation can be directly used in the extended Kalman filter design when we integrate the measurement solution with the propagation solution (such as inertial navigation). At this moment, we will use these equations to form a least square solution for Δr and, consequently, r can be estimated as

$$r = {}_0r + \Delta r$$

Obviously, the minimum number of the reference objects we need is 3. More measurements will provide the opportunity to reduce the contribution of measurement errors.

Stacking each of the equations in vector and matrix forms, we have

$$\begin{bmatrix} \Delta d_1({}_0r) \\ \Delta d_2({}_0r) \\ \vdots \\ \Delta d_n({}_0r) \end{bmatrix} = \begin{bmatrix} -\left(u_{LOS}({}_0r)_1\right)^T \\ -\left(u_{LOS}({}_0r)_2\right)^T \\ \vdots \\ -\left(u_{LOS}({}_0r)_n\right)^T \end{bmatrix} \Delta r + \begin{bmatrix} v_1 \\ v_2 \\ \vdots \\ v_n \end{bmatrix}$$

which can be written as

$$\Delta D = H\Delta r + V \tag{4.8}$$

where

$$\Delta D({}_0r) = \begin{bmatrix} \Delta d_1 \\ \Delta d_2 \\ \vdots \\ \Delta d_n \end{bmatrix}, \quad H({}_0r) = \begin{bmatrix} -\left(u_{LOS}({}_0r)_1\right)^T \\ -\left(u_{LOS}({}_0r)_2\right)^T \\ \vdots \\ -\left(u_{LOS}({}_0r)_n\right)^T \end{bmatrix}, \quad V = \begin{bmatrix} v_1 \\ v_2 \\ \vdots \\ v_n \end{bmatrix} \tag{4.9}$$

and a least square solution can be calculated using the following equation

$$\Delta D({}_0r) = H({}_0r)\Delta r + V$$

$$\Delta\hat{r} = \left(H({}_0r)^T H({}_0r)\right)^{-1} H({}_0r)^T \Delta D({}_0r) = H({}_0r)^+ \Delta D({}_0r)$$

and we can arrive at a new estimate of the position as

$$\hat{r} = {}_0r + \Delta\hat{r} - {}_1r$$

We have the opportunity to use the new estimate as the new point at which we linearize the equation and derive another solution. We can keep doing the iterations until we get the best results.

To be more precise, assuming that we start with *i*th iteration result,

$${}_ir$$

we can form the measurement vector and the H-matrix and the measurement equation becomes

$$\Delta D({}_ir) = H({}_ir)\Delta r + V$$

We solve the problem and we have

$$\Delta\hat{r} = \left(H\left({}_{i}r\right)^{T} H\left({}_{i}r\right)\right)^{-1} H\left({}_{i}r\right)^{T} \Delta D\left({}_{i}r\right) = H\left({}_{i}r\right)^{+} \Delta D\left({}_{i}r\right)$$

and

$$\hat{r} = {}_{i}r + \Delta\hat{r} = {}_{(i+1)}r$$

Assuming that the measurement noises from the different nodes are uncorrelated and have the same standard deviation:

$$\sigma\left(v_{k}\right) = \sigma, \quad (k = 1,2,\ldots,n)$$

After the convergence (i.e., when linearization errors can be ignored), the variance on the estimation error can be written as

$$\sigma^{2}(\hat{r} - r) = \left(H^{T}H\right)^{-1}\sigma^{2} \tag{4.10}$$

If each of the measurement noise terms has a different variance value, we can scale each of the equations by the standard deviation, and the linearized equation becomes

$$\Delta\bar{D}\left({}_{i}r\right) = \bar{H}\left({}_{i}r\right)\Delta r + \bar{V} \tag{4.11}$$

where

$$\Delta\bar{D} = \begin{bmatrix} \frac{1}{\sigma_{1}}\Delta d_{1} \\ \frac{1}{\sigma_{2}}\Delta d_{2} \\ \vdots \\ \frac{1}{\sigma_{n}}\Delta d_{n} \end{bmatrix}, \quad \bar{H} = \begin{bmatrix} -\frac{1}{\sigma_{1}}\left(u_{\mathrm{LOS1}}\right)^{T} \\ -\frac{1}{\sigma_{2}}\left(u_{\mathrm{LOS2}}\right)^{T} \\ \vdots \\ -\frac{1}{\sigma_{n}}\left(u_{\mathrm{LOS}n}\right)^{T} \end{bmatrix}, \quad \bar{V} = \begin{bmatrix} \frac{1}{\sigma_{1}}v_{1} \\ \frac{1}{\sigma_{2}}v_{2} \\ \vdots \\ \frac{1}{\sigma_{n}}v_{n} \end{bmatrix} \tag{4.12}$$

We have

$$\Delta\hat{r} = \left(\bar{H}\left({}_{i}r\right)^{T} \bar{H}\left({}_{i}r\right)\right)^{-1} \bar{H}\left({}_{i}r\right)^{T} \Delta\bar{D}\left({}_{i}r\right) = \bar{H}\left({}_{i}r\right)^{+} \Delta\bar{D}\left({}_{i}r\right) \tag{4.13}$$

$$\sigma^{2}(\hat{r} - r) = \left(\bar{H}^{T}\bar{H}\right)^{-1} \tag{4.14}$$

In addition to position solution, we can also derive velocity estimate from range rate measurements. It is easy to show that

$$\dot{D} = Hv + V_{\mathrm{rate}} \tag{4.15}$$

$$\hat{v} = H^{+}\dot{D} \tag{4.16}$$

where $\dot{D}$ is a vector of range rate measurements, v is the velocity, and $\hat{v}$ is the velocity estimate.

4.2.3 Range-Only Network Collaborative PNT

The results that we established in Sections 4.2.1 and 4.2.2 provide the foundations for the network collaborative PNT. Specifically:

1. By performing the two-way range/time transfer, we can allow multiple collaborative partners, each of them called a node, to measure the relative range and to synchronize the time between the two nodes.
2. Three or more nodes with known positions can establish a set of core nodes of genesis so that any other node that has range measurements against these core nodes will be able to solve for its position by using the range measurements against the core nodes.
3. Once the position of a node is established by the above method, this node can also serve as a core node and to be used to allow other nodes to derive their positions from it.

More specifically, we can list the following important but obvious conclusions. About relative positioning:

1. Relative range between collaborative partners or nodes can only help to determine relative positions. The cluster of partners can be translationally moved and rotated with 6 degrees of freedom without violating the constraints imposed by the relative ranges.
2. Three partners or nodes form a triangle. The size and shape of the triangle are completely defined by the distances of each side (ranges). Consequently, the shape and size of the cluster are fully defined by the relative ranges.
3. With $n = 3$ or more partners or nodes, $(n, 3)$ (choice of 3 from n) triangles can be formed that are interconnected; consequently, the shape and the size of the cluster are fully defined by the relative ranges.
4. In a fully deterministic case, the relative ranges are not necessarily independent. Take a quadrilateral, for example; all the four edges are independent. One of the diagonals is also independent. However, the four edges and one diagonal fully define two triangles. Consequently, the other diagonal is not independent.
5. However, when uncertainty exists, all measurements can help.

About absolute positioning:

1. For 3-point anchoring, in the 3-D space, if three partners or nodes know their absolute position, then with relative ranges between the partners, the whole cluster or network knows the positions of each of the nodes. This is because the three partners define a plane, which further defines a 3-axis frame. Once this frame is fixed, the entire network is defined in the space.

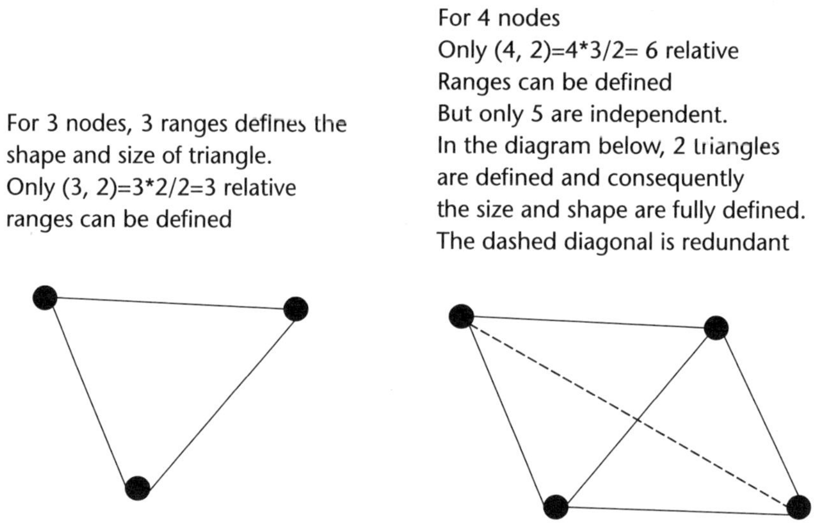

Figure 4.1 Range-only collaborative PNT.

2. For 2-point anchoring, in 2-D space, if two partners or nodes know their absolute position, then with relative ranges between the partners, the whole cluster or network knows the positions of the each of the nodes.
3. For anchoring by partial information, in general, if there are a number of partners or nodes with partial information available, as long as the results allow us to define two vectors (in the reference frame) in the 3-D space or 1 vector (in the reference frame) in the 2-D space, then the absolute position of the network can be determined.
4. For network-wide collaboration, in general, each of the nodes has its own position information with some uncertainty; by using the relative range measurements, the uncertainty of each of the nodes can be reduced.

Figure 4.1 illustrates some of the above principles.

4.3 PNT Using Pseudo-Range Measurement

4.3.1 Synchronized One-Way Ranging (Pseudo-Range) and PNT (GNSS)

GPS and other GNSS systems use one-way ranging instead of two-way ranging. This approach can also be used for the signal of opportunity positioning (SOOP) [1–4]. The biggest advantage of one-way ranging is that it does not require the infrastructure system to receive the signals from the users (which will limit how many users can be supported).

Given n objects (let us call them source nodes) with known positions located in a coordinate frame with their positions given as

$$r_k = \begin{bmatrix} x_k \\ y_k \\ z_k \end{bmatrix} \quad (k = 1,2,\cdots n)$$

and a navigator who wants to find its own position

$$r = \begin{bmatrix} x \\ y \\ z \end{bmatrix}$$

by measuring the pseudo-range to each of the objects, which can be measuring the TOA of the signal transmitted from each of the source nodes. Each of the source nodes uses the same time reference (synchronized source nodes):

$$\rho_k = \left\| r_k - r \right\| + b + v_k = \sqrt{\left(x - x_k\right)^2 + \left(y - y_k\right)^2 + \left(z - z_k\right)^2} + b + v_k \quad (4.17)$$

where b is the clock bias measured in distance unit (e.g., meters) and v_k is the range measurement error.

The same clock bias b is assumed for all the reference nodes. This is because all the source nodes are assumed to be synchronized, and the only time offset is the offset between the navigator's time and the time of the source nodes.

It will be useful to linearize the above equations around an approximate solution:

$${}_0r = \begin{bmatrix} {}_0x \\ {}_0y \\ {}_0z \end{bmatrix}$$

$$\begin{aligned} \rho_k &= \left\| r_k - r \right\| + b + v_k = \left\| \left(r_k - {}_0r\right) - \left(r - {}_0r\right) \right\| + b + v_k \\ &\approx \left\| r_k - {}_0r \right\| - \frac{\left(r_k - {}_0r\right)^T}{\left\| r_k - {}_0r \right\|}\left(r - {}_0r\right) + b + v_k \end{aligned}$$

or

$$\Delta\rho_k\left({}_0r\right) = \rho_k - \left\| r_k - {}_0r \right\| = -u_{\text{LOS}}\left({}_0r\right)_k^T \left(r - {}_0r\right) + b + v_k = -u_{\text{LOS}}\left({}_0r\right)_k^T \Delta r + b + v_k$$

where

$$u_{\text{LOS}}\left(r_0\right)_k = \frac{\left(r_k - {}_0r\right)}{\left\| r_k - {}_0r \right\|}$$

is the LOS unit vector from ${}_0r$ to r_k.

In summary, for each k, we have the following linearized measurement equation:

$$\Delta\rho_k\left({}_0r\right) = -\left(u_{\text{LOS}}\left({}_0r\right)_k\right)^T \Delta r + b + v_k \quad (4.18)$$

This equation can be directly used in an extended Kalman filter design when we integrate the measurement solution with the propagation solution (such as inertial navigation). At this moment, we will use these equations to form a least square solution for Δr and, consequently, r can be estimated as

$$r = {}_0r + \Delta r$$

Obviously, the minimum number of the source objects or nodes we need is 4. More measurements will provide opportunity to reduce the contribution of measurement errors.

Stacking each of the equations in vector and matrix forms, we have

$$\begin{bmatrix} \Delta\rho_1({}_0r) \\ \Delta\rho_2({}_0r) \\ \vdots \\ \Delta\rho_n({}_0r) \end{bmatrix} = \begin{bmatrix} -\left(u_{\text{LOS}}({}_0r)_1\right)^T & 1 \\ -\left(u_{\text{LOS}}({}_0r)_2\right)^T & 1 \\ \vdots & \vdots \\ -\left(u_{\text{LOS}}({}_0r)_n\right)^T & 1 \end{bmatrix} \begin{bmatrix} \Delta r \\ b \end{bmatrix} + \begin{bmatrix} v_1 \\ v_2 \\ \vdots \\ v_n \end{bmatrix}$$

which can be written as

$$\Delta \text{P} = H\Delta r + V \tag{4.19}$$

where

$$\Delta\text{P}({}_0r) = \begin{bmatrix} \Delta\rho_1({}_0r) \\ \Delta\rho_2({}_0r) \\ \vdots \\ \Delta\rho_n({}_0r) \end{bmatrix}, \quad H({}_0r) = \begin{bmatrix} -\left(u_{\text{LOS}}({}_0r)_1\right)^T & 1 \\ -\left(u_{\text{LOS}}({}_0r)_2\right)^T & 1 \\ \vdots & \vdots \\ -\left(u_{\text{LOS}}({}_0r)_n\right)^T & 1 \end{bmatrix}, \quad V = \begin{bmatrix} v_1 \\ v_2 \\ \vdots \\ v_n \end{bmatrix} \tag{4.20}$$

and a least square solution can be calculated using the following equation

$$\Delta\text{P}({}_0r) = H({}_0r)\begin{bmatrix} \Delta r \\ b \end{bmatrix} + V \tag{4.21}$$

$$\begin{bmatrix} \Delta\hat{r} \\ \hat{b} \end{bmatrix} = \left(H({}_0r)^T H({}_0r)\right)^{-1} H({}_0r)^T \Delta\text{P}({}_0r) = H({}_0r)^+ \Delta\text{P}({}_0r) \tag{4.22}$$

and we can arrive at a new estimate of the position as

$$\hat{r} = {}_0r + \Delta\hat{r} = {}_1r$$

We have the opportunity to use the estimate as the new point around which we linearize the equation and derive for another solution. We can keep doing the iterations until we get the best results.

To be more precise, assuming that we start with ith iteration results

$${}_ir$$

we can form the measurement vector and the H-matrix, and the measurement equation becomes

$$\Delta P\left({}_i r\right) = H\left({}_i r\right)\begin{bmatrix} \Delta r \\ b \end{bmatrix} + V \tag{4.23}$$

We solve the problem and we have

$$\begin{bmatrix} \Delta\hat{r} \\ \hat{b} \end{bmatrix} = \left(H\left({}_i r\right)^T H\left({}_i r\right)\right)^{-1} H\left({}_i r\right)^T \Delta P\left({}_i r\right) = H\left({}_i r\right)^+ \Delta P\left({}_i r\right) \tag{4.24}$$

and

$$\hat{r} = {}_i r + \Delta\hat{r} = {}_{(i+1)} r \tag{4.25}$$

Assuming that the measurement noises from the different nodes are uncorrelated and have the same standard deviation:

$$\sigma\left(v_k\right) = \sigma, \quad (k = 1,2,\ldots,n)$$

After the convergence (i.e., when linearization errors can be ignored), the variance on the estimation error can be written as

$$\sigma^2\left(\begin{bmatrix} \hat{r} \\ \hat{b} \end{bmatrix} - \begin{bmatrix} r \\ b \end{bmatrix}\right) = \left(H^T H\right)^{-1} \sigma^2 \tag{4.26}$$

If each of the measurement noise has different variance values, we can scale each of the equations by the standard deviation and the linearized equation becomes

$$\Delta\bar{P}\left({}_i r\right) = \bar{H}\left({}_i r\right)\Delta r + \bar{V} \tag{4.27}$$

where

$$\Delta\bar{P} = \begin{bmatrix} \frac{1}{\sigma_1}\Delta\rho_1 \\ \frac{1}{\sigma_2}\Delta\rho_2 \\ \vdots \\ \frac{1}{\sigma_n}\Delta\rho_n \end{bmatrix}, \quad \bar{H} = \begin{bmatrix} -\frac{1}{\sigma_1}\left(u_{\mathrm{LOS}}\left({}_0 r\right)_1\right)^T & \frac{1}{\sigma_1} \\ -\frac{1}{\sigma_2}\left(u_{\mathrm{LOS}}\left({}_0 r\right)_2\right)^T & \frac{1}{\sigma_2} \\ \vdots & \vdots \\ -\frac{1}{\sigma_n}\left(u_{\mathrm{LOS}}\left({}_0 r\right)_n\right)^T & \frac{1}{\sigma_n} \end{bmatrix}, \quad \bar{V} = \begin{bmatrix} \frac{1}{\sigma_1}v_1 \\ \frac{1}{\sigma_2}v_2 \\ \vdots \\ \frac{1}{\sigma_n}v_n \end{bmatrix} \tag{4.28}$$

We then have

$$\begin{bmatrix} \Delta\hat{r} \\ \hat{b} \end{bmatrix} = \left(\bar{H}\left({}_i r\right)^T \bar{H}\left({}_i r\right)\right)^{-1} \bar{H}\left({}_i r\right)^T \Delta\bar{P}\left({}_i r\right) = \bar{H}\left({}_i r\right)^+ \Delta\bar{P}\left({}_i r\right) \tag{4.29}$$

and

$$\sigma^2\left(\begin{bmatrix} \hat{r} \\ \hat{b} \end{bmatrix} - \begin{bmatrix} r \\ b \end{bmatrix}\right) = \left(\bar{H}^T \bar{H}\right)^{-1} \tag{4.30}$$

In addition to the position solution and the clock bias, we can also derive velocity estimates and clock rate estimates from pseudo-range rate measurements in a very similar manner.

4.3.2 Asynchronous One-Way Ranging (Pseudo-Range) and PNT (SOOP)

For carefully planned man-made PNT infrastructures such as GPS, GLONASS, Galileo, and Beidou, the system is designed to be on the solid foundation of the precision system-wide master clock, which ensures that all the broadcasting signals are synchronized with very small errors.

This is, in general, not true if we want to use SOOP as a ranging source such as radio stations, TV stations, cell phone towers, and communicating satellites. These systems typically are privately owned by different organizations, and it will be very hard to expect that any synchronization can ever be done among these signal sources.

It is therefore important to develop an approach that can use pseudo-ranges from asynchronous sources. The key to the solution is the use of a reference receiver that is located at a known surveyed location and we use the single difference of pseudo-ranges between a navigating receiver and a reference receiver to cancel out the contributions of the clocks of the source nodes.

Assuming that n signal sources with known positions given by

$$r_k = \begin{bmatrix} x_k \\ y_k \\ z_k \end{bmatrix} \quad (k = 1,2,\cdots n)$$

a navigator (user) who wants to find its own position

$$r = \begin{bmatrix} x \\ y \\ z \end{bmatrix}$$

and a reference receiver located at a known location

$$r_r = \begin{bmatrix} x_r \\ y_r \\ z_r \end{bmatrix}$$

the pseudo-range measurement of the navigator to the source nodes are given by

$$\rho_{uk} = \left\| r_k - r \right\| + b_u + b_k + v_k = \sqrt{(x - x_k)^2 + (y - y_k)^2 + (z - z_k)^2} + b_u + b_k + v_{uk} \qquad (4.31)$$

where b_u is the user receiver clock bias measured in distance unit (e.g., meters), b_k is the source k clock bias measured in distance unit (e.g., meters), and v_{uk} is the user pseudo-range measurement error.

The pseudo-range measurement of reference receiver to the source nodes are given by

$$\rho_{rk} = \|r_k - r_r\| + b_r + b_k + v_{rk} = \sqrt{(x_r - x_k)^2 + (y_r - y_k)^2 + (z_r - z_k)^2} + b_r + b_k + v_{rk} \quad (4.32)$$

where b_r is the reference receiver clock bias measured in distance unit (e.g., meters) and v_{rk} is the reference pseudo-range measurement error.

If we perform a single difference between the user pseudo-range and the reference pseudo-range, we have

$$SD_{urk} = \rho_{uk} - \rho_{rk} = \|r_k - r\| - \|r_k - r_r\| + (b_u - b_r) + (v_{uk} - v_{rk}) \quad (4.33)$$

Let

$$\xi_k \equiv SD_{urk} + \|r_r - r_k\|$$

We arrive at the same form of equations as the standard synchronized pseudo-range-based positioning problem:

$$\xi_k = \|r_k - r\| + b_{ur} + v_{urk} \quad (4.34)$$

where $b_{ur} = v_u - b_r$ is the relative clock offset between the user node and the reference node, and $v_{urk} = v_{uk} - v_{rk}$ is the difference of the pseudo-range measurement errors.

The important implications of the above equations are:

1. Most importantly, the source-specific clock biases are cancelled out.
2. The measurement equation is in the same form as the synchronized pseudo-range positioning problem. That means that all the equations on the solution methods and the error analysis methods can be reused and therefore not repeated here.
3. The clock bias that the solution will be able to solve for is the relative clock bias between the user receiver and the reference receiver. In other words, we are achieving time synchronization and time transfer between the reference node and the user node.
4. The pseudo-range errors in the single difference equation are the difference of the errors of the user receiver and reference receiver. This has two implications:
   a. The noise after difference, typically high-frequency and filterable, becomes higher due to the difference of two random variables.
   b. The signal propagation errors (such as ionosphere delays and troposphere delays), typically low-frequency, are partially cancelled.

Fundamentally, the use of a reference node or receiver helps to achieve two objectives:

1. Solve the time-synchronization problem.
2. Create similar effects as differential GPS to cancel out common mode errors to achieve better performance.

Recognizing the above two points helps us to make wise system architecture and system engineering decisions. For example, if we have the capability to maintain the time accuracy using a good clock for an extended period of time, for the user, the reference, and the source, it means during this period of time, we do not need the reference node to help us solve the time-synchronization problem if it was done earlier.

Similarly, if we can model the signal propagation delays and correct them from time to time, we may also maintain good performance without the reference receiver.

4.4 Position and Attitude Determination Using Angle Measurements

4.4.1 Introduction

The angle measurement represents the measured direction of some type of optical, RF, or other signals. Examples of angle measurements include but are not limited to:

1. Stars, planets, satellites, and other heavenly bodies as seen by devices such as cameras or star-trackers. In these cases, the angle measurements are provided as the space objects' position in the camera FOV, which can be converted into two angles or into a unit vector.
2. LOS measurements as measured by laser or RF terminals that communicate with a communication partner. The angle measurements are provided as gimbal angles and/or refined angle measurements derived from laser or RF phase information.

When the object emitting the signal is so far away that it can be considered as if it is at infinity, such as a star, the direction is independent of the location of the receiving device. In other words, fundamentally, angle measurements for objects that are located at infinity only contain attitude information.

When the angle measurements are with respect to an object that has finite distance from the observer, the angles (i.e. directions) will change as the receiver of the signal moves. The angles will also change when the attitude of the receiver changes. In other words, fundamentally, angle measurements for objects that are located at a finite distance contain both attitude and position information.

In Section 4.4.2, we will focus on attitude determination using angle measurements with respect to objects both at infinity and at finite distance [5–13].

In Section 4.4.3, we will focus on combined attitude and position determination using angle measurements with respect to objects at finite distance.

4.4.2 Attitude Determination Using Angle Measurements

Attitude can be determined from angle measurements from both objects at infinity and objects at a finite distance. For the purpose of this section, when finite distance object is used, we assume that the user location is perfectly known. In Section 4.4.3, we will deal with the case when the position is not perfectly known and the angle measurements are used to determine the position and attitude at the same time.

4.4.2.1 Attitude Determination from Two Pairs of Unit Vectors (Directions)

Given a body frame B and a reference frame R, the attitude of the body frame with respect to the reference frame can be represented as quaternions, DCMs, or Euler angles or other mathematical representations as we have discussed in Chapter 2.

Assuming that we know the unit vector pointing from a user to an object, both in body frame (such as measured by a star-tracker) and the reference frame (such as from a star catalog or computed based on the positions of the objects), then we will be able to determine the attitude from two pairs of such unit vectors provided that the two vectors are not parallel to one another.

Let ${}^{B}u_1, {}^{B}u_2$ be the coordinate vectors of u_1, u_2 in the B frame and let ${}^{R}u_1, {}^{R}u_2$ be the coordinate vectors of u_1, u_2 in the R frame. By definition, we have

$$
\begin{aligned}
{}^{B}u_1 &= C_R^B\,{}^{R}u_1 \\
{}^{B}u_2 &= C_R^B\,{}^{R}u_2
\end{aligned}
$$

The TRIAD algorithm [5–13] simply solves the above equations for the DCM C_R^B.

Construct the vector w_1 that is orthogonal to both unit vectors and we do that in both the B frame and the R frame:

$$
\begin{aligned}
{}^{B}w_1 &= \frac{\left({}^{B}u_1 \times {}^{B}u_2\right)}{\left\|{}^{B}u_1 \times {}^{B}u_2\right\|} \\
{}^{R}w_1 &= \frac{\left({}^{R}u_1 \times {}^{R}u_2\right)}{\left\|{}^{R}u_1 \times {}^{R}u_2\right\|}
\end{aligned}
\tag{4.35}
$$

and it is easy to show that

$$
{}^{B}w_1 = C_R^B\,{}^{R}w_1
$$

Construct another vector w_2 that is orthogonal to both u_1 and w_1. Again we do this in both frames:

$$
\begin{aligned}
{}^{B}w_2 &= \frac{\left({}^{B}u_1 \times {}^{B}w_1\right)}{\left\|{}^{B}u_1 \times {}^{B}w_1\right\|} \\
{}^{R}w_2 &= \frac{\left({}^{R}u_1 \times {}^{R}w_1\right)}{\left\|{}^{R}u_1 \times {}^{R}w_1\right\|}
\end{aligned}
\tag{4.36}
$$

and it is easy to show that

$$
{}^{B}w_2 = C_R^B\,{}^{R}w_2
$$

Putting the three unit vectors into a matrix, we have

$$
\begin{bmatrix} {}^{B}u_1 & {}^{B}w_1 & {}^{B}w_2 \end{bmatrix} = C_R^B \begin{bmatrix} {}^{R}u_1 & {}^{R}w_1 & {}^{R}w_2 \end{bmatrix}
$$

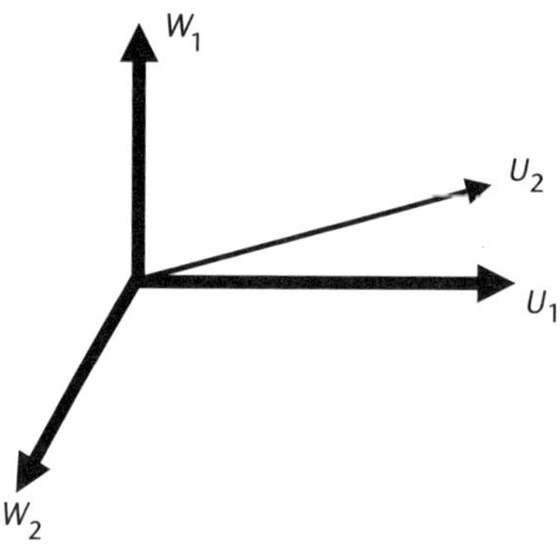

Figure 4.2 Vectors constructed in TRIAD.

Since the vectors are orthonormal with respect to one another, the two matrices are orthonormal matrices and we have

$$C_R^B = \begin{bmatrix} {}^B u_1 & {}^B w_1 & {}^B w_2 \end{bmatrix} \begin{bmatrix} {}^R u_1 & {}^R w_1 & {}^R w_2 \end{bmatrix}^T \tag{4.37}$$

Figure 4.2 illustrates the vectors we constructed.

4.4.2.2 Measurement Error Equation

Given the truth attitude DCM C_R^B and its estimate $\hat{C}_B^R$, using the symbols and equations from Section 3.4.4, we have

$$\hat{C}_B^R = C_R^{\hat{R}} C_B^R = \left(I - \left[\theta_R^{\hat{R}} \times\right]\right) C_B^R = \left(I - \left[\delta\, {}^R\theta_B^R \times\right]\right) C_B^R$$

and

$$\hat{C}_R^B = C_R^B \left(I + \left[\delta\, {}^R\theta_B^R \times\right]\right)$$

where $\delta\, {}^R\theta_B^R = \theta_R^{\hat{R}}$ is the error of the attitude (R with respect to B) expressed in the R frame. ($\delta\, {}^R\theta_B^R$ is a better symbol than $\theta_R^{\hat{R}}$ as it provides the context explicitly.)

Similarly, we can also view the difference between the two DCMs as an error of the knowledge of the body frame:

$$\hat{C}_B^R = C_B^R C_{\hat{B}}^B = C_B^R \left(I - \left[\theta_{\hat{B}}^B \times\right]\right) = C_B^R \left(I + \left[\theta_B^{\hat{B}} \times\right]\right) = C_B^R \left(I + \left[\delta\, {}^B\theta_R^B \times\right]\right)$$

and

$$\hat{C}_R^B = \left(I - \left[\delta\, {}^B\theta_R^B \times\right]\right) C_R^B$$

where $\delta\, {}^B\theta_R^B = \theta_B^{\hat{B}} = -\delta\, {}^B\theta_B^R$ is the error of the attitude (B with respect to R) expressed in the B frame. For the exercise below, we will use the B frame version of the error definition.

Since we have

$$^{B}u_k = C_R^B\,{}^{R}u_k$$

and

$$^{B}\hat{u}_k = \hat{C}_R^B\,{}^{R}u_k$$

we can write

$$^{B}\hat{u}_k = \hat{C}_R^B\,{}^{R}u_k = \left(I - \left[\delta\,{}^{B}\theta_R^B \times\right]\right) C_R^B\,{}^{R}u_k = \left(I - \left[\delta\,{}^{B}\theta_R^B \times\right]\right){}^{B}u_k$$

which can be rewritten as

$$\delta\,{}^{B}u_k = {}^{B}\hat{u}_k - {}^{B}u_k = -\left[\delta\,{}^{B}\theta_R^B \times\right]{}^{B}u_k = \left[{}^{B}u_k \times\right]\delta\,{}^{B}\theta_R^B \tag{4.38}$$

It is important to note that the symbols in the above equations can be interpreted in two different ways and consequently lead to different applications.

When we linearize a nonlinear measurement equation, we are interested in

$$\textit{Delta} = \textit{Truth} - (\textit{Point around which we linearize})$$

In most cases, the point around which we linearize can be the estimate. In other words,

$$\textit{Delta} = \textit{Truth} - \textit{Estimate}$$

When we perform error analysis, our convention has been

$$\textit{Error} = \textit{Estimate} - \textit{Truth}$$

which seems to lead to the confusion of the signs.

What is important is that the equation stays the same as long as the both sides of the equation use the same interpretation (i.e., either both sides are the errors or both sides are the deviation from nominal).

With the above, now we state the importance of the above equations:

1. They can be used to linearize the measurement equation.
2. They can be used as the linearized error analysis equation that relates the unit vector error to the attitude error.

Example 4.1: A Deeper Look at Attitude Determination Error from Two Pairs of Unit Vectors

Without loss of generality, let us choose the body frame axes such that the two unit vectors are in the *y*/*z* plane and the *z*-axis in the middle of the two vectors as illustrated in Figure 4.3.

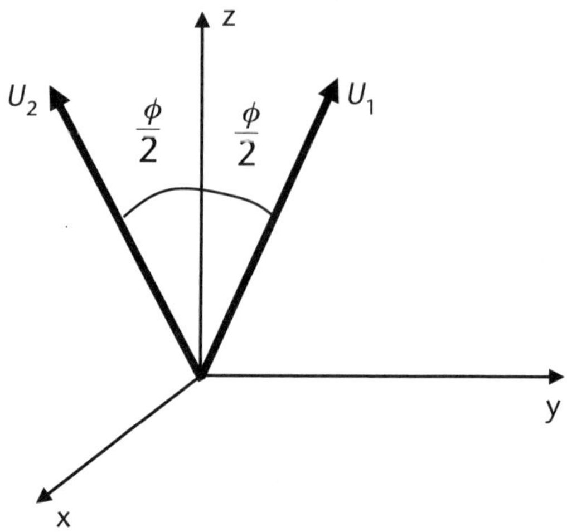

Figure 4.3 Coordinate Frame chosen to analyze TRIAD Errors

We have

$$^{B}u_1 \approx \begin{bmatrix} 0 \\ \sin\frac{\varphi}{2} \\ \cos\frac{\varphi}{2} \end{bmatrix}, \quad \left[{}^{B}u_1 \times\right] = \begin{bmatrix} 0 & -\cos\frac{\varphi}{2} & \sin\frac{\varphi}{2} \\ \cos\frac{\varphi}{2} & 0 & 0 \\ -\sin\frac{\varphi}{2} & 0 & 0 \end{bmatrix}$$

$$^{B}u_2 \approx \begin{bmatrix} 0 \\ -\sin\frac{\varphi}{2} \\ \cos\frac{\varphi}{2} \end{bmatrix}, \quad \left[{}^{B}u_2 \times\right] = \begin{bmatrix} 0 & -\cos\frac{\varphi}{2} & -\sin\frac{\varphi}{2} \\ \cos\frac{\varphi}{2} & 0 & 0 \\ \sin\frac{\varphi}{2} & 0 & 0 \end{bmatrix}$$

We can form two linearized equations (the delta's deviations)

$$\delta\,{}^{B}u_1 = \left[{}^{B}u_1 \times\right]\delta\,{}^{B}\theta_R^B$$
$$\delta\,{}^{B}u_2 = \left[{}^{B}u_2 \times\right]\delta\,{}^{B}\theta_R^B$$

Expand these equations and use the simplified symbol

$$\delta\,{}^{B}\theta_R^B = \begin{bmatrix} \theta_x & \theta_y & \theta_z \end{bmatrix}^T$$

We have the following expanded equations:

$$-\cos\frac{\varphi}{2}\theta_y + \sin\frac{\varphi}{2}\theta_z = \delta u_1(1)$$

$$\cos\frac{\varphi}{2}\theta_x = \delta u_1(2)$$

$$-\sin\frac{\varphi}{2}\theta_x = \delta u_1(3)$$

and

$$-\cos\frac{\varphi}{2}\theta_y - \sin\frac{\varphi}{2}\theta_z = \delta u_2(1)$$

$$\cos\frac{\varphi}{2}\theta_x = \delta u_2(2)$$

$$\sin\frac{\varphi}{2}\theta_x = \delta u_2(3)$$

From the above, we can derive a solution for θ_y

$$\theta_y = -\frac{\left(\delta u_1(1) + \delta u_2(1)\right)}{2\cos\frac{\varphi}{2}} \quad (4.39)$$

and a solution for θ_z

$$\theta_z = -\frac{\left(\delta u_1(1) - \delta u_2(1)\right)}{2\sin\frac{\varphi}{2}} \quad (4.40)$$

For θ_x (the x-axis is orthogonal to both of the two unit vectors), we have two redundant equations:

$$\theta_x = \delta u_1(2)\cos\frac{\varphi}{2} - \delta u_1(3)\sin\frac{\varphi}{2} \quad (4.41)$$

$$\theta_x = \delta u_2(2)\cos\frac{\varphi}{2} + \delta u_2(3)\sin\frac{\varphi}{2} \quad (4.42)$$

We have the following important conclusions:

1. For the two axes that are in the plane spanned by the two unit vectors, the attitude determination errors are amplified by $1/(2\cos(\varphi/2))$ and $1/(2\sin(\varphi/2))$.
2. For the one axis that is orthogonal to the plane spanned by the two unit vectors, attitude can be solved from the measurement from any or both of the unit vectors. The dilution of error will be 1 (when only one of the two equations is used) or 1/sqrt(2) (when both equations are used.)

Obviously, to achieve the best performance for all three axes, we want to choose the separation angle to be 90° so that both $1/(2\cos(\varphi/2))$ and $1/(2\sin(\varphi/2))$ are minimized.

4.4.2.3 Attitude Determination from More Than Two Pairs of Unit Vectors

In real applications, such as attitude determination using star-trackers, we have typically many more than just two pairs of unit vectors to work with. Based on what we discussed earlier, we can imagine that having more pairs of the unit vectors will provide at least two advantages:

1. Better geometry, which leads to smaller dilution of precision;
2. Averaging of measurement noise and errors.

Both lead to better performance.

Historically, this problem has been called the Wabba problem [6–13] and has attracted the attention of many researchers who have created algorithms such as QUEST (Quaternion Estimator). Practically, the problem is very simple and there is an easy two-step solution. For all practical purposes, there is no need to use any of the fancy Wabba problem solutions other than this practical method.

- *Step 1:* Pick a pair of the unit vectors and form an initial solution using the TRIAD algorithm we discussed earlier.
- *Step 2:* Use the solution to linearize the measurement equations and solve a linear least square problem. This step can use some iterations if the errors in the initial solution still lead to significant nonlinear terms.

Assuming that we have achieved step 1, as presented earlier, we then can linearize the unit vector measurement equation as

$$\delta\,{}^{B}u_{k} = \left[{}^{B}u_{k}\times\right]\delta\,{}^{B}\theta_{R}^{B}$$

Here, the deltas are interpreted as a deviation from nominal instead of errors.

It is typically simpler if we can transform the above equation into the sensor boresight frame:

$$C_{B}^{S}\delta\,{}^{B}u_{k} = C_{B}^{S}\left[{}^{B}u_{k}\times\right]\delta\,{}^{B}\theta_{R}^{B} = C_{B}^{S}\left[{}^{B}u_{k}\times\right]C_{B}^{S}C_{B}^{S}\delta\,{}^{B}\theta_{R}^{B}$$

Note that

$$C_{B}^{S}\left[{}^{B}u_{k}\times\right]C_{S}^{B} = \left[\left(C_{S}^{B}u_{k}\right)\times\right]$$

and we end up with

$$\delta\,{}^{S}u_{k} = \left[{}^{S}u_{k}\times\right]C_{B}^{S}\delta\,{}^{B}\theta_{R}^{B}$$

When the sensor frames are different for each one of the unit vector measurements, we just have to use the specific DCM individually. In other words, we have the following equation:

$$\delta\,{}^{S_k}u_{k} = \left[{}^{S_k}u_{k}\times\right]C_{B}^{S_k}\delta\,{}^{B}\theta_{R}^{B} \tag{4.43}$$

where S_k is the sensor frame for the kth unit vector.

Let

$$\delta^{S_k}u_k = \left[{}^{S_k}u_k \times\right] C_B^{S_k} \delta^B\theta_R^B = G_k \delta^B\theta_R^B$$

For most of the unit vector measurement sensors, we choose the sensor boresight axis to be the z-axis and typically the attitude deviations are reflected in the first two elements through the sine effect and the third element through cosine effect. As a result, if is often productive just to ignore the third element and we form

$$\delta^{S_k}u_k(1{:}2) = G_k(1{:}2,:)\delta^B\theta_R^B = H_{1k}\delta^B\theta_R^B$$

Including the sensor measurement noise, we have the following measurement equations

$$y_k = H_{1k}\delta^B\theta_R^B + v_k$$

The equally weighted least square solution is then given by

$$\delta\hat{\theta}_R^B = \left(\sum_k H_{1k}^T H_{1k}\right)^{-1}\left(\sum_k H_{1k}^T y_k\right) \tag{4.44}$$

Assuming that

$$\operatorname{cov}(v_k, v_j) = \sigma^2\delta_{kj}$$

then

$$E\left(\delta\theta\delta\theta^T\right) = \sigma^2\left(\begin{bmatrix} \cdots & H_{1k}{}^T & \cdots \end{bmatrix}\begin{bmatrix} \vdots \\ H_{1k} \\ \vdots \end{bmatrix}\right)^{-1} = \sigma^2\left(\sum_k H_{1k}{}^T H_{1k}\right)^{-1}$$

We can write the above as

$$E\left(\delta\theta\delta\theta^T\right) = \frac{\sigma^2}{n}\left(\frac{1}{n}\sum_k H_{1k}{}^T H_{1k}\right)^{-1} = \frac{\sigma^2}{n}Z \tag{4.45}$$

which shows the two effects of using multiple pairs of unit vectors:

1. The averaging effects that reduces σ^2 to σ^2/n.
2. The geometric dilution effect through the average matrix $Z = \left((1/n)\sum_k H_{1k}{}^T H_{1k}\right)^{-1}$.

4.4.2.4 R. (Ken) Li Formula for the Single Star-Tracker Attitude Determination Performance Assessment

As an application, we will use the above method to derive a simple and convenient approximate formula to estimate the attitude determination performance using a single star-tracker.

Given N star unit vectors in the star-tracker (ST) frame,

$$^{ST}U_k = \begin{bmatrix} x_k & y_k & z_k \end{bmatrix}^T, \quad k = 1,2,\cdots,N$$

We then have

$$H_{1k} = \left[{}^{ST}U_k \times \right](1{:}2,:) = \begin{bmatrix} 0 & -z_k & y_k \\ z_k & 0 & -x_k \end{bmatrix}$$

$$H_{1k}{}^T H_{1k} = \begin{bmatrix} 0 & z_k \\ -z_k & 0 \\ y_k & -x_k \end{bmatrix} \begin{bmatrix} 0 & -z_k & y_k \\ z_k & 0 & -x_k \end{bmatrix} = \begin{bmatrix} z_k^2 & 0 & -z_k x_k \\ 0 & z_k^2 & -z_k y_k \\ -z_k x_k & -z_k y_k & x_k^2 + y_k^2 \end{bmatrix}$$

and

$$\sum_k H_{1k}{}^T H_{1k} = \begin{bmatrix} \sum_k z_k^2 & 0 & -\sum_k z_k x_k \\ 0 & \sum_k z_k^2 & -\sum_k z_k y_k \\ -\sum_k z_k x_k & -\sum_k z_k y_k & \sum_k x_k^2 + y_k^2 \end{bmatrix} = N \begin{bmatrix} \frac{1}{N}\sum_k z_k^2 & 0 & -\frac{1}{N}\sum_k z_k x_k \\ 0 & \frac{1}{N}\sum_k z_k^2 & -\frac{1}{N}\sum_k z_k y_k \\ -\frac{1}{N}\sum_k z_k x_k & -\frac{1}{N}\sum_k z_k y_k & \frac{1}{N}\sum_k x_k^2 + y_k^2 \end{bmatrix}$$

When there are many stars, the above can be approximated by

$$\sum_k H_{1k}{}^T H_{1k} = N \begin{bmatrix} \frac{1}{N}\sum_k z_k^2 & 0 & -\frac{1}{N}\sum_k z_k x_k \\ 0 & \frac{1}{N}\sum_k z_k^2 & -\frac{1}{N}\sum_k z_k y_k \\ -\frac{1}{N}\sum_k z_k x_k & -\frac{1}{N}\sum_k z_k y_k & \frac{1}{N}\sum_k x_k^2 + y_k^2 \end{bmatrix} \approx N \begin{bmatrix} E(Z^2) & 0 & -E(ZX) \\ 0 & E(Z^2) & -E(ZY) \\ -E(ZX) & -E(ZY) & E(X^2 + Y^2) \end{bmatrix}$$

Assuming that the stars are uniformly distributed in the star-tracker FOV, we then have

$$\sum_k H_{1k}{}^T H_{1k} \approx N \begin{bmatrix} E(Z^2) & 0 & -E(ZX) \\ 0 & E(Z^2) & -E(ZY) \\ -E(ZX) & -E(ZY) & E(X^2 + Y^2) \end{bmatrix} \approx N \begin{bmatrix} 1 & 0 & 0 \\ 0 & 1 & 0 \\ 0 & 0 & \frac{2FOV^2}{12} \end{bmatrix} = N \begin{bmatrix} 1 & 0 & 0 \\ 0 & 1 & 0 \\ 0 & 0 & \frac{FOV^2}{6} \end{bmatrix}$$

Here we assume that the star-tracker has a square FOV and the side of the square has a length of FOV radians.

Consequently, the attitude covariance matrix can be written as

$$P \approx \frac{\sigma^2}{N}\begin{bmatrix} 1 & 0 & 0 \\ 0 & 1 & 0 \\ 0 & 0 & 1/\left(\dfrac{FOV^2}{6}\right) \end{bmatrix} \qquad P^{1/2} \approx \frac{\sigma}{\sqrt{N}}\begin{bmatrix} 1 & 0 & 0 \\ 0 & 1 & 0 \\ 0 & 0 & 1/\left(\dfrac{FOV}{\sqrt{6}}\right) \end{bmatrix} \tag{4.46}$$

This is the R. (Ken) Li Formula, which essentially says the following:

1. The number of star averaging effects is simply $1/\sqrt{N}$ as expected.
2. The geometric dilution effect is such that, for the two axes normal to the boresight, the effect is 1 to 1, and there is no amplification or reduction due to geometry. For the attitude around the boresight, the errors are amplified by $1/(FOV/\sqrt{6})$.

4.4.3 Position and Attitude Determination from Angle Measurements

As discussed before, angular measurements with respect to an object at a finite distance contain both position and attitude information. Fundamentally, both celestial navigation and vision-based navigation are based on angle measurements [14–33].

Let the location of a navigator in ECEF frame be given by ${}^{E}r$ and the location of a finite distance object (FDO), FDO j be given by ${}^{E}r_j$.

The LOS vector from the navigator to the FDO can then be computed as

$${}^{E}u_j = \frac{1}{\left\| {}^{E}r_j - {}^{E}r \right\|}\left({}^{E}r_j - {}^{E}r\right) = \frac{1}{\rho_j}\left({}^{E}r_j - {}^{E}r\right) \tag{4.47}$$

where ρ_j is the range to the FDO.

${}^{E}u_j$ can be transformed to the sensor frame (S) using the following equation:

$${}^{S}u_j = C_B^S C_E^B \, {}^{E}u_j \tag{4.48}$$

where C_E^B is the DCM that transforms the coordinates of a vector in the ECEF frame to the body frame. In other words, C_E^B represents the navigator's attitude. C_B^S is the DCM that transforms the coordinates of a vector in the body frame to the sensor (such as a star-tracker) frame.

The measured unit vector can be modeled as

$$m\,{}^{S}u_j = C_B^S C_E^B \, {}^{E}u_j + v \tag{4.49}$$

where v is the measurement noise.

There are a few interesting scenarios that can be discussed:

1. Position determination with a known attitude with respect to the ECEF frame (so that the unit vector is known in the ECEF frame) and known range. In this case, measurement with respect to single object will be able to determine the position of the navigator by the following equation:

$$ {}^{E}r = {}^{E}r_j - \rho_j {}^{E}u_j $$

2. Position determination with known attitude. In this case, measurement with respect to two objects will be able to allow the solution of the navigator to be determined. Intuitively, this is easy to understand. For each object, we can draw a line from the object along the direction of the unit vector ${}^{E}u_j$ and the navigator must be somewhere along this line. If we have two such objects and two unit vectors, we can draw two lines and the intersecting point of the two lines is the location of the navigator. Algebraically, each of unit vector equations provide two independent constraints, two unit vector equations provide four constraints that are sufficient to constraint the position solution (three unknowns) under nonsingular conditions. Singularity will occur if the two unit vectors are parallel, which is equivalent to a single unit vector.
3. When both position and attitude are unknown, we have 6 degrees of freedom; it will take at least three objects to be able to constrain the position and attitude of the navigator.

In the following, we will focus on the linearized measurement model that can be used to solve the problems through iterations.

The measurement residual for the unit vector can be written as

$$ \begin{aligned} \Delta u &= m^{S}u_j - {}^{S}\hat{u}_j = C^{S}_{B}C^{B}_{E}{}^{E}u_j - \hat{C}^{S}_{B}\hat{C}^{B}_{E}{}^{E}\hat{u}_j + v \\ &= \left(C^{S}_{B}C^{B}_{E}{}^{E}u_j - \hat{C}^{C}_{B}C^{B}_{E}{}^{E}u_j\right) + \left(\hat{C}^{S}_{B}C^{B}_{E}{}^{E}u_j - \hat{C}^{S}_{B}\hat{C}^{B}_{E}{}^{E}u_j\right) + \left(\hat{C}^{S}_{B}\hat{C}^{B}_{E}{}^{E}u_j - \hat{C}^{S}_{B}\hat{C}^{B}_{E}{}^{E}\hat{u}_j\right) + v \\ &= \left(C^{S}_{B} - \hat{C}^{S}_{B}\right)C^{B}_{E}{}^{E}u_j + \hat{C}^{S}_{B}\left(C^{B}_{E} - \hat{C}^{B}_{E}\right){}^{E}u_j + \hat{C}^{S}_{B}\hat{C}^{B}_{E}\left({}^{E}u_j - {}^{E}\hat{u}_j\right) + v \end{aligned} $$

The first term in the above equation accounts for the contribution of the sensor misalignment knowledge error.

Note that

$$ C^{S}_{B} = C^{S}_{\hat{S}}C^{\hat{S}}_{B} = \left(I - \left[\theta^{S}_{\hat{S}}\times\right]\right)C^{\hat{S}}_{B} $$

and we have

$$ \left(C^{S}_{B} - \hat{C}^{S}_{B}\right)C^{B}_{E}{}^{E}u_j \approx -\left[\theta^{S}_{\hat{S}}\times\right]C^{\hat{S}}_{B}C^{B}_{E}{}^{E}u_j \approx -\left[\theta^{S}_{\hat{S}}\times\right]{}^{\hat{S}}\hat{u}_j = \left[{}^{\hat{S}}\hat{u}_j\times\right]\theta^{S}_{\hat{S}} $$

where $\theta^{S}_{\hat{S}} = \delta\theta_{sma}$ is the sensor DCM knowledge deviation from the previous estimate represented in the sensor frame as three small angles.

The second term accounts for the contribution of the navigator attitude knowledge error.

Note that

$$ C^{E}_{B} = C^{E}_{\hat{E}}C^{\hat{E}}_{B} = \left(I - \left[\theta^{E}_{\hat{E}}\times\right]\right)C^{\hat{E}}_{B} \text{ and } C^{B}_{E} = \hat{C}^{B}_{E}\left(I + \left[\theta^{E}_{\hat{E}}\times\right]\right) $$

and we have

$$\hat{C}_B^S\left(C_E^B - \hat{C}_E^B\right){}^E u_j = \hat{C}_B^S\hat{C}_E^B\left[\theta_{\hat{E}}^E \times\right]{}^E u_j = -\hat{C}_B^S\hat{C}_E^B\left[{}^E u_j \times\right]\theta_{\hat{E}}^E$$
$$\approx -\hat{C}_E^S\left[{}^E\hat{u}_j \times\right]\theta_{\hat{E}}^E = -\left[{}^S\hat{u}_j \times\right]\hat{C}_E^S\theta_{\hat{E}}^E = -\left[{}^S\hat{u}_j \times\right]\hat{C}_E^S\delta^E\theta_B^E$$

where $\delta^E\theta_B^E = \theta_{\hat{E}}^E$ is the attitude deviation from the previous estimate represented as three small angles in the ECEF frame.

The third term accounts for the contribution of the navigator position knowledge error as well as the FSO position knowledge error.

We have

$$\left(\hat{C}_B^S\hat{C}_E^B{}^E u_j - \hat{C}_B^S\hat{C}_E^B{}^E\hat{u}_j\right) = \hat{C}_B^S\hat{C}_E^B\left({}^E u_j - {}^E\hat{u}_j\right) = \hat{C}_B^S\hat{C}_E^B\left(\delta^E u_j\right)$$

where

$$\delta^E u_j = {}^E u_j - {}^E\hat{u}_j,\ {}^E u_j = \frac{1}{l_j}\left({}^E r_j - {}^E r\right),\ {}^E u_j = \frac{1}{\hat{l}_j}\left({}^E\hat{r}_j - {}^E\hat{r}\right)$$

Ignoring the difference between $1/\hat{\rho}_j$ and $1/\rho_j$, we have

$$\left(\hat{C}_B^S\hat{C}_E^B{}^E u_j - \hat{C}_B^S\hat{C}_E^B{}^E\hat{u}_j\right) = \hat{C}_B^S\hat{C}_E^B\left(\delta^E u_j\right) = \frac{1}{\hat{\rho}}\hat{C}_B^S\hat{C}_E^B\left(\delta^E r_j - \delta^E r\right)$$

where $\delta^E r = {}^E r - {}^E\hat{r} \quad \delta^E r_j = {}^E r_j - {}^E\hat{r}_j$

Finally, we have

$$\Delta u_j = m^S u_j - {}^S\hat{u}_j = \left[{}^S\hat{u}_j \times\right]\delta\theta_{sma} - \left[{}^S\hat{u}_j \times\right]\hat{C}_E^S\delta^E\theta_B^E + \frac{1}{\hat{\rho}}\hat{C}_B^S\hat{C}_E^B\delta^E r_j - \frac{1}{\hat{\rho}}\hat{C}_B^S\hat{C}_E^B\delta^E r + v \quad (4.50)$$

The above linearized equation can also be written in the following way to make the unknowns appear more obviously:

$$\Delta u_j = m^S u_j - {}^S\hat{u}_j = \begin{bmatrix} \left[{}^S\hat{u}_j \times\right] & -\left[{}^S\hat{u}_j \times\right]\hat{C}_E^S & \frac{1}{\hat{\rho}}\hat{C}_B^S\hat{C}_E^B & -\frac{1}{\hat{\rho}}\hat{C}_B^S\hat{C}_E^B \end{bmatrix}\begin{bmatrix} \delta\theta_{sma} \\ \delta^E\theta_B^E \\ \delta^E r_j \\ \delta^E r \end{bmatrix}\delta^E r + v \quad (4.51)$$

If we know the location of each of the reference objects perfectly, then $\delta^E r_j = 0$, and the equation above can be simplified as

$$\Delta u_j = m^S u_j - {}^S\hat{u}_j = \begin{bmatrix} \left[{}^S\hat{u}_j \times\right] & -\left[{}^S\hat{u}_j \times\right]\hat{C}_E^S & \frac{1}{\hat{\rho}}\hat{C}_B^S\hat{C}_E^B \end{bmatrix}\begin{bmatrix} \delta\theta_{sma} \\ \delta^E\theta_B^E \\ \delta^E r \end{bmatrix} + v \quad (4.52)$$

This linearized measurement equations can be used in a couple of ways:

1. We can stack all the FDO equations together to form a least square solution problem if we have a sufficient number of objects available.
2. We can use this in the Kalman filter design.

Since we have demonstrated multiple times how least square solutions are used to solve the various types of position and attitude determination problems, we will assume that the readers can easily repeat the same exercise using the above equations.

4.5 Generic Terrain Reference

4.5.1 Simplified Case: The Sensor on the Perfect Platform

The Earth has many location-related physical phenomena that can be used to provide navigational information [34–42]. The following are a few examples that fall into the category that we call generic terrain reference (GTR):

1. The terrain constrains the 3-D position to a 2-D surface of the Earth.
2. The nominal Earth gravitational field changes its direction at different locations and is the most convenient source of local vertical reference.
3. The Earth gravity field changes with of the location and creates the gravity anomaly:
   a. A 1-axis gravity anomaly measurement constraint a 3-D position to a 2-D one.
   b. A 3-axis gravity anomaly has the possibility of mapping the gravity anomaly directly to the position if we can avoid ambiguity;
   c. A gravity gradient anomaly can provide even more constraints. The gravity gradients form a 3×3 matrix that is symmetric and has a zero trace (summation of diagonal elements). In other words, there are $9 - 3 - 1 = 5$ independent variables associated with gravity gradients.
4. Similarly, the Earth magnetic field can also be used for navigation purposes in a similar way as the Earth gravitational field.

The details of the navigation methods using the GTR will be discussed in later chapters. In this section, we will address some of the generic mathematical problems.

Let us first study a simpler problem. Let us assume that somehow we have a way to keep the sensor aligned in the ECEF frame perfectly, or the sensor measurement is simply independent of attitude and alignment errors (for example, the scalar magnetometer measures the whole field), and therefore we do not need to consider the attitude and sensor alignment in this first simplified exercise.

We have one equation that models the location sensitive variable, which is often derived from a database:

$$^{E}\gamma = g\left(^{E}r\right) \tag{4.53}$$

The above equation can be the Earth gravity anomaly model, the magnetic field model, or the terrain model and the output is typically the field vector. We will use another equation to model the sensor measurement that is typically a function of γ.

$$z = h\left({}^{E}\gamma\right) + v \tag{4.54}$$

Combining the two equations together, we have

$$z = h\left(g\left({}^{E}r\right)\right) + v \tag{4.55}$$

Most of the time, the above equation only provides partial knowledge of the position ${}^{E}r$, and the full observability will have to be realized over a trajectory when navigating with inertial navigation.

In the rare case when we do have sufficient information from a single sample, and we know an initial guess of the position, we can linearize the measurement equation as

$$\delta z = \left[\frac{\partial h}{\partial g}\frac{\partial g}{\delta\, {}^{E}r}\right]\delta\, {}^{E}r + v \tag{4.56}$$

and we can derive a position measurement using the following equation:

$$\delta\, {}^{E}r = \left[\frac{\partial h}{\partial g}\frac{\partial g}{\delta\, {}^{E}r}\right]^{-1}(\delta z - v) \tag{4.57}$$

The gradient matrix $[(\partial h/\partial g)(\partial g/\delta\, {}^{E}r)]$ characterizes the sensitivity of how the measurement changes with the change of the position. As can be seen from the above equation, it plays an important role in the performance of the GTR navigation.

Let use a simple 2-D example to illustrate a number of important issues associated with the GTR navigation.

Figure 4.4 illustrates a 2-D terrain, the x-axis is the horizontal position, and the z-axis is the vertical position. The terrain profile is a curve in this coordinate frame and the gradient is the slope of this curve. The TRN in this case simply means backing out the horizontal position from the vertical measurement.

The first observation using this simple example is that the inverse function that maps vertical position to horizontal position is multivalued (i.e., a single vertical position will map to several horizontal positions). Because of this nature, terrain referenced updates can only be used locally in the sense that we can only use one segment of the curve at a time using our current knowledge of position and current assessment of the maximum error. Each such segment must be monotonic.

A related issue is consequently the level of details of the GTR model (either in the form of database, analytical equations, or the combinations of both). We can imagine if we model all the details of the terrain to centimeter level, there will be lots of up and downs and the monotonic segments will be very small. If we are not interested in that level of details for the navigation problem, it will be wise to use a terrain reference model that ignores such details.

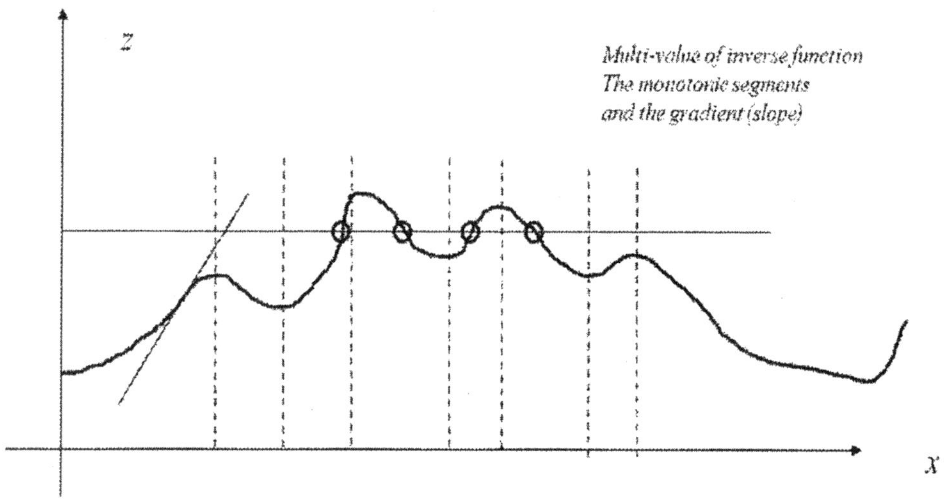

Figure 4.4 The 2-D TRN.

A smarter approach is that we can actually average the model or smooth the model so that the size of the monotonic segments matches with the position uncertainty. When the position uncertainty is large, we will use a lower-resolution terrain model that is an averaged-down version of the higher-resolution model. When the position uncertainty is small, we can use a higher-resolution database. Figure 4.5 illustrates this point.

Although we used a simple example to facilitate the above discussions, all these discussions are applicable in general and they are summarized here:

1. Mapping from measurement to position using the GTR model is, in general, multivalued. Consequently, GTR-based positioning requires sufficient *a priori* position knowledge such that the GTR model is monotonic in the neighborhood of the current position knowledge with uncertainty.
2. In order to support the locally monotonic requirement, the resolution of the GTR model can be chosen to match the expected position uncertainty. In addition, the GTR model can be dynamically averaged or smoothed to support different resolution levels according to the current level of position uncertainty.
3. The gradient of the combination of the measurement model and the GTR model is the sensitivity from position change to measurement change and is one of the most important contributors to performance.

4.5.2 Strapdown Sensors: Earth Gravity Anomaly and Earth Magnetic Field

We use the same gravity anomaly and/or magnetic field model as the platform case:

$$^{E}\gamma = g\left(^{E}r\right)$$

where $^{E}\gamma$ is the field related vector represented in the ECEF frame.

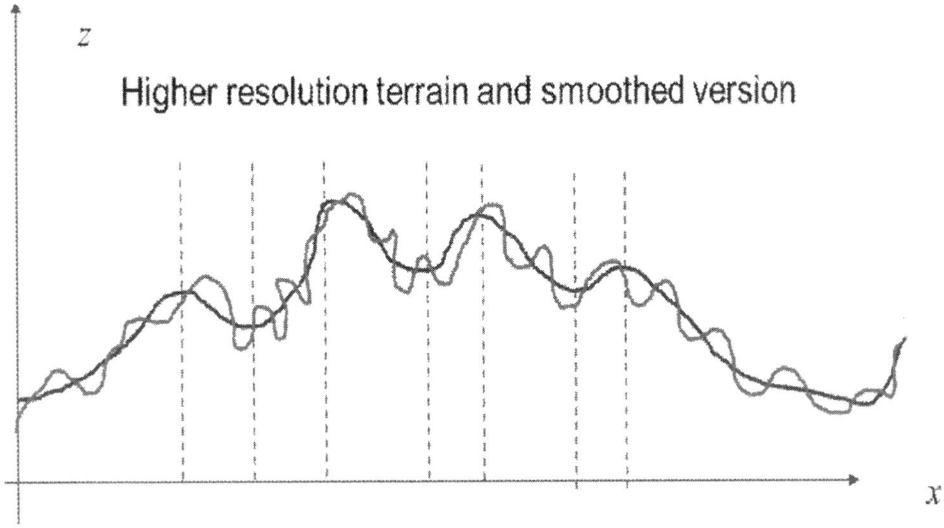

Figure 4.5 Dynamic adjustment of the terrain reference resolution.

The above equation can be the Earth gravity anomaly model, the magnetic field model, or the terrain model and the output is typically the field vector. For illustration purposes, we will simply assume that we have a 3-D field vector. Similar things (slightly more complicated) can be done when the output is not a 3-D field vector.

The field can be transformed to the sensor frame in order for it to be used in a strapdown sensor model:

$$^{S}\gamma = C_B^S C_E^B \, {}^{E}\gamma \tag{4.58}$$

We will use another equation to model the sensor measurement, which is typically a function of $^{S}\gamma$

$$z = h\left({}^{S}\gamma\right) + v \tag{4.59}$$

We would like to study the linearized version of the models:

$$\delta z = \frac{\partial h}{\partial^{s}\gamma}\delta^{s}\gamma + v$$
$$\delta^{S}\gamma = \delta C_B^S\left(C_E^B \, {}^{E}\gamma\right) + C_B^S\left(\delta C_E^B\right){}^{E}\gamma + \left(C_B^S C_E^B\right)\delta^{E}\gamma$$

Using similar techniques to what was done in Section 4.4, we have

$$\begin{aligned}\delta^{S}\gamma &= \delta C_B^S\left(C_E^B \, {}^{E}\gamma\right) + C_B^S\left(\delta C_E^B\right){}^{E}\gamma + \left(C_B^S C_E^B\right)\delta^{E}\gamma \\ &= \left[{}^{S}\gamma\times\right]\delta\theta_{sma} - \left[{}^{S}\gamma\times\right]C_E^S\delta^{E}\theta_B^E + C_E^S\delta^{E}\gamma\end{aligned}$$

and

$$\delta^{E}\gamma = \frac{\partial g}{\partial^{E}r}\delta^{E}r$$

Combining the above together, we have the following linearized equation:

$$\delta z = \left[\frac{\partial h}{\partial^s \gamma}\right]\left[{}^S\gamma\times\right]\delta\theta_{sma} - \left[\frac{\partial h}{\partial^s \gamma}\right]\left[{}^S\gamma\times\right]C_E^S\delta\,{}^E\theta_B^E + \left[\frac{\partial h}{\partial^s \gamma}\right]C_E^S\left[\frac{\partial g}{\partial^E r}\right]\delta\,{}^E r + v \tag{4.60}$$

which can be rewritten as

$$\delta z = \left[\left[\frac{\partial h}{\partial^s \gamma}\right]\left[{}^S\gamma\times\right] \quad -\left[\frac{\partial h}{\partial^s \gamma}\right]\left[{}^S\gamma\times\right]C_E^S \quad \left[\frac{\partial h}{\partial^s \gamma}\right]C_E^S\left[\frac{\partial g}{\partial^E r}\right]\right]\begin{bmatrix}\delta\theta_{sma}\\ \delta\,{}^E\theta_B^E\\ \delta\,{}^E r\end{bmatrix} + v \tag{4.61}$$

Again, the above equations can be used in two ways:

1. Form the foundation for the iterative least square solution.
2. Use in the Kalman filter design.

It is important to point out that the exercise above is mainly to illustrate the methodology. For practical applications, strapdown gravity anomaly referenced navigation (GARN) has a number of fundamental issues and scalar magnetometer is fundamentally not dependent on attitude and alignment (with a few exceptions). We will discuss these issues in detail in later chapters.

4.5.3 Strapdown Sensors: Slant Range Sensor

A strapdown range sensor and more broadly a slant range sensor makes the problem more complicated than typical GARN and magnetic anomaly referenced navigation (MARN). For this reason, we treat this problem separately. Figure 4.6 illustrates the need to determine the Earth-piercing point in this scenario.

Given the known LOS vector for the sensor in the Earth frame:

$${}^E u_S = C_B^E\,{}^B u_S$$

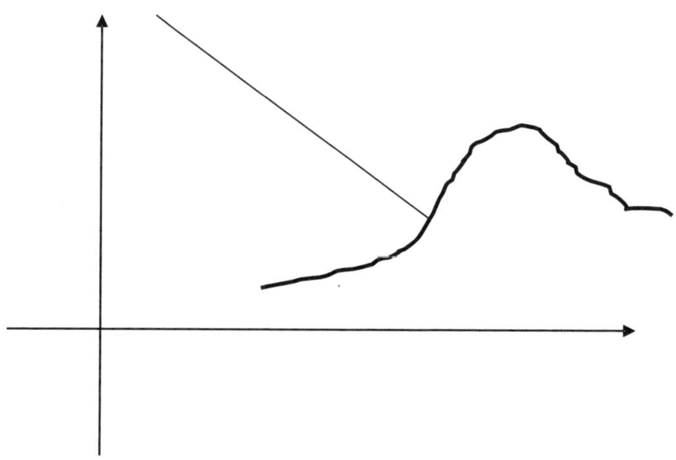

Figure 4.6 When the range measurement is from a slanted sensor, there is a need to determine the Earth-piercing point of the LOS vector.

where ${}^{B}u_S$ is the sensor boresight vector in the body frame.

The Earth-piercing point position vector can be written as

$$ {}^{E}R_{pp} = {}^{E}R + (d)\,{}^{E}u_S \tag{4.62} $$

where ${}^{E}R$ is the navigator's location and d is an unknown scalar that is the distance from the vehicle to the Earth-piercing point.

However, the Earth-piercing point satisfies the following terrain constraint equation:

$$ g\left({}^{E}R_{pp}\right) = 0 \tag{4.63} $$

As an example, let us first study the case when we ignore the terrain profile and use a simplified spherical Earth model to constrain the Earth-piercing point.

In this case, we have

$$ g\left({}^{E}R_{pp}\right) = \left\|{}^{E}R_{pp}\right\|^2 - \left(r_e + h\right)^2 = 0 $$

where r_e is the geocentric Earth radius and h is the altitude.

Expanding the above equation, we get

$$ \left\|{}^{E}R_{pp}\right\|^2 = {}^{E}R_{pp}^{\;T}\,{}^{E}R_{pp} = \left({}^{E}R + (d)\,{}^{E}u_S\right)^T\left({}^{E}R + (d)\,{}^{E}u_S\right) = \left(r_e + h\right)^2 $$

The above can be written as a quadratic equation for d:

$$ d^2 + 2\left(\left({}^{E}u_S\right)^T\left({}^{E}R\right)\right)d + \left(\left\|{}^{E}R\right\|^2\right) - \left(r_e + h\right)^2 = 0 \tag{4.64} $$

Solving the above equation, we consequently can calculate the piercing point as

$$ {}^{E}R_{pp} = {}^{E}R + (d)\,{}^{E}u_S \tag{4.65} $$

With general terrain constraints, things are more complicated. We typically start with an initial guess and then use iterations to solve the nonlinear equation. The above simplified solution, for example, can be used as the initial guess.

Let us assume that the initial guesses are given by

$$ \bar{d} \text{ and } {}^{E}\bar{R}_{pp} $$

We can then linearize the two equations we discussed before:

$$ {}^{E}\bar{R}_{pp} + \delta\,{}^{E}R_{pp} = {}^{E}R + \left(\bar{d} + \delta d\right){}^{E}u_S $$
$$ g\left({}^{E}\bar{R}_{pp}\right) + g_R\left({}^{E}R_{pp}\right)\delta\,{}^{E}R_{pp} = 0 $$

where

$$ g_R\left({}^{E}R_{pp}\right) = \frac{\partial g_R\left({}^{E}R_{pp}\right)}{\partial\,{}^{E}R_{pp}} \tag{4.66} $$

We can then form the following linear equations with four unknowns:

$$\begin{bmatrix} I & -{}^{E}u_S \\ g_R\left({}^{E}R_{pp}\right) & 0 \end{bmatrix}\begin{bmatrix} \delta\,{}^{E}R_{pp} \\ \delta d \end{bmatrix} = \begin{bmatrix} {}^{E}R + \bar{d}\,{}^{E}u_S - {}^{E}\bar{R}_{pp} \\ -g\left({}^{E}\bar{R}_{pp}\right) \end{bmatrix} \tag{4.67}$$

and the solution can be written as

$$\begin{bmatrix} \delta\,{}^{E}R_{pp} \\ \delta d \end{bmatrix} = \begin{bmatrix} I & -{}^{E}u_S \\ g_R\left({}^{E}R_{pp}\right) & 0 \end{bmatrix}^{-1}\begin{bmatrix} {}^{E}R + \bar{d}\,{}^{E}u_S - {}^{E}\bar{R}_{pp} \\ -g\left({}^{E}\bar{R}_{pp}\right) \end{bmatrix} \tag{4.68}$$

For the purpose of the least square solution or for the purpose of building a Kalman filter, we want to know the sensitivity of the range measurement against the navigator position and the LOS unit vector (which is further a function of attitude and sensor misalignments, and we already learned how to deal with that earlier).

Let

$$G = G(1{:}4,1{:}4) = \begin{bmatrix} I & -{}^{E}u_S \\ g_R\left({}^{E}R_{pp}\right) & 0 \end{bmatrix}^{-1} \tag{4.69}$$

The δd part of (4.68) can be written as

$$\delta d = G(4,1{:}3)\left({}^{E}R + \bar{d}\,{}^{E}u_S - {}^{E}\bar{R}_{pp}\right) - G(4,4)g\left({}^{E}\bar{R}_{pp}\right)$$

The partial against ${}^{E}R$ is easy

$$\frac{\partial d}{\partial\,{}^{E}R} = \frac{\partial \delta d}{\partial\,{}^{E}R} = G(4,1{:}3)\bullet I + G(4,4)\bullet 0 = G(4,1{:}3)$$

The partial against ${}^{E}u$ is slightly harder.

$$\frac{\partial d}{\partial^{E}\,{}^{E}u_S} = \frac{\partial \delta d}{\partial\,{}^{E}u_S} = G(4,1{:}3)\bar{d}\bullet I + \frac{\partial G(4,1{:}3)}{\partial\,{}^{E}u_S}\left({}^{E}R + \bar{d}\,{}^{E}u_S - {}^{E}\bar{R}_{pp}\right) - \frac{\partial G(4,4)}{\partial\,{}^{E}u_S}g\left({}^{E}\bar{R}_{pp}\right)$$

Since both ${}^{E}R + \bar{d}\,{}^{E}u_S - {}^{E}\bar{R}_{pp}$ and $g({}^{E}\bar{R}_{pp})$ are small (they are perfectly 0 when the solution converges), we therefore can avoid doing the partials on the matrix G and arrive at the following simple equation:

$$\frac{\partial d}{\partial^{E}\,{}^{E}u_S} = \frac{\partial \delta d}{\partial\,{}^{E}u_S} = G(4,1{:}3)\bar{d}$$

In summary, we have

$$G = G(1{:}4,1{:}4) = \begin{bmatrix} I & -{}^{E}u_S \\ g_R\left({}^{E}R_{pp}\right) & 0 \end{bmatrix}^{-1}$$

and

$$\delta d = G(4,1{:}3)\delta^E R_E^B - G(4,1{:}3) d\delta^E u_S \tag{4.70}$$

which can also be written as

$$\delta d = \begin{bmatrix} G(4,1{:}3) & -G(4,1{:}3)d \end{bmatrix} \begin{bmatrix} \delta^E R_E^B \\ \delta^E u_S \end{bmatrix} \tag{4.71}$$

4.6 Geo-Referenced Image-Based Navigation: Measurement Equations

4.6.1 Geo-Referenced Image as Collection of Image Anchor Points

Although the detailed image or vision-based navigation will be a topic of a later chapter, we will address some of the most fundamental concept and mathematical equations in this section. When we navigate by matching camera images against database images, we are fundamentally concentrating the information we derive from the image to one or more unit vectors pointing from the navigator to points on the Earth.

For example, the center of the image can be used as an anchor point. With the current position, attitude, and camera alignment knowledge, we can calculate the predicted Earth-piercing point corresponding to the center of the image. We can then retrieve the Earth image database around the piercing point and project the database image to the sensor frame. When the reference image and the sensor image are correlated, we can determine the difference between the actual camera image with respect to the predicted image which can be represented as the offsets of the center of the image.

More generally, the images can be divided into tiles or regions, and each of the tiles or regions can be treated as a separate image and the center points of the subimages can be used the same way as the whole image mentioned above. Consequently we will end up with a number of subimage center points associated with each of the subimages.

We call these subimage center points on the camera FOV the image anchor points. Each of the anchor points has corresponding Earth-piercing points on the Earth. These piercing points are located where the LOS vectors associated with each of the image anchor points will hit. For convenience, we will call both the anchor points and the corresponding Earth-piercing points the anchor points since they are fundamentally the same points in different coordinate frames (camera frame and the Earth frame).

It is easy to understand that once we can compute a prediction of the Earth-piercing points, we will be able to calculate a prediction of the unit vector in the ECEF frame. Together with the camera frame measured unit vectors, we will have the same problem as we have discussed before: determining attitude and position from angle measurements with respect to the finite distance objects (FDOs). Here the FDOs are simply the Earth-piercing points associated with the image anchor points. In other words, we essentially already have all the equations we need to model the measurements.

The following steps provide one of the many possible ways to perform image-based navigation:

1. Divide the image into a number of subimages. This step can be used to eliminate the part of image that is not expected to match the database due to weather or construction or other reasons. This step can also be used to select the subimages to optimize the dilution of precision.
2. For the subimages that are considered as useful, the center of the subimage will be used as an anchor point which is known in the camera frame. Using the current knowledge of the position, attitude and camera alignment matrix, the corresponding Earth-piercing point, also known as the anchor point in the Earth frame, can be solved using the equations in Section 4.4.
3. The reference images around the anchor points can be retrieved from the database.
4. The reference images will then be transformed to the camera frame, using the current knowledge of position, attitude, and camera alignment matrix.
5. The camera frame reference image can then be correlated against the camera image for each of the subimages; the correlation results will be the offset between the camera subimage and the reference subimage, which can be used as the measurement residual.
6. The measurement residual can then be used by a Kalman filter to generate corrections to navigation solutions and sensor calibration parameters.

4.6.2 Measurement Model for Image Anchor Points

Let the location of a navigator in the ECEF frame be given by ${}^{E}r$ and the location of an image anchor point j be given by ${}^{E}r_j$. (Section 4.5 provides the equations on how the anchor points can be solved.)

The LOS vector from the navigator to the anchor point can then be computed as

$$ {}^{E}u_j = \frac{1}{\left\| {}^{E}r_j - {}^{E}r \right\|}\left({}^{E}r_j - {}^{E}r \right) = \frac{1}{\rho_j}\left({}^{E}r_j - {}^{E}r \right) $$

where $\rho_j = \| {}^{E}r_j - {}^{E}r \|$ is the range to the anchor point.

${}^{E}u_j$ can be transformed to the sensor frame (camera frame) using the following equation:

$$ {}^{S}u_j = C_B^S C_E^B \, {}^{E}u_j $$

where C_E^B is the DCM that transforms the coordinates of a vector in ECEF frame to the body frame. In other words, C_E^B represents the navigator's attitude. C_B^S is the DCM that transforms the coordinates of a vector in the body frame to the sensor (camera) frame.

The measured unit vector can be modeled as

$$m^S u_j = C_B^S C_E^B {}^E u_j + v$$

where v is the measurement noise.

An astute reader will realize that the equations are identical to the FDO equations in Section 4.4; consequently, we also have the same linearized equations:

$$\Delta u = m^S u_j - {}^S\hat{u}_j = \left[{}^S\hat{u}_j \times\right]\delta\theta_{sma} - \left[{}^S\hat{u}_j \times\right]\hat{C}_E^S \delta^E\theta_B^E + \frac{1}{\hat{\rho}}\hat{C}_B^S\hat{C}_E^B\delta^E r_j - \frac{1}{\hat{\rho}}\hat{C}_B^S\hat{C}_E^B\delta^E r + v \quad (4.72)$$

which can also be written as

$$\Delta u = m^S u_j - {}^S\hat{u}_j = \begin{bmatrix} \left[{}^S\hat{u}_j \times\right] & -\left[{}^S\hat{u}_j \times\right]\hat{C}_E^S & -\frac{1}{\hat{\rho}}\hat{C}_B^S\hat{C}_E^B \end{bmatrix} \begin{bmatrix} \delta\theta_{sma} \\ \delta^E\theta_B^E \\ \delta^E r \end{bmatrix} + \frac{1}{\hat{\rho}}\hat{C}_B^S\hat{C}_E^B\delta^E r_j + v \quad (4.73)$$

where $\delta\theta_{sma}$ is the sensor (camera) misalignment error expressed in the sensor (camera) frame.

This linearized measurement equations can be used in several ways:

1. We can stack all the anchor point equations together to form a least square solution problem if we have a sufficient number of anchor points available,
2. We can use this in the Kalman filter design.

References

[1] Markley, F., and U. Naval, "Autonomous Navigation Using Landmark and Intersatellite Data," *AIAA/AAS Astrodynamics Conference*, Seattle, WA, 1984.

[2] Li, R., H. Qiu, and K. Xiong, "Autonomous Navigation for Constellation Based on Inter-Satellite Ranging and Directions," *IECON 2017—43rd Annual Conference of the IEEE Industrial Electronics Society*, Beijing, China, November 2017, pp. 2985–2990.

[3] Whelan, D., G. Gutt, and P. Enge, "Boeing Timing & Location: An Indoor Capable Time Transfer and Geolocation System," *Proc. of 5th Stanford University, Symposium on Position, Navigation, and Timing*, Menlo Park, CA, 2011.

[4] Tan, Z., et al., "New Method for Positioning Using IRIDIUM Satellite Signals of Opportunity," *IEEE Access*, Vol. 7, June 24, 2019, pp. 83412–83423.

[5] Black, H., "A Passive System for Determining the Attitude of a Satellite," *AIAA Journal*, Vol. 2 No. 7, July 1964, pp. 1350–1351.

[6] Wertz, J. R., *Spacecraft Attitude Determination and Control*, Boston, MA: D. Reidel Publishing, 1986.

[7] Shuster, M. D., "A Survey of Attitude Representations," *The Astronautical Science*, Vol. 41, 1993, pp. 439–517.

[8] Mekky, T., "A Comparative Study of Spacecraft Attitude Determination and Estimation Algorithms (A Cost-Benefit Approach)," *Aerospace Science and Technology*, Vol. 26, No. 1, April-May 2013, pp. 211–215.

[9] Mohd Zamri, H., et al., "Review on Attitude Estimation Algorithm of Attitude Determination System," *ARPN Journal of Engineering and Applied Sciences*, Vol. 11, 2016, pp. 4455–4460.

[10] Sidi, M. I., *Spacecraft Dynamics and Control*, Cambridge, U.K.: Cambridge University Press, 1997.

[11] Shuster, M. D., and S. D. Oh, "Three Axis Attitude Determination from Vector Observations," *Guid. Control*, Vol. 4, 1981, pp. 70–77.

[12] Simon, D., *Optimal State Estimation*, New York: John Wiley & Sons, 2006.

[13] Bar-Itzhack, Y., and R. R. Harman, "Optimized Triad Algorithm for Attitude Determination," *Guidance Control and Dynamics*, pp. 208–211.

[14] Chowdhary, G., et al., "GPS-Denied Indoor and Outdoor Monocular Vision Aided Navigation and Control of Unmanned Aircraft," *Journal of Field Robotics*, Vol. 30, No. 3, pp. 415–438.

[15] Won, D. H., et al., "Integration of Limited GNSS Signals with Monocular Vision Based Navigation," *12th International Conference on Control, Automation and Systems (ICCAS)*, 2012, pp. 1364–1367.

[16] Li, M., B. Xu, and L. Zhang, "Orbit Determination for Remote-Sensing Satellites Using Only Optical Imagery," *International Journal of Remote Sensing*, Vol. 38, No. 5, 2017, pp. 1350–1364.

[17] Li, M., and B. Xu, "Autonomous Orbit and Attitude Determination for Earth Satellites Using Images of Regular-Shaped Ground Objects," *Aerospace Science and Technology*, Vol. 80, 2018, pp. 192–202.

[18] George, M., and S. Sukkarieh, "Camera Aided Inertial Navigation in Poor GPS Environments," *2007 IEEE Aerospace Conference*, 2007, pp. 1–12.

[19] Giebner, M. G., "Tightly-Coupled Image-Aided Inertial Navigation System via a Kalman Filter," M.S. thesis, AFIT/GE/ENG/03-10, Graduate School of Engineering and Management, Air Force Institute of Technology (AU), Wright-Patterson AFB OH, March 2003 (ADA415246).

[20] Kauffman, K., et al., "Simulation Study of UWB-OFDM SAR for Navigation with INS Integration," *Proc. of 2011 International Technical Meeting of the Institute of Navigation*, 2011.

[21] Veth, M. J., "Fusion of Imaging and Inertial Sensors for Navigation," Ph.D. dissertation, Air Force Institute of Technology (AU), Wright-Patterson AFB OH, September 2006 (ADA468382).

[22] Chiu, H. -P., et al., "Precise Vision-Aided Aerial Navigation," *IEEE/RSJ International Conference on Intelligent Robots and Systems (IROS)*, 2014.

[23] Chiu, H. -P., et al., "Constrained Optimal Selection for Multi-Sensor Robot Navigation Using Plug-and-Play Factor Graphs," *IEEE International Conference on Robotics and Automation (ICRA)*, 2014.

[24] Patterson, T., et al., "Utilizing Geographic Information System Data for Unmanned Aerial Vehicle Position Estimation," *Proc. of Canadian Conf. on Computer and Robot Vision*, 2011.

[25] Sim, D., et al., "Integrated Position Estimation Using Aerial Image Sequences," *IEEE Transactions on Pattern Analysis and Machine Intelligence*, Vol. 24, No. 1, 2002.

[26] Conte, G., and P. Doherty, "Vision-Based Unmanned Aerial Vehicle Navigation Using Geo-Referenced Information," *EURASIP Journal of Advances in Signal Processing*, 2009.

[27] Mourikis, A., et al., "Vision-Aided Inertial Navigation for Spacecraft Entry, Descent, and Landing," *IEEE Transactions on Robotics*, Vol. 25, No. 2, April 2009.

[28] Chiu, H., et al., "Robust Vision-Aided Navigation Using Sliding-Window Factor Graphs," *Proc. IEEE Intl. Conf. on Robotics and Automation (ICRA)*, 2013.

[29] Kumar, R., et al., "Registration of Video to Geo-Referenced Imagery," *Proc. IEEE Intl. Conf. on Pattern Recognition (ICPR)*, 1998.

[30] Mourikis, A., and S. Roumeliotis, "A Multi-State Constraint Kalman Filter for Vision-Aided Inertial Navigation," *Proc. IEEE Intl. Conf. on Robotics and Automation (ICRA)*, 2007.

[31] Han, K., et al., "UAV Vision: Feature Based Accurate Ground Target Localization Through Propagated Initializations and Interframe Homographies," *Proc. IEEE Intl. Conf. on Robotics and Automation (ICRA)*, 2012.

[32] Kumar, R., et al., "Registration of Highly-Oblique and Zoomed in Aerial Video to Reference Imagery," *Proc. IEEE Intl Conf. on Pattern Recognition (ICPR)*, 2000.

[33] Kumar, R., et al., "Aerial Video Surveillance and Exploitation," *Proc. of the IEEE*, Vol. 89, No. 10, October 2001.

[34] Hollowell, J., "Heli/SITAN: A Terrain Referenced Navigation Algorithm for Helicopters," *IEEE Symposium on Position Location and Navigation. A Decade of Excellence in the Navigation Sciences*, March 1990.

[35] McGuffin, J. T., "Terrain Referenced Navigation Electromagnetic-Gravitational Correlation," U.S. 5272639A, 1992 (2019-11-14 Application status is Expired—Lifetime).

[36] Fellerhoff, J. R., "SITAN Implementation in the Saint System," *IEEE Journal*, 1986, pp. 89–95.

[37] Baird, C. A., and M. R. Abramson, "A Comparison of Several Digital Map-Aided Navigation Techniques," *IEEE Journal*, 1984, pp. 286–293.

[38] Hostetler, L. D., and R. D. Andreas, "Nonlinear Kalman Filtering Techniques for Terrain-Aided Navigation," *IEEE Transactions on Automatic Control*, Vol. AC-28, No. 3, March 1983, pp. 315–323.

[39] Priestley, N., "Terrain Referenced Navigation," *IEEE Symposium on Position Location and Navigation. A Decade of Excellence in the Navigation Sciences*, March 20, 1990.

[40] Henley, A. J., "Terrain Aided Navigation: Current Status, Techniques for Flat Terrain and Reference Data Requirements," *IEEE Symposium on Position Location and Navigation. A Decade of Excellence in the Navigation Sciences*, March 20, 1990.

[41] Jalving, B., et al., "Terrain Referenced Navigation of AUVs and Submarines Using Multi-Beam Echo Sounders," *Norwegian Defense Research*, 2004, http://citeseerx.ist.psu.edu/viewdoc/download?doi=10.1.1.459.3521&rep=rep1&type=pdf.

[42] Metzger, J., et al., "Sigma-Point Filter for Terrain Referenced Navigation," *AIAA Guidance, Navigation, and Control Conference and Exhibit*, San Francisco, CA, August 15–18, 2005.

CHAPTER 5

All Source PNT Solution: Integration Through Kalman Filters

5.1 Introduction

With the foundation that we laid in Chapters 2, 3, and 4, the integrations of various PNT sensors into an All Source PNT solution become easy. We basically just have to bring the propagation solutions from Chapter 3, and the measurement solutions from Chapter 4, into the framework of the extended Kalman filer (EKF) and unscented Kalman filer (UKF) from Chapter 2. It is worth noting that the approaches taken by various government agencies [1–4] and the developments in the industry and academia [5–21] are very much consistent with this approach.

5.2 Kalman Filter and All Source PNT Algorithm Architecture

As we discussed in Chapter 2, a Kalman filter is useful in integrating propagation solutions as discussed in Chapter 3, with a number of measurement and update solutions as discussed in Chapter 4.

Without loss of generality, we will use the EKF as an example to discuss the architecture. Similar discussions and architectures can be derived for other variations of the Kalman filters (such as UKFs and particle filters).

Figure 5.1 illustrates the architecture of a general EKF. Within this architecture, there are a few key elements:

1. Nonlinear state propagation module. In an All Source PNT system, this typically means a combination of the following:
    a. The IMU hardware, firmware, software, the IMU interface hardware and software, and the inertial navigation software;
    b. The clock propagation hardware, firmware, and software;
    c. The sensor and other parameter models that are often modeled as almost constant parameters as discussed in Chapter 3;
    d. We often also need to propagate or integrate some variables useful for the Kalman filter covariance propagation model and sometimes even for the measurement models.
2. Nonlinear measurement prediction module. In an All Source PNT system, there will be a number of sensors or information sources that will be used to update or correct the propagation solution. For each of the sensor types,

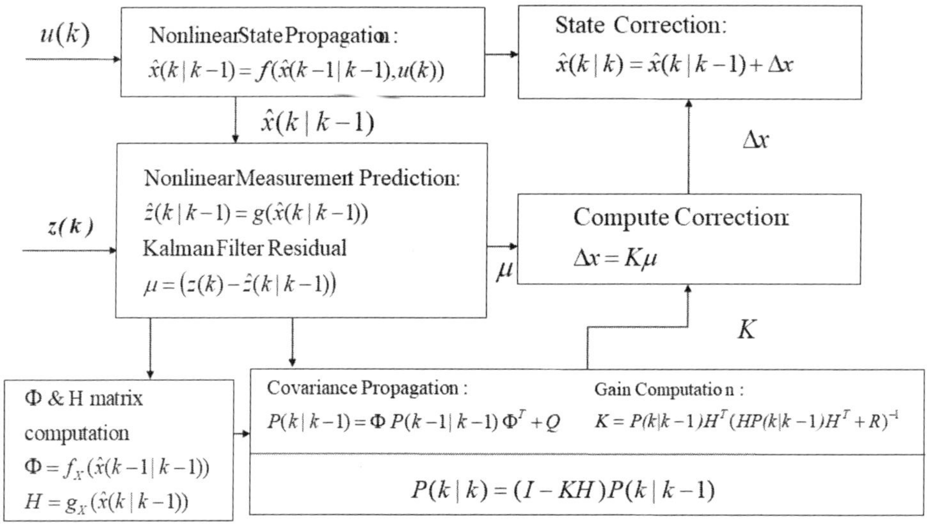

Figure 5.1 Architecture of generic EKF.

there will be an algorithm, typically implemented by software that will be used to predict the sensor output based on the current state estimate available from the nonlinear state propagation module.

3. Residual, measurement sensitivity matrix (H-matrix), and measurement variance calculation module:
   a. This is where the difference between the measurement and the prediction is calculated that produces the most important variable in the Kalman filter, the Kalman filter residual.
   b. The H-matrix can be calculated based on analytical linearized models discussed in Chapter 4 or can even be numerically calculated by numerically differentiating the measurement models.
   c. The R-matrix is the covariance matrix of the measurement noise that is often dependent on a number of parameters associated with the sensor.
4. Covariance propagation, gain computation, and covariance update module. This part is often called the narrow-sense Kalman filter module in an All Source PNT software where we perform the covariance propagation, the gain computation, and the covariance update for each type of the sources that are summarized in the form of the triplets: residual, the H-matrix, and the R-matrix.
5. Correction computation and application module. The product of the gain matrix and the residual vector produces the correction vector. The elements of the correction vector are then applied in the propagators and sensor-processing modules.

There are a few things that are not in the standard textbook Kalman filter and need to be included in the All Source PNT architecture. This is because the different sources of information often become available in an asynchronous manner. Specifically, there are three aspects:

1. Use of a time-matching buffer to store a short history of the propagated states around the current time.
2. Use of a time tag for all the measurements with respect to a common time reference and the mechanism to perform time synchronization or time transfer among the multiple sources.
3. Proper accounting in the H-matrix that the measurements are taken some time before they are made available.

With the above considerations, Figure 5.2 presents a naturally derived All Source PNT algorithm level architecture.

As illustrated in Figure 5.2, the inertial navigation and clock propagation typically represent the propagation-based solution. There could be many variations of the propagation-based solution. For example, for some applications, INS can be replaced by a solution based on models of vehicle dynamics and kinematics.

The propagation solution is then saved into a time-matching buffer such as those described in [19] to allow a short history of the estimated states to be stored in a circular buffer or anything that achieves a similar purpose.

For each of the update sensors, there is a basic three-module design pattern:

1. Sensor abstraction module to abstract the details of a specific sensor type to a generic mathematical abstraction of the sensor, which is capable of producing certain measurements and producing fault detection results;
2. Sensor output prediction module to predict the abstracted output based on the current estimate of the state;

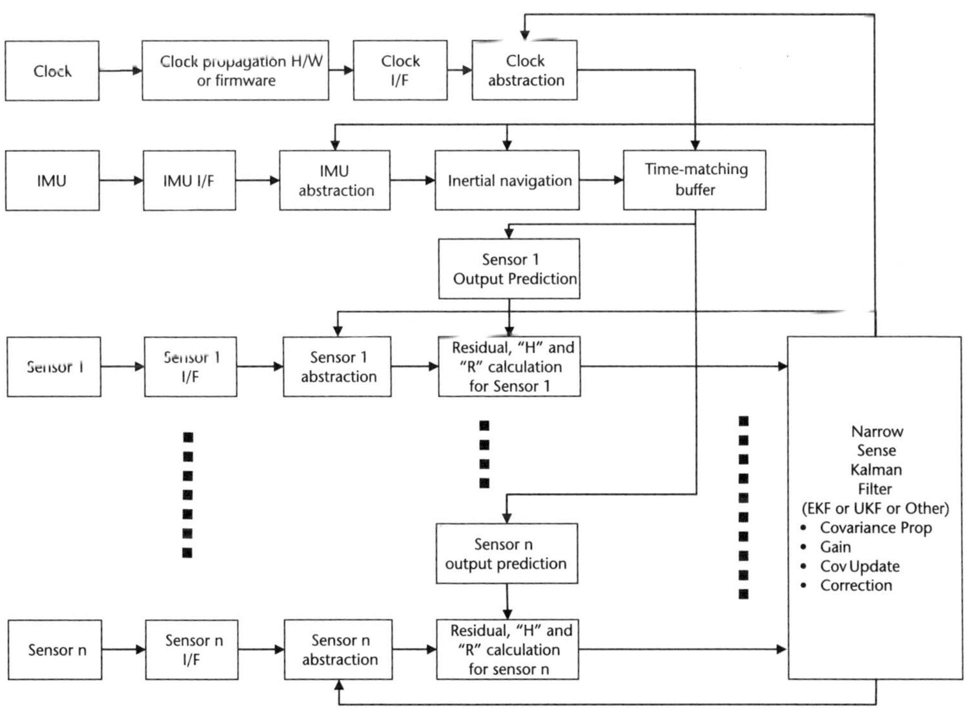

Figure 5.2 All Source PNT architecture.

3. Residual, H-matrix, and R-matrix (RHR) calculation module to produce the necessary inputs for a generic narrow-sense Kalman filter to calculate the gains and the corrections and to update the covariance matrices.

Finally, the narrow-sense Kalman filter serves as the center of fusion where all the update sensors are used, the system covariance matrices are maintained, and the gains and corrections are calculated.

5.3 Fault and Threat Management Architecture

What we did not discuss in the previous section is the important topic of fault and threat management, which ultimately addresses the topics of availability and access, integrity, and performance. This is especially true since an All Source PNT system is designed to take advantages of the redundant information available from multiple sensor and information sources to provide assured performance.

Fault and threat management consists of fault and threat detection, containment, isolation, and response (FDCIR).

Fault and threat detection is about determining if a fault has occurred, which typically are abnormal operational conditions associated with failures of components, or a threat produced by external factors such as jamming and spoofing. For fault and threat detection, we need to be concerned with the test criteria, test persistency, the probability of failure to detect, which is also known as the probability of missed detection (PMD), and the probability of false alarm (PFA). PMD is the probability of not detecting a fault when there is a fault or threat, and PFA is the probability of detecting a fault when there is no fault or threat.

In the rest of this section, we will often just refer to fault or threat as fault with the understanding that most of the discussions apply to both.

Fault containment is about doing immediate, typically local actions, to prevent a potential fault from propagating even further if a fault detection has not been confirmed yet. For example, if a single sample of measurement is out of the expected bound, the containment action is not to use that single sample. The mere fact that a single sample is bad will not necessarily be leading to a fault detection but an action of containment can help to reduce the probability of causing other problems.

Fault isolation is done after a fault is detected. This may not be always possible. The purpose of isolation is to identify the part of the system that caused the fault so that it may be taken out of the solution or be repaired or serviced when there is an opportunity.

Fault response is the action taken in response to a detected and possibly isolated fault. Typical action is to reconfigure the system to use a different combination of sensors to avoid the part that is causing the problem. If multiple parallel solutions are being produced all the time, the response may simply mean to choose a solution that is not dependent on the faulty part. For example, if multiple solutions are being produced with different combinations of sensors and a particular sensor has been isolated as the faulty one, then the solution, which does not use this particular sensor in its combination of sensor and information sources, will be chosen.

All these activities of detection, containment, isolation, and response can be done (and better be done) in a multilevel fashion. Specifically, for an All Source PNT system, FDCIR can be done at four levels:

1. At network level where information is collected and analyzed and decisions are made about the availability and integrity of PNT infrastructure systems such as various GNSS constellations, jamming and spoofing, augmentation and signal of opportunity information, and collaborative opportunities;
2. At the level of multiple parallel navigators within the platform;
3. At the level of a single navigator: consistency of individual sensors against the total navigator (the compound of other sensors);
4. At the level of local sensors: sensor-level FDCIR without using other sensors' information.

The best performance is achieved with the above four levels combined.

Figure 5.3 illustrates the multilevel FDCIR architecture.

Table 5.1 provides a summary of typical activities that can be done at each level for FDCIR.

In the following, we will provide some more elaboration on the fault or threat detection, containment, isolation, and response (F/TDCIR) approach. The reader is referred to [22–24] for general background in detection theory.

5.3.1 PFA and PMD

We will describe a generic fault detection scheme that can be used to support the detection, containment, and isolations at all of the four levels.

Let X be a random variable that behaves differently between two scenarios: H_0, when everything is normal, and H_1, when there is a fault.

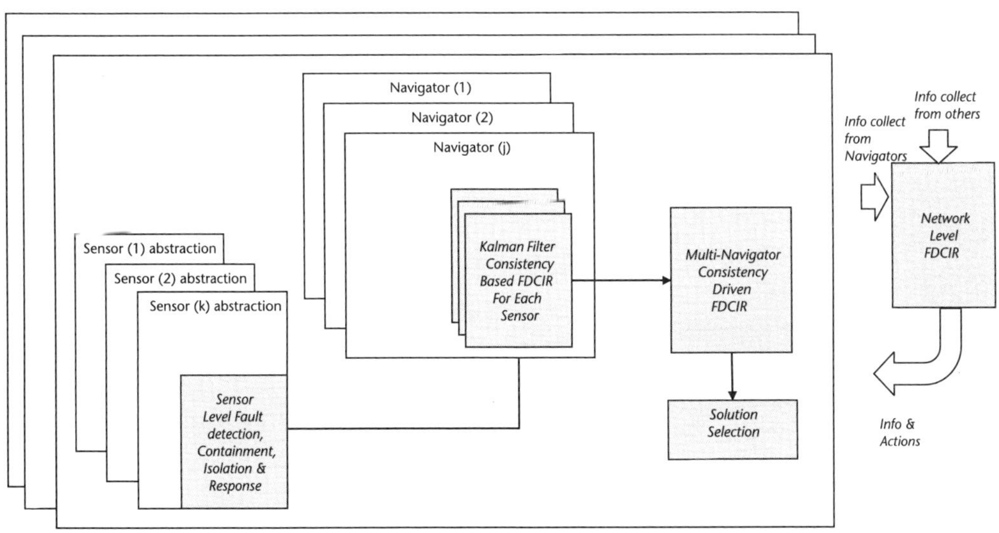

Figure 5.3 Generic four-level fault management architecture.

Table 5.1 Typical Fault Detection, Containment, Isolation, and Response Activities at Each of the Four Levels

	Typical Detection Activities	*Typical Containment Activities*	*Typical Isolation Activities*	*Typical Response Activities*
Local sensor level	Sensor built-in test; output reasonableness test; output consistency test; sudden jump test; persistency of test results	Not to use a bad sample; replace the bad sample by a predicted sample	We typically do not need to go below the sensor level; if a fault detection is confirmed, then the fault is isolated to this sensor	Declare the sensor to be invalid and not to use the sensor in the All Source PNT solution
Single navigator level	Check the residual of the Kalman filter for each sensor measurements	If residual is out of expected bound, the update by the Kalman filter using the residual is skipped	When this happens persistently to a sensor, it means that that sensor is being inconsistent with the rest of sensors, consequently a detection can be declared; again, since we only want to isolate to sensor level, in this case, detection means isolation	Declare the sensor to be invalid and not to use the sensor in the All Source PNT solution
Multinavigator level	Multiple navigators running in parallel using different combination of the sensors; the solutions from multiple navigators are voted and consistency-checked for a detection	Do not select the one that is inconsistent for output	Again, isolation is easy when a single solution acts inconsistently with the rest of the solutions; isolation may not be possible, when there are multiple faults at the same time, which happens with very low probability	Not to use the solutions that are contaminated
Network level	Network operation center collects information such as GNSS operational status, augmentation system operational status, jamming or spoofing threats, signal of opportunities, and collaboration opportunities	Declare certain PNT resource as suspicious before a detection is declared	Isolation is assumed to be done by lower levels but confirmation can be done with across-the-network checks	Declare certain PNT resources as unavailable or invalid; command PNT resource configuration actions such as deployment of augmentation services

Let the two conditional probability density function be given by

$$\begin{aligned} p(x \mid H_0) &= p_0(x) \\ p(x \mid H_1) &= p_1(x) \end{aligned} \tag{5.1}$$

Without loss of the generality, let us assume that the random variable X tends to take higher values when the fault happens and the density functions are such as what are depictured in Figure 5.4.

Assuming that we use a threshold x_{th} to determine if x is drawn from $p_0(x)$ or from $p_1(x)$, we can define

$$P_{fa} = P(x > x_{th} \mid H_0) = \int_{x_{th}}^{\infty} p(x \mid H_0)\,dx = \int_{x_{th}}^{\infty} p_0(x)\,dx \tag{5.2}$$

as the PFA. The PFA is the probability of the event when there is no fault in the system but we decided that there is a fault.

We can also define

$$P_{md} = P(x \le x_{th} \mid H_1) = \int_{-\infty}^{x_{th}} p(x \mid H_1)\,dx = \int_{-\infty}^{x_{th}} p_1(x)\,dx \tag{5.3}$$

as the PMD that is the probability of the event when we decided that there is no fault but, in reality, there is a fault.

Clearly, the value of the two important probabilities is determined by the choice of the threshold and, more importantly, by the two probability distribution functions.

If we increase the threshold, we will reduce the PFA. In the meantime, we are also increasing the PMD. In other words, with the probability distributions fixed, we cannot reduce the PFA and the PMD at the same time. If we reduce one, we must increase the other.

Obviously, we will want to minimize both the PFA and the PMD. There are limits to what we can do by adjusting the threshold to balance the need for PFA and PMD. To be able to have small values for both PFA and PMD, we must find ways to change the probability distribution directly or change the testing method, which is fundamentally equivalent to changing the probability distribution.

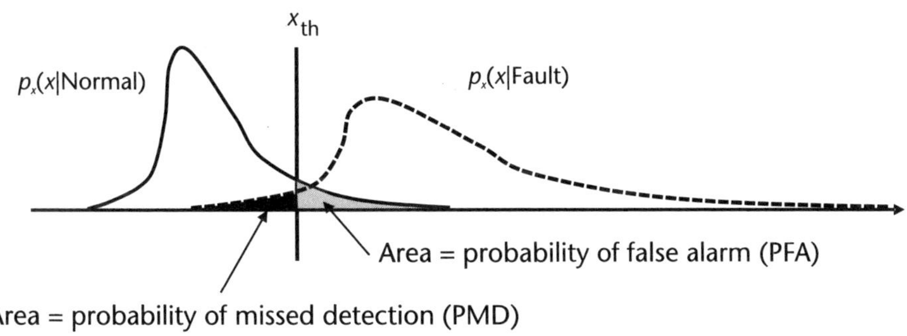

Figure 5.4 PFA and PMD.

For example, we can form a new random variable that is an average of the samples taken. Assuming that each of the samples are independent, the probability density functions of the new random variable for both hypotheses, similar to what are illustrated in Figure 5.4, will become narrower as the number of samples used for the average increases. Consequently, both PFA and PMD will be reduced.

In the following, instead of creating a new random variable with better distributions, we will simply change the testing method, that is, to repeat the test for multiple samples independently drawn. A decision is only made when the same criteria are maintained persistently for n times, where n is the persistency count threshold.

5.3.2 Reducing Both PFA and PMA at the Same Time by the Persistency Test

Assuming that there is no fault, the probability for all of the n tests to report fault is simply

$$P_{fa} = \left(P\left(x > x_{th} \mid H_0\right)\right)^n = \left(\int_{x_{th}}^{\infty} p_0(x)\,dx\right)^n \tag{5.4}$$

As n increase, the PFA will be reduced to small values very quickly.

Assuming that there is a fault, the probability for all of the n tests to report as normal is simply

$$P_{md} = \left(P\left(x \le x_{th} \mid H_1\right)\right)^n = \left(\int_{x_{th}}^{\infty} p_1(x)\,dx\right)^n \tag{5.5}$$

Again, as n increase, the PMD will be reduced to small values very quickly.

Since both the PFA and PMD reduce exponentially as n increases, the persistency test approach is very powerful. It greatly improves the performance of a fault detection algorithm.

One obvious question is what happens in between. In other words, if we perform n independent tests on n samples and the test trips m ($<n$) times, what should happen during this time? This is the situation when no decision can be made as far as fault detection is concerned. This is exactly the reason why we need to contain a potential fault before the fault is detected or declared. Only after a fault is detected, a fault response can be activated.

5.3.3 Containment Versus Response

The reason that we introduce the concept of containment is mostly because persistency is probably one of the most frequently used approach to improve fault detection performance. When a test is tripped once, because of the persistency requirement, we do not declare a detection yet. In the meantime, if a fault has indeed occurred, the fact that we are not declaring a fault yet may mean that we have to allow potentially bad data to enter the system that can cause collateral damages subsequently.

To avoid this, we will need to somehow handle this condition even if we are not ready to declare a fault yet.

This is where containment comes in. For example, if a test for a sensor fault is tripped, but we do not have enough data to say that it is happening persistently yet, we will want to not declare that this particular sensor is bad so that we should be switching to a backup sensor or an alternative sensor combination. Instead, we will simply contain the problem (i.e., stop the problem from propagating into the rest of the system) by providing a replacement for the data or by ignoring the data. The replacement can be derived from the previous samples or derived from the state of the system using some kind of sensor model. Typically, if the sensor is used to form the propagation solution, then a replacement has to be provided. If the sensor data is used to form the update solution, then we can simply just ignore the sample.

If the bad data was simply due to an outlier that happened briefly, the containment actions would have provided replacement data or ignored the data briefly and then everything would return to normal. No fault would have been detected. If the test tripped persistently, it would lead to the detection of a fault and then a response would be warranted and the response could be such that the sensor would be removed and a backup or replacement configuration would be used.

In summary, containment is the response to a tripped test that is not yet declared reliably as a fault to stop the potential fault from propagating through the system. The fault response is the action taken by the system after the test has tripped persistently and a fault is declared.

5.3.4 Fault Detection, Containment, Isolation, and Response at Each of the Four Levels

The general fault detection and persistency strategy is applicable to all four levels. Typically, at the sensor level, various combinations of tests can be used, and with the proper choice of persistency, a fault can be detected and declared. Typically, we only want to be able to isolate the fault to a particular sensor; at the sensor level, a fault is detected also means it is isolated.

Before the persistency criteria are satisfied, a tripped test means that we will need to contain the problem. For a propagation sensor such as an IMU, containment is critical since each sample of the sensor outputs is important to the solution. When a test is tripped, we will need to provide a replacement data by extrapolation and interpolation or by using a model. However, for update sensors, it is OK if we missed a sample. Containment will simply mean that we will skip a sample.

At a single-solution level, we can test the Kalman filter residuals and other variables. Again, persistency will be required to reliably detect a fault with low PFA and low PMD. Containment in this case can be just skipping the Kalman filter update for the particular piece of measurements produced by a particular sensor. It is important to realize that this containment action may not be a fair action, which assumes that the propagation solution (typically the inertial navigation solution based on IMU data) is not the one which has faulted.

At the single-solution level, fault detection may or may not lead to isolation. This is because a number of updater sources can be used and each one of the sources can lead to a fault.

That is why the isolation may have to be done at the multisolution level where multiple solutions with different sensor combinations are running in parallel and voting between the multiple solutions can be used to isolate the fault.

At the network level, fault detection and persistency algorithms can be used to detect faults or threats such as those associated with GNSS constellations. Those detections will lead to the distribution of the information to the individual solutions to take out sensor sources that would have otherwise not been detected by other means.

5.4 Steps to Build an All Source PNT System

In this section, we will provide some general guidelines on how one should proceed with building an All Source PNT system.

In general, there are typically 11 steps, as listed next, through which we go when we develop an All Source PNT system:

1. Develop the system concept.
2. Establish requirements and design objectives.
3. Define top-level hardware and software architecture.
4. Develop detailed algorithms.
5. Establish mathematical models.
6. Perform simulation and analysis.
7. Build the system hardware and software.
8. Perform the hardware test.
9. Perform the software test.
10. Integrate the system.
11. Perform the system test.

We will briefly discuss each one of the steps below.

Step 1: Concept Development

At this stage, we want to determine at the concept level:

1. What the propagation solutions are;
2. What the update sources are;
3. How all the sensor sources help synergistically the overall solutions in terms of availability, integrity, reliability, and performance;
4. What the faults and threats will likely be and how we want to manage them.

Step 2: Establish Requirements and Design Objectives

Once we have the system concept developed, we can proceed to analyze the use cases of the system and capture the quality attributes. It is important to note that:

1. The requirements and objectives not only deal with performance, availability, integrity, and operational environment and scenarios, but they also deal

with development cost, production cost, maintenance, service, and support cost, as well as quality.

2. In order to fully capture the requirements and design objectives, it is typically necessary to conduct use case workshops and quality attribute workshops to elaborate and elicit the requirements.

Step 3: Define Top-Level Hardware and Software Architecture

By using the architecture design patterns that we discussed in this book, it will not be difficult to develop and define the top-level hardware architecture and the top-level software architecture. Specifically:

1. We will need to identify all the sensor hardware, hardware I/F (Interface), software I/F, time-tagging, and time-synchronous mechanisms.
2. We will need to understand the sensor performance and error characteristics for each one of the sources as well as failures and fault scenarios.
3. We will need to identify the computing resources, such as an embedded computer with the electronic interfaces that can talk to each one of the sensor sources through the hardware I/F using certain software and communication protocol.
4. We will need to determine the software architecture that is similar to Figure 5.2 but with further elaborations on:
   a. What the operating system will be;
   b. What the device drivers that handle the sensor interfaces will be;
   c. How we want to use sensor abstraction modules or layers to make the rest of the PNT application software independent of the specific hardware;
   d. How software modules are interconnected and how they are dispatched;
   e. If any multithreading or multiprocess is required and, if so, how the inter-thread and interprocess communications are done;
   f. We will also need to design the fault architecture using the four-level approach mentioned in the previous section and decide how the fault architecture are integrated with the sunny day architecture.

Step 4: Develop Detailed Algorithms

Once the architecture is defined and elaborated, we can start to develop and define the detailed algorithms. Specifically,

1. We will need to define all the time-synchronization algorithms for each sensors and sources. The following are a few examples of a time-synchronization mechanism that are popular:
   a. Use of existing time-synchronization schemes: For example, the use of NTP or a more precise version of that, PTP. The advantage of NTP and PTP is that they are off-the-shelf proven designs available on many operating system environments.
   b. Use of a hardware timing pulse (such as 1 pulse per second (1PPS)). This typically requires specific hardware that has the capability to save the

system time at, for example, the rising edge of the timing pulse. This is the scheme how most of the GNSS receivers synchronize time with the receiver client system.

c. Use of a software time reset scheme. In this case, the PNT computer, for example, will send a message to a sensor, and the sensor will set its clock to a known value upon receiving the message. Assuming that the time delay from the command to the actual reset can be modeled to certain accuracy, then this allows the sensor to be synchronized to the PNT computer within that accuracy.

2. We will need to identify all the state variables and to define all the nonlinear propagation algorithms.
3. We will need to derive the linearized version of the propagation algorithms that can be used to support the Kalman filter covariance propagation.
4. We will need to define all the nonlinear sensor output prediction algorithms.
5. We will need to derive the linearized version of the output prediction algorithms that can be used to calculate the H-matrix.
6. We will need to elaborate the design details for fault detection, containment, isolation, and response for each sensors and other components that can fail, become unavailable, or become integrity-challenged.

Step 5: Establish Mathematical Models

In order to implement and analyze the detailed algorithm design, we must first develop the mathematical models. Specifically, we want to:

1. Establish trajectory models, operational scenarios, and operational parameters.
2. Establish all sensor models including sensor noise and error models.
3. Determine how nonwhite noises are to be modeled.
4. Identify the propagation noises and the Q-matrix.
5. Identify the measurement noises and the R-matrix.
6. Establish simulation models of the All Source PNT hardware and software.

Step 6: Perform Simulation and Analysis

With the mathematical models available, we will be able to perform simulation and analysis with the objectives of validating and verifying the following designs:

1. System architecture design;
2. System algorithm design;
3. System performance under nominal scenarios;
4. System performance under fault or challenged scenarios;
5. Availability;
6. Integrity;
7. Reliability;
8. Other design attributes.

Step 7: Build Hardware and Software for the System

This step is very specific to the nature of the system. How this step is done highly depends on the process and the quality standards that the system needs to comply to. There will be a huge difference between building a PNT system for a toy and building a PNT system that impacts the safety of human life.

Step 8: Perform the Hardware Test

Again, this highly depends on the nature of the system and the quality standards. Typical tests to ensure that the hardware meets the requirements and design objectives include the sensor unit test, the sensor interface test, and the hardware integration test. The hardware integration test will typically test the sensors, the sensor I/Fs, the computer hardware, and the device drivers in an integrated manner.

Step 9: Perform the Software Test

Typically, we want to perform unit tests both as a debug mechanism and as the tests to provide comprehensive coverage of the details of software implementation. In addition, we will want to integrate the software into the simulation environment and perform software-in-the-loop (SITL) tests and processor-in-the-loop (PITL) tests to reduce the risks for hardware and software integration.

Step 10: Perform the System Integration

Depending on the specific system, sometimes, it makes sense to integrate the real-time computing system, the software, and the simulated sensors as a first step to test the integrated system in a controlled environment. The second step is then to proceed with the integration of the sensor, the real-time computer, and the software. Sometimes, it just makes sense to go directly to the second step.

Step 11: Perform the System Test

Depending on the requirements and design objectives, the system-level tests will test the integrated system for its requirement compliance informally and formally depending on the process requirements.

Clearly, we will not be able to demonstrate the above 11 steps and their substeps within the scope of this book. Instead, in the subsequent chapters, we will illustrate some of Steps 1 to 5 through a number of examples.

References

[1] DARPA, "Brad Agency Announcement: All-Source Positioning and Navigation (ASPN)," Strategic Technology Office, DARPA-BAA-11-14, 2010.

[2] https://www.transportation.gov/pnt/national-positioning-navigation-and-timing-pnt-architecture.

[3] https://www.darpa.mil/program/adaptable-navigation-systems.

[4] Owano, N., "DARPA Seeks New Positioning, Navigation, Timing Solutions," *TechXplore*, March 28, 2015, https://techxplore.com/news/2015-03-darpa-positioning-solutions.html.

[5] Groves, P. D., "The Complexity Problem in Future Multi-Sensor Navigation and Positioning Systems: A Modular Solution," *The Journal of Navigation*, Vol. 67, 2014, pp. 311–326.

[6] Si, S., et al., "A New Underwater All Source Positioning and Navigation (ASPN) Algorithm Based on Factor Graph," *2019 Chinese Control and Decision Conference (CCDC)*, 2019, https://ieeexplore.ieee.org/document/8832700.

[7] Groves, P. D., *Principles of GNSS, Inertial, and Multisensor Integrated Navigation Systems*, Second Edition, Norwood, MA: Artech House, 2013.

[8] Penn, T. R., "All Source Sensor Integration Using an Extended Kalman Filter," Air Force Institute of Technology, Master's thesis, March 2012.

[9] Pullen, S., T. Walter, and P. Enge, "Integrity for Non-Aviation Users: Moving Away from Specific Risk," *GPS World*, July 2011, pp. 28–36.

[10] Thomas, M., et al., "Global Navigation Space Systems: Reliance and Vulnerabilities," Royal Academy of Engineering, March 2011.

[11] Wang, L., and S. Li, "Enhanced Multi-Sensor Data Fusion Methodology Based on Multiple Model Estimation for Integrated Navigation System," *Int. J. Control Autom. Syst.*, Vol. 16, 2018, pp. 295–305.

[12] Gong, X., J. Zhang, and J. Fang, "A Modified Nonlinear Two-Filter Smoothing for High-Precision Airborne Integrated GPS and Inertial Navigation," *IEEE Transactions on Instrumentation and Measurement*, Vol. 64, 2015, pp. 3315–3322.

[13] Gao, W., Y. Zhang, and J. Wang, "A Strapdown Interial Navigation System/Beidou/Doppler Velocity Log Integrated Navigation Algorithm Based on a Cubature Kalman Filter," *Sensors*, Vol. 14, 2014, pp. 1511–1527.

[14] Abdolkarimi, E. S., G. Abaei, and M. R. Mosavi, "A Wavelet-Extreme Learning Machine for Low-Cost INS/GPS Navigation System in High-Speed Applications," *GPS Solutions*, Vol. 22, No. 15, 2018.

[15] Chang, L., K. Li, and B. Hu, "Huber's M-Estimation-Based Process Uncertainty Robust Filter for Integrated INS/GPS," *IEEE Sens. J.*, Vol. 15, 2015, pp. 3367–3374.

[16] Gelb, M., *Applied Optimal Estimation*, Cambridge, MA: MIT Press, 1974.

[17] Brown, R. G., and P. Hwang, *Introduction to Random Signals and Applied Kalman Filtering*, New York: John Wiley & Sons, 1996.

[18] Li, R., and A. Wu, "Kalman Filtering & Applications," *UCLA Lecture Notes*, 2005–2019.

[19] Li, R., et al., "Spacecraft Attitude Determination System and Method," U.S. Patent 6,285,927, May 1999.

[20] Julier, S. J., J. K. Uhlmann, and H. Durrant-Whyte, "A New Approach for Filtering Nonlinear System," *Proc. of the American Control Conference*, 1995, pp. 1628–1632.

[21] Julier, S. J., and J. K. Uhlmann, "General Method for Approximating Nonlinear Transformation of Probability Distributions," Technical report RRG Department of Engineering Science, University of Oxford, November 1996.

[22] Izadi, I., et al., "A Framework for Optimal Design of Alarm Systems," *Proc. of 7th IFAC Symposium on Fault Detection, Supervision and Safety of Technical Processes*, Barcelona, Spain, June 30–July 3, 2009.

[23] https://en.wikipedia.org/wiki/Detection_theory.

[24] Wickens, T. D., *Elementary Signal Detection Theory*, New York: Oxford University Press, 2002.

CHAPTER 6

Integrated Clock

6.1 Introduction

Precise timing is critically important for many applications and plays a central role for many PNT systems. As the smallest example of an integrated PNT system, the integrated clock, without exception, is the integration of at least two types of the solutions: the propagation solution and the update solution. The propagation solution is typically based on an oscillator and most of the propagation solutions are actually done by hardware with the clock phase available for software to read.

There are a number of possible update solutions:

1. A propagator solution can be updated by another, typically more stable but noisier, oscillator. This happens frequently because most crystal oscillators (CXOs) have very good short-term noise but do not have very good long-term stability. Atomic clocks, such as Rubidium and Cesium atomic clocks, tend to have higher short-term noise but much better long-term stability. As a result, most atomic clocks use a combination of a CXO and an atomic clock to achieve optimized performance.
2. A propagation solution can be updated by another time source such as a GNSS receiver.
3. A propagation solution can be updated by a time-transfer mechanism that leads to a clock bias estimate that can be used to correct the time.

6.2 Architecture

Figure 6.1 illustrates the architecture of an integrated clock. In this generic architecture, the oscillator and the integration mechanism are often integrated and can be many different things. Some of the examples are listed here:

1. Earth self-rotational motion;
2. Moon rotation around the Earth;
3. Earth rotation around the Sun;
4. Mechanical pendulum;
5. Other types of mechanical oscillators;
6. CXO;
7. Atomic oscillator.

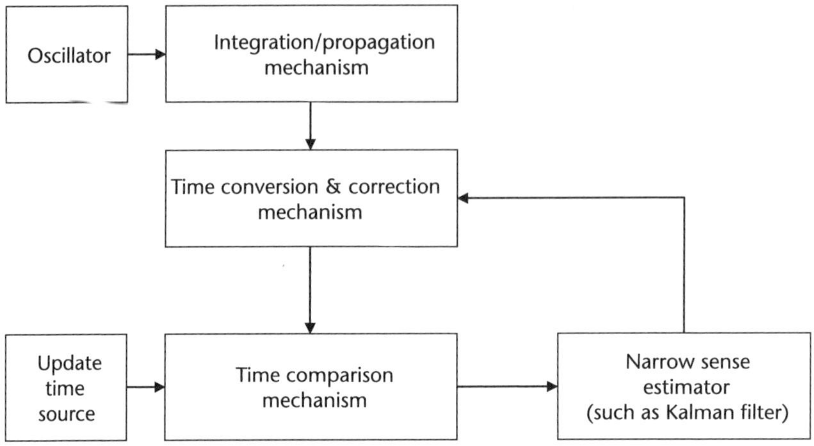

Figure 6.1 Integrated clock architecture.

The integration and propagation mechanism in the above examples can be natural motion, an analog or digital mechanical system, or analog or digital electronics.

It is important to realize that most of the architecture is done by hardware in order to achieve precision time propagation and precision time readout. The narrow-sense Kalman filter can still be done by software (or by hardware). Since the time-matching is done by hardware, there is often no longer a need for software-based time-matching.

Because the time-matching is done by hardware, each of the update sensors needs to use the current time (the hardware propagator time) output to drive the hardware to allow the update sensors or the interface electronics to directly measure the time difference between the clocks.

As an example, digital counter or register driven by an oscillator is probably among the most common oscillators or integrators in modern clocks. Similarly, triggering by an external clock to buffer a copy of the current time register content will allow the buffered time to be used as the time mark for the external clock event. The external clock at the same event can then be compared against the time mark to calculate the difference. With today's technology, these mechanisms are easily implemented by FPGAs(Field Programmable Gate Array) or ASICs (Application Specific Integrated Circuit).

6.3 Propagation Solution and Linearized Error Model

As discussed before, the propagation solution is mostly done by hardware. It is therefore only important to model the behavior of the propagation or, more usefully, to model the behavior of the error.

In this section, we will show three examples of the propagation error model. The reader is encouraged to review similar examples in [1–4].

Example 6.1

The clock error is modeled by the time error and the time rate error

$$\frac{dx}{dt} = b + w_{PRW} \tag{6.1}$$

$$\frac{db}{dt} = w_{RRW} \tag{6.2}$$

$$x_{out} = x + w_{pwn} \tag{6.3}$$

where $x(t) = t_{\text{raw}} - t$ is the difference between the raw time as the result of the integration of the oscillator output and true time, which we do not know. So $x(t)$ is the error of the raw time.

$b(t)$ is the raw clock frequency or rate error. $b(t)$ is normally a slow-changing, almost constant variable, which is modeled as a random walk process.

w_{PRW} is the white noise process that integrates into the phase random walk error.

$$E(w_{PRW}) = 0 \tag{6.4}$$

$$E(w_{PRW}(t)w_{PRW}(s)) = \sigma_{prw}^2 \delta(t - s) \tag{6.5}$$

w_{RRW} is the white noise process that integrates into the rate random walk, that is, the random walk in the clock rate bias $b(t)$.

$$E(w_{RRW}) = 0 \tag{6.6}$$

$$E(w_{RRW}(t)w_{RRW}(s)) = \sigma_{rrw}^2 \delta(t - s) \tag{6.7}$$

w_{PWN} is the white noise process that adds to the phase output.

$$E(w_{PWN}) = 0 \tag{6.8}$$

$$E(w_{PWN}(t)w_{PWN}(s)) = \sigma_{pwn}^2 \delta(t - s) \tag{6.9}$$

In vector form, we have

$$\frac{d}{dt}\begin{bmatrix} x \\ b \end{bmatrix} = \begin{bmatrix} 0 & 1 \\ 0 & 0 \end{bmatrix}\begin{bmatrix} x \\ b \end{bmatrix} b + \begin{bmatrix} w_{PRW} \\ w_{RRW} \end{bmatrix} \tag{6.10}$$

$$x_{out} = x + w_{pwn} = \begin{bmatrix} 1 & 0 \end{bmatrix}\begin{bmatrix} x \\ b \end{bmatrix} + w_{pwn} \tag{6.11}$$

The last equation models the readout of x.

The implementation of the Kalman filter using Example 6.1 actually is not trivial. It implies that we will need to be able to correct the raw time using estimated $x(t)$ and correct the oscillator frequency or the clock rate by the estimated $b(t)$. One can imagine that these two corrections will not be easily implemented when the clock propagation and integration are achieved by a hardware mechanism.

Example 6.2

In this second example, we will leave the raw time based on the oscillator alone and will create a corrected output by software using the raw time with a predefined correction formulae. We will use the external time source to estimate the parameters of the correction equation.

For example, we can use the following equation to correct the raw time into the corrected time

$$t_c = b_0 + b_1\left(t_{\text{raw}} - t_0\right) + \frac{1}{2}b_2\left(t_{\text{raw}} - t_0\right)^2 \tag{6.12}$$

where t_0 is some chosen starting epoch. The parameters b_0, b_1, b_2 are roughly the time bias, rate bias, and acceleration errors of the raw clock with respect to the external aiding clock.

We will model the parameters in the following way

$$\frac{d}{dt}\begin{bmatrix} b_0 \\ b_1 \\ b_2 \end{bmatrix} = \begin{bmatrix} 0 & 1 & 0 \\ 0 & 0 & 1 \\ 0 & 0 & 0 \end{bmatrix}\begin{bmatrix} b_0 \\ b_1 \\ b_2 \end{bmatrix} + \begin{bmatrix} w_{PRW} \\ w_{RRW} \\ w_{ARW} \end{bmatrix} \tag{6.13}$$

The readout of the corrected time can be modeled as

$$t_{c\_\text{read\_out}} = b_0 + b_1\left(t_{\text{raw}} - t_0\right) + \frac{1}{2}b_2\left(t_{\text{raw}} - t_0\right)^2 + w_{pwn} \tag{6.14}$$

Example 6.3

Along the same lines as Example 6.2, the correction formulae can be further augmented to include other terms that are believed to exist based on physics or test results. For example, one might want to include multiple sinusoidal terms with known frequencies

$$t_{c\_\text{read\_out}} = b_0 + b_1\left(t_{\text{raw}} - t_0\right) + \frac{1}{2}b_2\left(t_{\text{raw}} - t_0\right)^2 + \sum_{i=1}^{n} c_i \cos\omega_i\left(t - t_0\right) + s_i \sin\omega_i\left(t - t_0\right) + w_{wpn} \tag{6.15}$$

In this case, the propagation error model will include additional $2n$ terms and the model of the parameter states are therefore

$$\frac{d}{dt}\begin{bmatrix} b_0 \\ b_1 \\ b_2 \\ c_1 \\ s_1 \\ c_2 \\ s_2 \\ \vdots \\ c_n \\ s_n \end{bmatrix} = \begin{bmatrix} 0 & 1 & 0 & 0 & 0 & 0 & 0 & \cdots & 0 & 0 \\ 0 & 0 & 1 & 0 & 0 & 0 & 0 & \cdots & 0 & 0 \\ 0 & 0 & 0 & 0 & 0 & 0 & 0 & \cdots & 0 & 0 \\ 0 & 0 & 0 & 0 & 0 & 0 & 0 & \cdots & 0 & 0 \\ 0 & 0 & 0 & 0 & 0 & 0 & 0 & \cdots & 0 & 0 \\ 0 & 0 & 0 & 0 & 0 & 0 & 0 & \cdots & 0 & 0 \\ 0 & 0 & 0 & 0 & 0 & 0 & 0 & \cdots & 0 & 0 \\ \vdots & \vdots & \vdots & \vdots & \vdots & \vdots & \vdots & \ddots & \vdots & \vdots \\ 0 & 0 & 0 & 0 & 0 & 0 & 0 & \cdots & 0 & 0 \\ 0 & 0 & 0 & 0 & 0 & 0 & 0 & \cdots & 0 & 0 \end{bmatrix}\begin{bmatrix} b_0 \\ b_1 \\ b_2 \\ c_1 \\ s_1 \\ c_2 \\ s_2 \\ \vdots \\ c_n \\ s_n \end{bmatrix} + \begin{bmatrix} w_{PRW} \\ w_{PRW} \\ w_{ARW} \\ c_1 rw \\ s_1 rw \\ c_2 rw \\ s_2 rw \\ \vdots \\ c_n rw \\ s_n rw \end{bmatrix} \tag{6.16}$$

where the coefficients s_i and c_i (i = 1, 2, n) are modeled as random walks and $s_i rw$ and $c_i rw$(i = 1, 2, n) are the white noise processes that integrate into the random walk processes.

For the three examples above, the high-fidelity models are already linear and, consequently, we do not need to worry about developing the linearized models.

6.4 Updates for Integrated Clock

The external clocks that we use to correct the propagation solution could be a time source that is slaved to some high-accuracy time reference such as a GPS receiver or simply the clock of another vehicle.

The second type of external clock could be just another propagator-based solution that uses a better oscillator such as an atomic clock. In this case, the integrated clock will take advantage of the short-term better performance of a CXO and the long-term stability of the atomic clock.

Fundamentally, there is no difference between the two types of external update sources. Other than short-term (high-frequency) noise, we will simply treat the external clock as perfect over a long run and, consequently, the integrated clock will be slaved to the long-term performance of the external clock update.

Typically, we want to use precision hardware to measure the time error with respect to an external time source. The measurement becomes a time error. The good thing about the measurement being a time error and not the time is that the error itself changes slowly and, consequently, we will not need to worry about the precision time match for the corrections.

In Chapter 4, we discussed the two-way time transfer approach and the generation of the measured relative clock errors. That process will be one of the many ways by which the relative time error can be computed when the external time reference is not colocated with the propagation clock and the time of signal transmission must be considered. The widely used NTP and PTP [5] are based on this principle.

When the two time sources are colocated, the mechanism to generate the relative time error measurement typically relies on some electronics hardware to generate the precision measurements.

Here we will discuss two mechanisms:

1. Precision measurement of TOA of a pulse (such as 1 PPS);
2. Precision measurement of time difference by correlation of signals (which could be carrier, code, or data on the top of the code).

The two mechanisms are fundamentally the same. The first method uses a single signature as a time mark. The second mechanism uses a sequence of discrete or continuous signatures as time marks. We can imagine that, for the second mechanism, we have the opportunity to average the time synchronization errors over multiple time marks and, consequently, have the potential to achieve very high accuracy. With this understanding, one can easily be convinced that the 1-PPS type of mechanism is actually naïve and is not the best mechanism for high precision

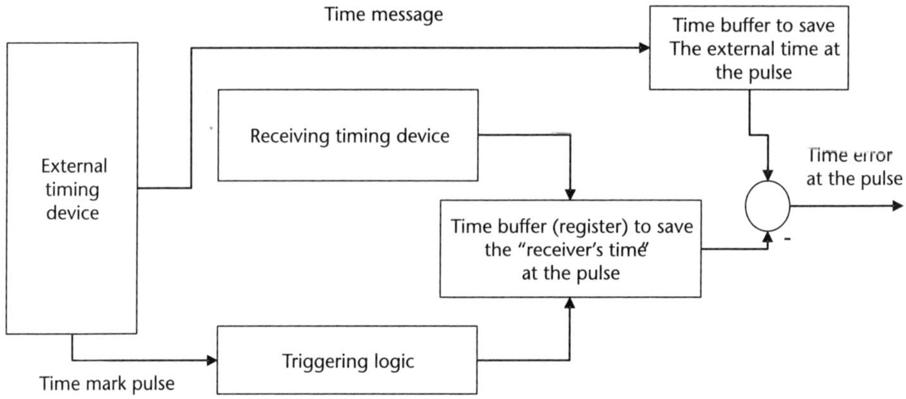

Figure 6.2 The 1-PPS style time difference measurement scheme.

time synchronization tasks. Figure 6.2 illustrates the 1-PPS approach to derive the measurement for the relative time difference.

This mechanism relies on three key design components:

1. The capability for the external clock to provide a pulse such as the popular 1-PPS to another device (receiving device);
2. The capability for the external clock to provide a message that has the actual time of the transmission at the pulse using the clock of the external device;
3. The mechanism to save the time of the receiving device's clock when the pulse is received.

With the above three components, the relative time error is simply the difference between the time for the pulse in the external clock's time and the time of the pulse in the receiver's clock.

Figure 6.3 illustrates the correlator approach for the time-difference measurement. In this case, we will look at the carrier or code or data modulated signals and compare them by correlation.

For example, let us assume that the external clock reference comes as a sine wave

$$z = A\sin\left(\omega t_e\right) + v_e \tag{6.17}$$

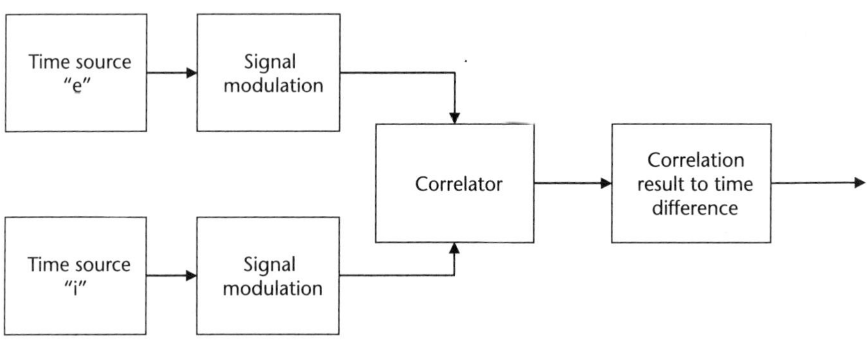

Figure 6.3 Correlation-based, time-difference measurement scheme.

where v_e is the noise on the signal.

The receiving device's time can also be represented by the phase of a sine wave

$$y = A_i \sin(\omega t_i) + v_i \tag{6.18}$$

where v_i is the noise on the signal.

In this simple case, the correlator can simply be classical multiplication-plus-lowpass-filtering mechanism, and the output can be written as

$$\begin{aligned} c &= LPF(y(t)z(t)) \\ &= \frac{A_e A_i}{2}\sin(\omega(t_e - t_i)) + LSF(v_i A_e \sin(\omega t_e) + v_e A_i \sin(\omega t_i)) \\ &= \frac{A_e A_i}{2}\sin(\omega(t_e - t_i)) + v_{ei} \end{aligned} \tag{6.19}$$

where LPF represents the lowpass filtering operator and

$$v_{ei} = LSF(v_i A_e \sin(\omega t_e) + v_e A_i \sin(\omega t_i)) \tag{6.20}$$

is the residual noise after the correlation process.

From the above, we can write

$$t_e - t_i = \frac{1}{\omega}\sin^{-1}\left(\frac{2c}{A_e A_i} - \frac{2v_{ei}}{A_e A_i}\right) \tag{6.21}$$

when the errors are small, the above can be approximated by

$$t_e - t_i = \frac{2}{\omega A_e A_i}c + v_t \tag{6.22}$$

where $v_t = -(2/\omega A_e A_i)v_{ei}$ is the residual noise on the time-difference measurement.

Now let us write the measurement equations for the Kalman filter.

For the purpose of Example 6.1, the measurement model can be written as

$$z = t_{\mathrm{raw}} - mt = x + w_{pwn} - v = \begin{bmatrix} 1 & 0 \end{bmatrix}\begin{bmatrix} x \\ b \end{bmatrix} + v_t \tag{6.23}$$

where $v_t = w_{pwn} - v$ is the combination of the phase white noise of the propagation solution clock and the measurement noise of the external clock being used.

For the purpose of Example 6.2,

$$\begin{aligned} z &= t_{c\_\mathrm{read\_out}} - mt = b_0 + b_1(t_{\mathrm{raw}} - t_0) + \frac{1}{2}b_2(t_{\mathrm{raw}} - t_0)^2 + w_{pwn} - v \\ &= \begin{bmatrix} 1 & (t_{\mathrm{raw}} - t_0) & \frac{1}{2}(t_{\mathrm{raw}} - t_0)^2 \end{bmatrix}\begin{bmatrix} b_0 \\ b_1 \\ b_2 \end{bmatrix} + v_t \end{aligned} \tag{6.24}$$

For the purpose of Example 6.3,

$$z = t_{c\_read\_out} - mt$$
$$= b_0 + b_1(t_{raw} - t_0) + \frac{1}{2}b_2(t_{raw} - t_0)^2 + \sum_{i-1}^{n} c_i \cos(\omega_i(t - t_0)) + s_i \sin\omega_i((t - t_0) + w_{wpn} - v \quad (6.25)$$

which can be written as

$$z = \begin{bmatrix} 1 & (t_{raw} - t_0) & \frac{1}{2}(t_{raw} - t_0)^2 & \cos\omega_1(t_{raw} - t_0) & \sin\omega_1(t_{raw} - t_0) & \cos\omega_2(t_{raw} - t_0) & \sin\omega_2(t_{raw} - t_0) & \cdots & \cos\omega_n(t_{raw} - t_0) & \sin\omega_n(t_{raw} - t_0) \end{bmatrix} \begin{bmatrix} b_0 \\ b_1 \\ b_2 \\ c_1 \\ s_1 \\ c_2 \\ s_2 \\ \vdots \\ c_n \\ s_n \end{bmatrix} + v_t \quad (6.26)$$

where $v_t = w_{pwn} - v$ is the combination of the phase white noise of the propagation solution clock and the measurement noise of the external clock being used.

6.5 Integrated Clock Simulation Example

The simulation result of an integrated clock with exaggerated errors is included here to illustrate an Example 6.1-style, classic, two-state integrated clock Kalman filter. For more examples, the readers are encouraged to review [1–4, 6–8].

Figure 6.4 shows behavior of the raw clock. The first strip in Figure 6.4 shows the slow-changing clock rate bias over time. The second strip shows the raw time error before any corrections are applied. Figure 6.5 shows the simulated update

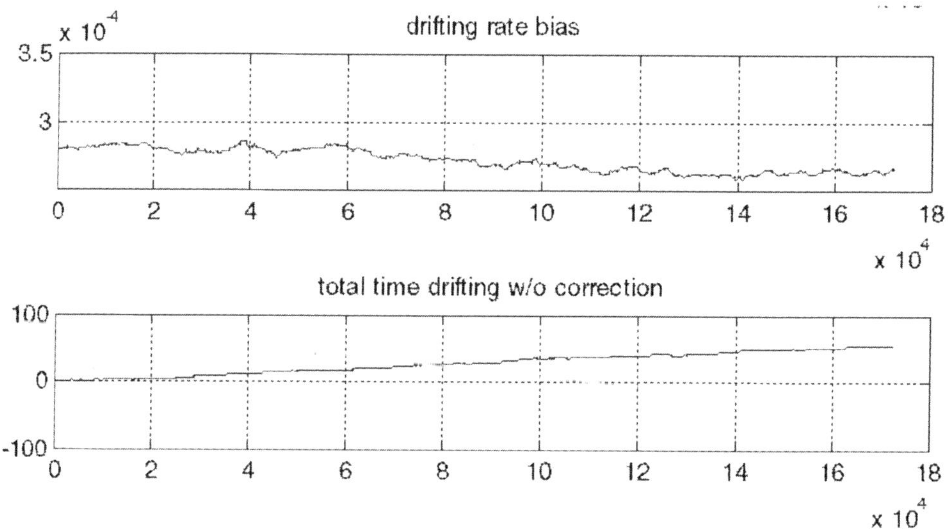

Figure 6.4 Behavior of the raw clock.

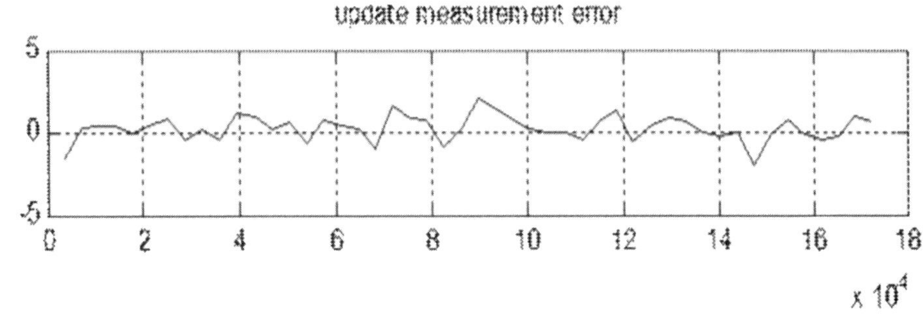

Figure 6.5 Measurement error of the external clock.

measurement error. Figure 6.6 shows the time error and the time rate error of the integrated clock plotted together with their +/–3 sigma upper and lower bounds computed from the Kalman filter covariance matrix. Figure 6.7 shows how the estimated rate bias is converging to the truth rate bias.

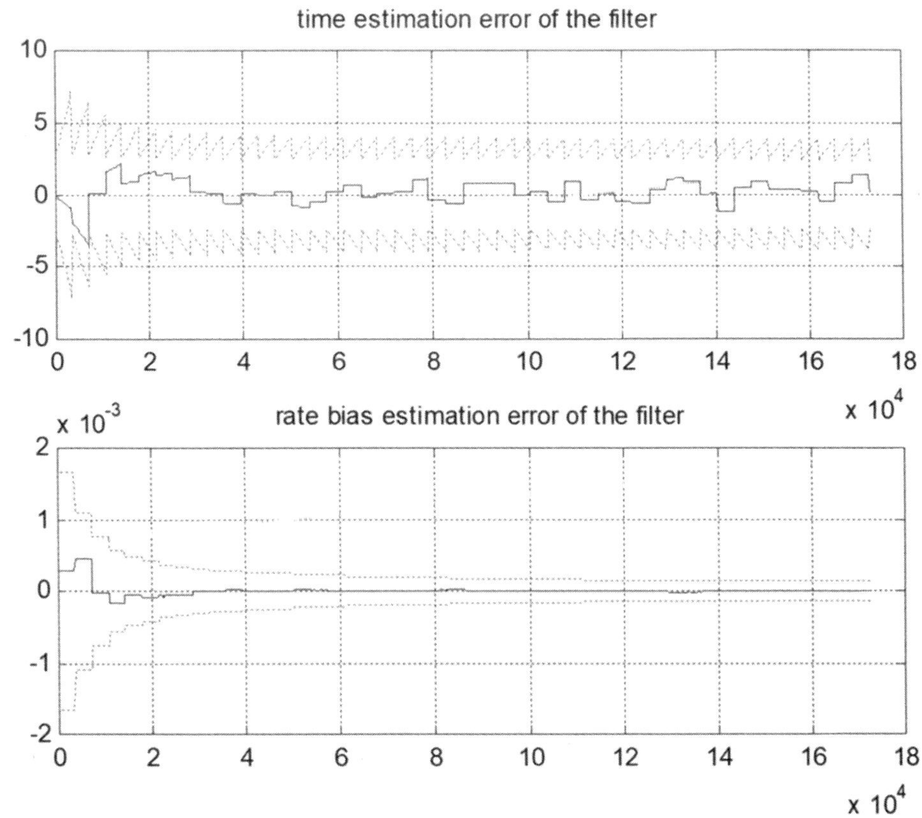

Figure 6.6 The simulated performance of the integrated clock.

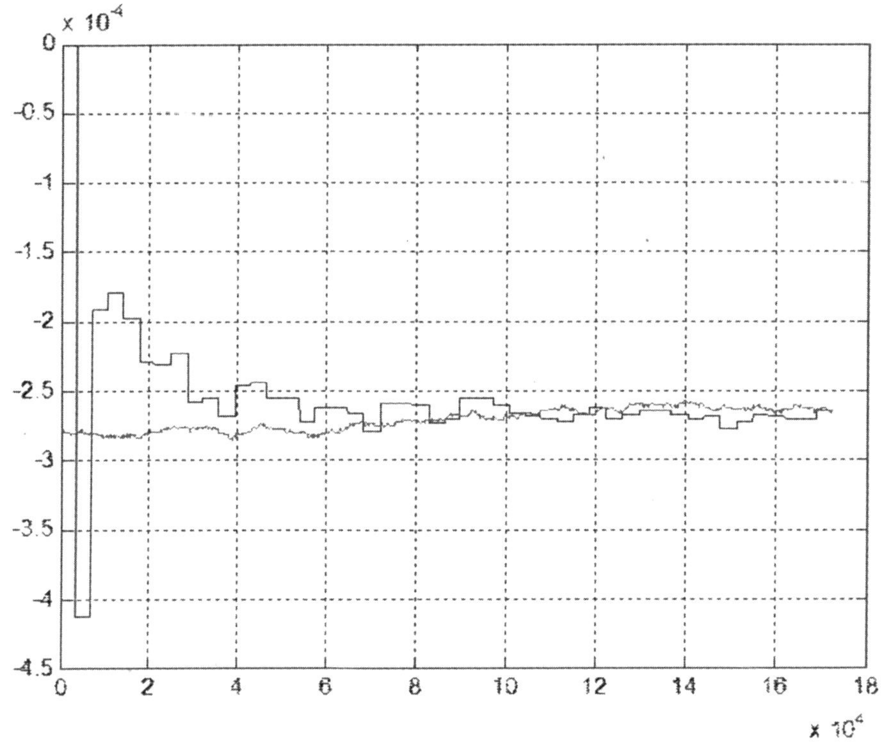

Figure 6.7 Convergence process of the estimated rate bias to the truth rate bias.

References

[1] Giorgi, G., and C. Narduzzi, "Performance Analysis of Kalman-Filter-Based Clock Synchronization in IEEE 1588 Networks," *IEEE Transactions on Instrumentation and Measurement*, Vol. 60, No. 8, August 2011.

[2] Stein, S. R., and R. L. Filler, "Kalman Filter Analysis for Real Time Applications of Clocks and Oscillators," *Proc. of 42nd Annual Frequency Control Symposium*, 1988.

[3] Kim, H., et al., "Tracking Low-Precision Clocks with Time-Varying Drifts Using Kalman Filtering," *Journal IEEE/ACM Transactions on Networking (TON)*, Vol. 20, No. 1, February 2012, pp. 257–270.

[4] Wang, B., J. Chen, and B. Wang, "Kalman Filter Simulation and Characterization of BDS Satellite Clock," *2018 European Frequency and Time Forum (EFTF)*, April 10–12, 2018.

[5] https://en.wikipedia.org/wiki/Precision_Time_Protocol.

[6] Gelb, M., *Applied Optimal Estimation*, Cambridge, MA: MIT Press, 1974.

[7] Brown, R. G., and P. Hwang, *Introduction to Random Signals and Applied Kalman Filtering*, New York: John Wiley & Sons, 1996.

[8] Li, R., and A. Wu, "Kalman Filtering & Applications," *UCLA Lecture Notes*, 2005–2019.

CHAPTER 7

Spacecraft Attitude Determination

7.1 Introduction

Since Chapter 6, we have used specific examples to further elaborate upon what we have presented in Chapters 3 to 5. Since spacecraft attitude determination is relatively simple, we will use spacecraft attitude determination as one of the early topics.

Although we are using spacecraft as an example, most of the algorithms, equations, and design principles apply to other vehicles when only attitude is to be determined.

Attitude determination is a widely covered topic in the literature, [1–10] provide some good coverage of attitude determination in general, with [11–15] providing the coverage of Kalman filtering and applications in areas such as attitude determination and navigation.

For spacecraft, Stellar Inertial Attitude Determination (SIAD) [16–64] is by far the most popular method of attitude determination.

Attitude determination is a critical capability and is required by most of spacecraft systems for at least three purposes:

1. To allow a spacecraft and its payload to be precision-pointed to a certain direction in order to provide the intended services [27, 31, 41]. For example, most of the Earth satellites are either communication service satellites or a sensor satellite. For both of the most popular services provided by an Earth orbiting spacecraft, precision attitude determination and precision pointing of the spacecraft are critical.
2. To allow a spacecraft to orient itself properly for its own survival. Specifically:
   a. Properly orient the spacecraft with respect to the Sun so that it is power-safe [28];
   b. Properly orient the spacecraft with respect to the Sun so that it is thermal-safe [28];
   c. Properly orient the spacecraft with respect to Earth ground stations so it is communication safe.
3. To allow the spacecraft to point to other directions to support spacecraft and payload calibration, spacecraft station-keeping or orbit control, and other operations [22, 27, 33].

Figure 7.1 illustrates the top-level architecture of a spacecraft attitude control system and the role of the attitude determination sensors and algorithms. Within the architecture, there are several key components and modules that are described here:

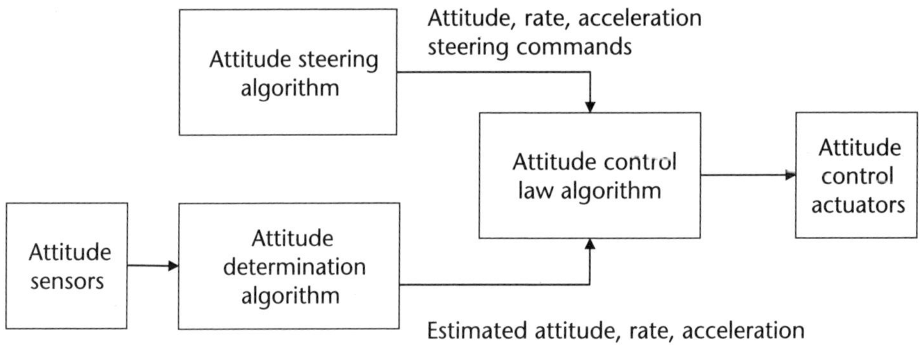

Figure 7.1 The role of attitude determination.

1. *Attitude steering algorithm:* This module determines how the spacecraft shall be oriented to achieve its operational objectives. The outputs of this subsystem are often a function of the operational modes and often contain the desired attitude quaternion, the desired attitude angular rate vector, and the desired attitude angular acceleration vector. These outputs are typically called steering commands.
2. *Attitude determination sensors and algorithm:* This module determines the current orientation of the spacecraft (i.e., the current spacecraft attitude, the current spacecraft attitude angular rate, and the current spacecraft angular acceleration). The outputs are called estimated states.
3. *Attitude control law and actuators:* This is where the differences between the steering commands and the current states are computed and the differences are then used to pass through control laws and generate actuator activities to null the differences.

As we discussed before, there are two types of sensors and information sources for PNT systems. Attitude determination, as a type of a PNT system, has no exception. The typical examples of propagation sensors and information sources used for spacecraft attitude determinations are listed here:

1. The three-axis gyros that produce angular rate outputs that can be used to integrate into attitude;
2. Spacecraft dynamic models that can be used to calculate the angular acceleration-based on the torques applied to the spacecraft and then integrate into angular rate and attitude.

Typical update sensors and information sources used for spacecraft attitude determinations are listed here:

1. Star-trackers that are becoming by far the most popular attitude determination sensors for spacecraft.
2. Sun sensors that are also popular due to the need to determine attitude with respect to the Sun for the purposes of power and thermal safety and because Sun sensors are typically inexpensive.

3. Earth-staring sensors that are capable of determining the LOS direction for the center of earth. Earth sensors were popular 15 to 30 years ago among Earth satellite systems since most of the satellites need to be pointing to the center of the Earth (nadir-pointing). Earth sensors are now mostly replaced by star-trackers for the benefits of cost reduction, SWAP reduction, performance improvements, and operational simplicity.
4. RF beacon sensors that measure the LOS direction of an RF transmitter on the Earth. Beacon sensors are still used in modern spacecraft when the payload to spacecraft bus thermal deformations and misalignments introduce too much uncompensated errors for the payload pointing. The beacon antenna and the S/C (Spacecraft) payload antennas can be colocated and can even be the same antenna.
5. Magnetometers that measure the local magnetic field vector. Compared to other sensors, three-axis magnetometers are typically cheaper but much less accurate, not because of the sensor but because the magnetic field can be easily distorted or contaminated by electronic or mechanical systems.
6. GPS receivers with multiple antennas that measure two or more baseline vectors in a reference frame can also be used to determine attitude. However, GPS receiver-based attitude determination has limited accuracy due to multipath effects and the limitations of using a long baseline.

Table 7.1 is a summary comparison of the above listed sensors.

7.2 Stellar Inertial Attitude Determination (SIAD)

7.2.1 System Concept

We will roughly follow the steps we outlined in Chapter 5 to develop and describe the SIAD design as one of the simplest examples of All Source PNT system.

SIAD is a synergistic integration of two types of attitude solutions: the IRU-based propagation solution and the star-tracker-based update solution.

We will use IRU (three-axis gyro unit) to provide rotational rate information. We will integrate the rotational rate into attitude. This will be the baseline propagation solution for the SIAD system. The IRU errors will be modeled and will be included in the Kalman design [39].

As discussed before, the propagation solution has two problems: it requires an initial value, and the error will grow over time. A star-tracker is therefore used to complement the IRU by providing LOS measurements for multiple stars. The stars will need to be identified to match with entries in a star catalog and then used to determine attitude.

When we do not have any attitude information at all, in other words, when we are in the state of being lost in space (LIS), a stellar attitude acquisition (SAA) algorithm [32, 44] will have to be used to match star patterns against star catalog patterns. The SAA algorithm will be able to provide initial attitude knowledge to initialize or reinitialize or reset the attitude state owned by the attitude propagator.

Table 7.1 Summary Comparison of Attitude Determination Sensors

	–	*Typical Performance (Rough Order of Magnitude)*	*Typical Recurring Cost (Rough Order of Magnitude)*	*Advantages*	*Limitations*
Propagate Attitude Determination Sensors					
1.1	3-axis gyro IRU (Inertial Reference Unit)	1e-3–10°/hour	$1,000–$1e7	Available all the time and at high rate	Error growing over time
1.2	S/C dynamics model	0.1%–10% of motion	0	All the above and no recurring cost	Growing error; low accuracy; requires torque and force measurements
Update Sensors					
2.1	Star-tracker	1e-4–1e-2 degrees (star position error)	$50,000 to $1 million	Very accurate	Sun, Moon, or planet intrusion concerns; accuracy may not be transferrable to payload due to thermal deformations
2.2	Sun sensor	0.1°–10°	$100 to $100,000	Low cost, low SWAP	Not very accurate; not useful in eclipse
2.3	Earth sensor	0.01°–1°	$100,000 to $1 million		Expensive, not very accurate, outdated
2.4	RF beacon sensor	0.001°–0.01°	$1,000 to $1 million (typically part of payload, may be free)	Very accurate; colocated with payload	Requires ground station. (which is typically OK, since there is a ground station for the services provided anyway)
2.5	3-axis magnetometer	0.1°–10°	$1,000 to $100,000	Low cost, low SWAP	Limited accuracy due to field distortions by other parts of S/C
2.6	Multi-antenna GNSS receiver	0.1°–1°	$10,000–$1 million	May be low cost if GNSS receivers are used anyway	Not very accurate

Most of the time during normal operations, the attitude propagator maintains a good attitude knowledge that can be used to predict the following:

1. What stars we are going to see.
2. What will be the positions of the stars in the FOV.

The above information can be used in two ways:

1. To steer or help the star-tracker to track stars that we expect to see.
2. To directly identify the stars by matching the predicted star position with the actual measured star position and to produce the measurement residuals that will be used by the Kalman filter to generate corrections to the states.

The Kalman filter properly models each of the error parameters including the IRU error parameters, the star-tracker error parameters, and the attitude error and will be able to properly attribute the star residuals into different correction parameters.

7.2.2 The SIAD Architecture

The system concept discussed in Section 7.2.1 can be solidified by an architecture diagram.

Figure 7.2 illustrates the architecture that is derived from the general All Source PNT architecture diagram from Chapter 5.

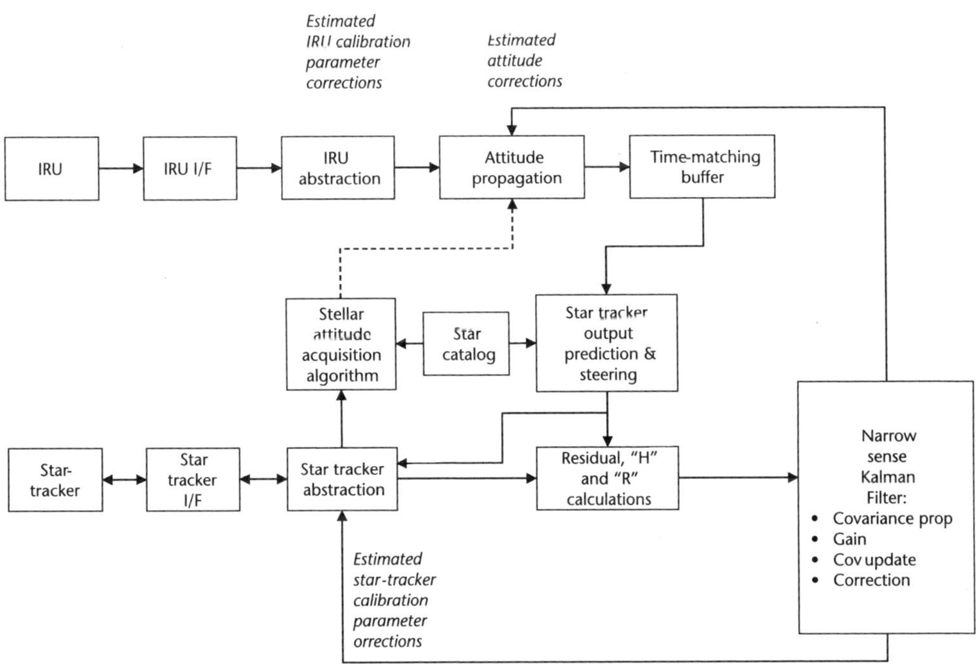

Figure 7.2 Stellar inertial attitude determination architecture.

Within the architecture, we expect the sensors (i.e., the IRU and the star-trackers that are separate hardware, firmware, or software entities) to be integrated into the system. The IRU and star-tracker interfaces consist of the hardware interfaces and the software interfaces. The rest of the architecture is naturally application software that runs on a computer processor.

7.2.3 Propagation Algorithms and Error Models

The attitude propagation algorithms have been discussed in detail in Chapter 3. Here we will simply use the third-order algorithm.

For spacecraft applications, we typically use the ECI frame (the I frame) as the reference frame. The differential equation is

$$\dot{q}_B^I = \frac{1}{2} q_B^I \otimes {}^B\omega_I^B = \frac{1}{2} q_B^I \, {}^B\omega_I^B \tag{7.1}$$

The third-order numerical integration algorithm is summarized as

$$q_B^I(t) = q_B^I(t - \Delta t) \begin{pmatrix} \left(1 - \frac{1}{6}\delta_0^2\right)\delta_1 - \frac{1}{6}\left(\delta_2\delta_3^p - \delta_3\delta_2^p\right) \\ \left(1 - \frac{1}{6}\delta_0^2\right)\delta_2 - \frac{1}{6}\left(\delta_3\delta_1^p - \delta_1\delta_3^p\right) \\ \left(1 - \frac{1}{6}\delta_0^2\right)\delta_3 - \frac{1}{6}\left(\delta_1\delta_2^p - \delta_2\delta_1^p\right) \\ 1 - \frac{1}{2}\delta_0^2 \end{pmatrix} \tag{7.2}$$

where

$$\begin{aligned} &\delta = \begin{bmatrix} \delta_1 \\ \delta_2 \\ \delta_3 \end{bmatrix} = \frac{1}{2} {}^B\omega_I^B(t)\Delta t, \quad \delta^P = \begin{bmatrix} \delta_1^P \\ \delta_2^P \\ \delta_3^P \end{bmatrix} = \frac{1}{2} {}^B\omega_I^B(t - \Delta t)\Delta t \\ &\delta_0^2 = \delta_1^2 + \delta_2^2 + \delta_3^2 \end{aligned} \tag{7.3}$$

The attitude error kinematics equation is given by

$$\delta^I \dot{\theta}_B^I = -C_B^I \delta \, {}^B\omega_I^B \tag{7.4}$$

We will use the following error model for the IRU

$$\delta\, {}^B\omega_I^B = m\, {}^B\omega_I^B - {}^B\omega_I^B = C_G^B \mathrm{diag}\left({}^G s\right) {}^G\omega_I^B + \left(\left[{}^B\theta\times\right] + L\left({}^B\beta\right)\right) {}^B\omega_I^B + {}^B b + C_G^B ARW \tag{7.5}$$

Here we introduced the gyro sensor frame (G), which is a coordinate frame with each of the axes aligned with each of the axis of the sensitivity of the 3 gyros. Both the scale factor error and the angle random walk (ARW) are terms associated with the sensor frame.

$$\operatorname{diag}\left({}^{G}s\right) = \begin{bmatrix} s_x & 0 & 0 \\ 0 & s_y & 0 \\ 0 & 0 & s_z \end{bmatrix}$$

represents the scale factor error in the sensor frame where

$${}^{G}s = \begin{bmatrix} s_x & s_y & s_z \end{bmatrix}^T$$

The DCM C_G^B is not necessarily an orthonormal matrix since the 3 gyro axes are not necessarily orthogonal to each other. It represents the transformation from the gyro sensor frame to the body frame.

${}^{B}b$ is the gyro bias vector in the body frame that is transformed from the sensor frame. In other words, this bias vector is a linear combination of the biases of the individual gyros. In the case that the 3 gyros have very different behaviors, we may want to choose to use

$${}^{B}b = C_G^B\,{}^{G}b$$

to model the bias directly in the gyro sensor frame.

The misalignment matrix (associated with the body frame) are decomposed into two transformations

$$M = (1 + \delta M) = \begin{bmatrix} 1 & -\theta_z & \theta_y \\ \beta_z + \theta_z & 1 & -\theta_x \\ -\left(\beta_y + \theta_y\right) & \beta_x + \theta_x & 1 \end{bmatrix} \approx \begin{bmatrix} 1 & 0 & 0 \\ \beta_z & 1 & 0 \\ -\beta_x & \beta_y & 1 \end{bmatrix} \begin{bmatrix} 1 & -\theta_z & \theta_y \\ \theta_z & 1 & -\theta_x \\ -\theta_y & \theta_x & 1 \end{bmatrix}$$

which can also be written additively as

$$M = (1 + \delta M) = \begin{bmatrix} 1 & -\theta_z & \theta_y \\ \beta_z + \theta_z & 1 & -\theta_x \\ -\left(\beta_y + \theta_y\right) & \beta_x + \theta_x & 1 \end{bmatrix} = I + \begin{bmatrix} 0 & -\theta_z & \theta_y \\ \theta_z & 0 & -\theta_x \\ -\theta_y & \theta_x & 0 \end{bmatrix} + \begin{bmatrix} 0 & 0 & 0 \\ \beta_z & 0 & 0 \\ -\beta_x & \beta_y & 0 \end{bmatrix} \tag{7.6}$$

Let

$$\left[{}^{B}\theta\times\right] = \begin{bmatrix} 0 & -\theta_z & \theta_y \\ \theta_z & 0 & -\theta_x \\ -\theta_y & \theta_x & 0 \end{bmatrix}, \quad L\left({}^{B}\beta\right) = \begin{bmatrix} 0 & 0 & 0 \\ \beta_z & 0 & 0 \\ -\beta_x & \beta_y & 0 \end{bmatrix} \tag{7.7}$$

where

$${}^{B}\theta = \begin{bmatrix} \theta_x & \theta_y & \theta_z \end{bmatrix}^T, \quad {}^{B}\beta = \begin{bmatrix} \beta_x & \beta_y & \beta_z \end{bmatrix}^T$$

We have

$$\delta M = \left[{}^{B}\theta\times\right] + L\left({}^{B}\beta\right) \tag{7.8}$$

It is important to realize that $[\theta\times]$ models the three rotational misalignment errors and $L(\beta)$ models the three nonorthogonality misalignment errors. As discussed in Chapter 3, the attitude errors due to rational misalignments do not accumulate over attitude maneuvers but those due to nonorthogonality misalignments do. In particular, the errors will accumulate over multi-axis coning-like motions.

In a SIAD system, the rotational misalignments are typically a combination of three parts, the rotational misalignments of the IRU due to calibration residuals and thermal deformations, the rotational misalignments between the IRU and the star-tracker due to calibration residuals and thermal deformations, and the rotational misalignments of the star-tracker due to calibration residuals and thermal deformations. When the IRU and the star-tracker are mounted over a relatively large baseline, the rotational misalignments due to thermal deformations can be significant. Fortunately, the attitude errors due this error do not accumulate.

The nonorthogonality misalignments only happen between the 3 gyros. The initial error can be calibrated and the thermal deformation typically will be very small due to the 3 gyros being mounted closely together.

The rotational rate error due to these misalignment error terms is given by

$$\delta M\omega = \left[\theta\times\right]\omega + L(\beta)\omega = -\left[\omega\times\right]\theta + \begin{bmatrix} 0 & 0 & 0 \\ 0 & 0 & \omega_x \\ \omega_y & -\omega_x & 0 \end{bmatrix}\beta = -\left[\omega\times\right]\theta + J(\omega)\beta \tag{7.9}$$

where

$$J(\omega) = \begin{bmatrix} 0 & 0 & 0 \\ 0 & 0 & \omega_x \\ \omega_y & -\omega_x & 0 \end{bmatrix} \tag{7.10}$$

The error equation can then be written in a way that is friendlier to the Kalman filter design:

$$\delta\,{}^{B}\omega_I^B = C_G^B \mathrm{diag}\left({}^{G}\omega_I^B\right){}^{S}s - \left[{}^{B}\omega_I^B\times\right]{}^{B}\theta + J\left({}^{B}\omega_I^B\right){}^{B}\beta + {}^{B}b + C_G^B ARW \tag{7.11}$$

Combining the above with the error kinematic equation, we end up with the following equations:

$$\begin{aligned} {}^{I}\dot{\theta}_B^I &= -C_B^I \delta\,{}^{B}\omega_I^B \\ &= -C_G^I \mathrm{diag}\left({}^{G}\omega_I^B\right)\begin{bmatrix} s_1 \\ s_2 \\ s_3 \end{bmatrix} - C_B^I J\left({}^{B}\omega_I^B\right)\begin{bmatrix} \beta_1 \\ \beta_2 \\ \beta_3 \end{bmatrix} + C_B^I\left[{}^{B}\omega_I^B\times\right]\begin{bmatrix} \theta_1 \\ \theta_2 \\ \theta_3 \end{bmatrix} - C_B^I\begin{bmatrix} b_1 \\ b_2 \\ b_3 \end{bmatrix} - C_G^I ARW \end{aligned} \tag{7.12}$$

If we model each one of the IRU error parameters as a random walk process and we can define the state vector and the system process noise vector as follows:

$$
\begin{aligned}
X^T &= \left[{}^I\theta_B^I, {}^Bb^T, {}^Ss^T, {}^B\beta^T, {}^B\theta^T \right] \\
W^T &= \left[-\left(C_G^I ARW\right)^T, \; RRW^T, \; SRW^T, \beta RW^T, \; \theta RW^T \right]
\end{aligned}
\tag{7.13}
$$

where

SRW is the 3×1 white noise process vector that generates the scale factor random walk.

βRW is the 3×1 white noise process vector that generates the nonorthogonality misalignment random walk.

θRW is the 3×1 white noise process vector that generates the rotational misalignment random walk.

We then have the following differential equations as the linearized error propagation model.

$$\dot{X} = AX + W$$

$$
A = \begin{bmatrix}
0_{3x3} & -C_B^I & -C_G^I \mathrm{diag}\left({}^G\omega_I^B\right) & -C_B^I J\left({}^B\omega_I^B\right) & C_B^I\left[{}^B\omega_I^B \times\right] \\
0_{3x3} & 0_{3x3} & 0_{3x3} & 0_{3x3} & 0_{3x3} \\
0_{3x3} & 0_{3x3} & 0_{3x3} & 0_{3x3} & 0_{3x3} \\
0_{3x3} & 0_{3x3} & 0_{3x3} & 0_{3x3} & 0_{3x3} \\
0_{3x3} & 0_{3x3} & 0_{3x3} & 0_{3x3} & 0_{3x3}
\end{bmatrix}
\tag{7.14}
$$

7.2.4 Measurement Equations and Error Models

Star-tracker measurements are typically provided in the star-tracker frame as the star position coordinates (h, v), which are related to the unit vector in the star-tracker sensor frame by

$$
{}^S u = \frac{1}{\sqrt{1 + h^2 + v^2}} \begin{bmatrix} h \\ v \\ 1 \end{bmatrix}
\tag{7.15}
$$

or we have

$$
\begin{bmatrix} h \\ v \end{bmatrix} = \begin{bmatrix} \dfrac{{}^S u(1)}{u(3)} \\ \dfrac{{}^S u(2)}{u(3)} \end{bmatrix}
\tag{7.16}
$$

and we can map the errors or deviations of the unit vector to the errors or deviations in (h, v) in the following way:

$$\begin{bmatrix} \delta h \\ \delta v \end{bmatrix} = \begin{bmatrix} \dfrac{\delta^S u(1)}{^S u(3)} - \dfrac{^S u(1)\delta^S u(3)}{^S u^2(3)} \\ \dfrac{\delta^S u(2)}{^S u(3)} - \dfrac{^S u(2)\delta^S u(3)}{^S u^2(3)} \end{bmatrix} = \begin{bmatrix} \dfrac{1}{^S u(3)} & 0 & -\dfrac{^S u(1)}{^S u^2(3)} \\ 0 & \dfrac{1}{^S u(3)} & -\dfrac{^S u(2)}{^S u^2(3)} \end{bmatrix} \delta^S u$$

Alternatively, the above equation can be rewritten as

$$\begin{bmatrix} \delta h \\ \delta v \end{bmatrix} = \frac{1}{\sqrt{1 + h^2 + v^2}} \begin{bmatrix} 1 & 0 & -h \\ 0 & 1 & -v \end{bmatrix} \delta^S u \tag{7.17}$$

For a star unit vector in the ECI frame (which can be retrieved from a star catalog), the corresponding star-tracker frame unit vector can be written as

$$^S u = C_B^S C_I^B\, {}^I u$$

From Chapter 4, we have

$$\delta^S u = \left[{}^S u\times\right]\delta^S\theta_I^S = \left[{}^S u\times\right]C_B^S\delta^B\theta_I^B = \left[{}^S u\times\right]C_I^S\delta^I\theta_I^B = -\left[{}^S u\times\right]C_I^S\delta^I\theta_B^I$$

Combining these equations, we have

$$\begin{bmatrix} \delta h \\ \delta v \end{bmatrix} = \frac{1}{\sqrt{1 + h^2 + v^2}} \begin{bmatrix} -1 & 0 & h \\ 0 & -1 & v \end{bmatrix} \left[{}^S u\times\right]C_I^S\delta^I\theta_B^I = H_1\delta^I\theta_B^I \tag{7.18}$$

where

$$H_1 = \frac{1}{\sqrt{1 + h^2 + v^2}} \begin{bmatrix} -1 & 0 & h \\ 0 & -1 & v \end{bmatrix} \left[{}^S u\times\right]C_I^S \tag{7.19}$$

which is a function of the star position.

The above equations map the attitude error in the ECI frame to a star position error in the star-tracker sensor frame.

The above equations assume that the star position has no errors. In reality, the star position typically contains three types of errors:

$$\begin{bmatrix} mh \\ mv \end{bmatrix} = \begin{bmatrix} h \\ v \end{bmatrix} + \begin{bmatrix} h_{LSF} + h_{HSF} + h_{TN} \\ v_{LSF} + v_{HSF} + v_{TN} \end{bmatrix} \tag{7.20}$$

where the low spatial frequency (LSF) errors are also called fixed pattern errors, which change slowly over the star-tracker FOV. When a star moves across the FOV, the errors will appear in the time domain as changing noises. They change fast if the star moves fast. They change slowly if the star moves slowly. Typical reasons for LSF errors are (but not limited to):

1. Effective focal length shift due to aging and thermal deformation of lens or other relevant part of the optical system.

2. Star-tracker focal plane response nonuniformity.
3. For the CCD-based focal plane array, the charge transfer efficiency (CTE) can cause a LSF error especially after the CCD array has been exposed to radiations over a long time. The active pixel matrix based focal plane array does not need to transfer the charges from a pixel to another pixel during readout and consequently does not have this problem.
4. Calibration residuals.

Some part of the LSF errors may be repeatably modeled as functions of star position (h, v). Sometimes it is beneficial to include the model parameters as part of Kalman filter states and consequently allow the system to continuously calibrate the star-tracker LSF errors [43].

The high spatial frequency (HSF) errors are errors that change fast over the FOV, typically with the frequency contents at pixel level. One good example of an HSF error is the centroiding error due to the finite resolution of the image that generates an S-curve (sine-wave-like curve) at 1 cycle per pixel [29]. The other source of an HSF error is the pixel-to-pixel response nonuniformity of the focal plane array.

The temporal noise is the random electronic noise that changes with time and with star position. Typical sources of such noise include the focal plane array noise (such as shot noise) and readout noise.

In summary, for each star in the FOV, the nonlinear measurement model is given by the following three equations:

$$\begin{aligned} {}^{S}u &= C_B^S C_I^B\, {}^{I}u \\ \begin{bmatrix} h \\ v \end{bmatrix} &= \begin{bmatrix} \dfrac{{}^{S}u(1)}{u(3)} \\ \dfrac{{}^{S}u(2)}{u(3)} \end{bmatrix} \\ \begin{bmatrix} mh \\ mv \end{bmatrix} &= \begin{bmatrix} h \\ v \end{bmatrix} + \begin{bmatrix} h_{LSF} + h_{HSF} + h_{TN} \\ v_{LSF} + v_{HSF} + v_{TN} \end{bmatrix} \end{aligned} \tag{7.21}$$

The linearized measurement model is given by the following equations

$$\begin{aligned} \begin{bmatrix} \delta mh \\ \delta mv \end{bmatrix} &= H_1 \delta^I \theta_B^I + \begin{bmatrix} h_{LSF} + h_{HSF} + h_{TN} \\ v_{LSF} + v_{HSF} + v_{TN} \end{bmatrix} \\ &\text{where} \\ H_1 &= \frac{1}{\sqrt{1 + h^2 + v^2}} \begin{bmatrix} -1 & 0 & h \\ 0 & -1 & v \end{bmatrix} \left[{}^{S}u \times \right] C_I^S \end{aligned} \tag{7.22}$$

Both the nonlinear model and the linearized model are used to construct the Kalman filter. The nonlinear model is used to generate the predicted star positions. The linear model provides the H-matrix and the measurement noise model that are necessary for the Kalman filter covariance update and gain computations.

7.2.5 Direct Star Identification and Star-Catalog Near-Neighbor Criteria

During normal operations, we can use the current attitude knowledge to predict what stars we should be seeing in the star-tracker FOV by the following steps:

1. Given the current attitude quaternion q_I^B, we will first calculate the attitude direction cosine matrix C_I^B.
2. We can then map the star-tracker boresight vector into the ECI frame as

$$^I u_{bs} = C_B^I C_S^B \begin{bmatrix} 0 \\ 0 \\ 1 \end{bmatrix} = C_B^I C_S^B(:,3)$$

3. Using this unit vector, we can find all the stars in the circular FOV that encloses the square or rectangular FOV (if the FOV is square or rectangular).
4. We then can calculate the (h, v) for each star by using the following equations:

$$^S u = C_B^S C_I^B \, {}^I u$$

$$\begin{bmatrix} h \\ v \end{bmatrix} = \begin{bmatrix} \dfrac{^S u(1)}{u(3)} \\ \dfrac{^S u(2)}{u(3)} \end{bmatrix}$$

5. We will use the (h, v) calculated above to check if the star is indeed in the FOV.
6. All the stars that pass the test are saved in a subcatalog, which contains stars that should be in the FOV at this moment.

It is important to point out that the attitude DCM must be generated for the time moment corresponding to the time tag of the star-tracker measurements.

With the predicted star positions in the subcatalog, we can now check each of the measured star in the FOV against each of the stars in the subcatalog by calculating the square of the distance:

$$d_k^2 = \left(mh - \hat{h}_k\right)^2 + \left(mv - \hat{v}_k\right)^2$$

where (mh, mv) is the measured position for a star of interest and $\left(\hat{h}_k, \hat{v}_k\right)$ is the predicted position for the stars in the current subcatalog. The criteria that we will use to identity a star is simply

$$d_k^2 = \left(mh - \hat{h}_k\right)^2 + \left(mv - \hat{v}_k\right)^2 < (\text{minsep})^2 \tag{7.23}$$

where the parameter minsep is the minimum near-neighbor separation angle used when we build the star catalog. In other words, if two stars visible to the star-tracker are separated for less than minsep, they are both dropped from the catalog.

Typically, the parameter minsep is chosen to be significantly larger than the attitude errors and the star measurement errors combined and small enough so that the near-neighbor criteria will not cause the dropping of too many stars from the catalog. A good example of minsep is 0.2°, which will satisfy the above criteria for most applications.

Clearly, the near-neighbor criteria for the star catalog is critically important to guarantee a unique correct match of the stars provided that:

1. The attitude knowledge errors are significantly smaller than minsep.
2. The star position errors are significantly smaller than minsep.

If, for any reason, the attitude knowledge error is so large that it violates the first rule above, it is highly likely that we may match measured stars with the incorrect catalog stars and lead to wrong attitude solution. It is very important to keep this in mind when designing the concept of operations for an SIAD system.

With all the above discussed, the direct-match criteria are very simple. If a predicted star in the subcatalog matches with the measured star such that

$$d_k^2 = \left(mh - \hat{h}_k\right)^2 + \left(mv - \hat{v}_k\right)^2 < (\text{minsep})^2$$

then the measured star is identified to be the star k in the subcatalog.

With the constraints and the near-neighbor criteria discussed above, the identification is guaranteed to be unique and correct.

Once the star is identified, the corresponding residual can then be used by the Kalman filter to produce corrections.

7.2.6 Aberration Error Correction

It is important to point out that the star-tracker measurements actually contain an important error that must be corrected when calculating the residual for the Kalman filter. That is the stellar aberration error due to the speed of the vehicle in the inertial space. The most important part of that speed is actually the Earth rotation around the Sun at roughly 107,000 km/hour, which causes about 20 arcsec of star position error.

The stellar aberration error is small enough to be ignored before the star is identified but must be corrected before the residuals are sent to the Kalman filter since most of the star-trackers are much more accurate than 20 arcsec.

The aberration error equation can be derived using simple classical arguments, although these easy-to-understand arguments are not entirely correct because relativistic effects are not included. However, the results are correct and sufficiently accurate for most applications.

Given a star unit vector u_{star} and an observer of the star traveling at velocity vector v, assuming that it takes δt for the star light to reach the observer, the distance vector from the star is then

$$r_{\text{star}} = (c\delta t)u_{\text{star}}$$

The observer has traveled in the mean time for a distance vector

$$(v\delta t)$$

As a result, the apparent distance vector observed by the observer is then

$$r_{\text{star\_observed}} = (c\delta t)u_{\text{star}} - v\delta t$$

The observed unit vector is then

$$u_{\text{star\_observed}} = \frac{(c\delta t)u_{\text{star}} - v\delta t}{\left\|(c\delta t)u_{\text{star}} - v\delta t\right\|}$$

A first-order approximation is given by

$$u_{\text{star\_observed}} = \frac{(c\delta t)u_{\text{star}} - v\delta t}{\left\|(c\delta t)u_{\text{star}} - v\delta t\right\|} \approx \frac{(c\delta t)u_{\text{star}} - v\delta t}{\left\|(c\delta t)u_{\text{star}}\right\|} = \frac{(c\delta t)u_{\text{star}} - v\delta t}{(c\delta t)} = u_{\text{star}} - \frac{1}{c}v \tag{7.24}$$

This first-order approximation is sufficient for most applications.

This correction can be applied to the catalog star positions or applied to the measured star position. Since there are always fewer measured stars than the candidate stars in the subcatalog, it is computationally more efficient to correct the measured star position.

For the measured unit vector, the measured plays the same role as the observed in the above discussion; consequently, the correction formula is given by

$$mu_{\text{star\_corrected}} = mu_{\text{star}} + \frac{1}{c}C_B^S C_I^B {}^I v \tag{7.25}$$

where the measured aberration-contaminated star unit vector is restored back to the original unit vector. The spacecraft velocity is transformed from the ECI frame to the star-tracker sensor frame.

Since we know that

$$\begin{bmatrix} \delta h \\ \delta v \end{bmatrix} = \frac{1}{\sqrt{1+h^2+v^2}} \begin{bmatrix} 1 & 0 & -h \\ 0 & 1 & -v \end{bmatrix} \delta^S u$$

the aberration correction in the (h, v) domain can be written as

$$\begin{aligned} \begin{bmatrix} mh \\ mv \end{bmatrix}_{\text{star\_corrected}} &= \begin{bmatrix} mh \\ mv \end{bmatrix}_{\text{star}} + \frac{1}{c}\frac{1}{\sqrt{1+h^2+v^2}} \begin{bmatrix} 1 & 0 & -h \\ 0 & 1 & -v \end{bmatrix} \begin{bmatrix} {}^S v(1) \\ {}^S v(2) \\ {}^S v(3) \end{bmatrix} \\ &= \begin{bmatrix} mh \\ mv \end{bmatrix}_{\text{star}} + \frac{1}{\sqrt{1+h^2+v^2}} \left(\frac{1}{c} \begin{bmatrix} {}^S v(1) - h^S v(3) \\ {}^S v(2) - v^S v(3) \end{bmatrix} \right) \end{aligned} \tag{7.26}$$

Because the correction is very small, most times, the following equation is sufficient:

$$\begin{bmatrix} mh \\ mv \end{bmatrix}_{star\_corrected} = \begin{bmatrix} mh \\ mv \end{bmatrix}_{star} + \frac{1}{c}\begin{bmatrix} {}^{S}v(1) - h^{S}v(3) \\ {}^{S}v(2) - v^{S}v(3) \end{bmatrix} \tag{7.27}$$

7.2.7 SAA Algorithm: An Example

As discussed before, with good knowledge of attitude, we can identify stars by direct match. This is not possible when we are lost in space. In this case, we will need to use the SAA to determine the attitude so that the attitude propagator can be initialized.

The basic principle of the SAA algorithm is to use star patterns to determine attitude. There are many algorithms that have been invented and are supported by different variations of star catalogs. In this section, we will describe one example algorithm, the pair-match-based algorithm for the SAA [32, 44, 50, 56].

The pair-match SAA algorithms are supported by two types of the catalogs:

1. *Regular star catalog:* This is where the unit vector or the right ascension and declination, the star magnitude for stars that are visible to the star-tracker are included. A number of criteria are applied to ensure that these criteria are satisfied by each of stars included in the catalog. Typically a star catalog contains a few thousand stars. The actual size is determined by the star-tracker FOV size and other considerations. For most applications, we want 4 to 10 stars in the FOV. For large FOV star-trackers, we can achieve that by using a fewer number of stars by cutting off at a brighter magnitude. For small FOV star-trackers, we will need to look deeper into the sky in order to maintain the same average number of stars in the FOV; as a result, we will need to use a larger catalog.
2. *Pair catalog:* This catalog contains star pairs. Since the detailed star information is already included in the regular catalog. Only star indices are included for each pair. The pairs are sorted according to the separation angle between the two stars in the pair. A pair catalog can be very big because there are $N(N - 1)/2$ pairs for an N-star catalog in general. Fortunately, we only care about star pairs with a separation angle limited by the maximum separation distance in a star-tracker FOV. In addition, we also do not care about pairs where stars are too close to each other. These restrictions significantly reduce the number of pairs and make it possible to store the catalog onboard even for a very old-fashioned computer.

The pair-matching LIS SAA algorithm can be described as follows:

1. We first select a pair of stars from the stars in the FOV. Typically, we want:
   a. The stars to be bright enough (so that we have fewer matches).
   b. The separation angle between the stars to be large so that the attitude determined from the pair has better accuracy (recall Chapter 4).
2. We then try to match the star pair against the pair catalog. With star-position measurement uncertainty (typically very small), we typically do not find too many candidate matching pairs.

3. Our next step is try each of the candidate matching pairs and decide which one is the correct match. Recall in Chapter 4, we can determine an attitude based on a pair of vectors in two frames, in this case, the star-tracker frame for the measurements and the ECI frame for the star-catalog. By using the TRIAD algorithm, for each candidate pair, we can determine an attitude.
4. We then verify if the attitude is actually correct by using this attitude to predict the positions of other stars in the star-tracker FOV. If we have good matches for two additional stars in the sense that the measured stars and the predicted stars match both in terms of position and magnitude, including the two in the pair, we then have four stars matched correctly in total. It can be shown that the probability for four stars to match, while the attitude is still wrong, is extremely small. For all practical purposes, the probability can be considered to be 0.
5. If we have a success, we are done and we can now use the attitude that we determined to initialize or reset the attitude propagator and begin normal operations.
6. If not, we will try the next candidate pair and so on until we have a match or we run out of the candidate pairs.
7. There is a possibility that we will not have a match after running out of all the candidate pairs. This does happen from time to time. The reason is that one or both of the stars that we selected for the pair are not in the catalog. In this case, we can repeat everything above by selecting another two stars in the FOV as the pair.

7.2.8 Star Catalog

7.2.8.1 Apparent Magnitude of Stars

When we use one object as a reference that is defined as having the magnitude m_{ref} with the power intensity as measured from the Earth in units of power per unit area (such as watts per square meter, W/m^2) I_{ref}, the magnitude of the object with the power intensity given by I can be calculated using the formula

$$m = m_{\text{ref}} - 2.5\log_{10}\left(\frac{I}{I_{\text{ref}}}\right) \tag{7.28}$$

This definition implies two things that are worth remembering:

1. Brighter stars have smaller magnitude (due to the minus sign in front of 2.5).
2. If a star is 10 times brighter (in terms of power), then the magnitude number will be smaller by 2.5.

7.2.8.2 Star Counts

The number of visible stars brighter than the visible magnitude m can be approximately predicted by the following formula [65]:

$$N(m) = \text{Number of stars brighter than } m = 41253 \times 10^{(-0.0003m^3+0.0019m^2+0.484m-3.82)} \tag{7.29}$$

Table 7.2 Typical Number of Stars in the FOV at Different Cutoff Magnitudes

Magnitude	*Actual Number of Stars Brighter Than*	*Prediction by Approximate Formula*	*Average Number in 8° × 8° FOV*	*Average Number in 14° × 14° FOV*
−1	2	2	0.0	0.0
0.5	8	11	0.0	0.0
1.5	22	33	0.0	0.1
2.5	93	103	0.1	0.4
3.5	283	316	0.4	1.3
4.5	893	965	1.4	4.2
5.5	2,822	2,918	4.4	13.4
6.5	8,768	8,696	13.6	41.7
7.5	26,533	25,455	41.2	126.1

Table 7.2 lists the number of stars brighter than magnitude −1 to 7.5 and provides an estimate of the average number of stars in the FOV for two popular star-tracker FOV sizes in the industry.

It can be seen that, for 8° × 8° FOV, we will want to use stars up to 5.5 to 6.5 magnitude and the number of stars to be included in the star catalog ranges from 2,800 to 8,800.

For 14° × 14° FOV, we will want to use stars up to 4.5 to 5.5 and the number of stars to be included in the catalog ranges from 900 to 2,800.

7.2.8.3 Catalog-Building Steps [19, 21, 36, 46]

Input Data

When we build a star catalog for a particular application, we typically start with two pieces of information:

1. A master catalog, typically, the NASA Sky 2000 Master Catalog, which now includes the data from the European Hipparcos mission that reduced most of the star position errors to the milli-arcsec level;
2. A model, typically a polynomial curve fitting, that allows one to convert the visual magnitude to the instrument magnitude (the magnitude as sensed by the star-tracker).

The Process

We typically take the following steps:

1. Take a subset of the master catalog based on generous considerations on the visibility of stars by the star-tracker. For example, if the star-tracker is designed to see up to magnitude 6 stars at beginning of life, we may want

to take a subset of the master catalog for stars brighter than 8.0 or 9.0 since some of the stars may appear brighter to the star-tracker than to human eyes.
2. Calculate the instrument magnitude for every star.
3. Retrieve or calculate quality attributes and eliminate bad stars. For example:
   a. Eliminate stars with the position error bigger than *max_star_pos_err*, for example, 1 arcsec.
   b. Proper motion bigger than *max_proper_motion*, for example, 0.5 arcsec/year.
   c. Near-neighbor criteria 1: eliminate stars within *maxsep_close_mag* (e.g., 0.2°) with magnitude difference < *mag_close* (e.g., 1 magnitude).
   d. Near-neighbor criteria 2: eliminate stars within *maxsep_far_mag* (e.g., 0.1°) with magnitude difference < *mag_far* (e.g., 2 magnitudes)
4. Perform instrument magnitude subsetting: It is important that this is done after the near-neighbor criteria are applied. Otherwise, some of the stars that should have been eliminated due to a dimmer star in the neighborhood would not have been eliminated.
5. Perform gap analysis and density analysis.
   a. For the parts of sky where there might be too many stars, eliminating some of them, for example, some dimmer ones in the area, may be a wise thing to do (and sometimes may not be, depending on the operational strategy);
   b. For the parts of sky where there are too few stars, try to allow some dimmer stars to fill the gap if possible. This typically only helps during the early part of the spacecraft life when the star-tracker can see dimmer stars.

The Building of a Pair Catalog [19, 21, 36, 46]

Since all the information required for a pair catalog can be derived from the regular catalog, it is often a good idea to generate the pair catalog from the regular star catalog, for example, at the startup of the flight software. There are several advantages of generating the pair catalog on the fly instead of generating that as part of the software and data build:

1. If the pair catalog is generated at the startup of the computer, it means that the pair catalog does not have to occupy any PROM (Programmable Read Only Memory) space, which typically is more expensive and has less capacity. Instead, only the regular catalog, which is smaller, has to be stored in the PROM. This is even more beneficial for many systems that actually copy the PROM contents to the RAM (Random Access Memory) at the beginning of operations in order to gain throughput improvements. In this case, there will be two copies of the regular star catalog, one in the ROM and the other in the RAM. The fact that the pair catalog is generated means that the pair catalog, which typically is much larger, only needs to exist in the RAM, consequently saving a significant amount of PROM space.
2. The pair catalog will be guaranteed to be consistent with the regular star catalog because pair catalog is generated using the regular star catalog.

The algorithm for the actual generation of the pair catalog is also very simple and can be summarized as the following:

1. The pair catalog is organized by a separation angle range and is sorted by the separation angle.
2. For every star, find all the other stars within the separation angle range from the regular catalog. The first star and each of the stars that we found form a pair and each of the pairs is added into the corresponding slot in the pair catalog data structure.
3. Since pair (i, j) and pair (j, i) are the same pair, we only have to create and evaluate pairs where the second index is bigger than the first index.

If the regular catalog has N stars, there are theoretically $N(N-1)/2$ pairs. However, because we are only interested in the pairs with a separation angle between, for example, 0.5 to 8*sqrt(2) degrees. The number of pairs is drastically reduced. Depending on the organization of the regular catalog, it is often easy to quickly eliminate the pairs beyond the maximum separation angle before a detailed evaluation of each of the candidate pairs is performed. For example, if the regular catalog is organized by declination strips, it will be easy to decide to not consider stars that are located in strips where stars are guaranteed to have separation angles larger than the maximum. Similarly, if the stars are sorted within the strip according to the right ascension angle, it will be easy to eliminate the stars out of range as well.

7.2.9 Star-Tracker Orientation and Configuration Considerations

7.2.9.1 More Than One Tracker for Normal Operation

Typically, the first decision that we have to make, regardless of redundancy strategy, is to decide how many star-trackers that we want to use during normal operations.

From the cost and operational convenience point of view, it is highly desirable to use only one star-tracker to support normal operations (regardless of how many are needed to provide backup for failures).

More than one star-tracker is selected to support normal operations often because of one or both of the following reasons:

1. The star-tracker FOV is too small to support reasonable attitude accuracy around the star-tracker boresight. When two star-trackers are used, the large separation angle between the two star-trackers will be able to provide much better dilution of precision.
2. Star-trackers may suffer from bright object intrusions such as the Sun or the Moon during part of the operations and there is no easy solution other than having two or more star-trackers oriented toward different directions and at least one of them not being intruded by a bright object when performance is required.

One easy way to develop an attitude determination error budget for two trackers is to create a virtual star by averaging all the stars in the FOV into an average star for each of the star-trackers and use the two average stars to determine attitude. We already derived how the 3-axis attitude errors are impacted by the separation angle between the two vectors in Chapter 4. The separation angle can be approximated by the boresight separation angle of the two star-trackers.

More specifically, we can define the vector, in the plane spanned by the two average star unit vectors, and in the middle of the separation angle, to be the z-axis. The x-axis is chosen to be normal to the plane, and the y-axis is such that x, y, z form a right-handed body frame defined by the two trackers. Using the equations from Chapter 4, we have the attitude errors as function of the unit vector errors as below

$$\theta_y = -\frac{\left(\delta u_1(1) + \delta u_2(1)\right)}{2\cos\frac{\varphi}{2}}$$

$$\theta_z = -\frac{\left(\delta u_1(1) - \delta u_2(1)\right)}{2\sin\frac{\varphi}{2}}$$

For θ_x (the x-axis is orthogonal to both of the two unit vectors), we have two redundant equations

$$\theta_x = \delta u_1(2)\cos\frac{\varphi}{2} - \delta u_1(3)\sin\frac{\varphi}{2}$$

$$\theta_x = \delta u_2(2)\cos\frac{\varphi}{2} + \delta u_2(3)\sin\frac{\varphi}{2}$$

Assuming that approximately we have

$$\sigma\left(\delta u_1(1)\right) = \sigma\left(\delta u_1(2)\right) = \frac{\sigma_{\text{star\_pos}}}{\sqrt{N_{\text{star}}}}$$

$$\sigma\left(\delta u_2(1)\right) = \sigma\left(\delta u_2(2)\right) = \frac{\sigma_{\text{star\_pos}}}{\sqrt{N_{\text{star}}}}$$

for simplicity, we assume that each of star-tracker has the same number of stars and has the same per-axis star position errors. We then have

$$\sigma\left(\theta_y\right) = \frac{\sigma_{\text{star\_pos}}}{\left(\sqrt{2N_{\text{star}}}\right)\cos\frac{\varphi}{2}}$$

$$\sigma\left(\theta_z\right) = \frac{\sigma_{\text{star\_pos}}}{\left(\sqrt{2N_{\text{star}}}\right)\sin\frac{\varphi}{2}}$$

For the x-axis, assuming that $u_2(3)$ is close to 1 and the deviation of that is a second-order effect (cosine of small angle) and will be ignored. Both of the solutions have the following standard deviation:

$$\sigma\left(\theta_x\right) = \frac{\sigma_{\text{star\_pos}}}{\sqrt{N_{\text{star}}}}\cos\frac{\varphi}{2}$$

Combining the two solutions by averaging, we can have a better estimate with the error reduced by factor of 1/sqrt(2)

$$\sigma(\theta_x) = \frac{\sigma_{\text{star\_pos}}}{\sqrt{2N_{\text{star}}}}\cos\frac{\varphi}{2}$$

The above can be summarized as the R (Ken) Li Formula for Two Star-Tracker Attitude Determination Performance:

$$\sigma(\theta_y) = \frac{\sigma_{\text{star\_pos}}}{\left(\sqrt{2N_{\text{star}}}\right)\cos\frac{\varphi}{2}}, \quad \sigma(\theta_z) = \frac{\sigma_{\text{star\_pos}}}{\left(\sqrt{2N_{\text{star}}}\right)\sin\frac{\varphi}{2}}$$
$$\sigma(\theta_x) = \frac{\sigma_{\text{star\_pos}}}{\sqrt{2N_{\text{star}}}}\cos\frac{\varphi}{2} \tag{7.30}$$

Together with the single star-tracker formula discussed in Chapter 4, one can quickly estimate the performance and make star-tracker configuration decisions.

7.2.9.2 Star-Tracker Orientation

When deciding on star-tracker orientation, typically the following has to be considered:

1. Attitude determination performance for the axis where the attitude knowledge is desired;
2. Intrusions of bright objects such as the Sun, the Moon, and the planets;
3. Intrusions, blockage, reflections, and glint by other equipment on the vehicle into the FOV;
4. Proximity to payloads and relative thermal deformation considerations;
5. Proximity to the IRU because star-trackers are integrated with IRU to provide the SIAD solution; relative thermal deformation between the two sensors is to be considered.

One interesting example is the orientation of the star-trackers for geostationary satellites. We know that the maximum elevation angle of the Sun, the Moon, and the planets is 23.5° above the GEO. Assuming 0.5° maximum inclination, 30° Sun exclusion angle, and 1° of margin, if the star-tracker boresight has an elevation angle of (23.5 + 0.5 + 30 + 1) = 55°, it will then be guaranteed that the star-tracker will not be intruded by the Sun, the Moon, or the planets during normal operations. Obviously, −55° will also do the trick.

The azimuth angle of the star-tracker boresight axis can be used to deal with other issues, mostly related to the clearance from other equipment on the vehicle.

The rotation around the boresight axis can also be used to optimize the performance. When the star-tracker h and v axes are rotated roughly 45° with respect to the orbit velocity direction, both h and v of a star will be changing when the S/C is rotating around the Earth orbit. This will cause the HSF errors to appear as changing errors in the time domain and consequently be attenuated by filtering techniques.

For the LEO, MEO, and HEO satellites, we typically are not that fortunate. The spacecraft attitude steering law will play a very important role when choosing the orientation of the star-trackers. For example, a frequently used steering law, the Sun-nadir steering law, ensures that the S/C z-axis points to the nadir, the solar wing gets maximum power, and there is always a dark side ($-x$ side) that will never see the Sun. It is often convenient to place the star-tracker on the dark side to avoid Sun intrusions.

In general, it is very difficult to avoid the Moon in the FOV. That is why some of the modern star-trackers are designed to continue to track stars even if the Moon is in the FOV. This has been possible when the CMOS (Complementary Metal-Oxide–Semiconductor) active pixel arrays instead CCD arrays are used as the focal plane array devices. The Moon intrusion often drives how accurate the IRU needs to be to keep the attitude accuracy during the period when no star-tracker updates are available.

7.3 Sun Sensor-Based Attitude Determination

7.3.1 The Concept

The Sun sensor measurement can be simply used as a single star measurement. If we know the time, the Sun sensor unit vector in the ECEF or the ECI frame can be calculated very accurately using the Sun ephemeris.

The use of the Sun sensor in this way can provide a much needed augmentation of accuracy, especially when the star-tracker is being intruded by bright objects from time to time. We can imagine that we can use the SIAD attitude determination algorithm that we developed and presented in earlier sections and treat the Sun as just another star and we are done. Typically, if the Sun sensor is used this way, a digital Sun sensor with a wide FOV will be desirable.

However, in many other applications, the use of a Sun sensor is primarily to help the spacecraft to find the Sun (Sun acquisition) and then maintain Sun-pointing to keep the spacecraft power and thermal safe.

In this case, we do not necessarily care about the absolute attitude; we probably also do not care about the attitude around the Sun-sensor boresight axis. Most likely, when we get into the situation that we have to perform Sun acquisition, we do not necessarily have the precise time to calculate the precise Sun unit vector in the ECI or the ECEF frame either. Fortunately, for the purposes of finding the Sun and maintaining Sun pointing, we do not need all that. This leads to the second use of the Sun sensor-based attitude determination: the Sun sensor boresight referenced attitude determination.

Typically, when we lose the Sun, the spacecraft will go through a sequence of maneuvers trying to find the Sun. One way to do that is to try to find the Sun in the Sun sensor FOV. Once the Sun is in the Sun sensor FOV, we can derive a measured Sun unit vector in the Sun sensor frame. If we simply choose the reference vector for the Sun to be ${}^{R}u_{sun} = [0,0,1]^T$, then the attitude of the Sun sensor frame with respect to the reference will be $q_R^S = [0,0,0,1]^T$ or expressed in DCM as $C_R^S = I$ (identity matrix) when the sun is at the boresight of the Sun sensor. At this moment, both the reference frame Sun unit vector and the Sun sensor frame Sun unit vector are $[0, 0, 1]^T$.

In order to do that, in the first moment that the Sun is in the FOV, we can calculate the Sun sensor frame attitude with respect to the Sun sensor boresight reference by using a pair of vectors, which will be used to reset the attitude quaternion.

The S/C will be trying to null the difference between the current attitude and the target attitude $q_R^S = [0,0,0,1]^T$ and in the end the boresight axis of the spacecraft will be Sun-pointing.

7.3.2 The Sun Sensor as an Additional Sensor for Normal Attitude Determination

As discussed before, the Sun sensor measurements can be used to augment an existing attitude determination system such as an SIAD system. We can simply use the existing SIAD architecture and treat the Sun as a star (and it is).

The Sun unit vector in the ECI frame will be calculated based on time and Sun ephemeris:

$$^{I}u_{sun} = \text{sun\_eph}(t)$$

The Sun vector in the Sun sensor frame and the Sun position in the FOV can be predicted using the following equations

$$^{SS}u_{sun} = C_B^{SS} C_I^{B}\,{}^{I}u_{sun}$$

$$\begin{bmatrix} h_{sun} \\ v_{sun} \end{bmatrix} = \begin{bmatrix} \dfrac{^{SS}u_{sun}(1)}{u_{sun}(3)} \\ \dfrac{^{SS}u_{sun}(2)}{u_{sun}(3)} \end{bmatrix}$$

where SS stands for the Sun sensor frame and $\begin{bmatrix} h_{sun} \\ v_{sun} \end{bmatrix}$ is the Sun position in the Sun sensor FOV.

The Sun sensor measurement is modeled by the following equation

$$\begin{bmatrix} mh_{sun} \\ mv_{sun} \end{bmatrix} = \begin{bmatrix} h_{sun} \\ v_{sun} \end{bmatrix} + \begin{bmatrix} v_{ssh} \\ v_{ssv} \end{bmatrix} + \begin{bmatrix} b_{ssh} \\ b_{ssv} \end{bmatrix}$$

where $\begin{bmatrix} v_{ssh} \\ v_{ssv} \end{bmatrix}$ is the Sun sensor measurement noise and $\begin{bmatrix} b_{ssh} \\ b_{ssv} \end{bmatrix}$ is the Sun sensor boresight offset errors.

The boresight offset errors are only important when Sun sensor is used together with other sensors such as a star-tracker. If we use star-tracker as the reference, then the misalignment due to the relative alignment knowledge between the star-tracker and the Sun sensor are accounted for by this term. This variable can be added as additional Kalman filter states so that when the attitude based on star-tracker has high accuracy, the bias can be estimated. The estimated values will be used when star-tracker becomes unavailable and the Sun sensor will continue to provide attitude corrections based on the same reference.

The linearized measurement model is given by the following equations:

$$\begin{bmatrix} \delta mh_{sun} \\ \delta mv_{sun} \end{bmatrix} = H_1 \delta^I \theta_B^I + \begin{bmatrix} b_{ssh} \\ b_{ssv} \end{bmatrix} + \begin{bmatrix} v_{ssh} \\ v_{ssv} \end{bmatrix}$$

where

$$H_1 = \frac{1}{\sqrt{1 + h^2 + v^2}} \begin{bmatrix} -1 & 0 & h \\ 0 & -1 & v \end{bmatrix} \left[{}^S u_{sun} \times \right] C_I^S$$

The state vector for SIAD is augmented to include the Sun sensor boresight offset vector:

$$b_{ss} = \begin{bmatrix} b_{ssh} \\ b_{ssv} \end{bmatrix}$$

$$X^T = \left[{}^I\dot{\theta}_B^I, {}^B b^T, {}^S s^T, {}^B \beta^T, {}^B \theta^T, b_{ss} \right]$$

$$W^T = \left[-\left(C_G^I ARW \right)^T, \ RRW^T, \ SRW^T, \ \beta RW^T, \ \theta RW^T, \ bssRW^T \right]$$

where $bssRW$ is the white noise process that generates the random walk process, which models the Sun sensor boresight offset vector.

The linear differential equations shown below are the linearized error propagation model.

$$\dot{X} = AX + W$$

$$A = \begin{bmatrix} 0_{3x3} & -C_B^I & -C_G^I \mathrm{diag}\left({}^G\omega_I^B \right) & -C_B^I J\left({}^B\omega_I^B \right) & C_B^I \left[{}^B\omega_I^B \times \right] & 0_{3x2} \\ 0_{3x3} & 0_{3x3} & 0_{3x3} & 0_{3x3} & 0_{3x3} & 0_{3x2} \\ 0_{3x3} & 0_{3x3} & 0_{3x3} & 0_{3x3} & 0_{3x3} & 0_{3x2} \\ 0_{3x3} & 0_{3x3} & 0_{3x3} & 0_{3x3} & 0_{3x3} & 0_{3x2} \\ 0_{3x3} & 0_{3x3} & 0_{3x3} & 0_{3x3} & 0_{3x3} & 0_{3x2} \\ 0_{2x3} & 0_{2x3} & 0_{2x3} & 0_{2x3} & 0_{2x3} & 0_{3x2} \end{bmatrix}$$

By now, we have sufficient information to build the Kalman filter.

7.3.3 The Sun Sensor for Attitude Determination to Support Sun-Pointing and Sun Acquisition

For Sun acquisition and Sun-pointing, accuracy is typically not a concern. In this case, it is more important to keep the design simple than to use fancy designs to optimize the performance. Because of this consideration, we will use a fixed gain filter to provide attitude update instead of using a Kalman filter used for the other attitude determination applications.

The basic idea is very simple. We will use the TRIAD algorithm discussed in Chapter 4 first to determine the attitude and we will then use the attitude determined as the measurement that will provide corrections to the attitude state.

For the simplicity, we will only keep gyro bias and gyro ARW in the gyro error model.

We will use the same exact attitude propagation algorithm and the corresponding third-order numerical integration algorithm as described in Section 7.2.2:

$$\dot{q}_{SS}^{I} = \frac{1}{2} q_{SS}^{I} \otimes {}^{SS}\omega_{I}^{SS} = \frac{1}{2} q_{SS}^{I} \, {}^{SS}\omega_{I}^{SS}$$

To implement a fixed gain filter, we will need to use the attitude error kinematics equation expressed in the Sun sensor body frame using (3.116)

$$\delta\, {}^{SS}\dot{\theta}_{I}^{SS} = -\left[{}^{SS}\omega_{I}^{SS} \times \right] \delta\, {}^{B}\theta_{I}^{SS} + \delta\, {}^{SS}\omega_{I}^{SS} \tag{7.31}$$

We will use the following error model for the IRU

$$\delta\, {}^{SS}\omega_{I}^{SS} = m\, {}^{SS}\omega_{I}^{SS} - {}^{SS}\omega_{I}^{SS} = {}^{SS}b_{g} + C_{G}^{SS} ARW$$

The attitude determination linearized error model can then be described as

$$\begin{gathered} X = \begin{bmatrix} {}^{SS}\theta_{I}^{SS} \\ {}^{SS}b_{G} \end{bmatrix}, \quad W = \begin{bmatrix} C_{G}^{B} ARW \\ RRW \end{bmatrix} \\ \dot{X} = AX + W, \quad A = \begin{bmatrix} -\left[{}^{SS}\omega_{I}^{SS} \times \right] & I_{3x3} \\ 0_{3x3} & 0_{3x3} \end{bmatrix} \end{gathered} \tag{7.32}$$

and the measurement equation is simply

$$mq_{SS}^{I} = q_{SS}^{I} + v$$

and the linearized model is given by

$$m\delta\, {}^{SS}\theta_{I}^{SS} = \delta\, {}^{SS}\theta_{I}^{SS} + v$$

We can design a fixed gain filter to update the state in the following way

$$\frac{d}{dt}\hat{X} - A\hat{X} + K\left(m\delta\, {}^{SS}\theta_{I}^{SS} - \delta\, {}^{SS}\hat{\theta}_{I}^{SS} \right) = A\hat{X} + K\left(\Delta\delta\, {}^{SS}\theta_{I}^{SS} \right)$$

One way to design the gain matrix K is to use Kalman filter steady state solution gains where $m\delta^{SS}\theta_{I}^{SS}$ can be calculated from

$$\begin{aligned} m\Delta q &= mq_{I}^{SS} \otimes \left(\hat{q}_{SS}^{I} \right) \\ m\delta\, {}^{SS}\theta_{I}^{SS} &= 2m\Delta q(1{:}3) \end{aligned} \tag{7.33}$$

and mq_{I}^{SS} is the measured attitude quaternion using TRIAD algorithm.

Note that the sensor-measured Sun unit vector and the boresight reference unit vector only form one pair of unit vectors. In order to use the TRIAD algorithm, we will need a second pair of unit vectors.

The essence of the problem that we are facing is that the Sun unit vector measurement only constrains the attitude around one axis. We have complete freedom to rotate around the Sun unit vector while still pointing to the Sun. In order not to add additional constraints, we will create a pair of a reference unit vector and a measured unit vector that is consistent with the current attitude estimate.

Since the Sun sensor unit vector in the reference frame has been chosen to be

$$^{I}u_{sun} = [0,0,1]^T$$

where I is the Sun sensor boresight reference frame (treated as an inertial frame).

We then choose a fictitious sensor such that

$$^{I}u_f = [0,1,0]^T$$

and the cross-product will be

$$^{I}u_f \times {}^{I}u_{sun} = [1,0,0]^T$$

Since we do not want the fictitious vectors to actually impose any new constraints, the SS frame unit vector is chosen to be transformed by the current attitude knowledge as

$$^{SS}u_f = C_I^{SS}\begin{bmatrix} 0 \\ 1 \\ 0 \end{bmatrix} = C_I^{SS}(:,2)$$

The new attitude DCM can then be calculated using the TRIAD algorithm as

$$mC_I^{SS} = \left[\frac{^{SS}u_f \times {}^{SS}u_{sun}}{\left\| ^{SS}u_f \times {}^{SS}u_{sun} \right\|}, {}^{SS}u_f, {}^{SS}u_{sun}\right]\left[\begin{bmatrix} 1 \\ 0 \\ 0 \end{bmatrix}, \begin{bmatrix} 0 \\ 1 \\ 0 \end{bmatrix}, \begin{bmatrix} 0 \\ 0 \\ 1 \end{bmatrix}\right]^{-1} = \left[\frac{^{SS}u_f \times {}^{SS}u_{sun}}{\left\| ^{SS}u_f \times {}^{SS}u_{sun} \right\|}, {}^{SS}u_f, {}^{SS}u_{sun}\right]$$

Because of our smart choices of the reference vectors, the computation of the DCM is greatly simplified.

We can then convert the DCM to a quaternion using algorithms discussed in Chapter 2.

$$mq_I^{SS} = dcm2quaterion\left(C_I^{SS}\right)$$

7.4 GPS Carrier-Phase-Based Attitude Determination

7.4.1 Introduction

The GPS carrier-phase-based attitude determination has been developed and implemented since the 1990s. A few commercial implementations were available. Some of the famous ones were Ashtech (Magellan) 3DF receiver and Trimble Navigation's

TANS vector receiver. The approach has been used by several spacecraft systems as a coarse attitude determination method.

The most important lesson learned from the past practices has been that the carrier-phase multipath is the most important error contributor. Because of the multipath error, the performance is often many times worse than advertised.

In general, the GPS carrier phase-based attitude determination does not provide the optimized balance between performance, cost, and SWAP and is inferior to other methods such as SIAD discussed earlier in this chapter. This topic is included here not because the technology is very useful, but because some of the ideals and methodology are still interesting and educational.

Figure 7.3 illustrates the basic principle of GPS carrier-phase-based attitude determination.

Three GPS antennas define a plane, which also defines a coordinate frame. The vectors connecting any two antennas in a body frame are known vectors in the body frame (which can be measured using a ruler, for example). The same vectors can also be measured using a GPS carrier phase in the ECEF frame or in the ECI frame.

Given two known pairs of vectors in the body frame and in the reference frame such as the ECEF or the ECI frame, the attitude of body frame with respect to the reference frame can be determined using the TRIAD algorithm discussed in Chapter 4.

With the above discussion, attitude determination using the GPS carrier phase is essentially equivalent to precise relative position determination. However, the biggest difference from a typical carrier-phase differential GPS-based relative positioning problem is that this relative positioning problem is much easier because we precisely know the length of the vectors. This is a very tight constraint and, consequently, the ambiguity resolution problem becomes much easier [13, 45].

The key contributors to the performance errors are listed here:

1. Receiver noise. Typically, this is not a problem and can be reduced by averaging or filtering.
2. Multipath. This is a performance driver, typically correlated over time and hard to average out. Multipath is also hard to model and estimate. The typical order of magnitude is 2 to 20 mm.

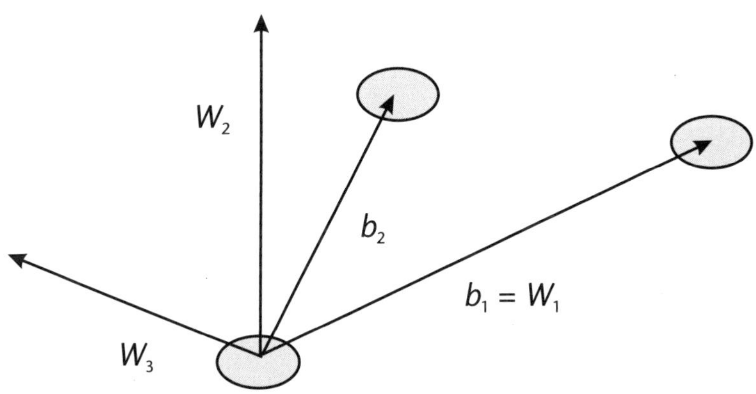

Figure 7.3 Three GPS antennas define a plane and a coordinate system.

3. Antenna phase center stability.
4. Thermal deformation between antennas.
5. Structural vibration between antennas.
6. Baseline distance.

The multipath error and the baseline distance are by far the most important two contributors and the following equation will provide a good rough estimate of the attitude error determined using the GPS carrier phase:

$$\sigma_{\text{att\_err}} = \frac{\sigma_{\text{multi-path}}}{L} \tag{7.34}$$

where L is the baseline length.

For example, if the baseline distance is 1m and the multipath error is 5 mm, the attitude error will be on the order 5 mrad. If the multipath is 20 mm, the attitude error will become 20 mrad, which is more than 1°. This is why GPS carrier-phase-based attitude determination is not very useful for precision applications.

7.4.2 Architecture

Figure 7.4 is an example architecture when GPS carrier-phase attitude determination is used to aid an inertial-based attitude determination system.

Within the architecture, there are three attitude determination solutions fused together:

1. The IRU (gyros)-based inertial attitude propagation (the propagation solution);
2. The GPS carrier-phase-based update through the Kalman filter (normal update solution) and carrier-phase ambiguity maintenance;

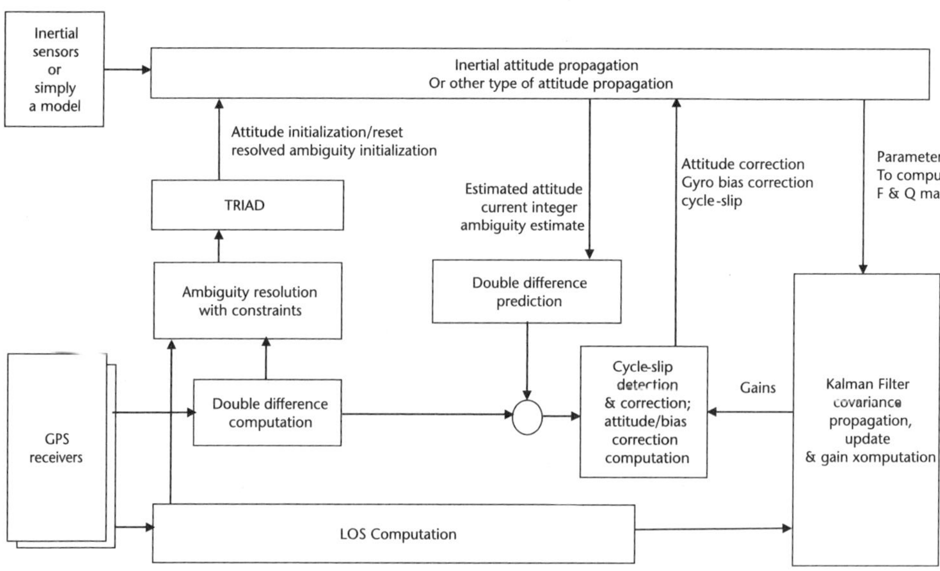

Figure 7.4 Architecture of integrated inertial and GPS carrier-phase attitude determination system.

3. The initial attitude determination (attitude acquisition) and initial carrier-phase integer ambiguity resolution.

Typically, the initial attitude determination and initial ambiguity resolution only have to be done once and we should be able to maintain the resolved integers across cycle slips and GPS outages because we can maintain the attitude knowledge well using the gyro data.

7.4.3 Attitude Initialization and Initial Ambiguity Resolution

We first show how attitude can be determined once we determined the baseline vectors between antenna phase centers.

With at least three antennas, we can construct at least two vectors known in the body frame:

$$^{B}b_1, \quad ^{B}b_2$$

The reference frame (ECEF or ECI) version of these two vectors can be measured using GPS carrier-phase relative positioning techniques and we get

$$^{R}b_1, \quad ^{R}b_2$$

By definition, we have

$$C_B^R \, ^{B}b_1 = {}^{R}b_1, \quad C_B^R \, ^{B}b_2 = {}^{R}b$$

We can then construct the attitude DCM C_R^B using the TRIAD algorithm. Construct

$$^{B}w_1 = {}^{B}b_1, \quad ^{R}w_1 = {}^{R}b_1, \quad ^{B}w_2 = \frac{\left(^{B}b_1 \times {}^{B}b_2\right)}{\left\|^{B}b_1 \times {}^{B}b_2\right\|}, \quad ^{R}w_2 = \frac{\left(^{R}b_1 \times {}^{R}b_2\right)}{\left\|^{R}b_1 \times {}^{R}b_2\right\|}$$

and

$$^{B}w_3 = \frac{\left(^{B}w_1 \times {}^{B}w_2\right)}{\left\|^{B}w_1 \times {}^{B}w_2\right\|}, \quad ^{R}w_2 = \frac{\left(^{R}w_1 \times {}^{R}w_2\right)}{\left\|^{R}w_1 \times {}^{R}w_2\right\|}$$

Form the matrices

$$^{B}W = \left[^{B}w_1, {}^{B}w_2, {}^{B}w_3\right], \quad ^{R}W = \left[^{R}w_1, {}^{R}w_2, {}^{R}w_3\right]$$

Then the attitude DCM can be computed as

$$C_B^R = \left(^{R}W\right)\left(^{B}W\right)^T$$

As a result, the attitude determination problem becomes the problem of relative position vector determination.

Let the carrier-phase measurement for satellite k and antenna j be given by

$$\varphi(i,j) = R(i,j) + ct_r(i) + ct_s(j) - I(i,j) + T(i,j) + v(i,j) + m(i,j) + \lambda N(i,j) \tag{7.35}$$

The meanings of the terms in the above equation are listed here:

$R(i, j)$: True range between antenna/receiver i to GPS satellite j;
$ct_r(i)$: Range error due to receiver i clock error;
$ct_s(j)$: Range error due to satellite j clock error;
$I(i, j)$: Ionosphere delay between antenna or receiver i and satellite j;
$T(i, j)$: Troposphere delay between antenna or receiver i and satellite j;
$v(i, j)$: Receiver carrier-phase noise;
$m(i, j)$: Carrier-phase multipath error;
$n(i, j)$: Carrier-phase integer ambiguity.

We can form the single difference of the carrier phase for the same satellite and different receivers (antennas) to cancel out ionosphere delays, troposphere delays, and GPS satellite clock errors:

$$SD_\varphi = \Delta\varphi\left(i_1, i_2, j\right) = R\left(i_1, j\right) - R\left(i_2, j\right) + \Delta ct_r + \Delta v + \Delta m + \Delta N\lambda \tag{7.36}$$

We then form the double difference, the difference of single differences for different satellites to cancel out receiver clock errors (for simplicity, the indices are omitted):

$$DD_\varphi = \delta\Delta\varphi = \delta\Delta R + \delta\Delta v + \delta\Delta m + \delta\Delta N\lambda \tag{7.37}$$

Using the techniques in Chapter 4, we can easily form the linearized equations for the single and the double differences:

$$SD_i = -l_i^T b + \Delta v + \Delta m + \Delta N\lambda + ct_r \tag{7.38}$$

$$DD_{i,j} = -\left(l_i^T - l_j^T\right)b + \delta\Delta v + \delta\Delta m + \delta\Delta N\lambda \tag{7.39}$$

The equations are geometrically meaningful and are illustrated by Figure 7.5.

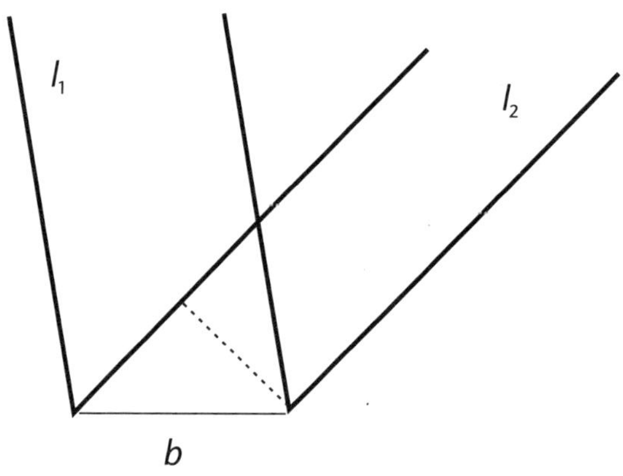

Figure 7.5 Geometric meaning of single and double-difference equations.

We can use the following standard vector form equation:

$$DD = Lb + V + M + N\lambda \tag{7.40}$$

where

DD: The vector formed by stacking the double differences together.

L: This is formed by differencing the LOS vectors and stacking them together.

N: This is the vector of unknown double-difference integer ambiguities.

$V + M$: This is formed by stacking the double-difference receiver noise and multipath errors.

The above equation forms the basis for the solution of the baseline vector by a search + least square solution approach. The search algorithm is described here:

1. Use the current knowledge of the baseline vector and determine the possible range of the integer ambiguity.
2. For each of the integer ambiguity candidates:
   a. Solve the equation $(DD - N_*\lambda) = Lb$ by least square method $\hat{b} = \left(L^T L\right)^{-1} L\left(DD - N_*\lambda\right)$ where N_* is the candidate integer ambiguity vector.
   b. The candidate integer vector is considered preliminarily verified if the following two conditions are verified:
      (i) The known vector length constraints are satisfied with sufficiently small errors that are consistent with the current dilution of precision of the least square problem.
      (ii) The residuals show that all the double-difference measurement equations are consistently satisfied.
   c. We typically want to continue to try other candidates to create more preliminarily verified solutions.
3. We carry this for several epochs of GPS data to ensure that the preliminarily verified integer solutions survive the checks over several epochs and the one that has the least error is selected.

By leveraging the known vector length constraints, the integer ambiguity resolution process is typically much faster than a generic relative positioning problem.

This is especially true when we already know some part of the attitude. For example, for most of terrestrial navigation problems, we can easily determine the roll and pitch attitude angles using accelerometer data through the coarse leveling process. In this case, what we really need is to determine the heading.

In 1996, the author published [65] for exactly this topic. This article provides an extremely fast ambiguity resolution approach for a special case of attitude determination, the heading determination. We will describe the key results of this article in this section.

Given two GPS antennas, a vector spanning the two antennas can be used to determine the heading. In the local-level navigation frame (Nav frame), the vector can be expressed as the function of the elevation angle and the heading angle in the local-level frame:

$$
\begin{aligned}
&{}^{N}b = \begin{bmatrix} b\cos\theta\sin\psi & b\cos\theta\cos\psi & b\sin\theta \end{bmatrix}^{T} \\
&\text{where} \quad b = \left\|{}^{B}b\right\| = \left\|{}^{N}b\right\|
\end{aligned} \tag{7.41}
$$

If the vector b is determined in the Nav frame, both the heading and the elevation angle are determined (see Figure 7.6). If neither the elevation angle nor the heading angle is known, then the vector b is constrained on the sphere. If the elevation angle is known, the vector b is constrained on a small circle (vs. the great circle) on the sphere.

We will assume that, for a period of time, the baseline vector in the Nav frame is not changing (for example, a period of initial alignment).

Stacking the double-difference equations together, we have

$$
DD = Lb + V + ZN\lambda \tag{7.42}
$$

where $Z = I$ for a single epoch and $Z = [I, I, \cdots, I]^T$ for multiple epochs. The use of the matrix Z is to recognize that the integer ambiguity remains unchanged across the epochs.

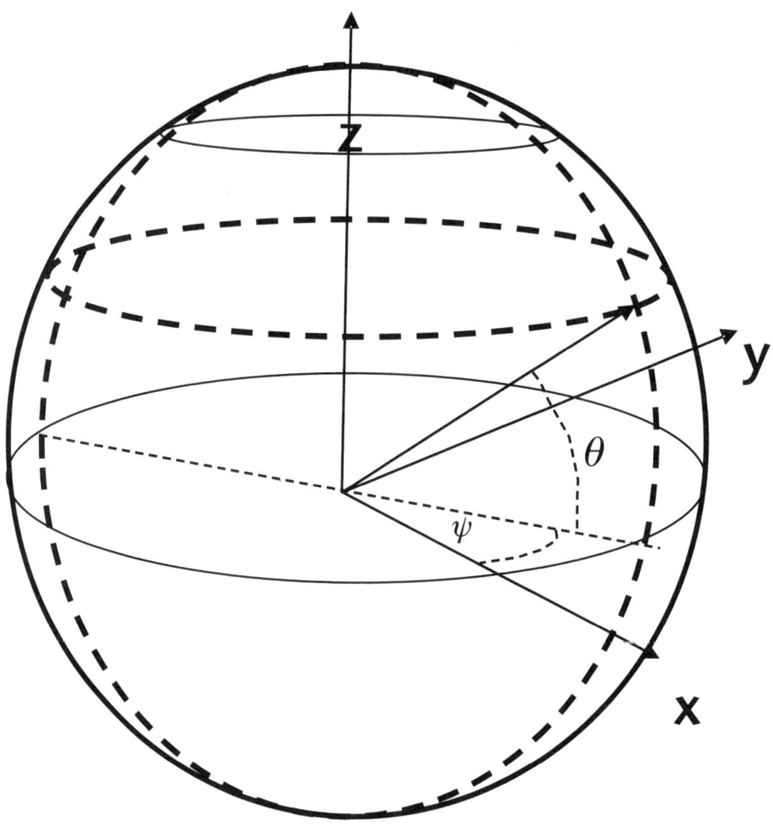

Figure 7.6 Heading determination problem.

Let $Z^{+} = (Z^{T}Z)^{-1}Z^{T}$. Z^{+} is simply the average operator. In other words, we will simply average multiple epoch of data.

The basic double-difference equation can be written as

$$Z^{+}DD = Z^{+}L^{N}b + Z^{+}V + N\lambda \tag{7.43}$$

Using

$${}^{N}b = \begin{bmatrix} b\cos\theta\sin\psi & b\cos\theta\cos\psi & b\sin\theta \end{bmatrix}^{T},$$

we can easily write the equation element-by-element:

$$\begin{aligned}\left(Z^{+}DD\right)(k) - \lambda N(k) &= b\left(Z^{+}L^{N}\right)(k,1)\cos\theta\sin\psi \\ &+ b\left(Z^{+}L^{N}\right)(k,2)\cos\theta\cos\psi + b\left(Z^{+}L^{N}\right)(k,3)\sin\theta + Z^{+}V\end{aligned}$$

Combining the $\sin\psi$ and $\cos\psi$ terms into a single term, we have

$$\left(Z^{+}DD\right)(k) - \lambda N(k) = bL_{12}(k)\cos\theta\sin\left(\psi + \psi_{0}(k)\right) + b\left(Z^{+}L^{N}\right)(k,3)\sin\theta + Z^{+}V \tag{7.44}$$

where

$$\begin{aligned} L_{12}(k) &= \sqrt{\left(\left(Z^{+}L^{N}\right)(k,1)\right)^{2} + \left(Z^{+}L^{N}\right)(k,2))^{2}}, \\ \psi_{0} &= \tan^{-1}\left(\left(Z^{\prime}L^{N}\right)(k,2),\left(Z^{+}L^{N}\right)(k,1)\right)\end{aligned} \tag{7.45}$$

Consequently, we have

$$\sin\left(\psi + \psi_{0}(k)\right) = \frac{\left(Z^{+}DD\right)(k) - \lambda N(k) - b\left(Z^{+}L^{N}\right)(k,3)\sin\theta}{bL_{12}(k)\cos\theta} \tag{7.46}$$

If the integer is known and the vector $(Z^{+}L^{N})(k)$ is not aligned with the local vertical axis, the heading can be determined from the double difference formed by a single pair of GPS satellites.

The basic equations have $M + 1$ unknowns, 1 heading angle, and M number of integer ambiguities. If we guess on only 1 of the M integer ambiguities, then we have M equations and M unknowns. In other words, we can solve for the heading angle and the rest of the integer ambiguities. We will then check if the solution for the integer ambiguity is close to integers; if any one of the solutions is not, then the guess of the integer ambiguity is invalidated. This gives us a 1-D search algorithm, which is extremely fast.

Since there are M integers, we can choose one of them as the search target. Since we have many choices, we can select the integer that allows the best observability of the heading angle. Based on (7.46), we want to choose k such that $|L_{12}(k)\cos\theta|$ is maximized.

In the following, we derive the upper and lower bounds for an integer ambiguity so that we can determine the search range.

Recall that we have

$$\left(Z^{+}DD\right)(k) - \lambda N(k) = bL_{12}(k)\cos\theta\sin\left(\psi + \psi_0(k)\right) + b\left(Z^{+}L^{N}\right)(k,3)\sin\theta + Z^{+}V$$

Solving for the integer ambiguity, we have

$$\lambda N(k) = \left(Z^{+}DD\right)(k) - bL_{12}(k)\cos\theta\sin\left(\psi + \psi_0(k)\right) - b\left(Z^{+}L^{N}\right)(k,3)\sin\theta + Z^{+}V$$

The upper and lower bounds can be easily found:

$$\begin{aligned} N(k) &< \frac{1}{\lambda}\left\{\left(Z^{+}DD\right)(k) - b\left(Z^{+}L^{N}\right)(k,3)\sin\theta + \left|bL_{12}(k)\cos\theta\right| + \max Z^{+}V\right\} = \max N(k) \\ N(k) &> \frac{1}{\lambda}\left\{\left(Z^{+}DD\right)(k) - b\left(Z^{+}L^{N}\right)(k,3)\sin\theta - \left|bL_{12}(k)\cos\theta\right| + \min Z^{+}V\right\} = \min N(k) \end{aligned} \tag{7.47}$$

Now we can summarize the 1-D integer ambiguity search algorithm:

1. Choose k as discussed before.
2. For each integer $N(k)$ in the search range $[\min N(k), \max N(k)]$:
   a. We can solve the following equation for two solutions of ψ:

$$\sin\left(\psi + \psi_0(k)\right) = \frac{\left(Z^{+}DD\right)(k) - \lambda N(k) - b\left(Z^{+}L^{N}\right)(k,3)\sin\theta}{bL_{12}(k)\cos\theta}$$

   b. For each ψ, compute the b vector as

$${}^{N}b = \begin{bmatrix} b\cos\theta\sin\psi & b\cos\theta\cos\psi & b\sin\theta \end{bmatrix}^{T}$$

   c. Since everything is known except for the rest of integers, we solve for an estimate of the rest of the integer ambiguity vector:

$$fN = \frac{1}{\lambda}Z^{+}\left(DD - L^{N}b\right)$$

   d. We then use hypothesis testing to check if each of the estimate of integers is close enough to integers; if not, then the candidate integer $N(k)$ is rejected (for details on how this is done, see [45]).
3. Continue the above for new GPS data until only one $N(k)$ is left.

We should point out that most of the algorithms before this approach was developed use multidimensional searches and, consequently, the search space is very large. A typical search algorithm requires $20^4 = 160{,}000$ searches. The 1-D search, in a typical case, only requires 20 searches and consequently is 8,000 times faster.

7.4.4 GPS Carrier-Phase Measurements as Update and Ambiguity Maintenance

Once we have the first solution and the initial ambiguities are resolved, we should not need to perform an integer ambiguity search again. Instead, we will just need to maintain them.

In addition, we no longer need to explicitly solve for attitude using the TRIAD algorithm; instead, we can simply use the double differences as measurements and use that as the updates to the Kalman filter.

We will use the same attitude propagation algorithm, the IRU models, and the linearized propagation equations to build the Kalman filter. In this section, we are only concerned with the measurement equations and their linearizations.

Given the current attitude knowledge $\hat{C}_B^R$ and the baseline vectors in the body frame, the two baseline vectors in the reference frame can then be predicted as

$$\hat{C}_B^{R\,B}b_1,\ \ \hat{C}_B^{R\,B}b_2$$

With the last resolved ambiguity vector N_1 and N_2, we can predict the double-difference measurements

$$\begin{aligned} {}^R\hat{D}D_1 &= L_1\hat{C}_B^{R\,B}b_1 + N_1\lambda, \\ {}^R\hat{D}D_2 &= L_2\hat{C}_B^{R\,B}b_2 + N_2\lambda \end{aligned} \tag{7.48}$$

The actual measurement is modeled as

$$\begin{aligned} m^R DD_1 &= L_1 C_B^{R\,B}b_1 + N_1 + \left(N_{1\text{cycle\_slip}}\right)\lambda + V_1 \\ m^R DD_2 &= L_2 C_B^{R\,B}b_1 + N_2 + \left(N_{2\text{cycle\_slip}}\right)\lambda + V_2 \end{aligned} \tag{7.49}$$

where $(N_{1\text{cycle\_slip}})$, $(N_{2\text{cycle\_slip}})$ are the number of cycles the carrier-tracking loop may have slipped.

The double-difference residual is modeled as

$$\delta^R DD_i = m^R DD_i - {}^R\hat{D}D_2 = L_i\left(\delta^R\theta_B^R\times\right)^b b_i + N_{i\text{cycle\_slip}}\lambda + V_i \tag{7.50}$$

which can be written as

$$\delta^R DD_i = -L_i\left[{}^R b_i\times\right]\delta^R\theta_B^R + N_{i\text{cycle\_slip}}\lambda + V_i \tag{7.51}$$

The contribution of the attitude error to the measurement after the attitude is initialized is $L_i(\delta^R\theta_B^R\times)^b b_i$. Since the attitude knowledge is maintained from frame to frame using the IRU data, we assume that the IRU-introduced error is significantly less than 1 wavelength (almost always true). The carrier-phase error V is typically much smaller than 1 cycle of the wavelength. Consequently, cycle slip can be easily detected and corrected.

The detection can be done by simply checking the value of $\delta^R DD_i$. If it is much smaller than λ, then no cycle slips has occurred. Otherwise, we will find an integer

N_{icycle_slip} such that value of $\delta^R DD_i$ is minimized (i.e., we will detect and correct the cycle slips).

After the cycle slips are corrected, the residual equation is then

$$\delta^R DD_i = -L_i\left[{}^R b_i \times\right]\delta^R \theta_B^R + V_i \tag{7.52}$$

which can be used by the Kalman filter to provide updates.

References

[1] Wertz, J. R., *Spacecraft Attitude Determination and Control*, Boston, MA: D. Reidel Publishing, 1986.

[2] Shuster, M. D., "A Survey of Attitude Representations," *The Astronautically Science*, Vol. 41, 1993, pp. 439–517.

[3] Mekky, T. A. H., "A Comparative Study of Spacecraft Attitude Determination and Estimation Algorithms (A Cost-Benefit Approach)," *Aerospace Science and Technology*, Vol. 26, No. 1, April-May 2013, pp. 211–215.

[4] Mohd, Z. H., et al., "Review on Attitude Estimation Algorithm of Attitude Determination System," *ARPN Journal of Engineering and Applied Sciences*, Vol. 11, 2016, pp. 4455–4460.

[5] Sidi, M. I., *Spacecraft Dynamics and Control*, Cambridge, U.K.: Cambridge University Press, 1997.

[6] Shuster, M. D., and S. D. Oh, "Three Axis Attitude Determination from Vector Observations," *Guid. Control*, Vol. 4, 1981, pp. 70–77.

[7] Simon, D., *Optimal State Estimation*, New York: John Wiley & Sons, 2006.

[8] Bar-Itzhack, I. Y., and R. R. Harman, "Optimized Triad Algorithm for Attitude Determination," *Guidance Control and Dynamics*, May 23, 2012, pp. 208–211.

[9] Savage, P. G., "Strapdown Inertial Navigation Integration Algorithm Design Part 1: Attitude Algorithms," *Journal of Guidance, Control, and Dynamics*, Vol. 21, No. 1, 1998, pp. 19–28.

[10] Savage, P. G., "Strapdown Inertial Navigation Integration Algorithm Design Part 2: Velocity and Position Algorithms," *Journal of Guidance, Control, and Dynamics*, Vol. 21, No. 2, March–April 1998.

[11] Gelb, M., *Applied Optimal Estimation*, Cambridge, MA: MIT Press, 1974.

[12] Brown, R. G., and P. Hwang, *Introduction to Random Signals and Applied Kalman Filtering*, New York: John Wiley & Sons, 1996.

[13] Li, R., and A. Wu, "Kalman Filtering & Applications," *UCLA Lecture Notes*, 2005–2019.

[14] Julier, S. J., J. K. Uhlmann, and H. Durrant-Whyte, "A New Approach For Filtering Nonlinear System," *Proc. of the American Control Conference*, 1995, pp. 1628–1632.

[15] Julier, S. J., and J. K. Uhlmann, *General Method for Approximating Nonlinear Transformation of Probability Distributions*, Technical Report RRG, Department of Engineering Science, University of Oxford, November 1996.

[16] Li, R., et al., "Modified Kalman Filter for Generation of Attitude Error Corrections," U.S. Patent 8,543,266, October 2011.

[17] Li, R., et al., "Real-Time Refinement Method of Spacecraft Star-Tracker Alignment Estimates," U.S. Patent 8,019,544, January 2005.

[18] Li, R., et al., "Method for Compensating Star Motion Induced Error in a Stellar Inertial Attitude Determination System," U.S. Patent 7,487,016, September 2005.

[19] Li, R., et al., "Fast Access, Low Memory, Pair Catalog," U.S. Patent 7,310,578, June 2004.

[20] Li, R., et al., "Spacecraft Gyro Calibration System," U.S. Patent 7,185,858, November 2003.

[21] Li, R., et al., "Method and Apparatus for On-Board Autonomous Pair Catalog Generation," U.S. Patent 7,136,752, June 2004.

[22] Li, R., et al., "Relative Attitude Estimator for Multi-Payload Attitude Determination," U.S. Patent 7,124,001, June 2004.

[23] Li, R., et al., "Refinement of Spacecraft Angular Velocity and Attitude Estimates Using Star Data," U.S. Patent 7,062,363, September 2004.

[24] Li, R., et al., "Attitude-Acquisition Methods and Systems for Controlled Spacecraft Attitude," U.S. Patent 6,766,227, November 2002.

[25] Li, R., et al., "System and Method for Calibrating Inter-Star-Tracker Misalignments in a Stellar Inertial Attitude Determination System," U.S. Patent 6,691,033, July 2000.

[26] Li, R., et al., "Attitude Determination System and Method with Outer-Loop Gyro Scale-Factor Non-Linearity Calibration," U.S. Patent 6,615,117, November 2001.

[27] Li, R., et al., "Attitude Control Methods and Systems for Multiple-Payload Spacecraft," U.S. Patent 6,595,469, October 2001.

[28] Li, R., et al., "Spacecraft Methods and Structures for Acquiring and Determining Power-Safe Attitudes" U.S. Patent 6,588,708, January 2001.

[29] Li, R., et al., "System and Method for Autonomous Correction of Star-Tracker High Spatial Frequency (HSF) Error," U.S. Patent 6,566,640, September 2001.

[30] Li, R., et al., "Lost-in-Space Stellar Attitude Acquisition Using Multiple Star Trackers," U.S. Patent 6,512,979, June 2002.

[31] Li, R., et al., "Method and Apparatus for Spacecraft Antenna Beam Pointing Correction," U.S. Patent 6,504,502, January 2000.

[32] Li, R., et al., "Lost-in-Space Stellar Attitude Acquisition Using Multiple Star Trackers," U.S. Patent 6,470,270, November 2001.

[33] Li, R., et al., "Method and Apparatus for Spacecraft Payload Pointing Registration," U.S. Patent 6,441,776, January 2000.

[34] Li, R., et al., "System for Stellar Inertial Attitude Determination," U.S. Patent 6,362,464, February 2000.

[35] Li, R., et al., "Stellar Attitude-Control Systems and Methods with Weighted Measurement-Noise Covariance Matrices," U.S. Patent 6,356,815, August 2000.

[36] Li, R., et al., "System for Star Catalog Equalization to Enhance Attitude Determination," U.S. Patent 6,330,988, July 2000.

[37] Li, R., et al., "Stable and Verifiable State Estimation Methods and Systems with Spacecraft Applications," U.S. Patent 6,317,662, July 2000.

[38] Li, R., et al., "Spacecraft Attitude Control Systems with Dynamic Methods and Structures for Processing Star-Tracker Signals," U.S. Patent 6,304,822, August 2000.

[39] Li, R., et al., "Autonomous Gyro Scale Factor and Misalignment Calibration," U.S. Patent 6,298,288, December 1998.

[40] Li, R., et al., "Attitude Determination System and Method," U.S. Patent 6,289,268, March 2000.

[41] Li, R., et al., "Beacon-Assisted Spacecraft Attitude Control Systems and Methods," U.S. Patent 6,288,671, April 2000.

[42] Li, R., et al., "Spacecraft Attitude Determination System and Method," U.S. Patent 6,285,927, May 1999.

[43] Li, R., et al., "System and Method for Correcting Star-Tracker Low Spatial Frequency Error in Stellar-Inertial Attitude Determination Systems," U.S. Patent 6,272,432, May 1999.

[44] Li, R., et al., "Autonomous Attitude Acquisition for a Stellar Inertial Attitude Determination System," U.S. Patent 6,108,594, May 1998.

[45] Kudva, P., and A. Throckmorton, "Preliminary Star Catalog Development for the Earth Observation Systems AMI (EOS-AMI) Mission," *AIAA Journal of Guidance, Control and Dynamics*, Vol. 19, No. 6, 1996, pp. 1332–1336.

[46] Bone, J. W., "On-Orbit Star Processing Using Multi-Star Star-Trackers," *SPIE*, Vol. 2221, April 5–8, 1994, pp. 6–14.

[47] Wicenec, A. J., and M. Albrecht, "Methods for Structuring and Searching Very Large Catalogs," *Astronomical Society of the Pacific*, 1998, pp. 512–515.

[48] Barrett, P., "Application of the Linear Quadtree to Astronomical Databases," *Astronomical Data Analysis Software and Systems IV, ASP Conference Series*, Vol. 77, 1995, pp. 1–4.

[49] Strikwerda, T. E., and J. L. Junkins, "Star Pattern Recognition and Spacecraft Attitude Determination," U.S. Army ETL, May 1981.

[50] Ju, G., and S. K. Lee, "Attitude Determination for Kompsat-2 Using Star Identification Technique," AAS 02-100, pp. 1–17, March 2002.

[51] Heide, E. J. v. d., et al., "Development and Validation of a Fast and Reliable Star Sensor Algorithm with Reduced Data Base," IAF-98.A.6.05, pp. 1–13, October 1998.

[52] Potteck, S., "A Star Recognition Algorithm Based on Star Groups and Associated Means," AAS 99-011, pp. 87–94, November 1999.

[53] Mortari, D., J. L. Junkins, and M. A. Samaan, "Lost-in-Space Pyramid Algorithm for Robust Star Pattern Recognition," AAS 01-004, pp. 1–20, January 2001.

[54] Mortari, D., "Search-Less Algorithm for Star Pattern Recognition," pp. 179–194. Journal of the Astronautical Sciences 45(2) · April 1997.

[55] Mortari, D., "A Fast On-Board Autonomous Attitude Determination System Based on a New Star-ID Technique for a Wide FOV Star-Tracker," AAS 96-158, pp. 893–903. January 1996.

[56] Schmidt, U., C. Elstner, and K. Michel, "ASTRO 15 Star-Tracker Flight Experience and Further Improvements Towards the ASTRO APS Star-Tracker," *AIAA Guidance, Navigation and Control Conference and Exhibit*, Honolulu, HI, August 18–21, 2008.

[57] Baldini, D., et al., "Star-Configuration Searching for Satellite Attitude Computation," *IEEE Transactions on Aerospace and Electronic Systems*, Vol. 31, No. 2, April 1995, pp. 768–777.

[58] Udomkesmalee, S., J. W. Alexander, and A. F. Tolivar, "Stochastic Star Identification," *Journal of Guidance, Control, and Dynamics*, Vol. 17, No. 6, November-December 1994.

[59] Ju, G., et al., "Lost-in-Space: A Star Pattern Recognition and Attitude Estimation Approach for the Case of No Prior Attitude Information," AAS-00-004, February 2–6, 2000, pp. 1–16.

[60] Samaan, M. A., D. Mortari, and J. L. Junkins, "Recursive Mode Star Identification Algorithms," AAS-01-149, February 2001, pp. 1–18.

[61] Shucker, B., and E. D. Fasse, "A Ground-Based Prototype of a CMOS Navigational Star Camera for Small Satellite Applications," *15th AIAA/USU Conference on Small Satellites*, April 2001.

[62] Ju, G., et al., "Digistar: A Low Cost Micro Star-Tracker," *AIAA Space Technology Conference & Exposition*, September 1999.

[63] https://spacemath.gsfc.nasa.gov/stars/6Page103.pdf.

[64] http://www.stargazing.net/david/constel/howmanystars.html.

[65] Li, R., "A Super-Fast Integer Ambiguity Resolution Approach to GPS Interferometer Heading Determination for the Application in a Low Cost Integrated GPS Inertial Navigation System," *Proc. of ION NTM-96*, 1996, pp. 83–88.

CHAPTER 8

Orbit Determination

8.1 Introduction

The orbit determination of heavenly bodies has been a human practice since 1801 when Gauss solved the problems of the orbit determination of Asteroid Ceres. The development in this field has been growing exponentially for the last 60 years after humans started to send artificial satellites to orbit around the Earth and for deep space missions. Today, orbit determination is a very mature field with a large amount of publications, a large amount of software tools, and a large amount of ground and spacecraft-based solutions that have been implemented and proven.

As we discussed before, there are two types of sensors and information sources for PNT systems. Orbit determination, as a type of a PNT system, is no exception. The typical examples of propagation sensors and information sources used for spacecraft orbit determinations are listed here:

1. Orbit propagation based on a gravity model and other force models is by far the most popular orbit propagation method. The information source, in this case, is not even a sensor but just the gravity model and other models of forces that are acting on the spacecraft.
2. Accelerometers. When a spacecraft is in orbit, the gravity and the spacecraft acceleration balance out in such a way that the accelerometers will output zeroes. However, accelerometers are still useful when the spacecraft is firing thrusters and doing orbit adjustments. This is especially true when traditional chemical thrusters are used that typically generate large forces and relatively large accelerations on the spacecraft. However, electrical thrusters typically generate much smaller forces and much smaller accelerations. The accelerations can be so small that they are near or below the resolution of typical accelerometers. In this case, it is typically not profitable to use accelerometers to measure the accelerations generated by thrusters. Instead, mathematical models with calibrations often work better and have zero recurring cost.

There are many types of update sensors that have been used to support orbit determination. Some of the sensors are ground-based, and some are spacecraft-based. The typical update sensors and information sources used for spacecraft orbit determinations are listed here [1–21]:

1. Ground-based two-way ranging is still by far the most popular measurements that are used to support spacecraft orbit determination. There are typically three variations:

a. Range and Doppler measurements from a single ground station;
b. Range and Doppler measurements, and LOS (elevation and azimuth angles, typically not very accurate) from a single ground station;
c. Range and Doppler measurements from multiple ground stations. Typically, two ground stations will provide much better performance than a single ground station.

2. GPS receivers onboard the spacecraft is probably the second most popular method.
3. Onboard intersatellite crosslinks that provide ranging information to another spacecraft with known ephemeris.
4. Ground-based telescopes that provide angle-only measurements to support orbit determination. Again, one can choose to use a single ground-based telescope or multiple telescopes located at different locations on the earth to provide better geometry. These telescopes typical observe both the spacecraft and the stars at the same time so that it can determine its attitude with respect to the ECI frame precisely while measuring the LOS unit vector for the spacecraft.
5. Optical sensors, such as cameras or star-trackers that track finite distance objects, such as another satellite with a known ephemeris, or planets and other heavenly bodies. Again, the star-tracker observes both the spacecraft and the stars in the background to eliminate knowledge errors associated with the star-tracker orientation.
6. Onboard imaging devices, such as visible, IR, or radar imaging devices that look at the Earth landmarks or just the Earth image in general.

In the above, item 6 is essentially vision or image-based navigation that will be discussed in detail in Chapter 11.

As mentioned before, there are a large amount of software tools available in the field of orbit determination. In the meantime, it is worth mentioning that a significant amount of these software items are both open-source and high-quality. The following are some examples:

1. NASA's open source orbit determination tool box (ODTBX);
2. NASA's General Mission Analysis Tool (GMAT);
3. NASA's JAT: Java Astrodynamics Toolkit;
4. Open-Source, Extensible Spacecraft Simulation And Modeling Environment (Open-SESSAME);
5. Orbit Reconstruction, Simulation, and Analysis (ORSA);
6. NASA's OpenFlightLinux.

Most of the orbit determination systems in the history have been ground-based and most of the orbit determination activities were not done in real time, but by postprocessing. The orbit determined for the past can be propagated forward into the future using the orbit propagator and the gravity model and other force models. Onboard orbit determination is typically done in real time and can be used to support autonomous operations of the spacecraft.

Because of the nonreal-time nature of the ground-based orbit determination, there are typically two types of the algorithms:

1. Batch algorithms that are typically based on the batch least square estimator;
2. Sequential processing algorithms that are typically based on the EKF.

The batch algorithms use the measurements collected for a segment of the orbit (an arc) and use the orbit propagation equations and force models to model all the measurements along the arc as a function of the initial states (or the states at another chosen time) and force model parameters. Consequently, the initial states and the force model parameters can be estimated using a batch least square solution.

There are several advantages of using the batch algorithms: First, the solutions can go through many iterations to improve the accuracy. By using an earlier iteration solution as the a priori knowledge to linearize the equations, another iteration can be done to achieve improved accuracy. In addition, the batch algorithms use all the data, past and future, with respect to the states that we want to estimate, and typically yield better results. Many of the popular postprocessing orbit determination tools, including open-source, private business-owned, and government-owned, provide batch processing-based orbit determination capabilities.

The sequential processing algorithms can be done both in nonreal time and in real time and is very much required for real-time processing. The sequential algorithms fit in our standard All Source PNT architecture. We will focus on the sequential algorithms for the rest of this chapter.

8.2 Orbit Determination Architecture

Figure 8.1 illustrates a typical generic architecture for orbit determination. Within the architecture, the propagation solution and the update solution are integrated through a Kalman filter.

The propagation solution uses the gravity model together with other force models. The force models can contain parameters that need to be calibrated or estimated by the Kalman filter.

The update sensors can be one or several of the sensor types that we discussed before. The processing and correction algorithms may process the information in a way that is more convenient for the Kalman filter. One example is the algorithm to combine the GNSS pseudo-range and carrier-phase measurements to eliminate ionosphere delays from the measurements. In addition, there could be additional parameters that can be used to perform corrections to the measurement and these parameters are also to be estimated by the Kalman filter.

8.3 Propagation Algorithms and Error Models

The most basic states for orbit have 6 degrees of freedom. They can be simply the position and velocity vector or the 6 Keplerian orbit elements. With the position

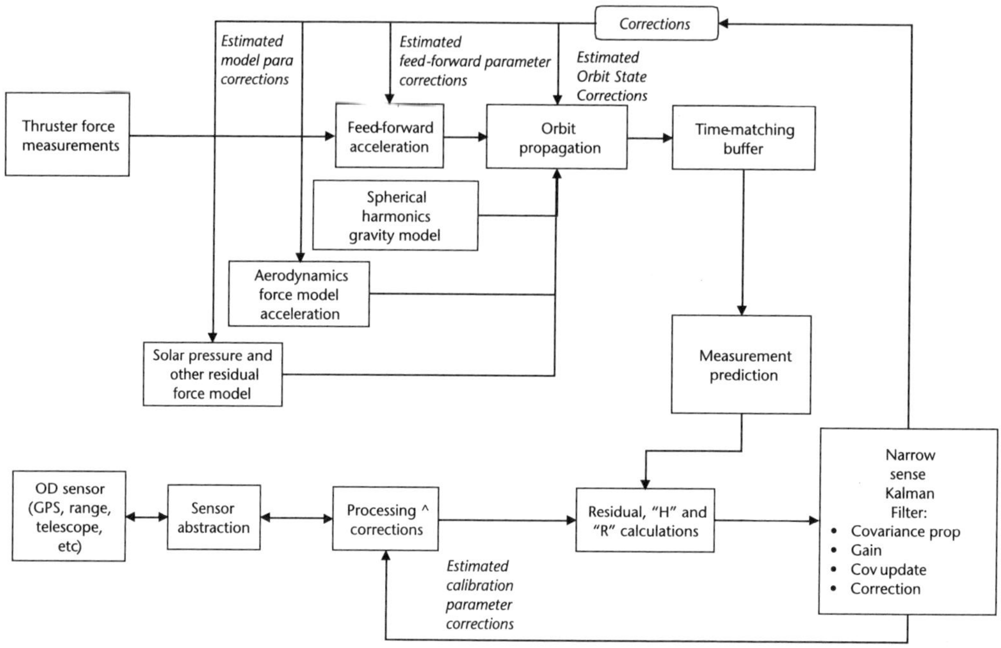

Figure 8.1 Orbit determination architecture.

and velocity and all the known external accelerations, the differential equations can be used to propagate the state to the future.

Depending on the force models that we use and the models of other sensors and information sources that we use, we may decide to augment the state vector with additional parameters.

The state propagation equations can be written as

$$
\begin{aligned}
\frac{d}{dt}{}^{I}R_{I}^{B} &= {}^{I}V_{I}^{B} \\
\frac{d}{dt}{}^{I}V_{I}^{B} &= {}^{I}a_{I}^{B} = {}^{E}g_{0}\left({}^{I}R_{I}^{B}\right) + f_{g}\left({}^{I}R_{I}^{B}\right) + f_{ngI}\left(p_{I}\right) + C_{B}^{I}\,{}^{B}f_{T} + C_{B}^{I}f_{ngb}\left(p_{B}\right) \\
\frac{d}{dt}p_{B} &= -\frac{1}{\tau_{B}}p_{B} + w_{pb} \\
\frac{d}{dt}p_{I} &= -\frac{1}{\tau_{B}}p_{I} + w_{pi}
\end{aligned}
\tag{8.1}
$$

where

${}^{E}g_{0}\left({}^{I}R_{I}^{B}\right) = -\left(\mu / \|{}^{I}R_{I}^{B}\|^{3}\right)R_{I}^{B}$ is the point mass portion of the gravity model.

$f_{g}({}^{I}R_{I}^{R})$ is the rest of the gravitational accelerations including a higher-order portion of the spherical harmonics gravity model (which includes the Sun and Moon effects), the Earth tide model, and accelerations due to planets.

$f_{ngI}(p_I)$ is the nongravitational acceleration model with the parameter vector p_I. This term is assumed to be changing slowly in inertial frame. The solar pressure-induced acceleration belongs here.

$f_{ngB}(p_B)$ is the nongravitational acceleration model with the parameter vector p_B. This term is assumed to be changing slowly in the body frame. RF radiation pressure, thermal radiation pressure, and aerodynamic forces can be included in this term.

${}^{B}f_T = C_T^B \begin{bmatrix} (1+s)f_t \\ \alpha f_t \\ \beta f_t \end{bmatrix}$ models the thruster force including the scale factor and the misalignment errors. When thruster is not firing, $f_t = 0$.

The total number of states in this nonlinear propagation model is

$n_{\text{states}} = 6$ (position & velocity) + 3 (thruster scale factor & misalignments)
+ n_{ngi} (non-gravitational terms, slow in inertial frame) +
+ n_{ngb} (non-gravitational terms, slow in body frame)

The linearized model for the deviations and errors can be written as

$$\frac{d}{dt}\delta^{I}R_I^B = \delta^{I}V_I^B$$

$$\frac{d}{dt}\delta^{I}V_I^B = \delta^{I}a_I^B = \frac{\partial^{E}g_0\left({}^{I}R_I^B\right)}{\partial^{I}R_I^B}\delta^{I}R_I^B + \frac{\partial f_{ngI}\left(p_I\right)}{\partial p_I}\delta p_I + C_B^I C_T^B \begin{bmatrix} \delta s_t \\ \delta\alpha_t \\ \delta\beta_t \end{bmatrix} f_t + C_B^I \frac{\partial f_{ngb}\left(p_B\right)}{\partial p_B}\delta p_B$$

$$\frac{d}{dt}\delta p_B = -\frac{1}{\tau_B}\delta p_B + w_{pb}$$

$$\frac{d}{dt}\delta p_I = -\frac{1}{\tau_B}\delta p_I + w_{pi}$$

Let $\gamma = \begin{bmatrix} s_t \\ \alpha_t \\ \beta_t \end{bmatrix}$ be the three-element vector that models the thruster scale factor and misalignment errors.

It will be modeled as a first-order Markov process:

$$\frac{d}{dt}\delta\gamma = -\frac{1}{\tau_\gamma}\delta\gamma + w_\gamma$$

We have the following equations for the linearized model:

State and process noise:

$$\delta X = \begin{bmatrix} \delta^I R_I^B \\ \delta^I V_I^B \\ \delta\gamma \\ \delta p_B \\ \delta p_I \end{bmatrix}, \quad W = \begin{bmatrix} 0 \\ 0 \\ w_\gamma \\ w_{pb} \\ w_{pI} \end{bmatrix} \tag{8.2}$$

The propagation model:

$$\frac{d\delta X}{dt} = A\delta X + W = \begin{bmatrix} 0 & I & 0 & 0 & 0 \\ 0 & \dfrac{\partial^I g_0\left({}^I R_I^B\right)}{\partial^I R_I^B} & C_B^I C_T^B f & C_B^I \dfrac{\partial f_{ngb}\left(p_B\right)}{\partial p_B} & \dfrac{\partial f_{ngi}\left(p_I\right)}{\partial p_I} \\ 0 & 0 & -\left(\dfrac{1}{\tau_\gamma}\right)I & 0 & 0 \\ 0 & 0 & 0 & -\left(\dfrac{1}{\tau_{pb}}\right)I & 0 \\ 0 & 0 & 0 & 0 & -\left(\dfrac{1}{\tau_{pi}}\right)I \end{bmatrix} \delta X + W \tag{8.3}$$

where

$$\begin{aligned} \frac{\partial^I g_0\left({}^I R_I^B\right)}{\partial^I R_I^B} &= {}^I R_I^B \frac{3\mu}{\left\|{}^I R_I^B\right\|^3} \left(\frac{1}{\left\|{}^I R_I^B\right\|^2} {}^I R_I^B\right)^T - \frac{\mu}{\left\|{}^I R_I^B\right\|^3} I \\ &= g_0 \left(\frac{3}{\left\|{}^I R_I^B\right\|^2} {}^I R_I^B\right)^T - \frac{\mu}{\left\|{}^I R_I^B\right\|^3} I \end{aligned} \tag{8.4}$$

8.4 Measurement Equations and Error Models

8.4.1 Range Measurement from Ground or Space Objects (Crosslinks)

Let the position of the spacecraft be given by

$${}^I R_I^B(t) = \begin{bmatrix} x(t) \\ y(t) \\ z(t) \end{bmatrix}$$

and the location of the ground station be given by

$${}^I R_k(t) = \begin{bmatrix} {}^I x_k(t) \\ {}^I y_k(t) \\ {}^I z_k(t) \end{bmatrix} = {}^I R_E^{G_k}(t) = C_E^I(t)\, {}^E R_E^{G_k} = C_E^I(t) \begin{bmatrix} {}^E x_k \\ {}^E y_k \\ {}^E z_k \end{bmatrix}$$

For the space object, ${}^{I}R_k(t)$ will be simply the location of the space object computed from its ephemeris:

$$ {}^{I}R_k(t) = \begin{bmatrix} {}^{I}x_k(t) \\ {}^{I}y_k(t) \\ {}^{I}z_k(t) \end{bmatrix} = {}^{I}R_I^{S_k}(t) $$

In both cases, the range measurement model is given by

$$ d_k = \left\| {}^{I}R_I^B(t) - {}^{I}R_k(t) \right\| + v_k = \sqrt{\left(x - x_k\right)^2 + \left(y - y_k\right)^2 + \left(z - z_k\right)^2} + v_k \quad (8.5) $$

where v_k is the range measurement error.

The linearized model is given by

$$ \Delta d_k = -\hat{u}_{\mathrm{LOS}k}{}^{T} \delta {}^{I}R_I^B(t) + v_k \quad (8.6) $$

where

$$ \hat{u}_{\mathrm{LOS}k} = \frac{{}^{I}R_k(t) - {}^{I}\hat{R}_I^B(t)}{\left\| {}^{I}R_k(t) - {}^{I}\hat{R}_I^B(t) \right\|} \quad (8.7) $$

This equation can be used to form the EKF H-matrix.

8.4.2 Angle Measurement from a Ground Telescope or a Spaceborne Camera

Similar to the previous section, we assume that the location of the spacecraft is given by

$$ {}^{I}R_I^B(t) = \begin{bmatrix} x(t) \\ y(t) \\ z(t) \end{bmatrix} $$

and the location of the ground station is given by

$$ {}^{I}R_k(t) = \begin{bmatrix} {}^{I}x_k(t) \\ {}^{I}y_k(t) \\ {}^{I}z_k(t) \end{bmatrix} = {}^{I}R_E^{G_k}(t) = C_E^I(t)\,{}^{E}R_E^{G_k} = C_E^I(t) \begin{bmatrix} {}^{E}x_k \\ {}^{E}y_k \\ {}^{E}z_k \end{bmatrix} $$

For the space object, ${}^{I}R_k(t)$ will be simply the location of the space object computed from its ephemeris:

$$ {}^{I}R_k(t) = \begin{bmatrix} {}^{I}x_k(t) \\ {}^{I}y_k(t) \\ {}^{I}z_k(t) \end{bmatrix} = {}^{I}R_I^{S_k}(t) $$

The LOS vector from spacecraft to the ground station or the space object is given by

$$^{I}u_j = \frac{1}{\left\| ^{I}R_j - {}^{E}R \right\|}\left(^{I}R_j - {}^{I}R_I^B \right) = \frac{1}{\rho_j}\left(^{I}R_j - {}^{I}R_I^B \right) \tag{8.8}$$

where $\rho_j = \| ^{I}R_j - {}^{I}R_I^B \|$ is the range to the object.

$^{I}u_j$ can be transformed to the sensor frame (S) using the following equation for the space-based camera

$$^{S}u_j = C_B^S C_I^B \, ^{I}u_j = C_I^S \, ^{I}u_j \tag{8.9}$$

For the ground-based telescope, we will have

$$^{S}u_j = C_E^S C_I^E \, ^{I}u_j = C_I^S \, ^{I}u_j \tag{8.10}$$

The measured unit vector can be modeled as

$$m^{c}u_j = C_I^S \, ^{I}u_j + v \tag{8.11}$$

Since, for spacecraft, the attitude determination and orbit determination are decoupled, for the purpose of this chapter, we will assume that C_I^S is known in both cases. One approach that will eliminate the need to worry about sensor misalignment errors is to use stars in the FOV of the telescope or the camera to determine the attitude while measuring the other finite distance objects.

We will derive the linearized model for the purpose of the Kalman filter H-matrix. The measurement residual for the unit vector can be written as

$$\delta m^{S}u_j = C_I^S \delta^{I}u_j + v$$

$$\delta^{I}u_j = -\frac{1}{\hat{\rho}}\delta^{I}R_I^B$$

where we are ignoring the sensitivity on the norm. Combining the above two equations, we have

$$\delta m^{S}u_j = -C_I^S \frac{1}{\hat{\rho}}\delta^{I}R_I^B + v \tag{8.12}$$

8.4.3 GPS-Based Orbit Determination

Although GPS was designed to serve users on Earth, the GPS signal is available for space use as well. The space where the GPS signal is available is called the GPS space service volume (SSV) and is illustrated in Figure 8.2.

For the LEO orbit, the service availability is similar to terrestrial applications with a slight degradation of a number of GPS satellites visible due to a raised altitude.

For the MEO orbit below the GPS orbit, when a GPS satellite is visible, we typically have very strong signals due to the proximity to the GPS orbit. However, because the MEO is so close to the GPS orbit, the same antenna that can see many

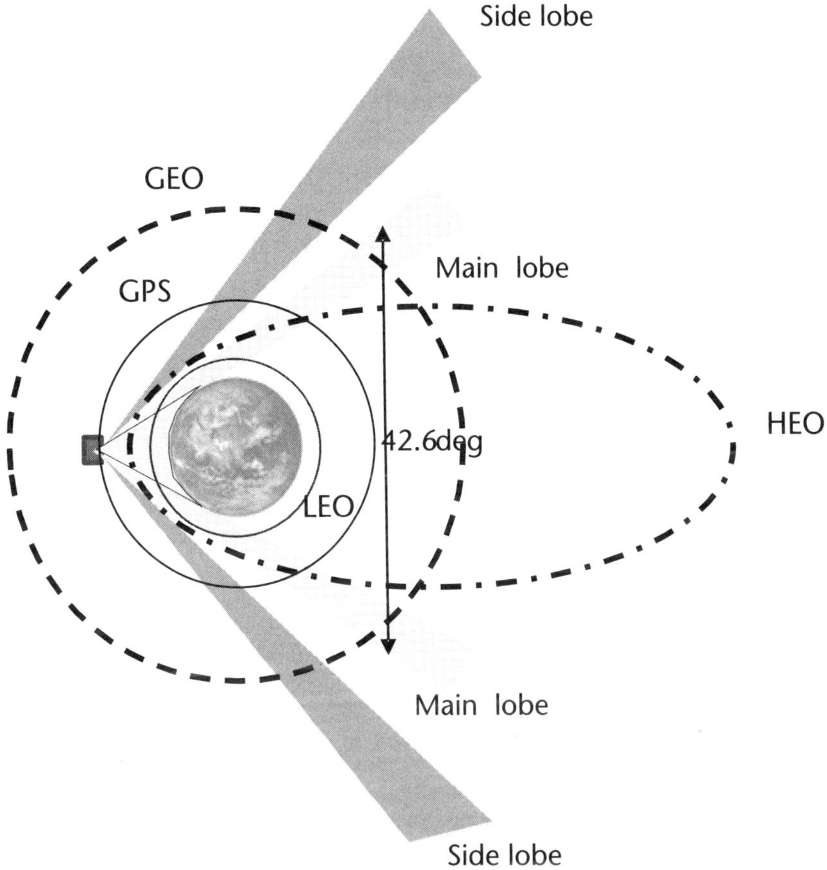

Figure 8.2 GPS SSV.

GPS satellites from the Earth or at the LEO orbit can no longer see as many at the MEO orbit. Consequently, even for a MEO satellite, there is often a need to use the GPS side-lobe power to expand the number of GPS satellites visible.

Figure 8.3 illustrates the fact that as the orbit becomes closer to the GPS orbit, the number of visible satellites by the same antenna FOV decreases.

For the GEO and HEO including lunar transfer orbits, we may be always or mostly above the GPS orbit and, as a result, the GPS satellite that is directly below the GEO and HEO is not visible because their antenna faces the Earth, not the space. Instead, what are visible are the GPS satellites that are from the other side of the Earth. Unfortunately, the strongest part of the main lobe signal will be blocked by the Earth, and only a small portion of the main lobe will be available for a higher orbit to use. Because of the Earth blockage, it will be necessary to use the side-lobes of the GPS signal in these orbits.

Figure 8.4 is the average of the publicly released GPS block IIR antenna patterns available at www.gps.gov. As can be seen, the main lobe is up to roughly 21° from the center with antenna gain ranges from 0 dB to 15 dB or so. The first side-lobe can extend the signal availability to 40° from the center and the gain ranges from −5 to −1 dB. The average gain around the center of the main lobe is about 15 dB higher than the side-lobe.

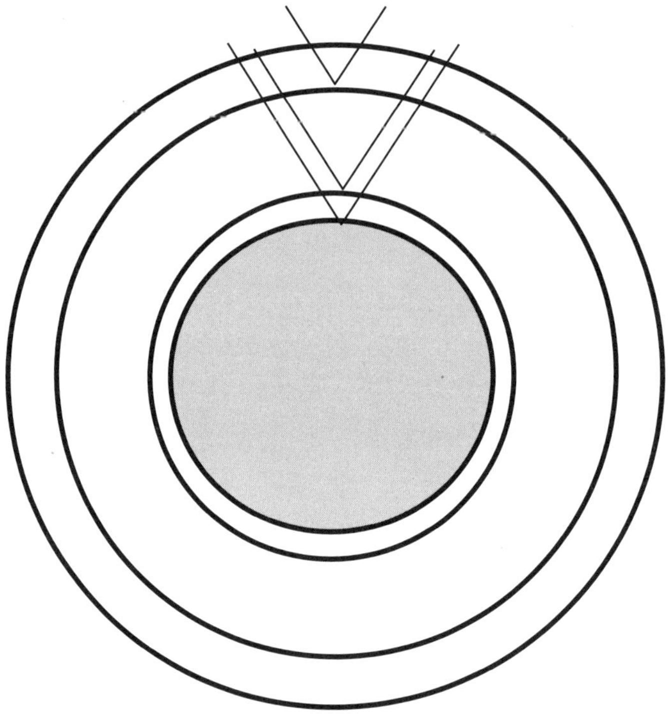

Figure 8.3 As the altitude increases, a fewer number of GPS satellites are visible.

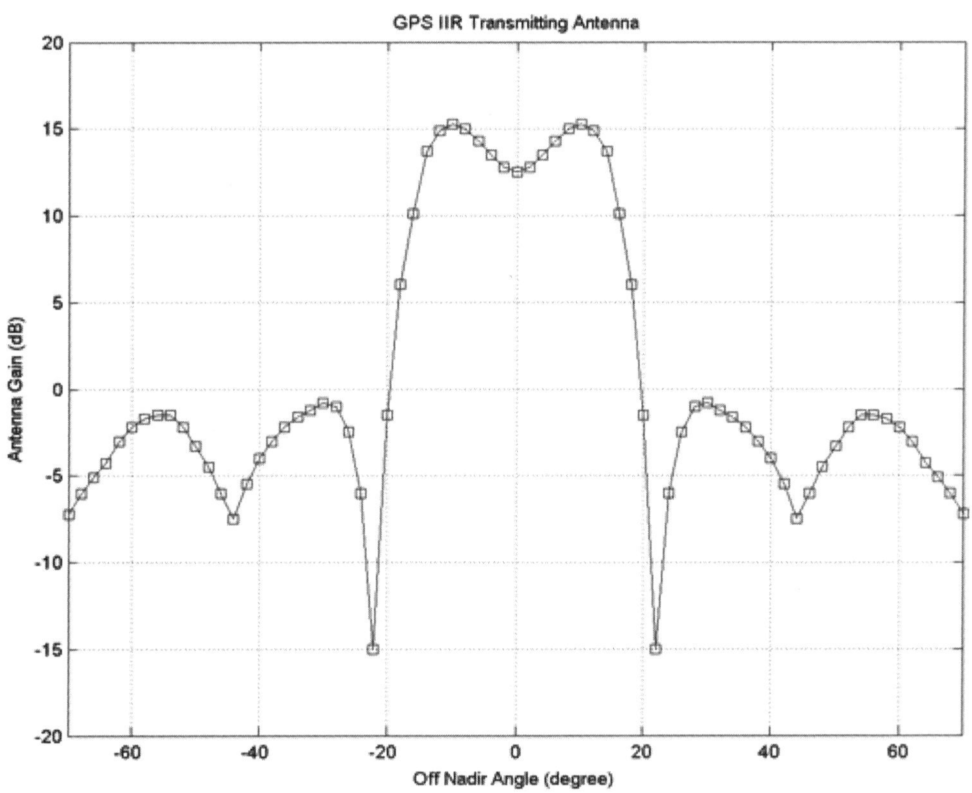

Figure 8.4 Average GPS block IIR antenna pattern.

Table 8.1 Minimum C/N0 to Acquire and Track and the Best C/N0 in Different Orbits

	Minimum C/N0 (dB-Hz) in Order to See Enough GPS Satellites	*Maximum C/N0 (dB-Hz) Typically Seen*	*Comments*
LEO	40	50	Main lobe
MEO	26	50	Main lobe and side-lobe
GEO and typical Earth service HEO	23	40	Main lobe and side-lobe
Lunar transfer HEO	*10*	*28*	*Main lobe and side-lobe*

The other thing that will hurt the GEO and HEO is the free space loss of the signal power.

Table 8.1 lists the typical minimum carrier-to-noise density ratio (C/N0) at which the GPS receiver must be able to acquire and track in order to operate in the corresponding orbits and the typical best C/N0s that can be observed. (Note that these are rough order-of-magnitude values. The actual values for specific applications depend on antenna specifics and orbit specifics.)

Because of the low C/N0s, the receiver often needs to be tightly coupled with the propagation-based solution in such a way that we will use our position, velocity, clock bias, and clock bias rate estimates to predict the pseudo-range, the Doppler, and even the I and Q signals to allow the receiver to coherently integrate the I/Q for extended period of time. In particular, for MEO and GEO and most HEO applications, coherent integration up to 20 ms will be required. For lunar transfer orbit, coherent integration could be required over 1 second in order to acquire and track the signals with C/N0 at low as 10 dB-Hz.

Since the classic GPS signal has 50-bit-per-second navigation data messages modulated on the top of the code, 20-ms coherent integration has to be managed carefully to deal with the data bit boundary issues. This is especially true for signal acquisition. For longer than 20 ms, coherent integration has to be done with data wipe-off schemes. In other words, we will need to use the navigation data message contents from some other sources to predict what the data will be and include the Nav data message modulation in the predicted I and Q signals. This approach effectively wipes off the data in the I/Q signals before coherent integration can be extended beyond 20 ms. Since GPS is not the focus of this book, the readers are referred to the references for further discussion of the low C/N0 acquisition and tracking problems.

Assuming that the GPS acquisition and tracking problems are solved for the orbit of interests, we may have the following types of measurements available:

1. Single-frequency pseudo-range;
2. Single-frequency carrier-phase;
3. Dual-frequency pseudo-range;
4. Dual-frequency carrier-phase.

Table 8.2 Precision GPS Orbit Determination Work Reported in the Literature

	Measurements Used	*Specials to Improve Performance*	*Position Accuracy Achieved*	*Reference*
1.1	Dual-frequency carrier-phase and code-phase measurements	Broadcast Ephemeris Only	0.3–0.5m	[1]
1.2	Dual-frequency carrier-phase and code-phase measurements	Precision ephemeris and clock data from IGS (International GNSS Service, http://www.igs.org/)	20–30 mm	[2, 4]
1.3	Dual-frequency carrier-phase and code-phase measurements	(1) Precision orbit and clock information from the Center for Orbit Determination in Europe (CODE); (2) modeling of antenna phase center variations; (3) ultra-precision, mission-specific gravity model (ITSG)	Better 10 mm	[5]
2.1	Single-frequency code and carrier phases	Broadcast ephemeris only	0.4–0.7m	[3]
2.2	Single-frequency code phase only	Broadcast ephemeris only	0.7–1m	[3]

For most of the space vehicles, the orbits are high enough such that the troposphere delay errors are not a concern. However, the ionosphere delay errors can be a contributor for most of the spacecraft orbit types, including GEO. This is because the GPS signal path from the GPS space vehicle to the GEO receiver can still pass through the space that is close to the Earth.

When dual-frequency data is available, it is often profitable to combine the data into ionosphere-error free pseudo-range and carrier-phase measurements.

When there is only single-frequency data available, it is often profitable to combine the code phase and the carrier phase into the Group and Phase Ionosphere Correction (GRAPHIC) to take advantage of the opposite signs of the ionosphere delay errors in the code phase and the carrier phase.

Precision LEO determination, in particular, has been widely studied by many authors [1–5] and very high precisions have been achieved. Table 8.2 summarizes what has been reported by various authors.

The integer ambiguity of the carrier phase can often be included as the states of the Kalman filter as floating point numbers and they will be estimated as increasingly more data are used to update the states.

Let the pseudo-range and carrier-phase measurement for satellite k be given by

$$\begin{aligned} PR_j(t) &= \left\| R(t) - R_j(t) \right\| + ct_r(t) + ct_{sj}(t) + I_j(t) + V_j(t) \\ C_j(t) &= \left\| R(t) - R_j(t) \right\| + ct_r(t) + ct_{sj}(t) - I_j(t) + v_j(t) + \lambda N_j \end{aligned} \quad (8.13)$$

where

R_j: the position of GPS SV j computed from ephemeris;

ct_t: the range error due to receiver clock bias;

ct_{sj}: the range error due to the satellite j clock;

I_j: the ionosphere delay error;

v_j: the receiver carrier-phase noise and multipath error;

V_j: the receiver code-phase noise and multipath error;

N_j: the carrier-phase integer ambiguity.

For dual-frequency receivers, the dominant errors from the GPS measurements will be receiver noise and multipath errors and GPS satellite ephemeris errors and clock errors.

If we use ionosphere delay corrected pseudo-range, we will typically just add the clock bias and clock bias rate into the Kalman filter states. If we use both pseudo-range and carrier phase, we will need to add the integer ambiguity as the states of the Kalman filter.

For single-frequency receivers, we can add the ionosphere delay for each satellite as the states of the Kalman filter. If we want to use carrier-phase measurements, we will also want to add the integer ambiguities as the states of the Kalman filter.

As mentioned before, for single-frequency receivers, we can also form the GRAPHIC measurements:

$$L_j(t) = \frac{PR_j(t) + C_j(t)}{2} = \left\| R(t) - R_j(t) \right\| + ct_r(t) + ct_{sj}(t) + \frac{V_j(t) + v_j(t)}{2} + \frac{\lambda}{2} N_j \tag{8.14}$$

In this case, since the ionosphere delays are canceled out, there is no need to add the ionosphere delay errors as part of the Kalman filter states.

In the following, we will use the single-frequency code and the carrier phase-based approach as an example to write out the states, the state propagation equations, the measurement equations, and the linearized measurement equations.

Recall that we have the following state vector

$$X = \begin{bmatrix} {}^{I}R_I^B \\ {}^{I}V_I^B \\ \gamma \\ p_B \\ p_I \end{bmatrix}$$

We will add the following additional elements into the state vector:

1. The clock bias and clock bias rate for the GPS receiver (two additional states);
2. The ionosphere delay for the GPS satellites $I = \begin{bmatrix} I_1 & I_2 & \cdots & I_M \end{bmatrix}^T$;

3. The integer ambiguity for the GPS satellites $N = \begin{bmatrix} N_1 & N_2 & \cdots & N_M \end{bmatrix}^T$.

where M is the number of GPS satellites currently used.

The state vector now looks like the following:

$$X = \begin{bmatrix} {}^{I}R_I^B \\ {}^{I}V_I^B \\ \gamma \\ p_B \\ p_I \\ b_r \\ d_r \\ I \\ N \end{bmatrix}$$

where b_r and d_r are the receiver clock bias and the clock rate bias.

The total number of states equals $(3 + 3 + 3 + Mb + Mi + 2 + 2M) = 11 + Mb + Mi + 2M$.

Both the ionosphere delay states and integer ambiguity states are to be reinitialized when new GPS satellites are used.

The nonlinear state propagation equations are therefore:

$$\begin{aligned}
&\frac{d}{dt}{}^{I}R_I^B = {}^{I}V_I^B \\
&\frac{d}{dt}{}^{I}V_I^B = {}^{I}a_I^B = {}^{E}g_0\left({}^{I}R_I^B\right) + f_g\left({}^{I}R_I^B\right) + f_{ngI}\left(p_I\right) + C_B^I\,{}^{B}f_T + C_B^I f_{ngb}\left(p_b\right) \\
&\frac{d}{dt}\gamma = -\frac{1}{\tau_\gamma}\gamma + w_\gamma \\
&\frac{d}{dt}p_B = -\frac{1}{\tau_B}p_B + w_{pb} \\
&\frac{d}{dt}p_I = -\frac{1}{\tau_B}p_I + w_{pi} \\
&\frac{d}{dt}b_r = d_r + w_{br} \\
&\frac{d}{dt}d_r = w_{dr} \\
&\frac{d}{dt}I = -\frac{1}{\tau_I}I + w_I \\
&\frac{d}{dt}N = w_N
\end{aligned} \tag{8.15}$$

The linearized propagation equation can be written as

$$\delta X = \begin{bmatrix} \delta^I R_I^B \\ \delta^I V_I^B \\ \delta\gamma \\ \delta p_B \\ \delta p_I \\ \delta b_r \\ \delta d_r \\ \delta I \\ \delta N \end{bmatrix}, \quad W = \begin{bmatrix} 0 \\ 0 \\ w_\gamma \\ w_{pb} \\ w_{pi} \\ w_{br} \\ w_{dr} \\ w_I \\ w_N \end{bmatrix} \tag{8.16}$$

$$\frac{d\delta X}{dt} = A\delta X + W =$$

$$\frac{d}{dt}\begin{bmatrix} \delta^I R_I^B \\ \delta^I V_I^B \\ \delta\gamma \\ \delta p_B \\ \delta p_I \\ \delta b_r \\ \delta d_r \\ \delta I \\ \delta N \end{bmatrix} = \begin{bmatrix} 0 & I & 0 & 0 & 0 & 0 & 0 & 0 & 0 \\ 0 & \frac{\partial^I g_0\left({}^I R_I^B\right)}{\partial^I R_I^B} & C_B^I C_T^B f & C_B^I \frac{\partial f_{ngb}(p_b)}{\partial p_b} & \frac{\partial f_{ngi}(p_i)}{\partial p_i} & 0 & 0 & 0 & 0 \\ 0 & 0 & -\left(\frac{1}{\tau_\gamma}\right) I & 0 & 0 & 0 & 0 & 0 & 0 \\ 0 & 0 & 0 & -\left(\frac{1}{\tau_{pb}}\right) I & 0 & 0 & 0 & 0 & 0 \\ 0 & 0 & 0 & 0 & -\left(\frac{1}{\tau_{pi}}\right) I & 0 & 0 & 0 & 0 \\ 0 & 0 & 0 & 0 & 0 & 0 & 1 & 0 & 0 \\ 0 & 0 & 0 & 0 & 0 & 0 & 0 & 0 & 0 \\ 0 & 0 & 0 & 0 & 0 & 0 & 0 & -\left(\frac{1}{\tau_I}\right) I & 0 \\ 0 & 0 & 0 & 0 & 0 & 0 & 0 & 0 & 0 \end{bmatrix} \begin{bmatrix} \delta^I R_I^B \\ \delta^I V_I^B \\ \delta\gamma \\ \delta p_B \\ \delta p_I \\ \delta b_r \\ \delta d_r \\ \delta I \\ \delta N \end{bmatrix} + \begin{bmatrix} 0 \\ 0 \\ w_\gamma \\ w_{pb} \\ w_{pi} \\ w_{br} \\ w_{dr} \\ w_I \\ w_N \end{bmatrix} \tag{8.17}$$

We have two measurements from each GNSS space vehicle. The nonlinear measurement equations are

$$\begin{aligned} PR_j(t) &= \left\| {}^I R_I^B(t) - R_j(t) \right\| + ct_r(t) + ct_{sj}(t) + I_j(t) + V_j(t) \\ C_j(t) &= \left\| {}^I R_I^B(t) - R_j(t) \right\| + ct_r(t) + ct_{sj}(t) - I_j(t) + v_j(t) + \lambda N_j \end{aligned} \tag{8.18}$$

The linearized equations can be written as

$$\begin{aligned} \delta PR_j(t) &= -\left(u_{\mathrm{LOS}j}\right)^T \delta^I R_I^B + c\delta t_r(t) + \delta I_j(t) + c\delta t_{sj}(t) + V_j(t) \\ \delta C_j(t) &= -\left(u_{\mathrm{LOS}j}\right)^T \delta^I R_I^B + c\delta t_r(t) + c\delta t_{sj}(t) - \delta I_j(t) + \lambda\delta N_j + c\delta t_{sj}(t) + v_j(t) \end{aligned} \tag{8.19}$$

Stacking the linearized measurement equations together, we have

$$Y = \begin{bmatrix} \delta PR \\ \delta C \end{bmatrix} = \begin{bmatrix} U_{\text{LOS}} & 0 & 0 & 0 & 0 & c & 0 & I & 0 \\ U_{\text{LOS}} & 0 & 0 & 0 & 0 & c & 0 & -I & \lambda I \end{bmatrix} \delta X + \begin{bmatrix} c\delta ts + V \\ c\delta ts + v \end{bmatrix} \tag{8.20}$$

where

$$U_{\text{LOS}} = \begin{bmatrix} (u_{\text{LOS}1})^T \\ (u_{\text{LOS}2})^T \\ \vdots \\ (u_{\text{LOS}m})^T \end{bmatrix} \tag{8.21}$$

With the above, we can construct the Kalman filter for orbit determination.

References

[1] Montenbruck, O., and P. Ramos-Bosch, "Precision Real-Time Navigation of LEO Satellites Using Global Positioning System Measurements," *GPS Solutions*, October 2009, https://www.dlr.de/rb/portaldata/38/resources/dokumente/gsoc_dokumente/rb-rft/gps_sol_0707.pdf.

[2] Cerri, L., et al., "Precision Orbit Determination Standards for the Jason Series of Altimeter Missions," *Marine Geodesy*, Vol. 33, December 2010, pp. 379–418.

[3] Gong, X., et al., "Precise Onboard Real-Time Orbit Determination with a Low-Cost Single-Frequency GPS/BDS Receiver," *Remote Sensing*, June 2018.

[4] Li, M., B. Xu, and J. Sun, "Autonomous Orbit Determination for a Hybrid Constellation," *International Journal of Aerospace Engineering*, Vol. 2018, Article ID 4843061, 2018.

[5] Zehentner, N., and T. Mayer-Gürr, "Precise Orbit Determination Based on Raw GPS Measurements," *J. Geod.*, Vol. 90, 2016, pp. 275–286.

[6] Lemoine, F. G., et al., "Precise Orbit Determination for Geosat Follow-On Using Satellite Laser Ranging Data and Intermission Altimeter Crossovers," *2001 Flight Mechanics Symposium*, Greenbelt, MD, 2001.

[7] Markley, F., and U. Naval, "Autonomous Navigation Using Landmark and Intersatellite Data," *AIAA/AAS Astrodynamics Conference*, Seattle, WA, 1984.

[8] Li, R., H. Qiu, and K. Xiong, "Autonomous Navigation for Constellation Based on Inter-Satellite Ranging and Directions," *IECON 2017—43rd Annual Conference of The IEEE Industrial Electronics Society*, Beijing, China, November 2017, pp. 2985–2990.

[9] Li, M., B. Xu, and L. Zhang, "Orbit Determination for Remote-Sensing Satellites Using Only Optical Imagery," *International Journal of Remote Sensing*, Vol. 38, No. 5, 2017, pp. 1350–1364.

[10] Li, M., and B. Xu, "Autonomous Orbit and Attitude Determination for Earth Satellites Using Images of Regular-Shaped Ground Objects," *Aerospace Science and Technology*, Vol. 80, 2018, pp. 192–202.

[11] Li, X., et al., "LEO Precise Orbit Determination with Inter-Satellite Links," *Remote Sensing*, Vol. 11, No. 18, 2019, p. 2117.

[12] Yoon, J. -C., et al., "Spacecraft Orbit Determination Using GPS Navigation Solutions," *Aerosp. Sci. Technol.*, Vol. 4, 2000, pp. 215–221.

[13] Rim, H. J., and B. E. Schutz, "Precision Orbit Determination," University of Texas Austin, Report, 2002.

[14] Montenbruck, O., and E. Gill, *Satellite Orbits*, New York: Springer, 2001.

[15] Bock, H., et al., "GRAS/Metop: Precise Orbit Determination Study Phase II," Terma Technical Note, 2001.

[16] Bock, H., et al., "Efficient Precise Orbit Determination of LEO Satellites Using GPS," *Advances in Space Research*, Vol. 30, No. 2, 2002, p. 295.

[17] Svehla, D., and M. Rothacher, "Kinematic Orbit Determination of LEOs Based on Zero- or Double-Difference Algorithms Using Simulated and Real SST GPS Data," *IAG Scientific Assembly*, Budapest, Hungary, 2001.

[18] Bisnath, S. B., and R. B. Langely, "Precise A Posteriori Geometric Tracking of Low Earth Orbiters with GPS," *Canadian Aeronautics and Space Journal*, Vol. 45, No. 3, 1999, p. 245.

[19] Bury, G., et al., "Multi-GNSS Orbit Determination Using Satellite Laser Ranging," *Journal of Geodesy*, pp 1–17.

[20] Paula, A., et al., "Onboard and Real-Time Artificial Satellite Orbit Determination Using GPS," *Mathematical Problems in Engineering*, Vol. 2013, Article ID 530516, 2013, http://dx.doi.org/10.1155/2013/530516.

[21] Gong, B., et al., "Angles-Only Initial Relative Orbit Determination Algorithm for Non-Cooperative Spacecraft Proximity Operations," *Astrodynamics*, Vol. 2, No. 3, September 2018, pp. 217–231.

CHAPTER 9

Terrain Referenced Navigation

9.1 Introduction

The Earth terrain changes as the geolocation changes. Consequently, it is possible to use the terrain profile as a geolocation reference, hence the terrain referenced navigation (TRN). The idea of TRN and several variations of TRN-based systems have been developed, implemented, and used since the 1950s. The technologies to realize and implement the idea have changed dramatically over this time, but the fundamental idea did not change much [1–10]. Among these systems, the TERrain COntour Matching (TERCOM) technique was developed in the 1960s. In particular, the design developed by Sandia Laboratories, the SITAN (Sandia Inertial Terrain Aided Navigation), which passes the terrain-matching errors to a Kalman filter to provide corrections to inertial navigation, is the closest to what will be described in this chapter.

For obvious reasons, the development of TRN paused for about 10 years when GPS/GNSS becomes the gold standard for PNT. The interest in TRN became renewed when both the military and commercial sectors realized the need to address the vulnerability of GNSS, due to signal blocking, jamming, spoofing, or simply the unavailability of the GNSS satellites. TRN has now become one of the key candidate technologies that can be included in an All Source PNT system that provides GNSS-denied or GNSS-challenged PNT solutions.

The TRN system has the advantage of being fully autonomous. It does not rely on any man-made infrastructure. However, it does rely on the terrain database, which needs to be created and maintained. One of the key disadvantages of the classical altimeter-based TRN is that it will not work if there are no terrain variations. This happens when a vehicle is flying over flat land and flat ocean. This is why other wider-sense TRNs such as MARN and GARN may be more attractive.

The wide-sense TRN includes the following variations:

1. The most basic TRN uses the difference of radar altimeter and baro altimeter as a measurement of the terrain height and uses the height to infer horizontal positions using a terrain database.
2. More generally, the radar altimeter can be replaced by one or more slant range sensors that measure the distance from an air vehicle to a point or a number of points on the ground. There are several advantages of this variation. First, for an air vehicle flying over the water but in the coastal area, it is possible that the slant range sensors can be pointing to the terrain along

the coastal line and the vehicle can navigate using TRN even when flying above the flat water. Second, when multiple slant range sensors are used, both the geometry and availability improve, which lead to better performance, availability, and integrity. Third, because the LOS directions do not have to be downward, it is possible that this technique can work for ground vehicles and marine/submarine vehicles.

3. When the measurement is actually a 2-D image of the terrain through a synthetic aperture radar (SAR), a LIDAR, or a camera.
4. When the sensor measures other geolocation-dependent variables, such as gravity anomaly and magnetic anomaly [11–19].

This chapter will focus on the first two types of TRN technology. Variation 3 will be discussed in Chapter 11 and Variation 4 will be discussed in Chapter 12.

As we discussed before, there are two types of sensors or information sources for PNT systems. TRN, as a type of a PNT system, is no exception. The typical examples of propagation sensors and information sources are listed here:

1. Inertial navigation is the most popular propagation solution. High-accuracy IMUs will need to be used if TRN and other updates may suffer from an extended period of unavailability. Lower-cost, lower-accuracy IMUs can be used if the combination of update sensors provides sufficient availability and integrity.
2. For some ultra-low-cost systems, the propagation solution can be simply the vehicle dynamics model. Due to input uncertainty and model uncertainty, the vehicle dynamic model-based propagations typically will not be able to maintain good accuracy for an extended period of time. As the result, if the vehicle dynamic model is used as the propagation solution, it is expected that the update solutions are such that they are available continuously.

The update sensors for terrain-based navigation are typically ranging devices, such as radar, sonar, or LIDAR. Some of these sensors use a large beam and have the capability to look for the shortest distance, consequently providing a good estimate of the altitude from the terrain. In other words, we have an altimeter. In this case, the altimeter update is mostly independent of the vehicle attitude.

Some of these devices use a very narrow beam and will provide a range measurement along the sensor boresight direction. In this case, the range measurements are coupled with both vehicle position and vehicle attitude. We call this type of devices slant range sensors.

Although it is not absolutely necessary, most of the TRN systems will also use a baro altimeter to provide altitude measurement with respect to the sea level.

It is worth mentioning that the Shuttle Radar Topography Mission (SRTM) in 2000 and the data processed and published thereafter have been a major breakthrough in digital global terrain mapping. The high-resolution, high-accuracy SRTM database that is publicly available has made TRN much easier, especially for nonmilitary applications [10].

9.2 TRN Architecture

Figure 9.1 illustrates a typical generic architecture for TRN. Within this architecture, the inertial navigation solution and the various altimeter/slant range sensor-based update solutions are integrated through a Kalman filter. The various types of sensors are all optional and there could be 0 to n sensors that need to be decided based on the requirements and the sensors available on the vehicle.

It is important to point out that, for the architecture described above, we assume that the initial navigation solution has a reasonable performance either through GNSS-based updates when GNSS is available or the solution has been initialized during initialization and alignment phase when the position knowledge is available. The reasonable performance is such that the terrain reference model around the current position with the position uncertainty remains in a monotonic region of the model. When the position uncertainty is too large, the inverse function of the terrain model will be multivalued and we will not be able to determine our position without using other information.

Similarly, we also need to realize that, in the classical altimeter-based TRN, the altitude is a 1-D variable and the derived horizontal position is a 2-D variable. In order to derive both axes of the horizontal position from the altitude measurements, the vehicle must experience terrain changes along multiple horizontal directions to create the observability. In between these direction changes, the inertial navigation plays the important role of allowing these observabilities generated by the different parts of the trajectory to be bridged together.

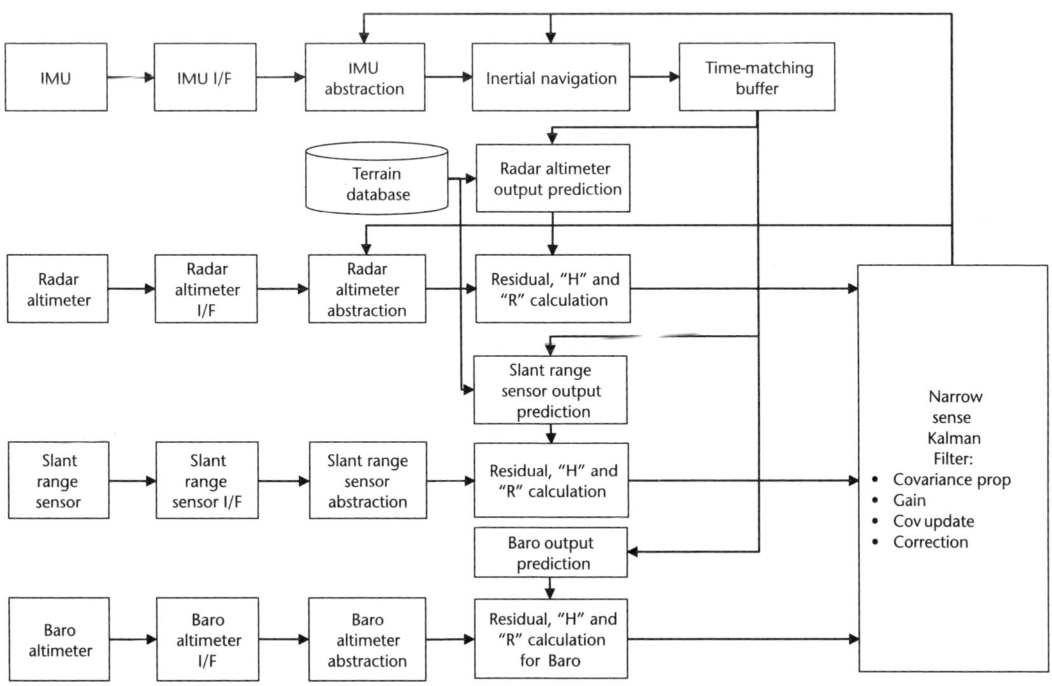

Figure 9.1 TRN architecture.

9.3 Propagation Algorithms and Error Models

The inertial navigation algorithms are discussed in Chapter 3 and there are several flavors of the algorithm depending on which frame in which the equations are written. For the purpose of this chapter, we use the INS algorithms presented in the ECEF frame, which are summarized here:

1. The attitude propagation equation in the quaternion form is simply

$$\dot{q}_B^E = \frac{1}{2}\left(q_B^E \otimes {}^B\omega_I^B - {}^E\omega_I^E \otimes q_B^E\right) \tag{9.1}$$

2. The DCM can be derived from the quaternion

$$C_B^E = q2DCM\left(q_B^E\right) \tag{9.2}$$

3. The velocity propagation equation is given by

$$\frac{d}{dt}{}^E v_E^B = C_B^E\,{}^B f_I^B + {}^E g - 2\,{}^E\omega_I^E \times {}^E v_E^B - \left[{}^E\omega_I^E \times\right]\left[{}^E\omega_I^E \times\right]{}^E r_E^B = C_B^E\,{}^B f_I^B + {}^E g_c - 2\,{}^E\omega_I^E \times {}^E v_E^B$$

where

$${}^E g_c = {}^E g - \left[{}^E\omega_I^E \times\right]\left[{}^E\omega_I^E \times\right]{}^E r_E^B \tag{9.3}$$

4. Finally, the position propagation equation is simply

$$\frac{d}{dt}{}^E r_E^B = {}^E v_E^B \tag{9.4}$$

For the covariance propagation of the Kalman filter, we will need the linearized error models for attitude, velocity, and position, which can be repeated from Chapter 3:

1. Attitude error propagation

$$\delta\,{}^E\dot{\theta}_B^E = -\left[{}^E\omega_I^E \times\right]\delta\,{}^E\theta_B^E - C_B^E\delta\,{}^B\omega_I^B \tag{9.5}$$

2. Velocity error propagation

$$\frac{d}{dt}\delta\,{}^E v_E^B = \left[\left(C_B^E m\,{}^B f_I^B\right)\times\right]\delta\,{}^E\theta_B^E + C_B^E\delta\,{}^B f_I^B + \left[\frac{\partial\,{}^E g}{\partial\,{}^E r_E^B}\right]\delta\,{}^E r_E^B - 2\left[\left({}^E\omega_I^E\right)\times\right]\delta\,{}^E v_E^B \tag{9.6}$$

3. Position error propagation

$$\frac{d}{dt}\delta\,{}^E r_E^B = \delta\,{}^E v_E^B \tag{9.7}$$

We will use the following gyro error models

$$\delta\,{}^{B}\omega_{I}^{B} = m^{B}\omega_{I}^{B} - {}^{B}\omega_{I}^{B} = C_{G}^{B}\mathrm{diag}\left({}^{G}s\right){}^{G}\omega_{I}^{B} + \left(\left[{}^{B}\theta_{G}\times\right] + L\left({}^{B}\beta_{G}\right)\right){}^{B}\omega_{I}^{B} + {}^{B}b_{G} + C_{G}^{B}ARW \tag{9.8}$$

where

$$\left[{}^{B}\theta\times\right] = \begin{bmatrix} 0 & -\theta_z & \theta_y \\ \theta_z & 0 & -\theta_x \\ -\theta_y & \theta_x & 0 \end{bmatrix}, \quad L\left({}^{B}\beta\right) = \begin{bmatrix} 0 & 0 & 0 \\ \beta_z & 0 & 0 \\ -\beta_x & \beta_y & 0 \end{bmatrix} \tag{9.9}$$

which can also be written in a more Kalman filter-friendly way as

$$\delta\,{}^{B}\omega_{I}^{B} = C_{G}^{B}\mathrm{diag}\left({}^{G}\omega_{I}^{B}\right){}^{G}s - \left[{}^{B}\omega_{I}^{B}\times\right]{}^{B}\theta_{G} + J\left({}^{B}\omega_{I}^{B}\right){}^{B}\beta_{G} + {}^{B}b_{G} + C_{G}^{B}ARW \tag{9.10}$$

where

$$J(\omega) = \begin{bmatrix} 0 & 0 & 0 \\ 0 & 0 & \omega_x \\ \omega_y & -\omega_x & 0 \end{bmatrix} \tag{9.11}$$

For the accelerometers, since we use the accelerometers as the reference to form the IMU frame, the accelerometers do not have any rotational misalignment errors but only have nonorthogonality misalignment errors:

$$\delta\,{}^{B}f_{I}^{B} = m^{B}f_{I}^{B} - {}^{B}f_{I}^{B} = C_{A}^{B}\mathrm{diag}\left({}^{A}s\right){}^{A}f_{I}^{B} + L\left({}^{B}\beta_{A}\right){}^{B}f_{I}^{B} + {}^{B}b_{A} + C_{A}^{B}VRW \tag{9.12}$$

which can also be written in a more Kalman filter-friendly way as

$$\delta\,{}^{B}f_{I}^{B} = C_{A}^{B}\mathrm{diag}\left({}^{A}f_{I}^{B}\right){}^{A}s + J\left({}^{B}f_{I}^{B}\right){}^{B}\beta_{A} + {}^{B}b_{A} + C_{A}^{B}VRW \tag{9.13}$$

where

$$J\left({}^{B}f_{I}^{B}\right) = \begin{bmatrix} 0 & 0 & 0 \\ 0 & 0 & {}^{B}f_{I\,x}^{B} \\ {}^{B}f_{I\,y}^{B} & -{}^{B}f_{I\,x}^{B} & 0 \end{bmatrix} \tag{9.14}$$

and VRW is the accelerometer velocity random walk, ${}^{A}s$ is the scale factor error in the accelerometer sensor frame, ${}^{B}\beta_{A}$ is the accelerometer nonorthogonality misalignments, and ${}^{B}b_{A}$ is the accelerometer bias vector in the body frame.

If we model each one of the IMU error parameters as a random walk process and let

$$\begin{aligned} \delta X_{I}^{T} &= \left[\delta\,{}^{E}\dot{\theta}_{B}^{E}, \delta\,{}^{B}b_{G}{}^{T}, \delta\,{}^{G}s^{T}, \delta\,{}^{B}\beta_{G}^{T}, \delta\,{}^{B}\theta_{G}^{T}, \left(\delta\,{}^{E}v_{E}^{B}\right)^{T}, \delta\,{}^{B}b_{A}^{T}, \delta\,{}^{A}s^{T}, \delta\,{}^{B}\beta_{A}^{T}, \left(\delta\,{}^{E}r_{E}^{B}\right)^{T}\right] \\ W_{I}^{T} &= \left[-\left(C_{G}^{E}ARW\right)^{T}, RRW^{T}, SRW_{G}^{T}, \beta RW_{G}^{T}, \theta RW_{G}^{T}, \left(C_{A}^{E}VRW\right)^{T}, ABRW^{T}, SRW_{A}^{T}, \beta RW_{A}{}^{T}, 0\right] \end{aligned} \tag{9.15}$$

we then have

$$\delta \dot{X}_{II} = A_{II}\delta X_I + W_I$$

$$A_{II} = \begin{bmatrix} -\left[{}^E\omega_I^E \times\right] & -C_B^E & -C_G^E \mathrm{diag}\left({}^G\omega_I^B\right) & -C_B^E J\left({}^B\omega_I^B\right) & C_B^E\left[{}^B\omega_I^B \times\right] & 0_{3x3} & 0_{3x3} & 0_{3x3} & 0_{3x3} & 0_{3x3} \\ 0_{3x3} & 0_{3x3} & 0_{3x3} & 0_{3x3} & 0_{3x3} & 0_{3x3} & 0_{3x3} & 0_{3x3} & 0_{3x3} & 0_{3x3} \\ 0_{3x3} & 0_{3x3} & 0_{3x3} & 0_{3x3} & 0_{3x3} & 0_{3x3} & 0_{3x3} & 0_{3x3} & 0_{3x3} & 0_{3x3} \\ 0_{3x3} & 0_{3x3} & 0_{3x3} & 0_{3x3} & 0_{3x3} & 0_{3x3} & 0_{3x3} & 0_{3x3} & 0_{3x3} & 0_{3x3} \\ 0_{3x3} & 0_{3x3} & 0_{3x3} & 0_{3x3} & 0_{3x3} & 0_{3x3} & 0_{3x3} & 0_{3x3} & 0_{3x3} & 0_{3x3} \\ \left[\left(C_B^E m\,{}^B f_I^B\right)\times\right] & 0_{3x3} & 0_{3x3} & 0_{3x3} & 0_{3x3} & -2\left[\left({}^E\omega_I^E\right)\times\right] & C_B^E & C_A^E \mathrm{diag}\left({}^A f_I^B\right) & C_B^E J\left({}^B f_I^B\right) & \left[\dfrac{\partial^E g}{\partial^E r_E^B}\right] \\ 0_{3x3} & 0_{3x3} & 0_{3x3} & 0_{3x3} & 0_{3x3} & 0_{3x3} & 0_{3x3} & 0_{3x3} & 0_{3x3} & 0_{3x3} \\ 0_{3x3} & 0_{3x3} & 0_{3x3} & 0_{3x3} & 0_{3x3} & 0_{3x3} & 0_{3x3} & 0_{3x3} & 0_{3x3} & 0_{3x3} \\ 0_{3x3} & 0_{3x3} & 0_{3x3} & 0_{3x3} & 0_{3x3} & 0_{3x3} & 0_{3x3} & 0_{3x3} & 0_{3x3} & 0_{3x3} \\ 0_{3x3} & 0_{3x3} & 0_{3x3} & 0_{3x3} & 0_{3x3} & I_{3x3} & 0_{3x3} & 0_{3x3} & 0_{3x3} & 0_{3x3} \end{bmatrix} \tag{9.16}$$

9.4 Measurement Equations and Error Models

9.4.1 Baro Altimeter Update

The baro altimeter measures the baro altitude and we typically use the following model to model the measurement of the baro altimeter:

$$mh_{baro} = h_e\left({}^E r_I^E\right) + \delta h\left({}^E r_I^E\right) + b_b + v_b \tag{9.17}$$

where h_e is the function that converts ECEF position to ellipsoidal altitude (see Chapter 2).

$\delta h\left({}^E r_I^E\right)$ is the spherical harmonics model that models the difference between the WGS84 ellipsoid and the geoid altitude.

b_b is the sensor bias that models the slow-changing part of the sensor error.

v_b is the measurement noise that is a combination of the sensor noise and the changes of measurement due to fluctuating pressure around the sensor.

It is important to point out that the difference between the sensor bias and the sensor noise becomes blurred in practice because the bias will unlikely be a pure constant and the measurement noise will unlikely be a pure white noise and it may take some more complicated mathematical models to model each of these variables.

For now v_b is modeled as a white noise process. In practice, the techniques described in Chapter 2 can be utilized to deal with the reality that the noise may not be actually white:

$$E\left(v_b(t)\right) = 0$$

$$E\left(v_b(t)v_b(s)\right) = \sigma_b^2\delta(t - s)$$

The prediction is simply done by

$$\hat{h}_{baro} = h_e\left({}^E\hat{r}_I^E\right) + \delta h\left({}^E\hat{r}_I^E\right) + \hat{b}_b \tag{9.18}$$

The model can be linearized as

$$\delta mh_{baro} = \frac{\partial h_e\left({}^E r_I^E\right)}{\partial {}^E r_I^E}\delta {}^E r_I^E + \delta b_b + v_b \tag{9.19}$$

where we ignored the sensitivity on the ellipsoid to the geoid correction.

The above equations can be used to form the H-matrix for the Kalman filter.

9.4.2 Radar Altimeter Update

The measured terrain altitude can be modeled as

$$mh_p = h_e\left({}^E r_E^B\right) - \eta(lat, lon) + b_r + v_r \tag{9.20}$$

where $\eta(lat, lon)$ is the terrain altitude above the WGS84 ellipsoid as a function of latitude and longitude (as typically provided by a terrain database), b_r is the sensor bias, and v_r is the radar altimeter measurement noise that is modeled as a white noise process. In practice, techniques described in Chapter 2 can be utilized to deal with the reality that the noise may not be actually white:

$$E\left(v_r(t)\right) = 0$$
$$E\left(v_r(t)v_r(s)\right) = \sigma_r^2\delta(t - s)$$

To predict the measurement, we first convert the ECEF position to (Lat, Lon, and Altitude) (see Chapter 2)

$$L\hat{L}A = \begin{bmatrix} l\hat{a}t \\ l\hat{o}n \\ \hat{h} \end{bmatrix} = ecef2lla\left({}^E \hat{r}_E^B\right)$$

then calculate

$$m\hat{h}_p = h_e\left({}^E \hat{r}_E^B\right) - \eta(l\hat{a}t, l\hat{o}n) + \hat{b}_r$$

We will derive the linearized measurement model for the purpose of the Kalman filter H-matrix

$$\begin{aligned}\delta mh_p &= \frac{\partial h_e\left({}^E r_E^B\right)}{\partial {}^E r_E^B}\delta {}^E r_E^B - \frac{\partial \eta(lat,lon)}{\partial lat}\frac{\partial lat}{\partial {}^E r_E^B}\delta {}^E r_E^B - \frac{\partial \eta(lat,lon)}{\partial lon}\frac{\partial lon}{\partial {}^E r_E^B}\delta {}^E r_E^B + \delta b_r + v_r \\ &= \left(\frac{\partial h_e\left({}^E r_E^B\right)}{\partial {}^E r_E^B} - \frac{\partial \eta(lat,lon)}{\partial lat}\frac{\partial lat}{\partial {}^E r_E^B} - \frac{\partial \eta(lat,lon)}{\partial lon}\frac{\partial lon}{\partial {}^E r_E^B}\right)\delta {}^E r_E^B + \delta b_r + v_r\end{aligned} \tag{9.21}$$

As an example of terrain database $\eta(lat, lon)$ that is publicly available, the NASA Shuttle Radar Topographic Mission (SRTM) provides 3-arcsec (approximately 90-m) resolution digital elevation data for over 80% of the globe. The availability of SRTM database has been a major breakthrough in the digital mapping of the world. This data is currently distributed free of charge by USGS (United States

Geological Survey) and is available for download from the National Map Seamless Data Distribution System [10].

9.4.3 Slant Range Update

As discussed before, the radar altimeter measures the distance from the local Earth surface to the vehicle that carries the altimeter along the local vertical direction. In order to do that, the radar altimeter has to either use a wider beam to look for the shortest distance or use a platform that allows a narrow radar beam to point along the local vertical direction.

A variation of the radar altimeter is the narrow beam slant range sensor that measures the range from the sensor to the piercing point on the Earth surface. Multiple slant range sensors can be used to provide improved observability through geometry augmentation and noise averaging.

To avoid complicated symbols, we will use

$$g(r) = h_e(r) - \eta\big(lat(r), lon(r)\big) = 0 \tag{9.22}$$

to represent the constraint imposed by the Earth terrain profile, where r is the 3-D ECEF position vector for a point on the Earth surface, and (lat, lon, h) are the corresponding latitude, longitude, and altitude for the same point. $g(.)$ is a 1-D function that constrains the position to a 2-D Earth terrain surface.

Similar to Chapter 4, we can define ${}^{B}u_s$, the sensor boresight axis in the body frame that can be transformed to the sensor boresight axis in the ECEF frame:

$${}^{E}u_s = C_B^E \, {}^{B}u_s$$

Given the vehicle position and the sensor boresight axis, the initial guess of the Earth-piercing point can be just the solution from the previous frame. When there is no previous frame solution available, we can calculate a first guess by using a spherical Earth model. In this case, the range $\bar{d}$ to the piercing point is the solution of the following quadratic equation:

$$d^2 + 2\left(\left({}^{E}u_s\right)^T \left({}^{E}R\right)\right)d + \left(\left\|{}^{E}R\right\|^2\right) - \left(r_e + h\right)^2 = 0 \tag{9.23}$$

The actual piercing point can then be calculated as

$${}^{E}\bar{R}_{pp} = {}^{E}R + (\bar{d})\,{}^{E}u_S \tag{9.24}$$

Once we have the first guess, the range and piecing point can then be calculated through iterations:

$$\begin{aligned}
\begin{bmatrix} \delta\,{}^{E}R_{pp} \\ \delta d \end{bmatrix} &= \begin{bmatrix} I & -{}^{E}u_s \\ g_R\left({}^{E}\bar{R}_{pp}\right) & 0 \end{bmatrix}^{-1} \begin{bmatrix} {}^{E}R + \bar{d}\,{}^{E}u_S - {}^{E}\bar{R}_{pp} \\ -g\left({}^{E}\bar{R}_{pp}\right) \end{bmatrix} \\
\begin{bmatrix} {}^{E}R_{pp} \\ d \end{bmatrix} &= \begin{bmatrix} {}^{E}\bar{R}_{pp} \\ \bar{d} \end{bmatrix} + \begin{bmatrix} \delta\,{}^{E}R_{pp} \\ \delta d \end{bmatrix}
\end{aligned} \tag{9.25}$$

where

$$g_R\left({}^E R_{pp}\right) = \frac{\partial g_R\left({}^E R_{pp}\right)}{\partial {}^E R_{pp}} \tag{9.26}$$

is the gradient of the terrain constraint function.

The linearized model for the range deviation is given by

$$\begin{aligned} G = G(1{:}4,1{:}4) &= \begin{bmatrix} I & -{}^E u_s \\ g_R\left({}^E R_{pp}\right) & 0 \end{bmatrix}^{-1} \\ \delta d &= G(4,1{:}3)\delta {}^E R_E^B - G(4,1{:}3) d\delta^E u_s \end{aligned} \tag{9.27}$$

Consequently, the nonlinear sensor model is given by

$$d = d\left({}^E R_E^B, {}^E u_S\right) + b_s + v_s \tag{9.28}$$

where the nonlinear function d represents the initialization and the iterative algorithm described above, b_s is the slant range sensor bias, and v_s is the slant range sensor measurement noise that is modeled as a white noise process. In practice, the techniques described in Chapter 2 can be utilized to deal with the reality that the noise may not be actually white:

$$\begin{aligned} E\left(v_s(t)\right) &= 0 \\ E\left(v_s(t)v_s(s)\right) &= \sigma_s^2 \delta(t - s) \end{aligned}$$

The linearized model can then be written as

$$\delta md = G(4,1{:}3)\delta {}^E R_E^B - G(4,1{:}3) d\delta^E u_s + \delta b_s + v_s \tag{9.29}$$

From Chapter 4, we also know that

$$\delta {}^E u_s = \left[{}^E u_s \times\right] \delta {}^E \theta_E^B \tag{9.30}$$

The linearized measurement equation can then be written as

$$\delta md = G(4,1{:}3)\delta {}^E R_E^B - G(4,1{:}3) d\left[{}^E u_s \times\right] \delta {}^E \theta_E^B + \delta b_s + v_s \tag{9.31}$$

Although, in reality, there are various combination of sensors used to achieve specific purposes of a system with various constraints, in the following, we will simply show the Kalman filter constructed with one of each of the three types of sensors mentioned above.

First, we will augment the Kalman filter state vector to include the bias for each of sensors, and we will simply model them as a random walk process:

$$\begin{aligned} X^T &= \left[X_I^T, b_b, b_r, b_s\right] \\ \delta X^T &= \left[X_I^T, \delta b_b, \delta b_r, \delta b_s\right] \end{aligned} \tag{9.32}$$

The linearized propagation error model is then given by

$$\frac{d}{dt}(\delta X) = \begin{bmatrix} A_{II} & 0_{30x1} & 0_{30x1} & 0_{30x1} \\ 0_{1x30} & 0 & 0 & 0 \\ 0_{1x30} & 0 & 0 & 0 \\ 0_{1x30} & 0 & 0 & 0 \end{bmatrix} \delta X + \begin{bmatrix} W_I \\ bbrw \\ brrw \\ bsrw \end{bmatrix} \tag{9.33}$$

where *bbrw*, *brrw*, and *bsrw* are the white noise processes that generate the random walk processes b_b, b_r, and b_s, respectively. Specifically, this means that each one of b_b, b_r, and b_s is modeled as a random walk process such that

$$\begin{aligned} &\frac{d}{dt} b_i = birw \\ &E\left(birw(t)\right) = 0 \\ &E\left(birw(t) birw(s)\right) = \left(\sigma_{birw}\right)^2 \delta(t-s) \\ &\text{where } i = b,\ r,\ s \end{aligned} \tag{9.34}$$

The total Q-matrix is then given by

$$Q = \begin{bmatrix} Q_I & 0 & 0 & 0 \\ 0 & \left(\sigma_{bbrw}\right)^2 & 0 & 0 \\ 0 & 0 & \left(\sigma_{brrw}\right)^2 & 0 \\ 0 & 0 & 0 & \left(\sigma_{bsrw}\right)^2 \end{bmatrix} \tag{9.35}$$

The linearized measurement equation can be written as

$$\begin{bmatrix} \delta mh_b \\ \delta mh_r \\ \delta md_s \end{bmatrix} = \begin{bmatrix} H_b & 1 & 0 & 0 \\ H_r & 0 & 1 & 0 \\ H_s & 0 & 0 & 1 \end{bmatrix} \delta X + \begin{bmatrix} v_b \\ v_r \\ v_s \end{bmatrix} \tag{9.36}$$

where

$$H_b = \left[0, 0, \ldots, 0, \frac{\partial h_e}{\partial {}^E r_E^B}\right] \tag{9.37}$$

$$H_b = \left[0, 0, \ldots, 0, \left(\frac{\partial h_e\left({}^E r_E^B\right)}{\partial {}^E r_E^B} - \frac{\partial \eta(lat, lon)}{\partial lat} \frac{\partial lat}{\partial {}^E r_E^B} - \frac{\partial \eta(lat, lon)}{\partial lon} \frac{\partial lon}{\partial {}^E r_E^B}\right)\right] \tag{9.38}$$

$$H_s = \left[-G(4{:},1{:}3) d\left[{}^E u_S \times\right], 0, \ldots, 0, G(4{:},1{:}3)\right] \tag{9.39}$$

and the total R-matrix is then given by

$$R = \begin{bmatrix} (\sigma_b)^2 & 0 & 0 \\ 0 & (\sigma_r)^2 & 0 \\ 0 & 0 & (\sigma_s)^2 \end{bmatrix} \tag{9.40}$$

It is worth mentioning that, in reality, we may not necessarily want to use all three measurements at the same time in the same equation and the total R-matrix as shown above. It is often profitable to apply each type of measurement separately using different update rates and different ways to deal with nonwhite measurement noise according to the reality of each of the individual sensors.

9.5 Simulation Examples

Figure 9.2 illustrates the trajectory and the terrain for this simple example. The simulation uses the SRTM terrain database. The shades in Figure 9.2 show the altitude variations and the box in the figure is the simple trajectory used in the simulation.

Example 9.1 Radar Altimeter

Figure 9.3 shows the simulation results for radar altimeter-based classical TRN. The upper and lower bounds are the 3-sigma bounds calculated from the diagonal elements of the Kalman filter covariance matrix. The error curve in the middle of the upper and lower bound curves is the result of a single instance of the Monte Carlo simulation run.

Example 9.2: Single Slant Range Sensor 45° Forward

Figure 9.4 shows the simulation results of TRN using a narrow-beam slant range sensor. The upper and lower bounds are the 3-sigma bounds calculated from the

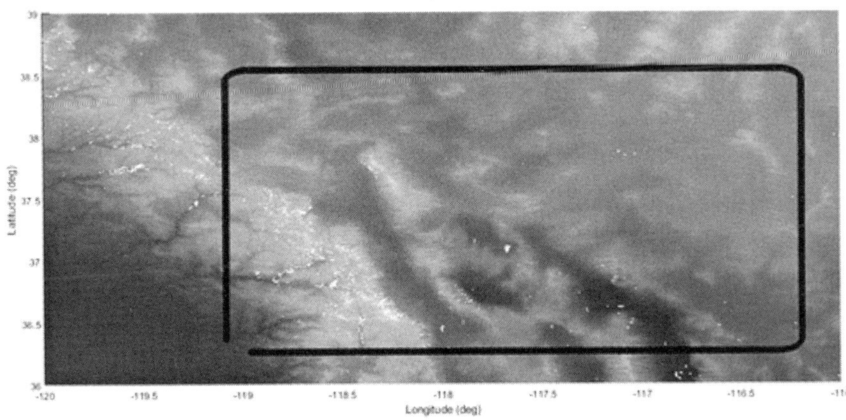

Figure 9.2 Terrain and trajectory for the simple examples.

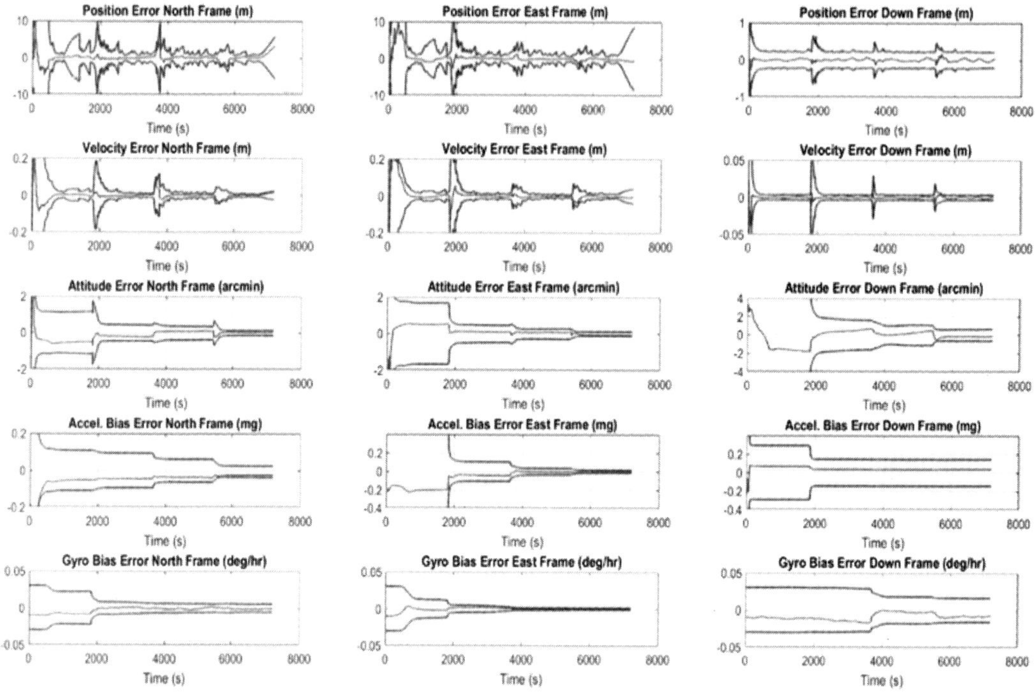

Figure 9.3 Radar altimeter-based classical TRN simulation results

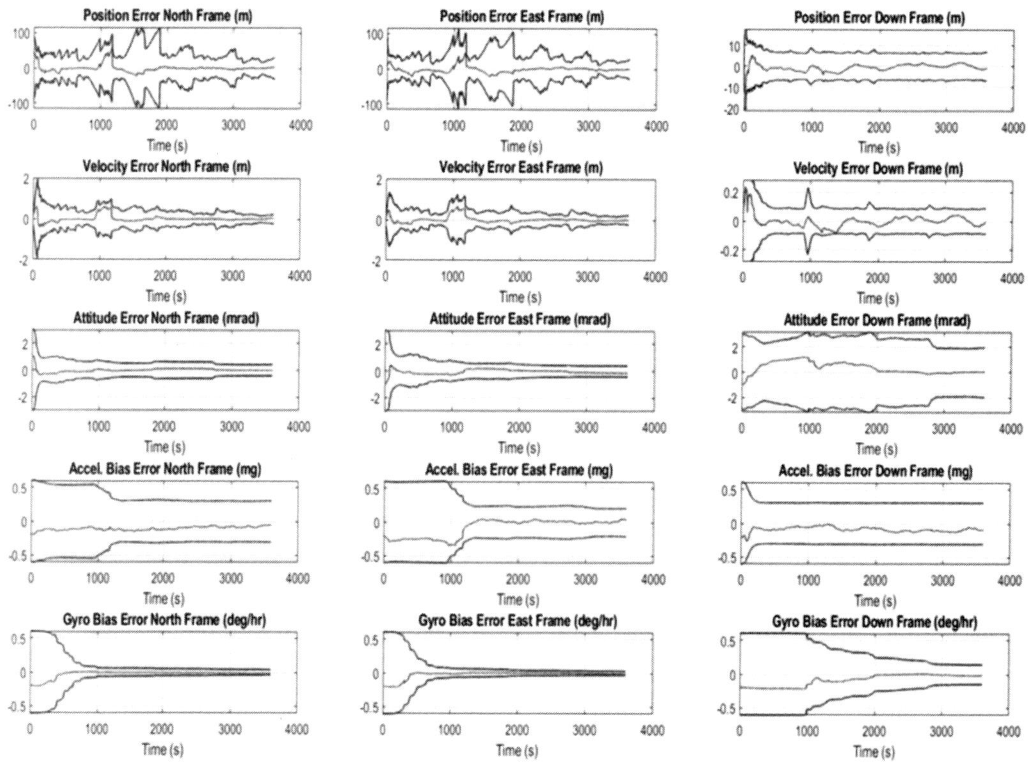

Figure 9.4 Single slant range sensor-based TRN: simulation results.

diagonal elements of the Kalman filter covariance matrix. The error curve in the middle of the upper and lower bound curves is the result of a single instance of the Monte Carlo simulation run.

References

[1] Hollowell, J., "Heli/SITAN: A Terrain Referenced Navigation Algorithm for Helicopters," *IEEE Symposium on Position Location and Navigation. A Decade of Excellence in the Navigation Sciences*, March 1990.

[2] McGuffin, J. T., "Terrain Referenced Navigation Electromagnetic-Gravitational Correlation," U.S. Patent 5272639A, 1992 (2019-11-14 Application status is Expired—Lifetime).

[3] Fellerhoff, J. R., "SITAN Implementation in the Saint System," *IEEE Journal*, 1986, pp. 89–95.

[4] Baird, C. A., and M. R. Abramson, "A Comparison of Several Digital Map-Aided Navigation Techniques," *IEEE Journal*, 1984, pp. 286–293.

[5] Hostetler, L. D., and R. D. Andreas, "Nonlinear Kalman Filtering Techniques for Terrain-Aided Navigation," *IEEE Transactions on Automatic Control*, Vol. AC-28, No. 3, March 1983, pp. 315–323.

[6] Priestley, N., "Terrain Referenced Navigation," *IEEE Symposium on Position Location and Navigation, A Decade of Excellence in the Navigation Sciences*, March 20, 1990.

[7] Henley, A. J., "Terrain Aided Navigation: Current Status, Techniques for Flat Terrain and Reference Data Requirements," *IEEE Symposium on Position Location and Navigation. A Decade of Excellence in the Navigation Sciences*, March 20, 1990.

[8] Jalving, B., et al., "Terrain Referenced Navigation of AUVs and Submarines Using Multi Beam Echo Sounders," Norwegian Defense Research, 2004, http://citeseerx.ist.psu.edu/viewdoc/download?doi=10.1.1.459.3521&rep=rep1&type=pdf.

[9] Metzger, J., et al., "Sigma-Point Filter for Terrain Referenced Navigation," *AIAA Guidance, Navigation, and Control Conference and Exhibit*, San Francisco, CA, August 15–18, 2005.

[10] https://cgiarcsi.community/data/srtm-90m-digital-elevation-database-v4-1/.

[11] Richeson, J. A., "Gravity Gradiometer Aided Inertial Navigation Within Non-GNSS Environments," Ph.D. Dissertation, University of Maryland, College Park, MD, 2008.

[12] Rogers, M. M., "An Investigation into the Feasibility of Using a Modern Gravity Gradient Instrument for Passive Aircraft Navigation and Terrain Avoidance," Master's Thesis, Air Force Institute of Technology, 2009.

[13] Wu, X., "Gravity Gradient Survey with a Mobile Atom Interferometer," Ph.D. Dissertation, Stanford University, 2009.

[14] Metzger, E. H., and A. Jircitano, "Inertial Navigation Performance Improvement Using Gravity Gradient Matching Techniques," *J. Spacecraft*, June 1976.

[15] Moryl, J., H. Rice, and S. Shinners, "The Universal Gravity Module for Enhanced Submarine Navigation," *IEEE 1998 Position Location and Navigation Symposium*, Palm Springs, CA, 1998.

[16] Wu, L., et al., "Performance Evaluation and Analysis for Gravity Matching Aided Navigation," *Sensors*, April 2017.

[17] Wang, H., et al., "Location Accuracy of INS/Gravity-Integrated Navigation System on the Basis of Ocean Experiment and Simulation," *Sensors*, December 2017.

[18] Wang, B., et al., "A Particle Filter-Based Matching Algorithm with Gravity Sample Vector for Underwater Gravity Aided Navigation," *IEEE/ASME Transactions on Mechatronics*, Vol. 21, No. 3, June 2016.

[19] Morales, G., "Magnetometer Aided Navigation Filters for Improved Observability and Estimation on Ground Vehicles," Master's Thesis, Auburn University, 2016.

CHAPTER 10

Modern Celestial Navigation

10.1 Introduction

Celestial navigation uses the angular measurements of heavenly bodies (e.g., stars, Sun, Moon, planets, satellites or RSOs) to determine attitude, position, and time. Celestial navigation is both a classical and a modern navigation approach. Celestial navigation is one of the earliest PNT methods of human being. It dates back at least to 1000 B.C. when humans started to use angular observations of the Sun, the Moon, the stars, and the observation of the horizon to determine position and local time.

Celestial navigation has been extensively used for marine navigation as well as the navigation of aircraft and missiles. It experienced a (very short) period of low demand when GPS became fully operational and available as a low-cost, high-accuracy, high-availability navigation method between the early 1990s and 2013 or so. Today, celestial navigation is receiving renewed attention due to concerns about the unavailability of GPS in a day without space or star-wars scenarios and the vulnerability of GPS due to jamming and spoofing.

It is worthwhile to say that spacecraft attitude determination went through a slightly different path. The optical sensor-based approach has always been the primary means of spacecraft attitude determination. Some of the early spacecraft used Sun sensors and Earth sensors for attitude determination. The star-tracker-based approach became the approach of choice for most spacecraft since the 1990s due to its high accuracy and the availability of relatively low-cost star-trackers. This is even more so after several European vendors have developed multiple product lines of a low-cost, high-performance, competitive offering of space-qualified star-trackers since the late 1990s. GPS position and velocity (or pseudo-range and delta range) measurements cannot help the determination of attitude for a spacecraft in orbit because the coupling of attitude errors into velocity and position errors in a terrestrial strapdown inertial navigation system is no longer true for orbital navigation. Multi-antenna carrier-phase differential GPS-based attitude determination has not been a competitive solution for spacecraft applications. This is mainly due to the significant errors caused by multipath and the difficulty to find clearance for the FOV of the GPS antennas to be Earth facing.

We will refer to the pre-1990 celestial navigation technology as the classic celestial navigation and the technology that is currently being developed as the modern celestial navigation. Table 10.1 summarizes the key differences between the classic and modern methods of celestial navigation.

Table 10.1 Modern Versus Classic Celestial Navigation

	Modern	*Classic*
Star-tracker technology	Wide FOV short-wave IR (SWIR) daytime capable star tracker [1, 2]	Very narrow FOV visible light star-tracker (also daytime capable)
Platform	Mostly strapdown, gimbals, or other mechanisms may be used to realize larger effective FOV	Gimbal platform is used to stabilize
Reference objects used	Stars and satellites or in general RSOs	Stars mostly, Sun and Moon sometimes
Cost	Much lower	Typically high

10.2 Fundamental Principles of Celestial Navigation (and Related Applications)

Fundamentally, celestial navigation is based on the following principles:

1. Attitude determination:
   a. The LOS measurements of a single star (a unit vector, two angles, or h and v coordinates of stars in the star-tracker FOV) provide information to constrain 2 degrees of freedom of the observer's attitude in the inertial frame (e.g., the ECI frame).
   b. Consequently, attitude with respect to the inertial frame can be determined using two or more stars (see Figure 10.1).
2. Attitude and position determination:
   a. The LOS measurements of a single finite distance object (FDO) (such as the Sun, the Moon, the planet, and the satellites and RSOs) provide

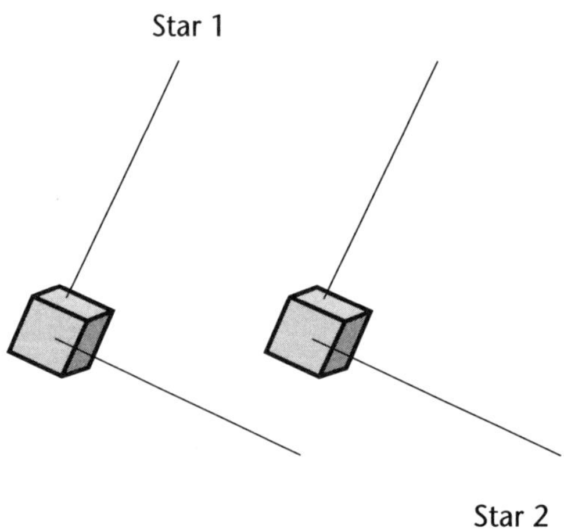

Figure 10.1 Attitude determination using stars.

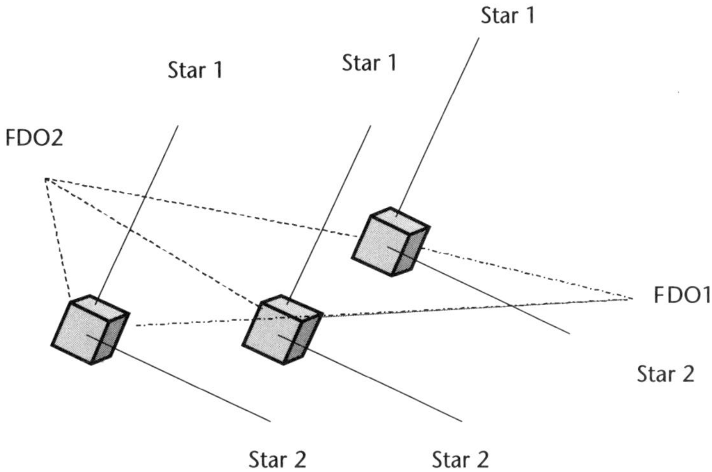

Without including the special cases of singularity
2 Stars and 1 FDO is not sufficient: the object can still move translational alone the LOS of FDO
2 FDOs are not sufficient 4 DOF information is not sufficient to determine 6DFO information
2 Stars with 2 FDO is sufficient to determine the attitude and position
1 Star with 2 FDO is also sufficient

Figure 10.2 Navigation using stars and FDOs.

2 degrees of freedom measurements that are functions of the observer's position and attitude.

b. Consequently, the combination of measurements of stars and FDOs from three to four or more sources with at least one FDO can determine both the position and attitude of the observer (see Figure 10.2).
c. For Earth navigation, horizon or local vertical measurements provide 2 degrees of freedom measurements with respect to the Earth. (One way to understand this is to consider the local vertical measurements as the LOS vector towards the center of the heavenly body Earth.) The altimeter measurement constrains another degree of freedom.
d. Consequently, combination star measurements, local vertical measurements, and altimeter can determine both position and attitude (see Figure 10.3).
e. Schuler tuned inertial navigation on Earth implicitly contain information about local vertical in the sense that the position error will translate into attitude error with respect to the local vertical, which then translates into gravity projection error for inertial navigation, which is then integrated into velocity and position error (hence, the Schuler tuned inertial navigation).

3. Importance of time and time determination:
   a. For Earth navigation, we need to know the difference between the inertial frame (e.g., ECI) and the Earth frame (e.g., ECEF).
   b. The relative orientation of the ECI and ECEF are accurately modeled by the Greenwich hour angle (GHA) and the Earth precession and nutation angles and all of them are accurately predicted using the absolute time.

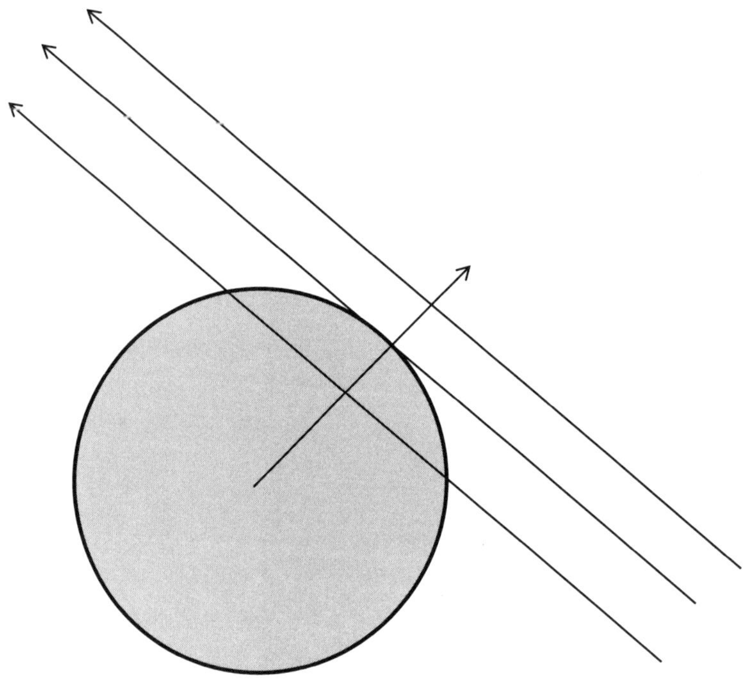

Figure 10.3 Navigation using stars, the local vertical, and an altimeter.

   c. In addition, if RSOs are used, accurate time is also required to calculate the RSO position using the RSO ephemeris for a particular time moment.
   d. Conversely, if we have both celestial measurements and other sources of position information (such as vision based navigation) available, we will have the capability to calibrate time knowledge errors. It is important to know that 1 second of time error only causes 15 arcsec of error in predicting the ECEF and ECI relative orientation. It will be very difficult to achieve very high time accuracy based on attitude or position error derived from other sources.
4. Attitude and position propagation and measurements of star and FDOs over time:
   a. Attitude and position information available at one time moment can be propagated to another time moment based on knowledge of rotation rate and acceleration that can be provided by a model (such as a gravity model-based orbit propagator) or by the measurements of gyros and accelerometers (such as an INS).
   b. With such a mechanism in place, the requirements of combinations of stars and FDOs can be relaxed such that we will have combination of

the measurements over time instead of at a single time epoch. As an example, the LOS measurements of Earth at different time moments are in different inertial directions and, consequently, the measurements of Earth LOS alone over time will be sufficient to determine the attitude of spacecraft when gyro data is used to propagate the attitude.

With the above principles, many different types of attitude determination and navigation systems can be defined and built. The following are a number of examples.

Example 10.1: Spacecraft Attitude Determination Using Star-Tracker Only

Two or more stars can be used to determine the inertial attitude of spacecraft. We discussed the topic of SIAD in Chapter 7 and there is also much literature in this field [3–60]. It is important to note that most of the contents in Chapter 7, such as the star-tracker error model, the building of the star catalog, and the SAA algorithms, are all still important for the purposes of celestial navigation and they are not repeated in this chapter.

Example 10.2: Strapdown Earth Navigation Using INS, Altimeter, and Star Measurements Only

In this case, the attitude information from star measurements will be bounding the attitude error to very small values (consequently, we do not need very high-quality gyros for such applications). The velocity and position errors will be Schuler-tuned and bounded. In this case, the INS is essentially providing the local vertical reference, which enables the total system to provide full attitude, velocity, and position solutions with bounded errors. The altimeter is required to bound error drift along the local vertical axis.

Example 10.3: Strapdown Earth Navigation Using INS and Star or RSO Measurements.

The addition of the RSO measurements will provide greatly improved position and velocity performance on the top of the star-only-based solution [61–64].

The details of Examples 10.2 and 10.3 are the focus of this chapter.

Example 10.4: Spacecraft Deep Space Optical Navigation.

By observing FDOs such as planet, Moon, or the Sun with a background stars, we will be able to determine both the attitude and position.

10.3 Observation of the Stars and the RSO: Difficulties and Solutions

Although modern celestial navigation offers the promise of GPS-independent navigation solution, there are a number of problems that must be solved in order to provide an operationally feasible solution. Some of the most important difficulties and solutions are discussed in this section.

Table 10.2 Probability of Clear Sky Versus Altitude

	Altitude	*Probability of Clear Sky*
1	Surface of Earth	50%
2	2,000m	60%
3	7,000m	70%
4	13,000m	95%

10.3.1 Cloud and Cloud Avoidance

Cloud presents one of the biggest challenges to celestial navigation. Fortunately, the probability to see clouds becomes much lower at a high altitude; most of the military aircraft and the commercial jet liners will be flying at altitudes where there is very little cloud coverage above the vehicles.

We are not that fortunate when it comes to marine navigation and ground navigation; as shown in Table 10.2, we only have 50% of probability of seeing the clear sky when we are at ground level.

Consequently, low-altitude celestial navigation must consider the impact of cloud on the availability of measurements. Since there are plenty of stars in the sky, one solution is to use a star-tracker that has steerable FOV either by installing the star-tracker on a gimbal platform or by using an optical beam director to allow us to avoid the clouds during operations.

One easy strategy for cloud avoidance is to use an inexpensive full-sky camera to map the clouds in real time and steer the star-tracker FOV towards the part of the sky where there is no clouds.

10.3.2 Daytime Star Observations/Day Time Capable Star Tracker

Another challenge to celestial navigation is the sky radiance during the day. That is exactly why human eyes cannot see any stars or satellites during the daytime. A number of techniques can be used to reduce the contribution of the sky radiance so that a star-tracker can reliably see stars and satellites during the day:

1. *Use near-IR or IR band:* The blue sky contains mostly short-wavelength visible light that is blue. As the wavelength increases, the atmospheric bright background becomes very dim. However, there are more near-IR and IR-band bright stars than visible light band stars. In addition, satellite typically remains bright in the near-IR and IR band. Consequently, using near-IR or IR-band detectors will make things much easier. SWIR band detectors seem to be the technology of choice for daytime star trackers.
2. *Simply limit the FOV:* The atmospheric bright background acts like a continuum of omnidirectional light sources. At a small area on the focal plane around the star, we have the summation of the energy from the light from the star (or a satellite or other objects of interest) and the integration of the

light from all of the sky. The second item will be reduced proportionally as the FOV is reduced. In other words, a small enough FOV will allow the star to stand out. This is in fact the technique used by the last generation of celestial navigation star-trackers, which use visible light detectors, with a very narrow FOV, to allow stars to be visible during the day.

3. *Use of long and well-designed light shade:* Using a long and well-designed light shade prevents the stray light from different sources hitting the detector array.
4. *Coherent integration over time:* There are two components of the bright background: the constant part and the part that changes over time. The constant part is not really a problem, since the star plus the constant background will make the pixels for the star image brighter. However, the changing part causes the background to fluctuate and leads to the burial of the star. (Imagine a light-colored dot on a white photo paper. It would be pretty easy to spot that dot despite the bright background. It will be very hard if there are many similar colored dots on the paper.) By integration of the light over time, we can reduce the level of the noise. This is easy if the star or satellite is not moving in the FOV. When the star is moving, we will need to transform the images to undo the motion and stack the motion-compensated images together to achieve the integration of the images. This technique is called the motion-compensated integration (MCI).
5. *Using interferometry:* Star lights are coherent light. Sky background light is incoherent, and an interferometer approach can eliminate or reduce the impact of the inherent background light.

Different combinations of the above techniques lead to different levels of performance, maturity, SWAP (size, weight, and power), and cost. The optimal decision for different applications will likely be different. It is reasonable to believe that the future daytime capable star-trackers will be based on SWIR detectors and the FOV will not be too small to allow more stars and RSOs to be used, and to allow room to keep the stars or RSOs in the FOV when the FOV is allowed to move in a range.

10.3.3 LEO RSO Observation: Scarcity and Midnight Gap

As discussed before, measurements of RSOs (e.g. satellites) provide observability for both the attitude and the position and have the potential to greatly improve the accuracy of a celestial navigation system design. However, it is fairly difficult for a star-tracker to see RSOs for the following reasons:

1. Except LEO RSOs, most RSO are very dim and very difficult to track. As a result, based on today's technology, it is a good idea to only use LEO RSOs. There are not that many bright LEO RSOs. For example, typically, there is one visual magnitude 5 or brighter RSOs passing through the sky once every 2 minutes.
2. For a ground observer of LEO RSOs, only a small fraction (around 0.58%) of the orbit is visible to a ground observer simply because of the low altitude of

LEO satellite. This point is illustrated in Figure 10.4 and is further elaborated upon in the rest of this section.

3. Unlike stars, RSOs are not light sources; they only reflect the sunlight. It turns out that there are about 6 hours of the midnight gap when the LEO satellites are not illuminated by the Sun and consequently are not visible to a local ground observer.

Because of the above concerns, it is almost impossible for a small FOV star-tracker, or even a number of small FOV star-trackers, to find a satellite with a high probability at any time. As a result, if we want to use RSO measurements, we will have to use either a gimbaled star-tracker or a star-tracker with a beam director so that we can practically place the star-tracker FOV anywhere in the already small portion of the LEO sky.

We will now further explain why we see only a very small portion of the LEO. This is illustrated in Figure 10.4. The angle θ_s is the sky coverage angle that represents how much of the LEO sky an observer is seeing. The angle θ_f is the effective star-tracker FOV angle, which is either the FOV of the star-tracker or the effective FOV achieved by allowing the star-tracker FOV to be steered. Since stars are practically at an infinity distance, the star-tracker effective FOV and the sky coverage angles are the same for stars. However, for the LEO sky, the sky coverage angle is much smaller than the effective FOV angle of the star-tracker.

We can show that

$$\frac{\theta_s}{2} = \frac{\theta_f}{2} - \sin^{-1}\left(\sin\left(\frac{\theta_f}{2}\right)\frac{r_e}{\left(r_e + h_o\right)}\right) \tag{10.1}$$

where r_e is the Earth radius and h_o is the orbit altitude. Table 10.3 provides some numerical examples that demonstrate that a ground observer can only see 0.58% of the LEO sky at a typical LEO altitude.

The derivation of the above equation is given here.

Table 10.3 Average Number of Stars Seen in Different Sizes of FOVs with Different Numbers of Stars in the Sky

R_e	6,370						
H_o	700						
$R_e/(R_e + h_o)$	0.90						
Theta FOV (deg)	Theta Sky (deg)	Solid Angle in Steradian	Percentage of Sky Coverage	Average Number of RSOs in FOV (200 in Sky	Average Number in FOV (500 in Sky)	Average Number in FOV (1,000 in Sky)	Average Number in FOV (2,000 in Sky)
120	17.428	0.072523	0.5771%	1.1542	2.886	5.771	11.542
8	0.793	0.000151	0.0012%	0.0024	0.006	0.012	0.024

For satellite at altitude h_o, θ_s is much smaller. Using the sine law, we have

$$\frac{\sin\left(\frac{\theta_s}{2}\right)}{s} = \frac{\sin\left(\pi - \frac{\theta_f}{2}\right)}{r_e + h_o} \Rightarrow \sin\left(\frac{\theta_s}{2}\right) = \sin\left(\frac{\theta_f}{2}\right)\frac{s}{r_e + h_o}$$

Separately, we also have

$$r_e + s\cos\left(\frac{\theta_f}{2}\right) = (r_e + h_o)\cos\left(\frac{\theta_s}{2}\right) \Rightarrow s = \frac{(r_e + h_o)\cos\left(\frac{\theta_s}{2}\right) - r_e}{\cos\left(\frac{\theta_f}{2}\right)}$$

Combining the above two basic equations, we yield

$$\sin\left(\frac{\theta_s}{2}\right) = \sin\left(\frac{\theta_f}{2}\right)\frac{1}{r_e + h_o}\frac{(r_e + h_o)\cos\left(\frac{\theta_s}{2}\right) - r_e}{\cos\left(\frac{\theta_f}{2}\right)} = \tan\left(\frac{\theta_f}{2}\right)\left(\cos\left(\frac{\theta_s}{2}\right) - \frac{r_e}{r_e + h_o}\right)$$

Noting that

$$\tan\left(\frac{\theta_f}{2}\right) = \frac{\sin\left(\frac{\theta_f}{2}\right)}{\cos\left(\frac{\theta_f}{2}\right)}$$

we have

$$\sin\left(\frac{\theta_s}{2}\right)\cos\left(\frac{\theta_f}{2}\right) = \sin\left(\frac{\theta_f}{2}\right)\cos\left(\frac{\theta_s}{2}\right) - \sin\left(\frac{\theta_f}{2}\right)\frac{r_e}{r_e + h_o}$$

or

$$\sin\left(\frac{\theta_f}{2}\right)\cos\left(\frac{\theta_s}{2}\right) - \sin\left(\frac{\theta_s}{2}\right)\cos\left(\frac{\theta_f}{2}\right) = \sin\left(\frac{\theta_f}{2}\right)\frac{r_e}{r_e + h_o}$$

Using the sine of difference formulae, the above can be written as

$$\sin\left(\frac{\theta_f - \theta_s}{2}\right) = \sin\left(\frac{\theta_f}{2}\right)\frac{r_e}{r_e + h_o} \Rightarrow \frac{\theta_f - \theta_s}{2} = \sin^{-1}\left(\sin\left(\frac{\theta_f}{2}\right)\frac{r_e}{r_e + h_o}\right)$$

and we arrive at the conclusion:

$$\frac{\theta_s}{2} = \frac{\theta_f}{2} - \sin^{-1}\left(\sin\left(\frac{\theta_f}{2}\right)\frac{r_e}{r_e + h_o}\right)$$

When the angles are small, the above can be approximated by

$$\frac{\theta_s}{2} \approx \frac{\theta_f}{2} - \left(\frac{\theta_f}{2}\right)\frac{r_e}{r_e + h_o} = \frac{\theta_f}{2}\left(\frac{h_o}{r_e + h_o}\right)$$

Although the above approximation is not accurate enough to support engineering analysis, it does provide an intuitively useful conclusion. The FOV angle is roughly scaled down by a factor of $h_o/(r_e + h_o)$. This ratio will converge to 1 when the orbit altitude is very high. For the LEO, the factor is unfortunately only around 10%. What makes the situation even worse is that the number of RSOs that can be found is proportional to the area of the FOV, which is proportional to the square of the effective FOV angle. If the angle is 10% of the FOV angle of the star-tracker, the area will be around 1% of the FOV of the star-tracker.

10.4 Example Configurations and General Architecture

10.4.1 Example Designs and Choices of Technology

Based on the discussions in the previous sections, we provide a few example configurations of the celestial navigation design.

Example 10.5: High Altitude Airborne Celestial Navigation System Using Stars Only

For this application, we can use fixed-head SWIR star-trackers integrated with an IMU that has high-accuracy accelerometers and low-cost gyros. We will need to use altimeters to bound vertical axis drift.

High altitude means that we do not need to be concerned with clouds. Stars-only means that we do not need to worry about finding RSOs. These two considerations allow us to use a fixed-head star-tracker.

The SWIR star-tracker is chosen to provide daytime availability of stars. Because of the high availability of stars, we do not need to use very good gyros to maintain attitude performance. Since star-only updates will leave the position and velocity to be Schuler-tuned, it is important to use high-accuracy accelerometers to maintain good position and velocity performance.

Example 10.6: High-Altitude Airborne Celestial Navigation System Using Stars and RSOs

In this case, even if we do not need to be concerned with cloud avoidance, we will need a steerable FOV either through a beam director or by using a gimbal platform to provide the star-tracker with the capability to move to an orientation where an RSO can be found.

Because of the RSO updates, we will be able to achieve an improved performance. However, the high-accuracy accelerometers are still important to deal with midnight gap scenarios.

Example 10.7: Low-Altitude Ground Celestial Navigation System Using Stars and RSOs

This turns out to be the most demanding application. We will need to use a beam director or a gimbal platform to allow the star-tracker to have a steerable FOV. In addition, we most likely have to support the star-tracker with MCI capability. A SWIR star-tracker that can perform coherent integration over several frames will be required.

We will also want to use a full sky camera to help to determine the cloud distribution. Again, low-cost gyros should be sufficient. High-accuracy accelerometers are required to maintain performance during the midnight gap.

10.4.2 General Architecture

We will describe a typical stellar or inertial celestial navigation system. Figure 10.5 is an architecture diagram.

The system consists of the following key components:

- The inertial measurement unit (IMU);
- The star-tracker;
- The celestial navigation software running on a real-time computer.

The celestial navigation software further consists of the following software components:

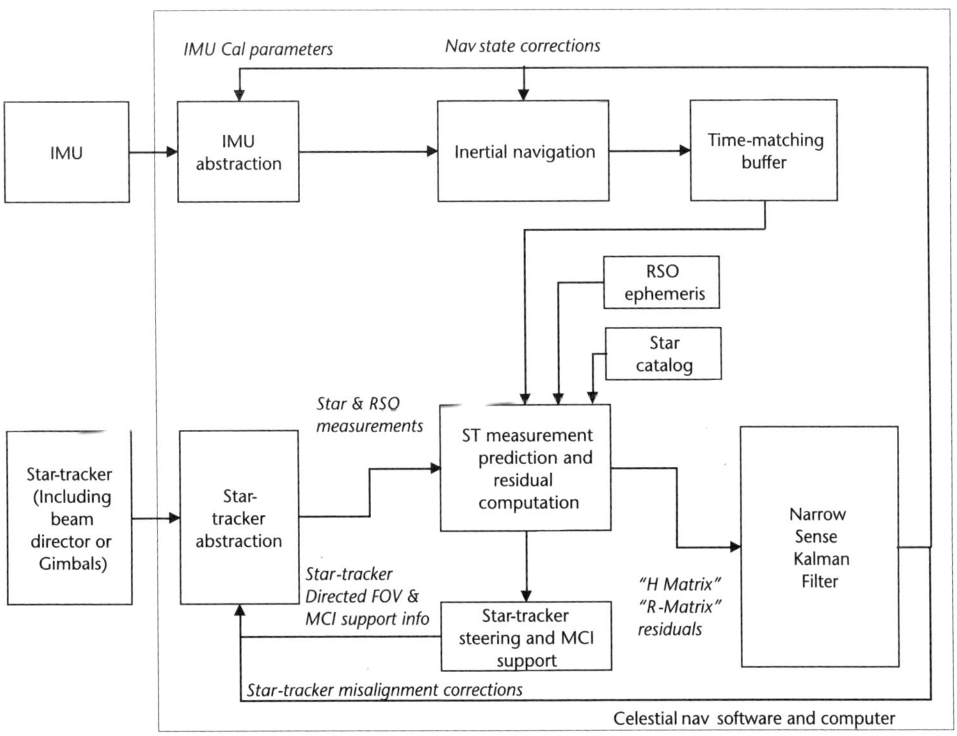

Figure 10.5 Celestial navigation architecture.

- The IMU and star-tracker abstractions that abstract specific hardware devices to generic software abstractions of the IMU and the star-tracker so that the rest of the celestial navigation software does not need to change with the changes of specific sensor hardware;
- The inertial navigation component;
- The time-matching buffer component that allows star and RSO measurements and IMU measurements with different time tags to be used to compute time-matched Kalman filter residuals;
- The star and RSO measurement processing, identification, steering, and MCI that support software component;
- The star catalog and RSO ephemeris that support the above processing;
- The narrow-sense Kalman filter that uses the star and RSO measurements to produce corrections for the IMU, the inertial navigation, and the star-tracker.

The IMU and inertial navigation software provide the baseline navigation solution on a continuous basis.

A circular buffer in the time-matching buffer component stores a short history of the most recent inertial navigation outputs.

The star tracker produces the star and RSO measurements which are then corrected for star-tracker misalignment corrections.

The time-tag of the star-tracker measurements is used to retrieve the time-matched inertial navigation states which are used to generate predicted star-tracker outputs using the star catalog and RSO ephemeris.

The star and RSO identification processing matches the star and RSO measurements against predicted star and RSO measurements. The difference between the measurements and the predictions are the Kalman filter residuals. We also compute the H-matrices using the predicted star and RSO information.

The star-tracker steering and MCI support module has several responsibilities:

1. Plan and compute the LOS pointing direction of the star-tracker for cloud avoidance and for RSO locating and tracking.
2. Compute the directed FOV information for the star-tracker to find the stars and RSOs in a small portion of the FOV around each of the stars and RSOs.
3. Compute motion compensation information to support motion compensated integration by the star-tracker.

The narrow-sense Kalman filter component uses the residual and H-matrices to compute corrections as well as updated covariance matrices. The corrections are sent to the IMU abstraction, the star-tracker abstraction, and inertial navigation to correct the state estimates.

10.5 Propagation Algorithms and Error Models

The inertial navigation algorithms are discussed in Chapter 3 and there are several flavors of the algorithm depending in which frame the equations are written. For the purpose of this chapter, we use the INS algorithms presented in the ECEF frame, which are summarized here:

1. The attitude propagation equation in quaternion form is simply

$$\dot{q}_B^E = \frac{1}{2}\left(q_B^E \otimes {}^B\omega_I^B - {}^E\omega_I^E \otimes q_B^E\right) \tag{10.2}$$

2. The DCM can be derived from the quaternion

$$C_B^E = q2DCM\left(q_B^E\right) \tag{10.3}$$

3. The velocity propagation equation is given by

$$\frac{d}{dt}{}^E v_E^B = C_B^E {}^B f_I^B + {}^E g - 2{}^E\omega_I^E \times {}^E v_E^B - \left[{}^E\omega_I^E \times\right]\left[{}^E\omega_I^E \times\right]{}^E r_E^B = C_B^E {}^B f_I^B + {}^E g_c - 2{}^E\omega_I^E \times {}^E v_E^B$$

where

$${}^E g_c = {}^E g - \left[{}^E\omega_I^E \times\right]\left[{}^E\omega_I^E \times\right]{}^E r_E^B \tag{10.4}$$

4. Finally, the position propagation equation is simply

$$\frac{d}{dt}{}^E r_E^B = {}^E v_E^B \tag{10.5}$$

For covariance propagation of the Kalman filter, we will need the linearized error models for attitude, velocity, and position repeated from Chapter 3.

1. Attitude error propagation

$$\delta{}^E\dot{\theta}_B^E = -\left[{}^E\omega_I^E \times\right]\delta{}^E\theta_B^E - C_B^E\delta{}^B\omega_I^B \tag{10.6}$$

2. Velocity error propagation

$$\frac{d}{dt}\delta{}^E v_E^B = \left[\left(C_B^E m{}^B f_I^B\right)\times\right]\delta{}^E\theta_B^E + C_B^E\delta{}^B f_I^B + \left[\frac{\partial{}^E g}{\partial{}^E r_E^B}\right]\delta{}^E r_E^B - 2\left[\left({}^E\omega_I^E\right)\times\right]\delta{}^E v_E^B \tag{10.7}$$

3. Position error propagation

$$\frac{d}{dt}\delta{}^E r_E^B = \delta{}^E v_E^B \tag{10.8}$$

We will use the following gyro error models

$$\delta{}^B\omega_I^B = m{}^B\omega_I^B - {}^B\omega_I^B = C_G^B \text{diag}\left({}^G s\right){}^G\omega_I^B + \left(\left[{}^B\theta_G \times\right] + L\left({}^B\beta_G\right)\right){}^B\omega_I^B + {}^B b_G + C_G^B ARW \tag{10.9}$$

where

$$\left[{}^B\theta\times\right] = \begin{bmatrix} 0 & -\theta_z & \theta_y \\ \theta_z & 0 & -\theta_x \\ -\theta_y & \theta_x & 0 \end{bmatrix}, \quad L\left({}^B\beta\right) = \begin{bmatrix} 0 & 0 & 0 \\ \beta_z & 0 & 0 \\ -\beta_x & \beta_y & 0 \end{bmatrix} \tag{10.10}$$

which can also be written in a more Kalman filter-friendly way as

$$\delta\,{}^{B}\omega_{I}^{B} = C_{G}^{B}\text{diag}\left({}^{G}\omega_{I}^{B}\right){}^{G}s - \left[{}^{B}\omega_{I}^{B}\times\right]{}^{B}\theta_{G} + J\left({}^{B}\omega_{I}^{B}\right){}^{B}\beta_{G} + {}^{B}b_{G} + C_{G}^{B}ARW \tag{10.11}$$

where

$$J(\omega) = \begin{bmatrix} 0 & 0 & 0 \\ 0 & 0 & \omega_{x} \\ \omega_{y} & -\omega_{x} & 0 \end{bmatrix} \tag{10.12}$$

and the following accelerometer model:

$$\delta\,{}^{B}f_{I}^{B} = m\,{}^{B}f_{I}^{B} - {}^{B}f_{I}^{B} = C_{A}^{B}\text{diag}\left({}^{A}s\right){}^{A}f_{I}^{B} + L\left({}^{B}\beta_{A}\right){}^{B}f_{I}^{B} + {}^{B}b_{A} + C_{A}^{B}VRW \tag{10.13}$$

which can also be written in a more Kalman filter-friendly way as

$$\delta\,{}^{B}f_{I}^{B} = C_{A}^{B}\text{diag}\left({}^{A}f_{I}^{B}\right){}^{A}s + J\left({}^{B}f_{I}^{B}\right){}^{B}\beta_{A} + {}^{B}b_{A} + C_{A}^{B}VRW \tag{10.14}$$

where

$$J\left({}^{B}f_{I}^{B}\right) = \begin{bmatrix} 0 & 0 & 0 \\ 0 & 0 & {}^{B}f_{I\,x}^{B} \\ {}^{B}f_{I\,y}^{B} & -{}^{B}f_{I\,x}^{B} & 0 \end{bmatrix} \tag{10.15}$$

If we model each one of the IMU error parameters as a random walk process and let

$$\begin{aligned} \delta X_{I}^{T} &= \left[\delta\,{}^{E}\dot{\theta}_{B}^{E},\ \delta\,{}^{B}b_{G}^{T},\ \delta\,{}^{G}s^{T},\ \delta\,{}^{B}\beta_{G}^{T},\ \delta\,{}^{B}\theta_{G}^{T},\ \left(\delta\,{}^{E}v_{E}^{B}\right)^{T},\ \delta\,{}^{B}b_{A}^{T},\ \delta\,{}^{A}s^{T},\ \delta\,{}^{B}\beta_{A}^{T},\ \left(\delta\,{}^{E}r_{E}^{B}\right)^{T}\right] \\ W_{I}^{T} &= \left[-\left(C_{G}^{E}ARW\right)^{T},\ RRW^{T},\ SRW_{G}^{T},\ \beta RW_{G}^{T},\ \theta RW_{G}^{T},\ \left(C_{A}^{E}VRW\right)^{T},\ ABRW^{T},\ SRW_{A}^{T},\ \beta RW_{A},\ {}^{T}0\right] \end{aligned} \tag{10.16}$$

we then have

$$\delta\dot{X}_{II} = A_{II}\delta X_{I} + W_{I}$$

$$A_{II} = \begin{bmatrix} -\left[{}^{E}\omega_{I}^{E}\times\right] & -C_{B}^{E} & -C_{G}^{E}\text{diag}\left({}^{G}\omega_{I}^{B}\right) & -C_{B}^{E}J\left({}^{B}\omega_{I}^{B}\right) & C_{B}^{E}\left[{}^{B}\omega_{I}^{B}\times\right] & 0_{3x3} & 0_{3x3} & 0_{3x3} & 0_{3x3} & 0_{3x3} \\ 0_{3x3} & 0_{3x3} & 0_{3x3} & 0_{3x3} & 0_{3x3} & 0_{3x3} & 0_{3x3} & 0_{3x3} & 0_{3x3} & 0_{3x3} \\ 0_{3x3} & 0_{3x3} & 0_{3x3} & 0_{3x3} & 0_{3x3} & 0_{3x3} & 0_{3x3} & 0_{3x3} & 0_{3x3} & 0_{3x3} \\ 0_{3x3} & 0_{3x3} & 0_{3x3} & 0_{3x3} & 0_{3x3} & 0_{3x3} & 0_{3x3} & 0_{3x3} & 0_{3x3} & 0_{3x3} \\ 0_{3x3} & 0_{3x3} & 0_{3x3} & 0_{3x3} & 0_{3x3} & 0_{3x3} & 0_{3x3} & 0_{3x3} & 0_{3x3} & 0_{3x3} \\ \left[\left(C_{B}^{E}m\,{}^{B}f_{I}^{B}\right)\times\right] & 0_{3x3} & 0_{3x3} & 0_{3x3} & 0_{3x3} & -2\left[\left({}^{E}\omega_{I}^{E}\right)\times\right] & C_{B}^{E} & C_{A}^{E}\text{diag}\left({}^{A}f_{I}^{B}\right) & C_{B}^{E}J\left({}^{B}f_{I}^{B}\right) & \left[\dfrac{\partial\,{}^{E}g}{\partial\,{}^{E}r_{E}^{B}}\right] \\ 0_{3x3} & 0_{3x3} & 0_{3x3} & 0_{3x3} & 0_{3x3} & 0_{3x3} & 0_{3x3} & 0_{3x3} & 0_{3x3} & 0_{3x3} \\ 0_{3x3} & 0_{3x3} & 0_{3x3} & 0_{3x3} & 0_{3x3} & 0_{3x3} & 0_{3x3} & 0_{3x3} & 0_{3x3} & 0_{3x3} \\ 0_{3x3} & 0_{3x3} & 0_{3x3} & 0_{3x3} & 0_{3x3} & 0_{3x3} & 0_{3x3} & 0_{3x3} & 0_{3x3} & 0_{3x3} \\ 0_{3x3} & 0_{3x3} & 0_{3x3} & 0_{3x3} & 0_{3x3} & I_{3x3} & 0_{3x3} & 0_{3x3} & 0_{3x3} & 0_{3x3} \end{bmatrix} \tag{10.17}$$

10.6 Measurement Equations and Error Models

As discussed before, what is special for each type of aiding sensors is its measurement model, and that is true for celestial navigation as well. The purpose of this section is to describe and discuss both the nonlinear and linear versions of the measurement model.

We will start by looking at the measurement models of a FDO such as an RSO. The models for star measurement can then be simplified by letting the distance of the RSO to go to infinity.

We will be introducing many mathematical equations and to make these equations and derivations easy to understand, we will break this down to into steps:

1. We will first look at the measurement model when the measurement is the LOS unit vector.
2. We will then show how the unit vector can be transformed to the standard star-tracker h and v measurements.
3. We will then combine the results of steps 1 and 2 to derive the nonlinear and linearized measurement equations. The latter gives us the H-matrices for the Kalman filter.
4. We will look at the modeling of RSO ephemeris errors.
5. We can then put all the equations together for RSOs.
6. We treat the star as a special case for the RSO by letting the object range to go to infinity.

Step 1: Unit Vector and Unit Vector Error Model

Let the location of vehicle star-tracker in the ECEF frame be given by ${}^E r$ and the location of the RSO (RSO j) be given by ${}^E r_j$.

The LOS vector from the vehicle to the RSO can then be computed as

$$ {}^E u_j = \frac{1}{\left\| {}^E r_j - {}^E r \right\|}\left({}^E r_j - {}^E r \right) = \frac{1}{\rho_j}\left({}^E r_j - {}^E r \right) \tag{10.18} $$

where $\rho_j = \| {}^E r_j - {}^E r \|$ is the range to the RSO.

${}^E u_j$ can be transformed to the star-tracker frame (ST) using the following equation

$$ {}^{ST} u_j = C_B^{ST} C_E^B \, {}^E u_j \tag{10.19} $$

where C_E^B is the DCM that transforms the coordinates of a vector in the ECEF frame to the body frame. In other words, C_E^B represents the vehicle attitude. C_B^{ST} is the DCM that transforms the coordinates of a vector in the body frame to the star-tracker frame.

The measured unit vector can be modeled as

$$ m^{ST} u_j = C_B^{ST} C_E^B \, {}^E u_j + v \tag{10.20} $$

where v is the measurement noise.

The measurement residual for the unit vector can be written as

$$\begin{aligned}\Delta u &= m^{ST}u_j - {}^{ST}\hat{u}_j = C_B^{ST} C_E^{B\,E} u_j - \hat{C}_B^{ST} \hat{C}_E^{B\,E} \hat{u}_j + v \\ &= \left(C_B^{ST} C_E^{B\,E} u_j - \hat{C}_B^{ST} C_E^{B\,E} u_j\right) + \left(\hat{C}_B^{ST} C_E^{B\,E} u_j - \hat{C}_B^{ST} \hat{C}_E^{B\,E} u_j\right) + \left(\hat{C}_B^{ST} \hat{C}_E^{B\,E} u_j - \hat{C}_B^{ST} \hat{C}_E^{B\,E} \hat{u}_j\right) + v \\ &= \left(C_B^{ST} - \hat{C}_B^{ST}\right) C_E^{B\,E} u_j + \hat{C}_B^{ST}\left(C_E^{B} - \hat{C}_E^{B}\right){}^{E} u_j + \hat{C}_B^{ST} \hat{C}_E^{B}\left({}^{E}u_j - {}^{E}\hat{u}_j\right) + v\end{aligned}$$

The first term in the above equation accounts for the contribution of star-tracker misalignment knowledge error.

Note that

$$C_B^{ST} = C_{S\hat{T}}^{ST} C_B^{S\hat{T}} = \left(I - \left[\theta_{S\hat{T}}^{ST} \times\right]\right) C_B^{S\hat{T}}$$

We have

$$\left(C_B^{ST} - \hat{C}_B^{ST}\right) C_E^{B\,E} u_j \approx -\left[\theta_{S\hat{T}}^{ST} \times\right] C_B^{S\hat{T}} C_E^{B\,E} u_j \approx -\left[\theta_{S\hat{T}}^{C} \times\right] {}^{S\hat{T}}\hat{u}_j = \left[{}^{S\hat{T}}\hat{u}_j \times\right] \theta_{S\hat{T}}^{ST}$$

where $\theta_{S\hat{T}}^{ST} = \delta\theta_{stma}$ is the star-tracker DCM knowledge deviation from the previous estimate represented in the star-tracker frame as three small angles.

The second term accounts for the contribution of the vehicle attitude knowledge error.

Note that

$$C_B^{E} = C_{\hat{E}}^{E} C_B^{\hat{E}} = \left(I - \left[\theta_{\hat{E}}^{E} \times\right]\right) C_B^{\hat{E}} \text{ and } C_E^{B} = \hat{C}_E^{B}\left(I + \left[\theta_{\hat{E}}^{E} \times\right]\right)$$

We have

$$\begin{aligned}\hat{C}_B^{ST}\left(C_E^{B} - \hat{C}_E^{B}\right){}^{E}u_j &= \hat{C}_B^{ST} \hat{C}_E^{B}\left[\theta_{\hat{E}}^{E} \times\right]{}^{E}u_j = -\hat{C}_B^{ST} \hat{C}_E^{B}\left[{}^{E}u_j \times\right]\theta_{\hat{E}}^{E} \\ &\approx -\hat{C}_E^{ST}\left[{}^{E}\hat{u}_j \times\right]\theta_{\hat{E}}^{E} = -\left[{}^{ST}\hat{u}_j \times\right]\hat{C}_E^{ST}\theta_{\hat{E}}^{E} = -\left[{}^{ST}\hat{u}_j \times\right]\hat{C}_E^{ST}\delta^{E}\theta_B^{E}\end{aligned}$$

where $\delta^{E}\theta_B^{E} \equiv \theta_{\hat{E}}^{E}$ is the attitude deviation from previous estimate represented as 3 small angles in the ECEF frame.

The third term accounts for the contribution of the vehicle position knowledge error as well as the RSO position knowledge error (RSO ephemeris error).

We have

$$\left(\hat{C}_B^{ST} \hat{C}_E^{B\,E} u_j - \hat{C}_B^{ST} \hat{C}_E^{B\,E} \hat{u}_j\right) = \hat{C}_B^{ST} \hat{C}_E^{B}\left({}^{E}u_j - {}^{E}\hat{u}_j\right) = \hat{C}_B^{ST} \hat{C}_E^{B}\left(\delta^{E}u_j\right)$$

where

$$\delta^{E}u_j = {}^{E}u_j - {}^{E}\hat{u}_j, {}^{E}u_j = \frac{1}{l_j}\left({}^{E}r_j - {}^{E}r\right), {}^{E}u_j = \frac{1}{\hat{l}_j}\left({}^{E}\hat{r}_j - {}^{E}\hat{r}\right)$$

where the $l_j = \|{}^{E}r_j - {}^{E}r\|$ and $\hat{l}_j = \|{}^{E}\hat{r}_j - {}^{E}\hat{r}\|$.

Ignoring the difference between $1/\hat{l}_j$ and $1/l_j$, we have

$$\left(\hat{C}_B^{ST}\hat{C}_E^{B\,E}u_j - \hat{C}_B^{ST}\hat{C}_E^{B\,E}\hat{u}_j\right) = \hat{C}_B^{ST}\hat{C}_E^{B}\left(\delta^E u_j\right) = \frac{1}{\hat{l}_j}\hat{C}_B^{ST}\hat{C}_E^{B}\left(\delta^E r_j - \delta^E r\right)$$

where $\delta^E r = {}^E r - {}^E\hat{r}$ $\delta^E r_j = {}^E r - {}^E\hat{r}_j$

Finally, we have

$$\Delta u = m^{ST}u_j - {}^{ST}\hat{u}_j = \left[{}^{ST}\hat{u}_j\times\right]\delta\theta_{stma} - \left[{}^{ST}\hat{u}_j\times\right]\hat{C}_E^{ST}\delta^E\theta_B^E + \frac{1}{\hat{l}_j}\hat{C}_B^{ST}\hat{C}_E^{B}\delta^E r_j - \frac{1}{\hat{l}}\hat{C}_B^{ST}\hat{C}_E^{B}\delta^E r + v \tag{10.21}$$

Step 2: Relating the Unit Vector to the Star-Tracker *h* and *v*

Although the unit vector-based equations at which we arrived in Step 1 are clean, the reality is that the LOS measurement is a 2-degree-of-freedom (DOF) measurement and it is more conveniently represented by the star-tracker h and v measurements, which are the angularized star position coordinates in the star-tracker FOV. Figure 10.6 illustrates the definition of h and v.

In Figure 10.6, the coordinates of the star in the star-tracker focal plane are given by x and y, the focal length is given by d, and we have the following relationships:

$$h = \frac{x}{d} = \tan\left(\theta_h\right)$$

$$v = \frac{y}{d} = \tan\left(\theta_v\right)$$

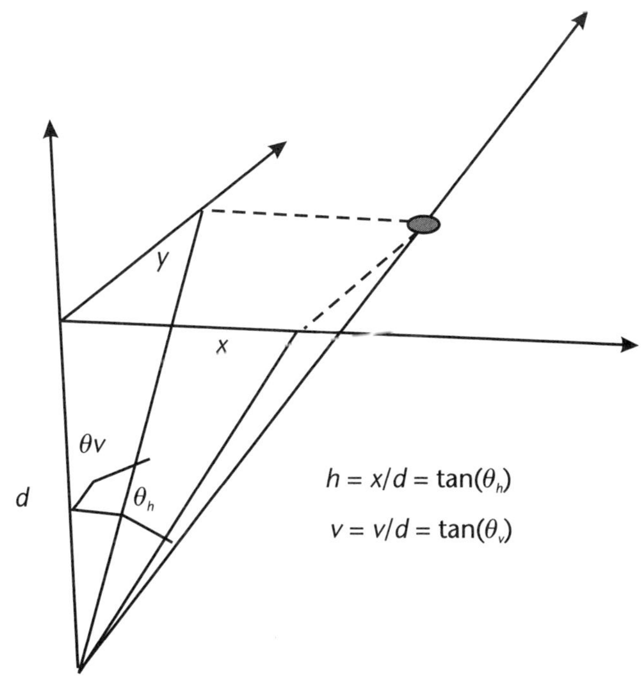

Figure 10.6 Star-tracker measurement *h* and *v* definitions.

The LOS unit vector is therefore given by

$$^{ST}u = \frac{1}{\sqrt{x^2 + y^2 + d^2}}\begin{bmatrix} x \\ y \\ d \end{bmatrix} = \frac{1}{\sqrt{h^2 + v^2 + 1^2}}\begin{bmatrix} h \\ v \\ 1 \end{bmatrix}$$

The h and v coordinates of the star position in the star-tracker FOV can be calculated using

$$h = \frac{^{ST}u(1)}{^{ST}u(3)}, \quad v = \frac{^{ST}u(2)}{^{ST}u(3)}$$

Conversely, we can also calculate the unit vector from h and v:

$$^{ST}u(1) = \frac{h}{\sqrt{1 + h^2 + v^2}},\ ^{ST}u(2) = \frac{v}{\sqrt{1 + h^2 + v^2}},\ ^{ST}u(3) = \frac{1}{\sqrt{1 + h^2 + v^2}}$$

We will use the above nonlinear equations as the measurement prediction equations. To derive the H-matrices for the Kalman filter, we will need to linearize the above equations:

$$\delta h = \frac{\delta\, ^{ST}u(1)}{^{ST}u(3)} - \frac{^{ST}u(1)}{^{ST}u^2(3)}\delta\, ^{ST}u(3) = \left(\frac{1}{^{ST}u(3)}\right)\left(\delta\, ^{ST}u(1) - h\delta\, ^{ST}u(3)\right)$$

$$\delta v = \frac{\delta\, ^{ST}u(2)}{^{ST}u(3)} - \frac{^{ST}u(2)}{^{ST}u^2(3)}\delta\, ^{ST}u(3) = \left(\frac{1}{^{ST}u(3)}\right)\left(\delta\, ^{ST}u(2) - v\delta\, ^{ST}u(3)\right)$$

which can be written as

$$\begin{bmatrix} \delta h \\ \delta v \end{bmatrix} = \frac{1}{^{C}u(3)}\begin{bmatrix} 1 & 0 & -h \\ 0 & 1 & -v \end{bmatrix}\delta\, ^{C}u = \frac{1}{\sqrt{1 + h^2 + v^2}}\begin{bmatrix} 1 & 0 & -h \\ 0 & 1 & -v \end{bmatrix}\delta\, ^{ST}u \quad (10.22)$$

Note that for the above equations to be true, h and v do not need to be small.

Step 3: Combine Steps 1 and 2 for Nonlinear and Linear or H-Submatrix Equations

The nonlinear equations are trivial since all we need to do is to put them together.
After we combine (10.21) with (10.22), we get

$$\begin{bmatrix} \delta h \\ \delta v \end{bmatrix} = \frac{1}{\sqrt{1 + h^2 + v^2}}\begin{bmatrix} 1 & 0 & -h \\ 0 & 1 & -v \end{bmatrix}\delta\, ^{ST}u = H_{stma}\delta\theta_{stma} + H_{att}\delta\, ^{E}\theta_{B}^{E} + H_{pos}\delta\, ^{E}r + H_{ppe}\delta\, ^{E}r_j + v \quad (10.23)$$

Each of the H-submatrices is calculated below.
For star-tracker frame misalignment errors, we have

$$H_{stma} = \frac{1}{^{ST}u(3)}\begin{bmatrix} 1 & 0 & -h \\ 0 & 1 & -v \end{bmatrix}\left[^{ST}\hat{u}_j \times\right] = \frac{1}{^{ST}u(3)}\begin{bmatrix} 1 & 0 & -h \\ 0 & 1 & -v \end{bmatrix}\begin{bmatrix} 0 & -^{ST}u_j(3) & ^{ST}u_j(2) \\ ^{ST}u_j(3) & 0 & -^{ST}u_j(1) \\ -^{ST}u_j(2) & ^{ST}u_j(1) & 0 \end{bmatrix}$$

$$= \begin{bmatrix} 1 & 0 & -h \\ 0 & 1 & -v \end{bmatrix} \begin{bmatrix} 0 & -1 & v \\ 1 & 0 & -h \\ -v & h & 0 \end{bmatrix} = \begin{bmatrix} hv & -(1+h^2) & v \\ 1+v^2 & -hv & -h \end{bmatrix}$$

In summary, we have

$$H_{stma} = \begin{bmatrix} hv & -\left(1+h^2\right) & v \\ 1+v^2 & -hv & -h \end{bmatrix} \tag{10.24}$$

Similarly, we have the H-submatrix for the ECEF frame attitude error stating

$$H_{att} = \frac{-1}{{}^{C}u(3)} \begin{bmatrix} 1 & 0 & -h \\ 0 & 1 & -v \end{bmatrix} \left[{}^{ST}\hat{u}_j \times \right] \hat{C}_E^{ST} = -\begin{bmatrix} hv & -\left(1+h^2\right) & v \\ 1+v^2 & -hv & -h \end{bmatrix} \hat{C}_E^{ST} \tag{10.25}$$

For the ECEF position states, we have

$$H_{pos} = -\frac{1}{{}^{ST}u(3)} \begin{bmatrix} 1 & 0 & -h \\ 0 & 1 & -v \end{bmatrix} \frac{1}{\hat{\rho}} \hat{C}_B^{ST} \hat{C}_E^B = \frac{-1}{\hat{\rho}\sqrt{1+h^2+v^2}} \begin{bmatrix} 1 & 0 & -h \\ 0 & 1 & -v \end{bmatrix} \hat{C}_E^{ST} \tag{10.26}$$

For the RSO position error (ephemeris error) in the ECEF,

$$H_{rppe} = \frac{1}{{}^{ST}u(3)} \begin{bmatrix} 1 & 0 & -h \\ 0 & 1 & -v \end{bmatrix} \frac{1}{\hat{\rho}} \hat{C}_B^{ST} \hat{C}_E^B = \frac{1}{\hat{\rho}\sqrt{1+h^2+v^2}} \begin{bmatrix} 1 & 0 & -h \\ 0 & 1 & -v \end{bmatrix} \hat{C}_E^{ST}$$
$$H_{rppe} = -H_{pos} \tag{10.27}$$

Step 4: Modeling the RSO Ephemeris Error

As mentioned before, it is often much casier to look at the ephemeris errors in the orbit frame of the RSO (j) (note that the orbits and orbit frames of RSOs are different) and we can then transform the error into the ECEF frame

$$\delta^E r_j = C_{Oj}^E \delta^{Oj} r_j$$

The H-submatrix for the orbit frame ephemeris error is then

$$H_{rppo} = \frac{1}{\hat{\rho}\sqrt{1+h^2+v^2}} \begin{bmatrix} 1 & 0 & -h \\ 0 & 1 & -v \end{bmatrix} \hat{C}_E^{ST} C_{Oj}^E - H_{pos} C_{Oj}^E \tag{10.28}$$

Step 5: Putting Everything Together for the RSO

Nonlinear Measurement Prediction Model

$$ {}^E u_j = \frac{1}{\left\| {}^E r_j - {}^E r \right\|} \left({}^E r_j - {}^E r \right) = \frac{1}{\rho_j} \left({}^E r_j - {}^E r \right)$$
$${}^{ST}u_j = C_B^{ST} C_E^B {}^E u_j \tag{10.29}$$
$$h = \frac{{}^{ST}u(1)}{{}^{ST}u(3)}, \quad v = \frac{{}^{ST}u(2)}{{}^{ST}u(3)}$$

Linear Model and Kalman Filter H-Submatrices (Sensitivity Matrices)

Star-tracker frame misalignment sensitivity matrix:

$$H_{stma} = \begin{bmatrix} hv & -\left(1+h^2\right) & v \\ 1+v^2 & -hv & -h \end{bmatrix} \tag{10.30}$$

Attitude sensitivity matrix:

$$H_{att} = -\begin{bmatrix} hv & -\left(1+h^2\right) & v \\ 1+v^2 & -hv & -h \end{bmatrix} \hat{C}_E^{ST} = -H_{stma}\hat{C}_E^{ST} \tag{10.31}$$

Position sensitivity matrix:

$$H_{pos} = \frac{-1}{\hat{\rho}\sqrt{1+h^2+v^2}} \begin{bmatrix} 1 & 0 & -h \\ 0 & 1 & -v \end{bmatrix} \hat{C}_E^{ST} \tag{10.32}$$

RSO ephemeris error sensitivity matrix:

$$H_{rppo} = \frac{1}{\hat{\rho}\sqrt{1+h^2+v^2}} \begin{bmatrix} 1 & 0 & -h \\ 0 & 1 & -v \end{bmatrix} \hat{C}_E^{ST} C_{Oj}^{E} = -H_{pos} C_{Oj}^{E} \tag{10.33}$$

Step 6: Taking the Limiting Case for Stars

Let the range go to infinity and we will arrive at the equations for stars:

Nonlinear Measurement Prediction Model

$$\begin{aligned} {}^{E}u_j &= \frac{1}{\rho_j}{}^{E}r_j = C_{ECI}^{E}\,{}^{ECI}u_j, \quad {}^{ECI}u_j \text{ is from star-catalog} \\ {}^{ST}u_j &= C_B^{ST} C_E^{B}\,{}^{E}u_j \\ h &= \frac{{}^{ST}u(1)}{{}^{ST}u(3)}, \quad v = \frac{{}^{ST}u(2)}{{}^{ST}u(3)} \end{aligned} \tag{10.34}$$

Linear Model and Kalman Filter H-Submatrices (Sensitivity Matrices)

The position and ephemeris related term go to zero. The star-tracker misalignment (*stma*) and attitude terms remain the same:

$$\begin{aligned} H_{stma} &= \begin{bmatrix} hv & -\left(1+h^2\right) & v \\ 1+v^2 & -hv & -h \end{bmatrix} \\ H_{att} &= -\begin{bmatrix} hv & -\left(1+h^2\right) & v \\ 1+v^2 & -hv & -h \end{bmatrix} \hat{C}_E^{ST} = H_{stma}\hat{C}_E^{ST} \end{aligned} \tag{10.35}$$

Augmented States

Although, in reality, there are various combinations of sensors used to achieve the specific purposes of a system with various constraints, in the following we will simply show the Kalman filter constructed with one star-tracker included.

We will need to augment the Kalman state vector to include star-tracker misalignment error states and the RSO ephemeris error states in the orbit frame for the currently used RSO. If more than one RSO is used at the same time, there needs to be three additional states for each of the RSOs. The following shows just one RSO, which is very typical.

$$\begin{aligned} X^T &= \left[X_I^T,\ \left(\delta\theta_{stma}\right)^T,\ \left(\delta^{O_j} r_j\right)^T \right] \\ \delta X^T &= \left[\delta X_I^T,\ \left(\delta\delta\theta_{stma}\right)^T,\ \left(\delta\delta^{O_j} r_j\right)^T \right] \end{aligned} \tag{10.36}$$

The linearized propagation error model is then given by

$$\frac{d}{dt}(\delta X) = \begin{bmatrix} A_{II} & 0_{30x3} & 0_{30x3} \\ 0_{3x30} & 0 & 0 \\ 0_{3x30} & 0 & 0 \end{bmatrix} \delta X + \begin{bmatrix} W_I \\ stmarw \\ rsooerw \end{bmatrix} \tag{10.37}$$

where *stmarw* and *rsooerw* are the white noise processes that integrate into the star-tracker misalignment random walk and the RSO orbit error random walk.

10.7 Example Simulation Results

As an illustration, we show some of the typical simulation results of a celestial navigation system. The simulation is run for 12 hours and the middle 6 hours are during the midnight gap. In Figures 10.7, 10.8, and 10.9, the upper and lower bound curves are the plots of the +3 and −3 sigma error bounds derived from the diagonal elements of the Kalman filter covariance matrix. The error curve in between is one instance of the Monte Carlo simulation results. Figure 10.7 shows the position and velocity estimation errors. Figure 10.8 shows the attitude and accelerometer bias estimation errors. Figure 10.9 shows the gyro bias and RSO ephemeris estimation errors.

During the day, dusk and dawn, we are able to use satellites to keep the position and velocity error small.

During "midnight gap" when the LEO satellites are not illuminated by the sun, we use stars to keep attitude error bounded and the position and velocity become "Schuler tuned". The inertial quality accelerometer helps to keep the error small

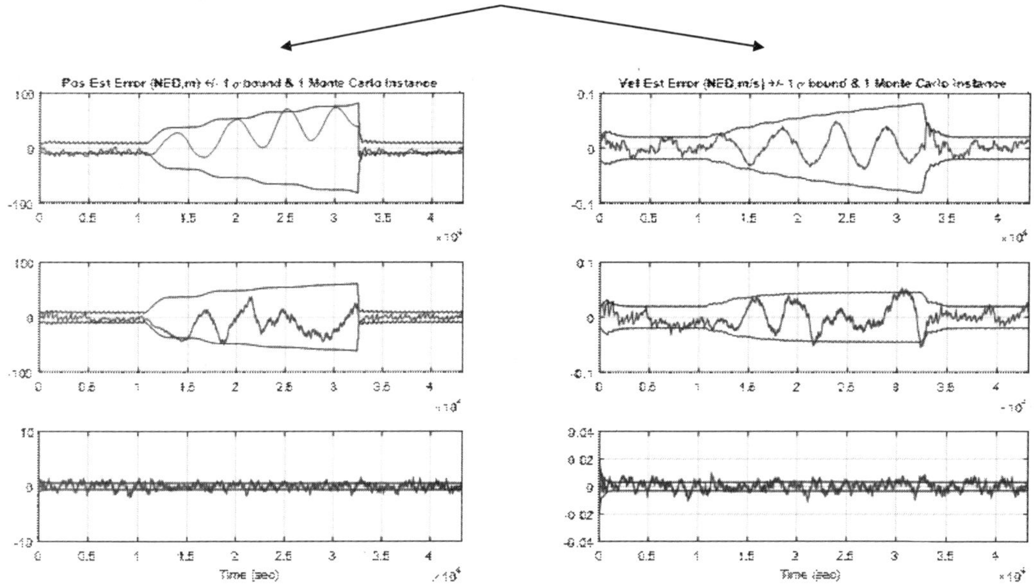

Figure 10.7 Position and velocity performance.

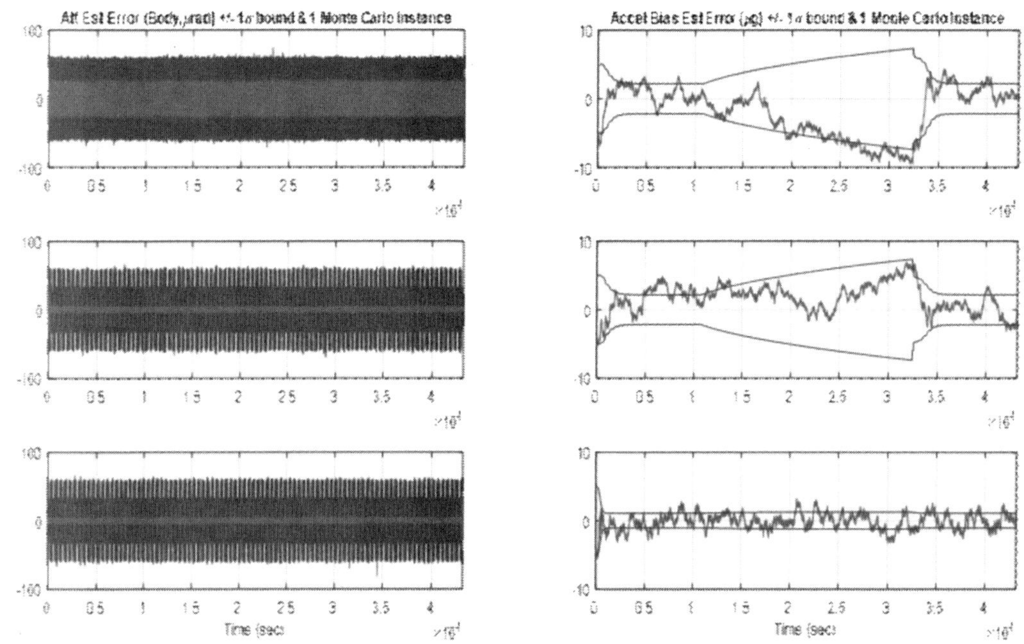

Figure 10.8 Attitude and accelerometer bias estimation errors.

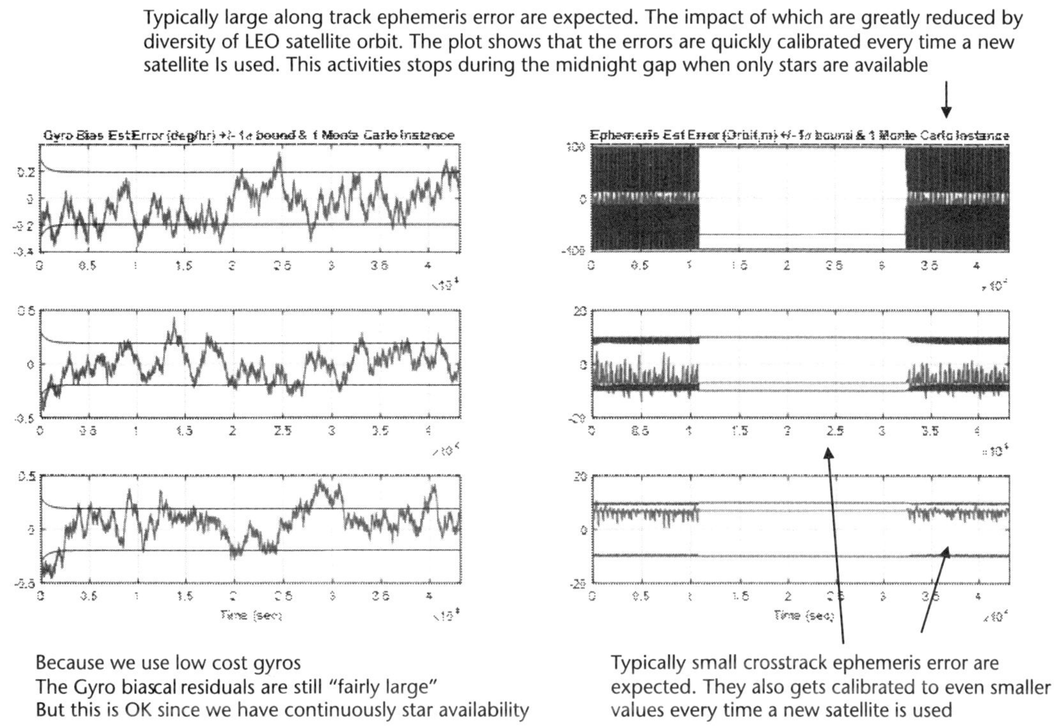

Figure 10.9 Gyro bias and RSO ephemeris estimation errors.

References

[1] Hansen, M. P., and D. S. Malchow, "Overview of SWIR Detectors, Cameras, and Applications," *Proc. of SPIE Defense and Security Symposium*, Orlando, FL, March 16–20, 2008.

[2] Truesdale, N. A., et al., "DayStar: Modeling and Test Results of a Balloon-Borne Daytime Star Tracker," *Proc. of 2013 IEEE Aerospace Conference*, Big Sky, MT, March 2–9, 2013, pp. 1–12.

[3] Liebe, C. C., "Accuracy Performance of Star Trackers—A Tutorial," *Aerosp. Electron. Syst.*, Vol. 38, February 2001, pp. 587–599.

[4] Vedder, J. D., "Star Trackers, Star Catalogs, and Attitude Determination- Probabilistic Aspects of System Design," *J. Guidance Control Dyn.*, Vol. 16, 1003, pp. 498–504.

[5] Wertz, J. R., *Spacecraft Attitude Determination and Control*, Boston, MA: D. Reidel Publishing, 1986.

[6] Shuster, M. D., "A Survey of Attitude Representations," *Astronautical Science*, Vol. 41, 1993, pp 439–517.

[7] Mekky, T. A. H., "A Comparative Study of Spacecraft Attitude Determination and Estimation Algorithms (A Cost-Benefit Approach)," *Aerospace Science and Technology*, Vol. 26, No. 1, April-May 2013, pp. 211–215.

[8] Mohd, Z. H., et al., "Review on Attitude Estimation Algorithm of Attitude Determination System," *ARPN Journal of Engineering and Applied Sciences*, Vol. 11, 2016, pp. 4455–4460.

[9] Sidi, M. I., *Spacecraft Dynamics and Control*, Cambridge, U.K.: Cambridge University Press, 1997.

[10] Shuster, M. D., and S. D. Oh, "Three Axis Attitude Determination from Vector Observations," *Guid. Control*, Vol. 4, 1981, pp. 70–77.

[11] Bar-Itzhack, Y., and R. R. Harman, "Optimized Triad Algorithm for Attitude Determination," *Guidance Control and Dynamics*, pp. 208–211.

[12] Li, R., and A. Wu, "Kalman Filtering & Applications," *UCLA Lecture Notes*, 2005–2019.

[13] Li, R., et al., "Modified Kalman Filter for Generation of Attitude Error Corrections," U.S. Patent 8,543,266, October 2011.

[14] Li, R., et al., "Real-Time Refinement Method of Spacecraft Star Tracker Alignment Estimates," U.S. Patent 8,019,544, September 2001.

[15] Li, R., et al., "Method for Compensating Star Motion Induced Error in a Stellar Inertial Attitude Determination System," U.S. Patent 7,487,016, September 2005.

[16] Li, R., et al., "Fast Access, Low Memory, Pair Catalog," U.S. Patent 7,310,578, June 2004.

[17] Li, R., et al., "Spacecraft Gyro Calibration System," U.S. Patent 7,185,858, November 2003.

[18] Li, R., et al., "Method and Apparatus for On-Board Autonomous Pair Catalog Generation," U.S. Patent 7,136,752, November 2003.

[19] Li, R., et al., "Relative Attitude Estimator for Multi-Payload Attitude Determination," U.S. Patent 7,124,001, June 2004.

[20] Li, R., et al., "Refinement of Spacecraft Angular Velocity and Attitude Estimates Using Star Data," U.S. Patent 7,062,363, September 2004.

[21] Li, R., et al., "Attitude-Acquisition Methods and Systems for Controlled Spacecraft Attitude," U.S. Patent 6,766,227, November 2002.

[22] Li, R., et al., "System and Method for Calibrating Inter-Star-Tracker Misalignments in a Stellar Inertial Attitude Determination System," U.S. Patent 6,691,033, July 2000.

[23] Li, R., et al., "Attitude Determination System and Method with Outer-Loop Gyro Scale-Factor Non-Linearity Calibration," U.S. Patent 6,615,117, November 2001.

[24] Li, R., et al., "Attitude Control Methods and Systems for Multiple-Payload Spacecraft," U.S. Patent 6,595,469, October 2001.

[25] Li, R., et al., "Spacecraft Methods and Structures for Acquiring and Determining Power-Safe Attitudes," U.S. Patent 6,588,708, January 2011.

[26] Li, R., et al., "System and Method for Autonomous Correction of Star Tracker High Spatial Frequency (HSF) Error," U.S. Patent 6,566,640, September 2001.

[27] Li, R., et al., "Lost-in-Space Stellar Attitude Acquisition Using Multiple Star Trackers," U.S. Patent 6,512,979, June 2002.

[28] Li, R., et al., "Method and Apparatus for Spacecraft Antenna Beam Pointing Correction," U.S. Patent 6,504,502, January 2000.

[29] Li, R., et al., "Lost-in-Space Stellar Attitude Acquisition Using Multiple Star Trackers," U.S. Patent 6,470,270, November 2001.

[30] Li, R., et al., "Method and Apparatus for Spacecraft Payload Pointing Registration," U.S. Patent 6,441,776, January 2000.

[31] Li, R., et al., "System for Stellar Inertial Attitude Determination," U.S. Patent 6,362,464, February 2000.

[32] Li, R., et al., "Stellar Attitude-Control Systems and Methods with Weighted Measurement-Noise Covariance Matrices," U.S. Patent 6,356,815, August 2000.

[33] Li, R., et al., "System for Star Catalog Equalization to Enhance Attitude Determination," U.S. Patent 6,330,988, July 2000.

[34] Li, R., et al., "Stable and Verifiable State Estimation Methods and Systems with Spacecraft Applications," U.S. Patent 6,317,662, July 2000.

[35] Li, R., et al., "Spacecraft Attitude Control Systems with Dynamic Methods and Structures for Processing Star Tracker Signals," U.S. Patent 6,304,822, August 2000.

[36] Li, R., et al., "Autonomous Gyro Scale Factor and Misalignment Calibration," U.S. Patent 6,298,288, December 1998.

[37] Li, R., et al., "Attitude Determination System and Method," U.S. Patent 6,289,268, March 2000.

[38] Li, R., et al., "Beacon-Assisted Spacecraft Attitude Control Systems and Methods," U.S. Patent 6,288,671, April 2000.

[39] Li, R., et al., "Spacecraft Attitude Determination System and Method," U.S. Patent 6,285,927, May 1999.

[40] Li, R., et al., "System and Method for Correcting Star Tracker Low Spatial Frequency Error in Stellar-Inertial Attitude Determination Systems," U.S. Patent 6,272,432, May 1999.

[41] Li, R., et al., "Autonomous Attitude Acquisition for a Stellar Inertial Attitude Determination System," U.S. Patent 6,108,594, May 1998.

[42] Kudva, P., and A. Throckmorton, "Preliminary Star Catalog Development for the Earth Observation Systems AMI (EOS-AMI) Mission," *AIAA Journal of Guidance, Control and Dynamics*, Vol. 19, No. 6, 1996, pp. 1332–1336.

[43] Bone, J. W., "On-Orbit Star Processing Using Multi-Star Star Trackers," *SPIE*, Vol. 2221, April 5–8, 1994, pp. 6–14.

[44] Wicenec, A. J., and M. Albrecht, "Methods for Structuring and Searching Very Large Catalogs," *Astronomical Society of the Pacific*, 1998, pp. 512–515.

[45] Barrett, P., "Application of the Linear Quadtree to Astronomical Databases," *Astronomical Data Analysis Software and Systems IV, ASP Conference Series*, Vol. 77, 1995, pp. 1–4.

[46] Strikwerda, T. E., and J. L. Junkins, "Star Pattern Recognition and Spacecraft Attitude Determination," U.S. Army ETL, May 1981.

[47] Ju, G., and S. K. Lee, "Attitude Determination for Kompsat-2 Using Star Identification Technique," AAS 02-100, pp. 1–17. March 2002.

[48] Heide, E. J., et al., "Development and Validation of a Fast and Reliable Star Sensor Algorithm with Reduced Data Base," IAF-98.A.6.05, pp. 1–13. October 1998.

[49] Potteck, S., "A Star Recognition Algorithm Based on Star Groups and Associated Means," AAS 99-011, pp. 87–94. November 1999.

[50] Mortari, D., J. L. Junkins, and M. A. Samaan, "Lost-In-Space Pyramid Algorithm for Robust Star Pattern Recognition," AAS 01-004, pp. 1–20. January 2001.

[51] Mortari, D., "Search-Less Algorithm for Star Pattern Recognition," pp. 179–194. Journal of the Astronautical Sciences 45(2), April 1997.

[52] Mortari, D., "A Fast On-Board Autonomous Attitude Determination System Based on a New Star-ID Technique for a Wide FOV Star Tracker," AAS 96-158, pp. 893–903. January 1996.

[53] Elstner, C., and D. Wunder, "Autonomous Attitude Determination," ESA Workshop, ESTEC, Noordwijk, October 1992

[54] Baldini, D., et al., "Star-Configuration Searching for Satellite Attitude Computation," IEEE Transactions on Aerospace & Electronics Systems, Vol. 31, No. 2, April 1995, pp 768–777,

[55] Udomkesmalee, S., J. W. Alexander, and A. F. Tolivar, "Stochastic Star Identification," J of Guidance, Control & Dynamics, Vol. 17, N0 6, November–December 1994

[56] Ju, G., et al., "Lost-in-Space: A Star Pattern Recognition and Attitude Estimation Approach for the Case of No Prior Attitude Information," AAS-00-004, February 2–6, 2000, pp. 1–16.

[57] Samaan, M. A., D. Mortari, and J. L. Junkins, "Recursive Mode Star Identification Algorithms," AAS-01-149, February 2001, pp. 1–18.

[58] Shucker, B., and E. D. Fasse, "A Ground-Based Prototype of a CMOS Navigational Star Camera for Small Satellite Applications," *15th AIAA/USU Conference on Small Satellites*, April 2001.

[59] Ju, G., et al., "Digistar: A Low Cost Micro Star Tracker," *AIAA Space Technology Conference & Exposition*, September 1999.

[60] Kim, H., J. L. Junkins, and D. Mortari, "A New Star Pair Pattern Recognition Method: Star Pair Axis and Image Template Matrix Method," Texas A & M University Internal Presentation, 2001.

[61] Rad, A. M., J. H. Nobari, and A. A. Nikkhah, "Optimal Attitude and Position Determination by Integration of INS, Star Tracker, and Horizon Sensor," *IEEE Aerosp. Electron. Syst. Mag.*, Vol. 29, 2014, pp. 20–33.

[62] Garvin, M., "Future of Celestial Navigation and the Ocean-Going Military Navigator" *OTS Master's Level Projects & Papers*, Vol. 41, 2010, https://digitalcommons.odu.edu/ots_masters_projects/41.

[63] Alshahin, R., "Effect of the Use of Celestial Navigation Technology in Marine Navigation," *The International Journal of Engineering and Science (IJES)*, Vol. 6, No. 4, 2017, pp. 63–73.

[64] Qiuying, W., et al., "Integrated Navigation Method Using Marine Inertial Navigation System and Star Sensor Based on Model Predictive Filtering," *Proc. of 2018 IEEE/ION Position, Location and Navigation Symposium (PLANS)*, Monterey, CA, April 23–26, 2018, pp. 850–857.

CHAPTER 11

Image and Vision-Based Navigation

11.1 Introduction

Image and vision-based navigation has been an active field of research and development since the 1980s with a significant amount of results publicly available. However, most of the image and vision-based navigation development was not done to solve the problem of PNT for aircraft, ships, and ground vehicles (the needs that drove the development of PNT technology). Instead, most of the image and vision-based navigation development was done by researchers in robotics and computer vision. It was done during the era when GPS became the most popular and least expensive means of PNT for human society. It is fair to say that we are very fortunate to be able to leverage the vision-based navigation development in robotics now that both the governments and industries have realized the important needs for GPS, GNSS-denied or challenged PNT.

The terminology used by robotics researchers for positioning is localization. The readers are advised that positioning equals localization and the terms can be used interchangeably. Another term that is popular in robotics is pose, which means relative attitude and relative position.

There are at least four major areas of image processing, computer vision, and image and vision-based navigation methods that can be found in computer vision and robotics literature that will be used for our purposes:

1. Image preprocessing, correction, and enhancement methods [1–10] that are necessary before any image and vision-based algorithms are exercised.
2. Image and vision-based navigation using pre-established geo-registered database (i.e., map-based localization (MBL), where camera images are matched against geo-registered database images, either directly, or indirectly by deriving or extracting features or landmarks from the image). Depending on how the database is organized and what matching or registration algorithms are used, there are many variations and subcategories of this method [11–35].
3. Image and vision-based navigation using simultaneous localization and mapping (SLAM) [36–40] where the landmarks are estimated and added into the database as they are being discovered. This is an area that have received significant amount of attention in robotics and there are many variations of this methods. One can imagine that SLAM is very difficult in general. However, in an All Source PNT operational scenario, other PNT sensor sources can be leveraged to achieve better performance when the system is still trying to discover and establish the map.

4. Image and vision-based navigation using the visual odometry method [41–58] where the camera images are not used to derive absolute information but are used to derive change of attitude and change of position information. At least two major variations of this approach exist. One derives change of position and change of attitude information by comparing features extracted from a sequence of images. The other is the optical flow method where the sequence of images is directly compared without performing feature extraction first.

Note that we use the terms "feature" and "landmark" interchangeably. Landmarks are not necessarily landmarks in the traditional sense such as the Golden Gate Bridge in San Francisco or the Empire State Building in New York. They are simply entities with consistent geographical locations that present themselves in images that can be described and identified through feature extraction image processing algorithms. For examples, a landmark or feature can be a corner of a building or a piece of rock on the side of a road. As another example, celestial navigation, which was presented in Chapter 10, can be viewed as a special case of vision-based navigation using landmarks or features, where the landmarks are simply the stars and RSOs. In this case, the star-tracker image processing algorithms are the specialized feature extraction algorithms that extract the simple features of stars and RSOs. If there is any difference between the two terms, it will be that we tend to use "landmark" when we pay more attention to the physical world and we use "feature" when we pay more attention to the image world. For vision-based navigation, since there is always the physical world and the image of the physical world, we tend to use "landmark" and "feature" interchangeably.

It is important to realize that there have been a tremendous amount of results that can now be viewed as vision-based navigation that were developed in the field of robotics and computer vision and it requires at least a whole book to capture the important results that are publicly available in the literature, not to mention that this is still an active field of research with results being published on a daily basis. What we will attempt to do is to show some of the results that are consistent with the methodology of the rest of the book and show these results as the foundation for further reading, research, and development in this very active field.

11.2 Image and Vision-Based Navigation Architecture

Figure 11.1 illustrates a typical generic architecture for image and vision-based navigation when a map database is used. When a map is used, the navigation performance is expected to have a bounded performance, in the sense that the navigation errors do not continuously grow with the time or distance traveled.

Within the architecture, the propagation solution and the update solution are integrated through a Kalman filter. In this architecture, the propagation solution is provided by inertial navigation and the camera is shown as the update sensor. What is implied and not shown is that, in an All Source PNT system, there will be other update sensors and the algorithms and the software associated with those sensors. In particular, the other sensors become very important when the image and vision-based navigator operates in the SLAM mode.

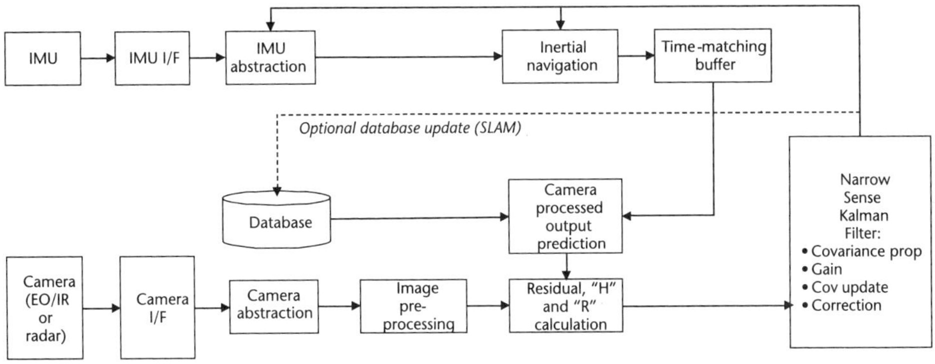

Figure 11.1 Architecture of MBL and SLAM.

The MBL and SLAM mode of the navigator are shown in the architecture diagram in a combined manner. In both cases, there will be a database that either stores the images (large database) or the information about the landmarks and features (small database); the information extracted from the database together with the current position and attitude information will be used to predict what the camera will see, either in terms of the image or in terms of the landmarks and features.

The camera images are time-tagged. The time tag is used to retrieve a time-matched navigation solution (position, velocity, and attitude) and camera alignments to support the output prediction algorithm mentioned above.

The camera image will typically be preprocessed in some or all of the following ways:

1. Correction by calibration parameters;
2. Histogram equalization (HE) or automatic gain control (AGC);
3. Extraction of features or landmarks and calculation of their corresponding LOS measurements.

The outputs of preprocessing are matched against the predicted outputs to determine one or several LOS measurement errors that can then be used by the Kalman filter to estimate corrections to the navigation solution and sensor error parameters.

In the case that the matching is done at the image level, the predicted image and the camera image are correlated to determine the offsets. This is often done by dividing the image into multiple tiles and performing the correlation for each of the tiles. This is because the transformation caused by attitude and position leads to different offsets for different parts of the image.

Alternatively, the matching can be done at the landmark or feature level. In this case, there are typically several (sometimes a lot) of landmarks or feature points predicted by the prediction algorithm. The preprocessing algorithm will attempt to generate the matched features by performing the feature or landmark extraction algorithm around the predicted area of the image. The landmarks or features produced from the camera image are then compared against the predicted landmarks or features to determine the offsets in the camera frame. The offsets become LOS measurement residuals to be used by the Kalman filter.

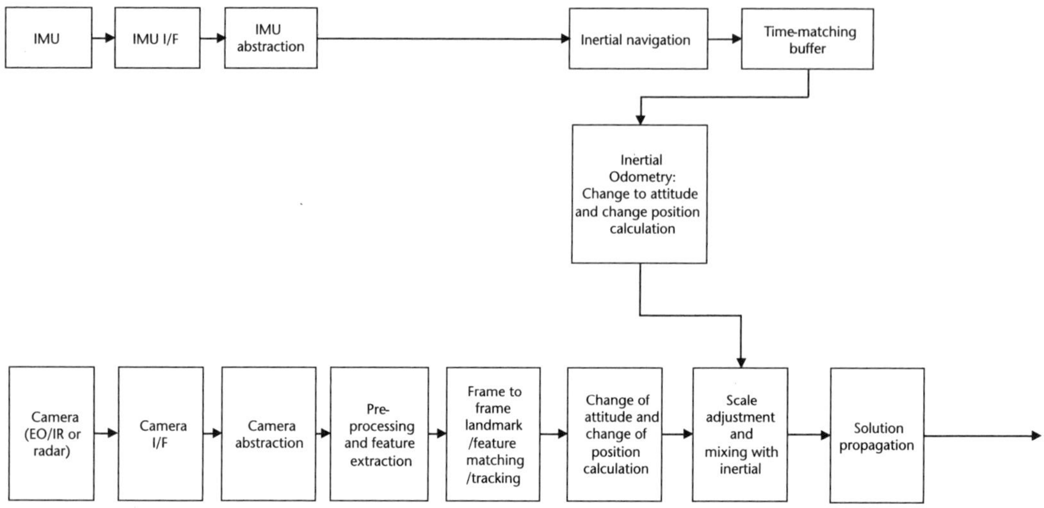

Figure 11.2 Candidate 1 architecture for visual odometry.

In the case of SLAM, we will typically include the landmark and feature locations as the states of the Kalman filter and as contents of the database. The Kalman filter will be able to provide updates to the landmark and feature locations when these variables are observable.

Figures 11.2 and 11.3 illustrate two of the many possible architectures for visual odometry-based approaches that do not require a map. Instead, visual odometry derives both attitude change and position change information by comparing a sequence of images over time and accumulating these changes over time to derive the attitude and position with respect to some reference. Clearly, one would expect

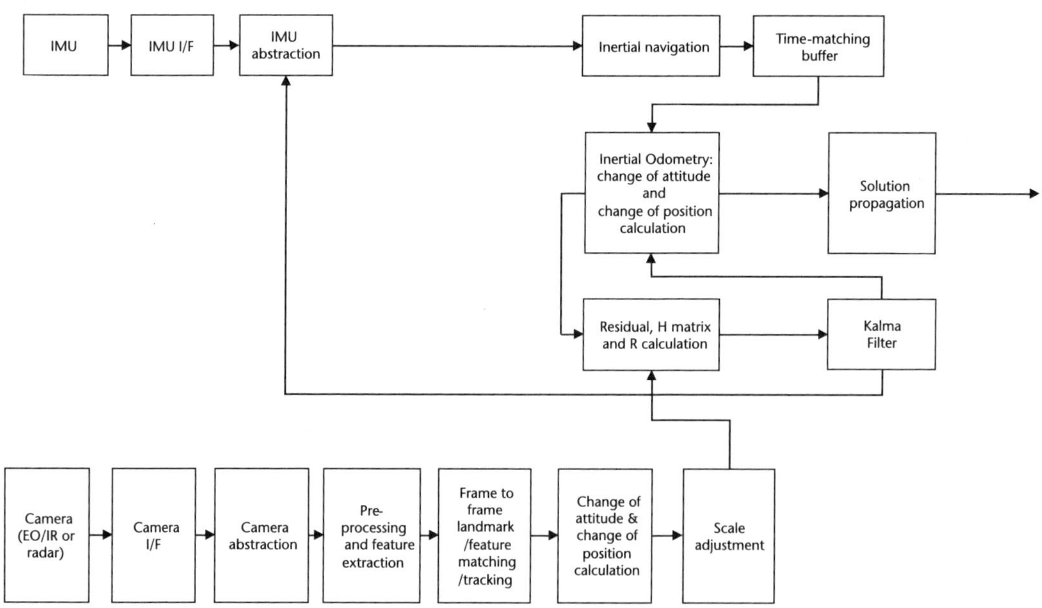

Figure 11.3 Candidate 2 architecture for visual odometry-aided inertial navigation.

that the navigation errors based on visual odometry will typically be growing as a function of distance traveled.

There are many ways in which the visual odometry information can be combined with the same information derived inertially. These different ways will lead to many variations of the architecture. What is illustrated in Figure 11.2 is one possible example approach where the visually derived attitude and position changes are used to propagate a navigation solution. In this architecture, the inertial-based solution is used for two purposes:

1. To provide the absolute scale information when such information is not available otherwise;
2. To act as a gap filler when visual data is not available.

Figure 11.3 illustrates another possible architecture where the inertially derived attitude and position changes are viewed as the primary solution that is updated by the visually derived solution with the help of a small Kalman filter. The Kalman filter will be able to generate corrections of the inertially derived odometry information and will also generate corrections to the IMU parameters.

11.3 Camera Image Preprocessing Algorithms

11.3.1 Overview

Before we start to process the image for image-based navigation using often expensive and sophisticated algorithms, it is often necessary to perform some basic processing so that the subsequent processing can be much more effective. We call these types of processing the preprocessing. The most commonly found preprocessing falls into two categories:

1. Image intensity dynamic range adjustment and equalization;
2. Image geometric distortion correction using calibration parameters.

To support category 2, we will also need to deal with calibration parameter estimation algorithms or simply calibration algorithms.

11.3.2 Image Intensity Dynamic Range Adjustment and Equalization

The most popular image intensity dynamic range adjustments and equalization algorithms are AGC and HE algorithms.

The AGC algorithm simply removes extreme values from both the maximum and minimum sides and linearly maps the dynamic range onto the full scale of a new dynamic range. Two things can happen during the AGC transformation:

1. We may change the storage required for each pixel, typically from bigger to smaller (such as a 16-bit integer to an 8-bit integer).
2. We maximize the usage of the dynamic range of the output image storage.

Other than eliminating the outliers on the bright and dark sides, the AGC algorithm does not change the shape of the histogram or distribution of the pixel intensity levels. It only gets shifted and rescaled.

AGC is very effective if the original image storage supports a large dynamic range but the actual image only uses a small dynamic range. AGC will eliminate the waste of storage by storing the image in a smaller memory without reducing the quality of the image.

HE is achieved by normalizing the intensity distribution using its cumulative distribution function so that the resulting image intensity is uniformly distributed.

HE typically increases contrast in the dominating intensity range but may increase the noise level and may wash out some of the details in areas where image intensities are almost the same.

There are a number of variations of the basic HE. Among them there are:

1. Plateau HE (PHE) improves the results by suppressing the enhancement of homogeneous regions with a plateau threshold value.
2. Adaptive HE (AHE) improves the performance by adaptively treating local areas differently.
3. More advanced methods are those such as contrast limited AHE (CLAHE), balanced CLAHE and contrast enhancement (BCCE), and bilateral filter and dynamic range partitioning (BF&DRP).

Since general image processing is not the focus of this book, the reader is referred to [1–7] for a more in-depth study of this topic. In the following we will describe the most basic AGC algorithm and the standard HE algorithm to serve as the foundation for further exploration of this topic.

We will first describe the AGC and HE algorithms for a general random variable X with its probability distribution density function given by $p_X(x)$.

Without loss of generality, let us assume that X takes value in the range $[0, x_{max}]$.

Assuming that the probabilities for $(X < A)$ and $(X > B)$ are really small, that is,

$$\int_0^A p_X(x)dx < \frac{\varepsilon}{2} \text{ and } \int_B^{x_{max}} p_X(x)dx < \frac{\varepsilon}{2} \tag{11.1}$$

the AGC algorithm simply shifts and scales the variable to new dynamic range $[0, y_{max}]$ by using the following transformation:

$$Y = \begin{cases} 0 & \text{if } X \le A \\ \dfrac{X - A}{(B - A)} y_{max} & \text{if } A < X < B \\ 0 & \text{if } X \ge B \end{cases} \tag{11.2}$$

Figure 11.4 illustrates how the AGC transformation saves the unused dynamic range by shifting and scaling and how it efficiently uses the new dynamic range. Because of this, it is often possible not to lose resolution while reducing the word length used to store the value (for example, converting from a 16-bit integer to an 8-bit integer). It is easy to see that the processing discussed above applies to all types of physical variables; it does not have to be just image.

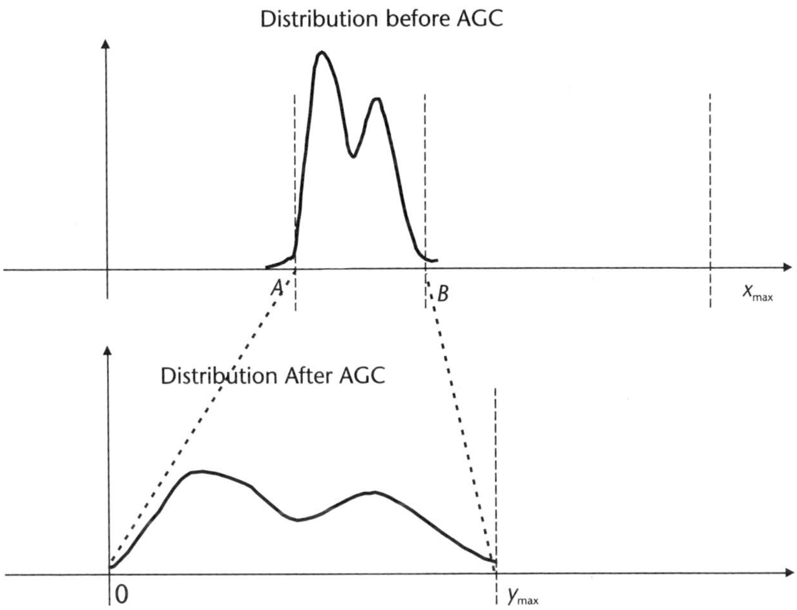

Figure 11.4 AGC transformation shifts and scales the distribution density function without changing the shape.

Histogram is really just another name for probability distribution density function in discrete case. The purpose of the HE is simply to perform a transformation of a random variable such that the resulting random variable is uniformly distributed. The advantage of uniform distribution is that each value in the dynamic range of the variable has an equal opportunity to be expressed in the new output. For image, it means that each intensity level has equal opportunity to be seen in the output image, which often leads to better contrast.

For the same random variable X mentioned earlier, the cumulative probability distribution function is defined to be

$$F(x) = \int_0^x p_X(x)\,dx = P(X \le x) \tag{11.3}$$

Define a new random variable

$$Y = y_{max}F(X) \tag{11.4}$$

Since $p_X(x) \ge 0$, $F(x)$ is monotonically increasing (or, more rigorously, nondecreasing).

Let the inverse function of $F(x)$ be given by $F^{-1}(y)$, and then the cumulative probability distribution function for the random variable Y can be written as

$$G(y) = P(Y \le y) = P\left(y_{max}F(X) \le y\right) = P\left(F(X) \le \frac{y}{y_{max}}\right) =$$
$$= P\left(X \le F^{-1}\left(\frac{y}{y_{max}}\right)\right) = F\left(F^{-1}\left(\frac{y}{y_{max}}\right)\right) = \frac{y}{y_{max}}$$

The probability density function of Y is then given by

$$g(y) = \frac{1}{y_{\max}} \tag{11.5}$$

which is uniformly distributed over the interval $[0, y_{\max}]$. In other words, we have just shown that the transformation $Y = y_{\max}F(x)$ transforms a random variable to a new random variable that is uniformly distributed over the interval $[0, y_{\max}]$. Figure 11.5 illustrates this process.

Now let us look at how this is done specifically to an image. Assuming that the image pixel intensities are integer valued in the interval $[0, N - 1]$, to simplify the discussion, we will do equalization without changing the dynamic range. In other words, the intensities of the output image have the same dynamic range of $[0, N - 1]$. Therefore, we have

$$x_{\max} = y_{\max} = N - 1$$

The probability distribution is provided by a normalized histogram as

$$p_k = \frac{\text{number of pixels with intensity } k}{\text{total number of pixels}}, \quad k = 0,1,2,\cdots N - 1 \tag{11.6}$$

The cumulative probability distribution function is given by

$$F(x) = \sum_{0}^{x} p_n, \quad x \in [0,1,2,\cdots N - 1] \tag{11.7}$$

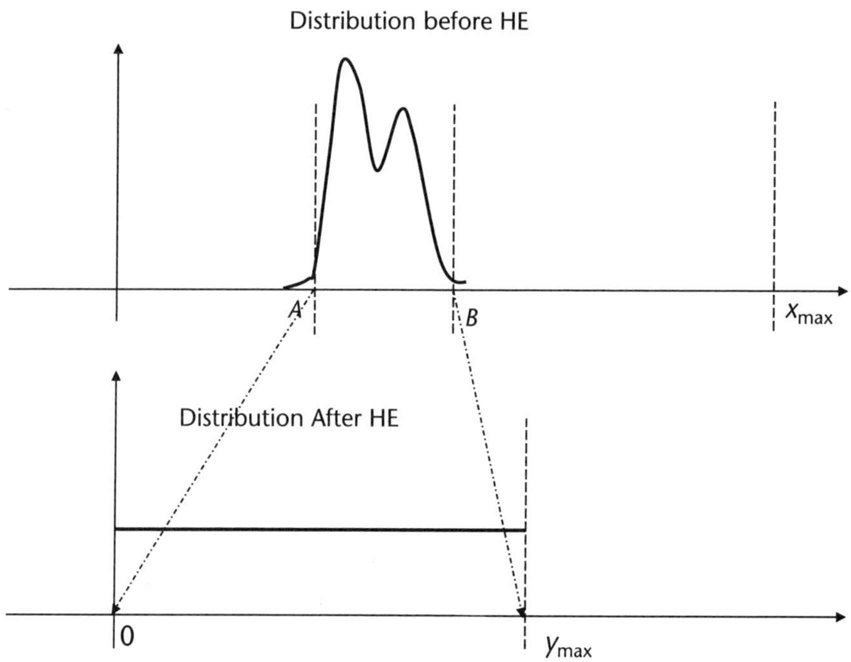

Figure 11.5 HE transforms a random variable of any distribution to a random variable of uniform distribution.

The transformation to transform the source image

$$X = \left[x_{i,j}\right] \quad x_{i,j} \in [0,1,2,\cdots N-1]$$

to a histogram equalized output image

$$Y = \left[y_{i,j}\right] \quad y_{i,j} \in [0,1,2,\cdots N-1]$$

is then given by

$$y = y_{\max} F(x)$$

where

$$y_{\max} = N - 1$$

and $F(x)$ is given by (11.7).

Consequently, the algorithm to transform the intensity of the pixels is therefore given by

$$y_{i,j} = (N-1)\left(\sum_{n=0}^{x_{i,j}} p_n\right) \tag{11.8}$$

Note that the above transformation is done on a pixel-by-pixel basis. Once the histogram is produced, the transformation is done to change the image intensity one pixel at a time.

Figure 11.6 is a demonstration of the effectiveness of the HE.

11.3.3 Camera Calibration and Image Correction by Calibration Parameters

In this section, we are concerned about the geometric distortion of the image. Specifically, we are concerned about the locations of pixels being distorted and consequently causing the image to appear distorted.

Figure 11.6 An example of the HE.

More specifically, we are concerned about the following three aspects of the problems:

1. How the distortions shall be parameterized in a mathematical model;
2. How the parameters shall be calibrated or estimated through a calibration procedure;
3. How to use the distortion model to correct the image.

11.3.3.1 Parameterization of the Problem

Given the distorted image and undistorted image as

$$I\left(x_{d,}y_{d}\right) \text{ and } I\left(x_{u,}y_{u}\right)$$

among the various types of image distortions, radial distortion is seen in most applications. One of the most effective models for radial distortion is the division model that was proposed by Fitzgibbon et al. [8–10]:

$$\begin{aligned} x_u &= \frac{x_d}{\left(1+\lambda_1 r_d^2+\lambda_2 r_d^4+\cdots\right)} \\ y_u &= \frac{y_d}{\left(1+\lambda_1 r_d^2+\lambda_2 r_d^4+\cdots\right)} \end{aligned} \tag{11.9}$$

where

$$r_d^2=\left(x-x_0\right)^2+\left(y-y_0\right)^2 \tag{11.10}$$

and (x_0, y_0) is the center of the distortion.

Even the single parameter model can be very effective:

$$\begin{aligned} x_u &= \frac{x_d}{\left(1+\lambda_1 r_d^2\right)} \\ y_u &= \frac{y_d}{\left(1+\lambda_1 r_d^2\right)} \end{aligned} \tag{11.11}$$

In general, to model other types of distortions, we can use the following two-variable polynomial model or simply use a table lookup correction:

$$\begin{aligned} x_u &= \sum_{k,j} a_{k,j} x_d^k y_d^j \\ y_u &= \sum_{k,j} b_{k,j} x_d^k y_d^j \end{aligned} \tag{11.12}$$

In many cases, we may choose a specific function type for each type of errors and we end up with

$$\begin{aligned} x_u &= \sum_k a_k f_k(x_d, y_d) \\ y_u &= \sum_k b_k g_k(x_d, y_d) \end{aligned} \tag{11.13}$$

It is easy to recognize that the last type actually is generic enough to include all the other types of parameterizations.

11.3.3.2 Calibration or Estimation of the Model Parameters

It is often a good idea to use an image with known coordinates of the key feature points. By determining the location of the feature points of the distorted image, we will have the data to estimate the model parameters.

One example is to use a checkerboard pattern and we can determine the coordinates of the grid points.

Another example is to use a star pattern (i.e., images with a number of dots located at different parts of the image). The nice thing about the star pattern is that the image processing to determine the location of the center of the stars of the distorted image will be the very simple centroiding algorithm that calculates the weighted center of the mass of the star image. Figure 11.7 illustrates the checkerboard pattern and the star pattern.

Once we have both the desired, ideal, or undistorted coordinates and the distorted coordinates, using the generic parameterization provided in (11.13), we have:

$$\begin{bmatrix} x_{u1} \\ x_{u2} \\ \vdots \\ x_{um} \end{bmatrix} = \begin{bmatrix} f_1(x_{d1}, y_{d1}) & f_2(x_{d1}, y_{d1}) & \cdots & f_n(x_{d1}, y_{d1}) \\ f_1(x_{d2}, y_{d2}) & f_2(x_{d2}, y_{d2}) & \cdots & f_n(x_{d2}, y_{d2}) \\ \vdots & \vdots & \vdots & \vdots \\ f_1(x_{dm}, y_{dm}) & f_2(x_{dm}, y_{dm}) & \cdots & f_n(x_{dm}, y_{dm}) \end{bmatrix} \begin{bmatrix} a_1 \\ a_2 \\ \vdots \\ a_n \end{bmatrix} \tag{11.14}$$

$$\begin{bmatrix} y_{u1} \\ y_{u2} \\ \vdots \\ y_{um} \end{bmatrix} = \begin{bmatrix} g_1(x_{d1}, y_{d1}) & g_2(x_{d1}, y_{d1}) & \cdots & g_n(x_{d1}, y_{d1}) \\ g_1(x_{d2}, y_{d2}) & g_2(x_{d2}, y_{d2}) & \cdots & g_n(x_{d2}, y_{d2}) \\ \vdots & \vdots & \vdots & \vdots \\ g_1(x_{dm}, y_{dm}) & g_2(x_{dm}, y_{dm}) & \cdots & g_n(x_{dm}, y_{dm}) \end{bmatrix} \begin{bmatrix} b_1 \\ b_2 \\ \vdots \\ b_n \end{bmatrix} \tag{11.15}$$

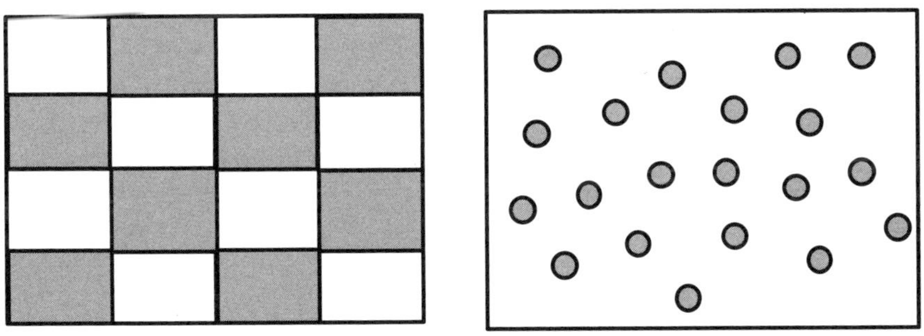

Figure 11.7 Typical image patterns used to calibrate the camera.

Solving the above two least square problems, we arrive at the estimates of the calibration parameters

$$a_k, b_k$$

It is important to note that what is described here is just a generic example. For specific applications, one is encouraged to properly model and transform the problem and find the most effective parameterization and solution.

Taking the two-parameter radial distortion only example, we can write the division model as

$$\begin{aligned} 1 + \lambda_1 r_d^2 + \lambda_2 r_d^4 &= \frac{x_d}{x_u} \\ 1 + \lambda_1 r_d^2 + \lambda_2 r_d^4 &= \frac{y_d}{y_u} \end{aligned} \tag{11.16}$$

If we simply average the two equations, we can write the following equations:

$$\lambda_1 r_d^2 + \lambda_2 r_d^4 = \frac{1}{2}\left(\frac{x_d}{x_u} + \frac{y_d}{y_u}\right) - 1 \tag{11.17}$$

With two points or pairs of points , we can easily estimate the two parameters by solving a linear least square problem.

11.3.3.3 Correction of the Image

Given the distorted image and the undistorted image as

$$\left\{I\left(x_{u,}y_u\right)\right\} \text{ and } \left\{I\left(x_{d,}y_d\right)\right\}$$

If (x_d, y_d) falls on preselected grid points, then (x_u, y_u) will generally not be.

Assuming that there is a nonlinear correction function that will map (x_d, y_d) to (x_u, y_u) such that

$$\left(x_{u,}y_u\right) = \left(f_x\left(x_d, y_d\right), f_y\left(x_d, y_d\right)\right) \tag{11.18}$$

in order to create a corrected image from the distorted image, it is often important to work on a super set of the image; in other words, we will need to deal with an image that is a super set of the area of interests.

In order to construct a new image on the preselected grid, we will need to use interpolations. Among many possibilities, bilinear interpolation is one of the most popular.

Let us assume that the image has sufficient resolution such that in a neighborhood of each of the pixels, the intensity appears to be a bilinear surface, which can be modeled as

$$I(x,y) = I_0 + \alpha x + \beta y + \lambda xy \tag{11.19}$$

For a given grid point

$$(x_{k,}y_k)$$

we will find the four or more nearest undistorted pixel locations that were calculated using the correction functions. Assuming that they are

$$\begin{array}{c}(x_{kn1,}y_{kn1})\\(x_{kn2,}y_{kn2})\\(x_{kn3,}y_{kn3})\\(x_{kn4,}y_{kn4})\end{array}$$

we then have

$$\begin{aligned}I_{kn1} &= I(x_{kn1}, y_{kn1}) = I_0 + \alpha x_{kn1} + \beta y_{kn1} + \gamma x_{kn1} y_{kn1}\\I_{kn2} &= I(x_{kn2}, y_{kn2}) = I_0 + \alpha x_{kn2} + \beta y_{kn2} + \gamma x_{kn2} y_{kn2}\\I_{kn3} &= I(x_{kn3}, y_{kn3}) = I_0 + \alpha x_{kn3} + \beta y_{kn3} + \gamma x_{kn3} y_{kn3}\\I_{kn4} &= I(x_{kn4}, y_{kn4}) = I_0 + \alpha x_{kn4} + \beta y_{kn4} + \gamma x_{kn4} y_{kn4}\end{aligned}$$

or

$$\begin{bmatrix} I_{kn1} \\ I_{kn2} \\ I_{kn3} \\ I_{kn4} \end{bmatrix} = \begin{bmatrix} 1 & x_{kn1} & y_{kn1} & x_{kn1}y_{kn1} \\ 1 & x_{kn2} & y_{kn2} & x_{kn2}y_{kn2} \\ 1 & x_{kn3} & y_{kn3} & x_{kn3}y_{kn3} \\ 1 & x_{kn4} & y_{kn4} & x_{kn4}y_{kn4} \end{bmatrix} \begin{bmatrix} I_0 \\ \alpha \\ \beta \\ \gamma \end{bmatrix} \tag{11.20}$$

which allows us to solve for (I_0, α, β, γ) and then we calculate the image intensity at each of the grid points using $I(x, y) = I_0 + \alpha x + \beta y + \lambda xy$.

Obviously, there could be many variations of the method described above. For example:

1. We could use more points to fit into the bilinear plane model, solve a least square problem for the coefficients, and use the same model for a few more grid points. This has the potential to reduce computational burden and reduce the impact of the noise at the same time.
2. We could also use a higher-order nonlinear model and use that model to cover a bigger area.

11.4 Map-Based Image and Vision Navigation

11.4.1 Architecture Variations

As discussed earlier, there are at least two classes of approaches to perform the registration or matching of the images: the direct matching of the images at the pixel level and the indirect matching of the images by matching landmarks and features.

Figure 11.8 illustrates the two types of processing. Both approaches comply with the architecture diagram in Figure 11.1. The differences are the algorithms and the databases used.

When direct-image matching method is used, the architecture has the following features:

1. The database contains geo-registered images and is typically very large. Image compression will be an important requirement.
2. In addition, a terrain database is needed to allow the calculation of the piercing points of the LOS vectors from the camera to the ground to allow the projection of the Earth frame image to the camera frame.
3. The image prediction algorithm will typically generate a predicted image that is larger (for example, 2×2 times larger) than the camera image to allow the correlation algorithms to perform searches during the correlation process.
4. The image preprocessing algorithm is typically very simple. It normally just includes HE and corrections by known calibration parameters such as lens deformation and focal-length shift corrections.
5. The image correlation algorithm will directly match the images. Typically, it means sliding the camera image against the bigger predicted image and evaluating the correlation score. The offset generated by the correlation process will be used as the residual between the measured image LOS vector and the predicted image LOS vector.
6. Because the position and attitude errors introduce a transformation to the image that is more than just offset, we often divide the image into tiles and

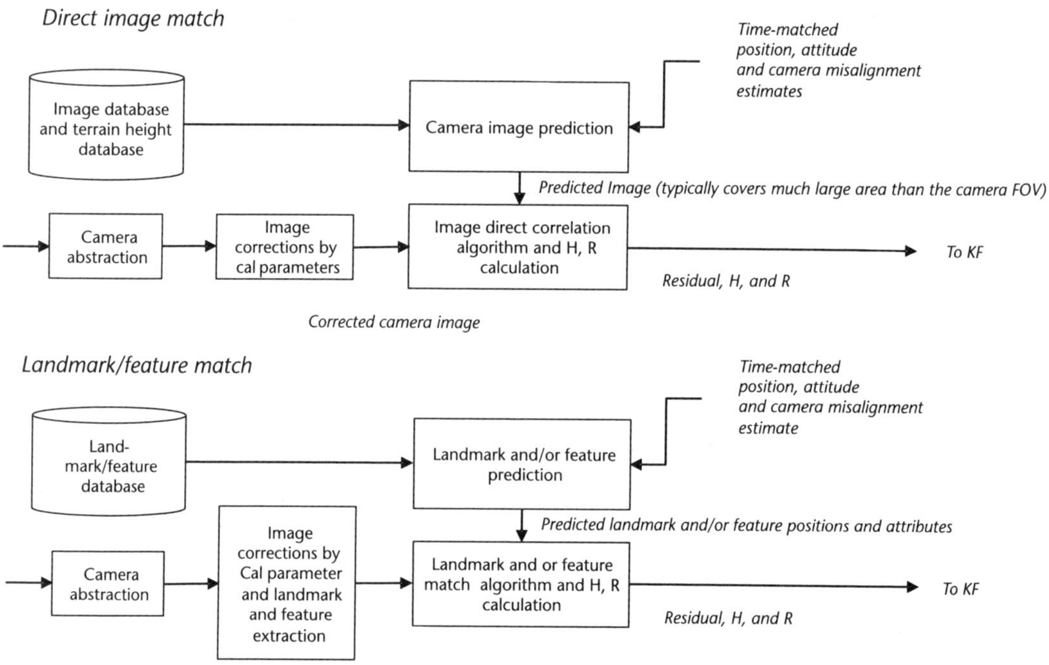

Figure 11.8 Direct image matching versus landmark or feature matching.

do correlation on a tile-by-tile basis. This will lead to multiple LOS residual measurements.

When the landmark or feature matching method is used, the architecture has the following features:

1. The database contains landmarks and features that typically have the following information:
   a. The landmark and feature location in the ECEF or geodetic coordinate frame, for example, the location of the corner point of a corner feature;
   b. The type of feature (which helps to determine the algorithm to be used to extract the feature and feature position);
   c. Other parameters that help the identification or the processing of the landmarks or features.
2. Typically, the database will be much smaller since the raw image is not necessary for the database.
3. The landmark or feature prediction algorithm will typically generate a list of predicted landmarks or features with their parameters such as the locations and the types of the features.
4. The predicted landmarks and features will be used to guide the image preprocessing algorithm to perform image processing in the neighborhood of the predicted landmarks and features to determine the locations and other parameters of the features. The features extracted are then compared against the predicted features to derive the landmark and feature residuals.
5. The residuals are then fed into the Kalman filter to generate corrections.

By now, we should realize that the celestial navigation algorithm discussed in Chapter 10 is actually a special version of the image and vision-based navigation with the landmarks and features being the stars and RSOs. In this case, the feature database is simply the combination of the star catalog and the RSO ephemeris database.

It is important to realize that the end results of the two types of the matching algorithms are the same in terms of producing the following output:

1. The positions in the ECEF or geodetic frame of the landmark and feature points or the positions of the Earth-piercing points of the LOS vectors that pass through the centers of tiles of interest. Note that, in the case of landmark and feature matching, this information is directly available from the database. In the case of direct matching, this information is derived by mapping the center points of the tiles in the camera frame to the ground (which requires the solution of nonlinear equations as discussed in Chapter 9).
2. The residuals corresponding to the landmark and feature points or the centers of the tiles.

With the above, the rest of the equations for processing are the same for the two approaches.

11.4.2 Output Prediction, Registration Matching, and the Correlation Algorithm

11.4.2.1 Image Prediction Algorithm

To reduce the processing burden, we typically create the grid in the camera frame first. We can calculate the LOS unit vectors in the camera frame that go through the grid points. The LOS vectors can then be transformed to the ECEF frame, which will be used to solve for the piercing points on the Earth.

By using interpolations, we can derive the piercing points on Earth for each of the pixels of the image. Let the coordinates of a grid point k be given by

$$\begin{bmatrix} h_k \\ v_k \end{bmatrix}$$

The unit vector in the camera frame can then be calculated as

$$^C u_k = \frac{1}{\sqrt{1 + h_k^2 + v_k^2}} \begin{bmatrix} h_k \\ v_k \\ 1 \end{bmatrix} \tag{11.21}$$

which can be transformed to the body frame and then the ECEF frame by

$$^E u_k = C_B^E C_C^B \, {}^C u_k \tag{11.22}$$

which can then be used to solve the following two equations (as discussed in Chapter 9)

$$\begin{aligned} {}^E R_{pp,k} &= {}^E R + (d)\,{}^E u_k \\ g\left({}^E R_{pp,k}\right) &= 0 \end{aligned} \tag{11.23}$$

where ${}^E R_{pp,k}$ is the position of the Earth-piercing point in ECEF frame, ${}^E R$ is the camera location in the ECEF frame, and d is the range from the camera to the ground-piercing point and $g({}^E R_{pp,k}) = 0$ is the terrain constraint function that defines the terrain surface.

The image prediction algorithm can then be summarized as:

1. For the image tile of the interest (typically larger than the actual image tile of the camera to allow search), create the grid points in the camera frame.
2. For each of the grid points, solve for the corresponding Earth-piercing points using (11.22) and (11.23) and the algorithms described in Chapter 9.
3. For each pixel in the camera frame, use linear interpolation to calculate the approximate Earth-piercing point corresponding to the pixel.
4. Use the Earth-piercing point to retrieve from the corresponding pixel the brightness value (when grayscale) or the RGB values (when colored) from the database. Since the Earth-piercing point will not be exactly at a database grid point, interpolations from the database values of the nearby points will be used.

11.4.2.2 Landmark and Feature Prediction Algorithm

The prediction of the landmark and feature points is, in general, easier than predicting the image. It contains four steps:

1. Determine an FOV that is bigger than the camera FOV based on position and attitude uncertainty.
2. Calculate the corresponding Earth-piercing points in the ECEF of the corner points of the FOV.
3. Use the corner points to retrieve all the feature points and landmarks with their ECEF location.
4. Transform the ECEF location to the camera frame and calculate the coordinates h and v using the following equations

$$
\begin{aligned}
{}^{E}u_k &= \frac{\left({}^{E}R_{pp,k} - {}^{E}R\right)}{\left\|{}^{E}R_{pp,k} - {}^{E}R\right\|} \\
{}^{C}u_k &= C_B^C C_E^B \, {}^{E}u_k \\
\begin{bmatrix} h_k \\ v_k \end{bmatrix} &= \frac{1}{{}^{C}u_k(3)} \begin{bmatrix} {}^{C}u_k(1) \\ {}^{C}u_k(2) \end{bmatrix}
\end{aligned} \tag{11.24}
$$

11.4.2.3 Direct Matching by Normalized Cross-Correlation (NCC)

Two algorithms are particularly popular for direct image registration and correlation. They are the NCC and the mutual information (MI) algorithms.

Given two images centered around (u_1, v_1) and (u_2, v_2)

$$
\begin{aligned}
&I_1\left(i + u_1, j + v_1\right) \quad (i = -n,\ldots n, \ \ j = -n,\ldots,n) \\
&I_2\left(i + u_2, j + v_2\right) \quad (i = -n,\ldots n, \ \ j = -n,\ldots,n)
\end{aligned}
$$

the NCC can be calculated using the following equation:

$$
NCC\left(u_1,v_1,u_2,v_2\right) = \frac{\sum_{i=-n}^{n}\sum_{j=-n}^{n} I_1\left(i+u_1,j+v_1\right) I_2\left(i+u_2,j+v_2\right)}{\sqrt{\left(\sum_{i=-n}^{n}\sum_{j=-n}^{n} I_1^2\left(i+u_1,j+v_1\right)\right)\left(\sum_{i=-n}^{n}\sum_{j=-n}^{n} I_2^2\left(i+u_2,j+v_2\right)\right)}} \tag{11.25}
$$

The NCC serves as a score of similarity between the two pictures. The higher the NCC value is, the more similar the two pictures are. Assuming that $I_1(i + u_1, j + v_1)$ is the camera image (or a portion of the camera image) and $I_2(i + u_2, j + v_2)$ is the predicted camera image, we would want to do is to scan through values of (u_2, v_2) and evaluate the NCC for each value. The result of this exercise is that we produce a matrix of NCC scores, and $(u_2, v_2) = (\bar{u}_2, \bar{v}_2)$, the center of the image that achieves the maximum score, is considered to be the center of the predicted image that leads to a match of the two images.

If we consider (u_1, v_1) as the measured center of the image, then the predicted image center will be $(u_2, v_2) = (\bar{u}_2, \bar{v}_2)$ and the difference between the two can be viewed as the residual, which is

$$(\delta u, \delta v) = \left(u_1 - \bar{u}_2, u_2 - \bar{v}_2\right) \tag{11.26}$$

11.4.2.4 Direct Matching by MI

MI is a statistical measure from the information theory. Given two random variables A and B, the MI of two random variables is defined to be

$$MI(A,B) = \sum_{a,b} p_{A,B}(a,b) \ln \frac{p_{A,B}(a,b)}{p_A(a)p_A(b)} \tag{11.27}$$

where $p_{A,B}(a, b)$ is the joint probability density function for A and B and $p_A(a)$, $p_B(b)$ are the marginal probability density function of A and B, respectively.

The above equation can also be written in terms of marginal and joint entropy of the random variables A and B:

$$MI(A,B) = H(A) + H(B) - H(AB) \tag{11.28}$$

where

$$\begin{aligned} H(A,B) &= -\sum_{a,b} p_{A,B}(a,b) \ln p_{A,B}(a,b) \\ H(A) &= -\sum_{a} p_A(a) \ln p_A(a) \end{aligned} \tag{11.29}$$

When we try to match two images, each image is treated as a random variable, and the 1-D and 2-D histograms after normalization are used as an approximation of the probability density functions.

Since the MI approach pays more attention to the distribution instead of the individual pixel intensities, the MI-based method is often more useful when matching images of different modalities. For example, MI-based matching is typically more useful when matching an IR image against the visible light image database.

To illustrate how MI is calculated, the following is a simple numerical example. We have two identical images, each having $3 \times 5 = 15$ pixels:

$$A = B = \begin{bmatrix} 0 & 0 & 1 & 0 & 0 \\ 0 & 1 & 2 & 1 & 0 \\ 0 & 0 & 1 & 0 & 0 \end{bmatrix}$$

There are only three gray levels (0, 1, 2).

The 1-D histograms are

$$hist(A) = hist(B) = \begin{bmatrix} 10 & 4 & 1 \end{bmatrix}$$

and the corresponding marginal density function is then

$$p(A) = p(B) = \frac{1}{15}\begin{bmatrix} 10 & 4 & 1 \end{bmatrix}$$

The 2-D mutual histogram is a 3 × 3 matrix

$$hist(A,B) = \begin{bmatrix} 10 & 0 & 0 \\ 0 & 4 & 0 \\ 0 & 0 & 1 \end{bmatrix}$$

and the corresponding mutual density function is

$$p(A,B) = \frac{1}{15}\begin{bmatrix} 10 & 0 & 0 \\ 0 & 4 & 0 \\ 0 & 0 & 1 \end{bmatrix}$$

and the final score is

$$MI(A,B) = 10\ln\frac{\left(\frac{10}{15}\right)}{\left(\frac{10}{15}\right)\times\left(\frac{10}{15}\right)} + 4\ln\frac{\left(\frac{4}{15}\right)}{\left(\frac{4}{15}\right)\times\left(\frac{4}{15}\right)} + 1\ln\frac{\left(\frac{1}{15}\right)}{\left(\frac{1}{15}\right)\times\left(\frac{1}{15}\right)} = 12.0497$$

As mentioned before, MI has the advantage of being able to match images of different modalities. The following are some examples to illustrate the point.

Example 11.1: Intensity Shift and Noise Insensitivity

The camera image is a 64 × 64 image with a 10 × 10 square feature in the middle and the noise level is at 50%. The predicted image has shifted intensity as shown in Figure 11.9. The camera image is shown in Figure 11.10. The correlation result shows the correct match of the image and is shown in Figure 11.11.

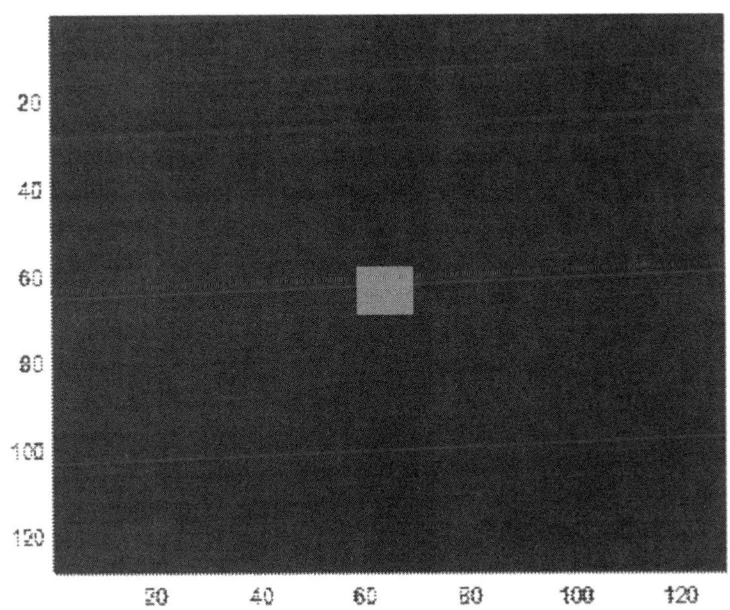

Figure 11.9 Predicted image with an intensity shift.

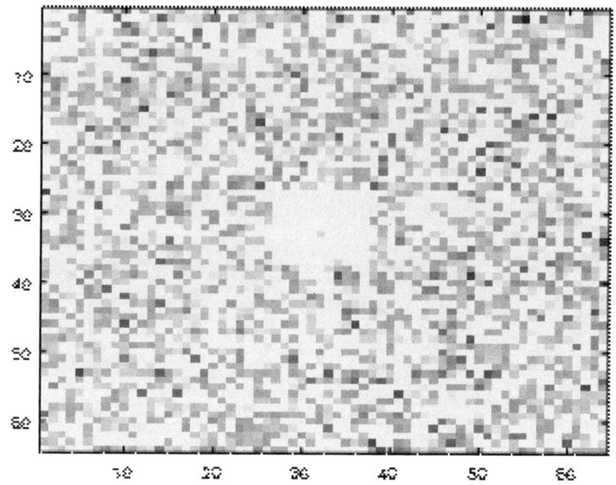

Figure 11.10 Camera image with large noise.

Example 11.2: Inverted Image

In this example, the camera image and the predicted image are inverted against each other and, as can be seen from the results, we still have a good-quality correlation result.

The predicted image is identical to Example 11.1 and consequently is not shown here. The camera image is shown in Figure 11.12. The correlation results are shown in Figure 11.13.

11.4.2.5 Landmark or Feature Matching

In order to match the camera-sensed landmark or features with the predicted landmarks and features, we must preprocess the image and generate landmarks or

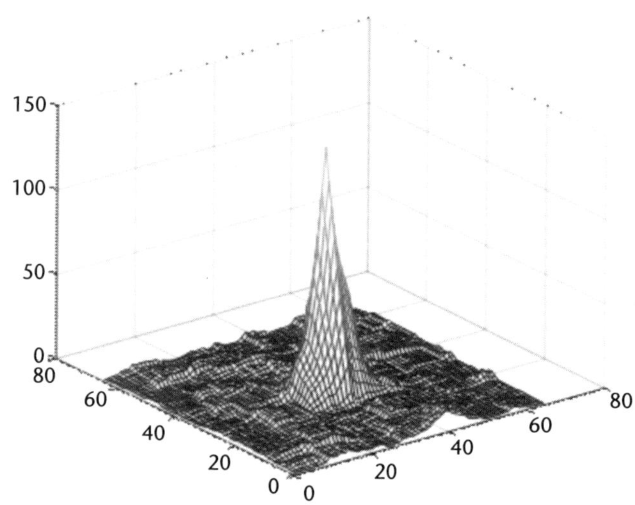

Figure 11.11 Good-quality correlation results.

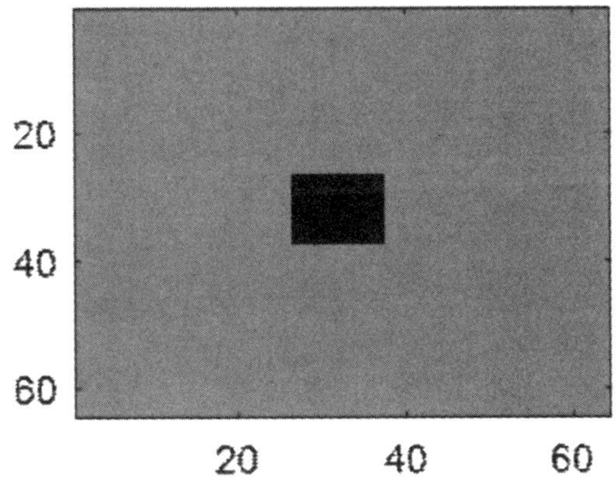

Figure 11.12 Inverted camera image.

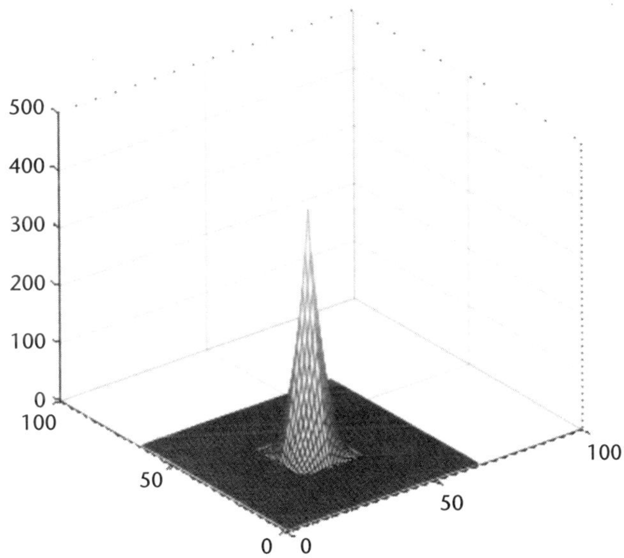

Figure 11.13 Quality correlation results with the inverted image.

features first. That means that we must detect the features and describe the features so that they can be matched later.

Obviously, the same feature extraction algorithms and feature descriptors have to be used for matching features between the database and the camera image. Each feature in the database can be generated using the same or different techniques depending on the nature of the features.

Feature extraction is a heavily studied topic in computer vision and there are thousands of methods available. Table 11.1 lists some of the popular algorithms and their corresponding feature descriptors. The selection of particular algorithms and descriptors depends on the specific application. The readers are encouraged to review these and other feature detectors or descriptors to decide on the best choice for their application.

Table 11.1 Examples of Feature Descriptors or Algorithms and Their Attributes

	Algorithm Types	Key Idea	Key Advantages	Key Disadvantages
1	SIFT descriptor and detector (Blobs) (D. G. Lowe, 2004	Scale invariant feature transform (SIFT) is one of the most robust feature extraction algorithms available; SIFT detects a blob; the SIFT detector is based on the difference-of-Gaussians (DoG) operator; feature points are detected by searching local maxima using DoG at various scales of the images	SIFT features are invariant to rotation, scale, illumination, and viewpoint changes; some limited invariance to affine transformations	Computationally expensive
2	SURF descriptor and detectors (blobs) (H. Bay, 2008)	Speeded-up robust features (SURF) are partially motivated by SIFT and were declared by the inventors of the method to be more robust and computationally cheaper; the SURF detector is based on the determinant of Hessian matrix and uses integral images to improve speed	Similar to SIFT and potentially more robust and computationally cheaper	Some believe SIFT is more robust
3	KAZE descriptor and detector (blobs) (P. F. Alcantarilla, 2012)	KAZE detector is based on scale-normalized determinant of Hessian matrix computed at multiple scale levels; the maxima of detector response are chosen as feature-points with a moving window	KAZE and AKAZE features are rotation and scale invariant and have limited affine transformation invariability	Computationally most expensive; can be 2 times as much or more expensive than SIFT
4	AKAZE descriptor and detector (Blobs) (P. F. Alcantarilla, 2013)	Accelerated KAZE		Computationally more efficient than KAZE, SIFT, and SURF
5	ORB descriptor and detector (Corners) (E. Rublee et al., 2011)	Oriented FAST and Rotated BRIEF (ORB) is derived from modified FAST (Features from Accelerated Segment Test) and BRIEF (Binary Robust Independent Elementary Features) description methods; FAST corners are detected in each layer of the scale pyramid and the results are evaluated using Harris Corner score to produce the high-quality points; a modified version of BRIEF descriptor has been employed to improve its stability with respect to rotations	ORB features are corners and are invariant to scale, rotation and limited affine changes; ORB is computationally efficient; [45] ranked this approach as the least expensive method	
6	BRISK descriptor and detectors (Corners) (S. Leutenegger et al., 2011)	Binary Robust Invariant Scalable Key (BRISK) points detect corners using the AGAST (Adaptive and Generic Accelerated Segment Test) algorithm, filters them with FAST Corner score, and searches for maxima in the scale space pyramid; BRISK identifies the characteristic direction of each feature to achieve rotation invariance	BRISK features are invariant to scale, rotation, and limited affine changes; computationally efficient, slightly worse than ORB	

In order to use feature or landmark-based methods against a known database, we have to go through the following steps:

1. Establish the landmarks or features database:
   a. We typically start with geo-referenced image database and use some of the methods to detect and describe the features including the geolocation associated with these features.
   b. These candidate features are evaluated against weather stability, seasonal stability, cloud cover stability, and the standard scale and rotation invariance to be selected to be included in the database.
2. Real-time feature detection and description:
   a. The camera image is processed to generate the landmarks and features using the chosen algorithm.
   b. This can be done completely in the open loop without using a priori knowledge.
   c. This can also be done with a prediction step by using the database features and the current knowledge of the position, attitude, and camera alignment to predict what features to find and where to find them.
3. Feature matching against the database features.
4. Outlier rejection: Each of the matched features will generate a measurement for the PNT solution. We will need to check each of the measurements before they are used. One way to do this is to check the Kaman filter residuals against the covariance matrix.

11.4.3 Propagation Algorithms and Error Models

The inertial navigation algorithms were discussed in Chapter 3 and there are several options of the algorithm depending on the frame in which the equations are written. For the purpose of this chapter, we use the INS algorithms presented in the ECEF frame that are summarized here:

1. The attitude propagation equation in quaternion form is simply

$$\dot{q}_B^E = \frac{1}{2}\left(q_B^E \otimes {}^B\omega_I^B - {}^E\omega_I^E \otimes q_B^E\right) \tag{11.30}$$

2. The DCM can be derived from the quaternion

$$C_B^E = q2DCM\left(q_B^E\right) \tag{11.31}$$

3. The velocity propagation equation is given by

$$\begin{aligned}\frac{d}{dt}{}^E v_E^B &= C_B^E\,{}^B f_I^B + {}^E g - 2\,{}^E\omega_I^E \times {}^E v_E^B - \left[{}^E\omega_I^E \times\right]\left[{}^E\omega_I^E \times\right]{}^E r_E^B \\ &= C_B^E\,{}^B f_I^B + {}^E g_c - 2\,{}^E\omega_I^E \times {}^E v_E^B \\ &\text{where} \\ {}^E g_c &= {}^E g - \left[{}^E\omega_I^E \times\right]\left[{}^E\omega_I^E \times\right]{}^E r_E^B\end{aligned} \tag{11.32}$$

4. Finally, the position propagation equation is simply

$$\frac{d}{dt}{}^{E}r_{E}^{B} = {}^{E}v_{E}^{B} \tag{11.33}$$

For the covariance propagation of the Kalman filter, we will need the linearized error models for attitude, velocity, and position repeated from Chapter 3.

1. Attitude error propagation

$$\delta^{E}\dot{\theta}_{B}^{E} = -\left[{}^{E}\omega_{I}^{E}\times\right]\delta^{E}\theta_{B}^{E} - C_{B}^{E}\delta^{B}\omega_{I}^{B} \tag{11.34}$$

2. Velocity error propagation

$$\frac{d}{dt}\delta^{E}v_{E}^{B} = \left[\left(C_{B}^{E}m^{B}f_{I}^{B}\right)\times\right]\delta^{E}\theta_{B}^{E} + C_{B}^{E}\delta^{B}f_{I}^{B} + \left[\frac{\partial^{E}g}{\partial^{E}r_{E}^{B}}\right]\delta^{E}r_{E}^{B} - 2\left[\left({}^{E}\omega_{I}^{E}\right)\times\right]\delta^{E}v_{E}^{B} \tag{11.35}$$

1. Position error propagation

$$\frac{d}{dt}\delta^{E}r_{E}^{B} = \delta^{E}v_{E}^{B} \tag{11.36}$$

We will use the following gyro error models

$$\delta^{B}\omega_{I}^{B} = m^{B}\omega_{I}^{B} - {}^{B}\omega_{I}^{B} = C_{G}^{B}\text{diag}\left({}^{G}s\right){}^{G}\omega_{I}^{B} + \left(\left[{}^{B}\theta_{G}\times\right] + L\left({}^{B}\beta_{G}\right)\right){}^{B}\omega_{I}^{B} + {}^{B}b_{G} + C_{G}^{B}ARW \tag{11.37}$$

where

$$\left[{}^{B}\theta\times\right] = \begin{bmatrix} 0 & -\theta_z & \theta_y \\ \theta_z & 0 & -\theta_x \\ -\theta_y & \theta_x & 0 \end{bmatrix}, \quad L\left({}^{B}\beta\right) = \begin{bmatrix} 0 & 0 & 0 \\ \beta_z & 0 & 0 \\ -\beta_x & \beta_y & 0 \end{bmatrix} \tag{11.38}$$

which can also be written in a more Kalman filter-friendly way as

$$\delta^{B}\omega_{I}^{B} = C_{G}^{B}\text{diag}\left({}^{G}\omega_{I}^{B}\right){}^{G}s - \left[{}^{B}\omega_{I}^{B}\times\right]{}^{B}\theta_{G} + J\left({}^{B}\omega_{I}^{B}\right){}^{B}\beta_{G} + {}^{B}b_{G} + C_{G}^{B}ARW \tag{11.39}$$

where

$$J(\omega) = \begin{bmatrix} 0 & 0 & 0 \\ 0 & 0 & \omega_x \\ \omega_y & -\omega_x & 0 \end{bmatrix} \tag{11.40}$$

We will use the following error model for the accelerometers:

$$\delta^{B}f_{I}^{B} = m^{B}f_{I}^{B} - {}^{B}f_{I}^{B} = C_{A}^{B}\text{diag}\left({}^{A}s\right){}^{A}f_{I}^{B} + L\left({}^{B}\beta_{A}\right){}^{B}f_{I}^{B} + {}^{B}b_{A} + C_{A}^{B}VRW \tag{11.41}$$

which can be written in a more Kalman filter-friendly way as

$$\delta\,{}^{B}f_{I}^{B} = C_{A}^{B}\text{diag}\left({}^{A}f_{I}^{B}\right){}^{A}s + J\left({}^{B}f_{I}^{B}\right){}^{B}\beta_{A} + {}^{B}b_{A} + C_{A}^{B}VRW \tag{11.42}$$

and

$$J\left({}^{B}f_{I}^{B}\right) = \begin{bmatrix} 0 & 0 & 0 \\ 0 & 0 & {}^{B}f_{I\,x}^{B} \\ {}^{B}f_{I\,y}^{B} & -{}^{B}f_{I\,x}^{B} & 0 \end{bmatrix} \tag{11.43}$$

where

VRW is the accelerometer velocity random walk.
${}^{A}s$ is the scale factor error in the accelerometer sensor frame.
${}^{B}\beta_{A}$ is the accelerometer nonorthogonality misalignments.
${}^{B}b_{A}$ is the accelerometer bias vector in the body frame.

If we model each one of the IMU error parameters as a random walk process and let

$$\begin{aligned} \delta X_{I}{}^{T} &= \left[\delta\,{}^{E}\dot{\theta}_{B}^{E}, \delta\,{}^{B}b_{G}^{T}, \delta\,{}^{G}s^{T}, \delta\,{}^{B}\beta_{G}^{T}, \delta\,{}^{B}\theta_{G}^{T}, \left(\delta\,{}^{E}v_{E}^{B}\right)^{T}, \delta\,{}^{B}b_{A}^{T}, \delta\,{}^{A}s^{T}, \delta\,{}^{B}\beta_{A}^{T}, \left(\delta\,{}^{E}r_{E}^{B}\right)^{T}\right] \\ W_{I}{}^{T} &= \left[-\left(C_{G}^{E}ARW\right)^{T}, RRW^{T}, SRW_{G}^{T}, \beta RW_{G}^{T}, \theta RW_{G}^{T}, \left(C_{A}^{E}VRW\right)^{T}, ABRW^{T}, SRW_{A}^{T}, \beta RW_{A},{}^{T} 0\right] \end{aligned} \tag{11.44}$$

we then have

$$\delta\dot{X}_{II} = A_{II}\delta X_{I} + W_{I}$$

$$A_{II} = \begin{bmatrix} -[{}^{E}\omega_{I}^{E}\times] & -C_{B}^{E} & -C_{G}^{E}\text{diag}({}^{G}\omega_{I}^{B}) & -C_{B}^{E}J({}^{B}\omega_{I}^{B}) & C_{B}^{E}[{}^{B}\omega_{I}^{B}\times] & 0_{3x3} & 0_{3x3} & 0_{3x3} & 0_{3x3} & 0_{3x3} \\ 0_{3x3} & 0_{3x3} & 0_{3x3} & 0_{3x3} & 0_{3x3} & 0_{3x3} & 0_{3x3} & 0_{3x3} & 0_{3x3} & 0_{3x3} \\ 0_{3x3} & 0_{3x3} & 0_{3x3} & 0_{3x3} & 0_{3x3} & 0_{3x3} & 0_{3x3} & 0_{3x3} & 0_{3x3} & 0_{3x3} \\ 0_{3x3} & 0_{3x3} & 0_{3x3} & 0_{3x3} & 0_{3x3} & 0_{3x3} & 0_{3x3} & 0_{3x3} & 0_{3x3} & 0_{3x3} \\ 0_{3x3} & 0_{3x3} & 0_{3x3} & 0_{3x3} & 0_{3x3} & 0_{3x3} & 0_{3x3} & 0_{3x3} & 0_{3x3} & 0_{3x3} \\ [(C_{B}^{E}m\,{}^{B}f_{I}^{B})\times] & 0_{3x3} & 0_{3x3} & 0_{3x3} & 0_{3x3} & -2[({}^{E}\omega_{I}^{E})\times] & C_{B}^{E} & C_{A}^{E}\text{diag}({}^{A}f_{I}^{B}) & C_{B}^{E}J({}^{B}f_{I}^{B}) & \left[\dfrac{\partial\,{}^{E}g}{\partial\,{}^{E}r_{E}^{B}}\right] \\ 0_{3x3} & 0_{3x3} & 0_{3x3} & 0_{3x3} & 0_{3x3} & 0_{3x3} & 0_{3x3} & 0_{3x3} & 0_{3x3} & 0_{3x3} \\ 0_{3x3} & 0_{3x3} & 0_{3x3} & 0_{3x3} & 0_{3x3} & 0_{3x3} & 0_{3x3} & 0_{3x3} & 0_{3x3} & 0_{3x3} \\ 0_{3x3} & 0_{3x3} & 0_{3x3} & 0_{3x3} & 0_{3x3} & 0_{3x3} & 0_{3x3} & 0_{3x3} & 0_{3x3} & 0_{3x3} \\ 0_{3x3} & 0_{3x3} & 0_{3x3} & 0_{3x3} & 0_{3x3} & I_{3x3} & 0_{3x3} & 0_{3x3} & 0_{3x3} & 0_{3x3} \end{bmatrix} \tag{11.45}$$

11.4.4 Measurement Equations and Error Models

When we match image tiles, we end up with determining the Earth-piercing points corresponding to the center of the titles and the difference between the measured point in the camera frame (the tile center) and the predicted matching point in the predicted image. The difference is actually the Kalman filter residual.

When we match the landmarks or feature points, the Earth-piercing point is already known in the database and the predicted point in the camera frame and the actual measured points are all available. The difference between the measured feature points and the predicted feature points are the Kalman filter residuals.

In other words, although we used different methods for the two different approaches, we end up with the same mathematical variables that we can use to model the measurements and errors.

Following the terminology we used in Chapter 4, we will call both the landmark or feature point and center of tile and its corresponding Earth-piercing point the image anchor point.

Let the location of a navigator in the ECEF frame be given by ${}^{E}r$ and the location of an image anchor point j in the ECEF frame be given by ${}^{E}r_j$.

The LOS vector from the navigator to the anchor point can then be computed as

$$ {}^{E}u_j = \frac{1}{\left\| {}^{E}r_j - {}^{E}r \right\|}\left({}^{E}r_j - {}^{E}r \right) = \frac{1}{\rho_j}\left({}^{E}r_j - {}^{E}r \right) $$

where $\rho_j = \| {}^{E}r_j - {}^{E}r \|$ is the range to the anchor point.

${}^{E}u_j$ can be transformed to the sensor frame (camera frame) using the following equation:

$$ {}^{C}u_j = C_B^C C_E^B \, {}^{E}u_j $$

where

C_E^B is the DCM that transforms the coordinates of a vector in the ECEF frame to the body frame. In other words, C_E^B represents the navigator's attitude.

C_B^C is the DCM that transforms the coordinates of a vector in the body frame to the sensor (camera) frame. In other words, C_B^C is the camera alignment matrix.

The measured unit vector can be modeled as

$$ m\,{}^{C}u_j = C_B^C C_E^B \, {}^{E}u_j + v $$

where v is the measurement noise.

From Chapter 4, we have the unit vector measurement residual given by

$$ \Delta u = m\,{}^{C}u_j - {}^{C}\hat{u}_j = \left[{}^{S}\hat{u}_j \times \right] \delta\theta_{sma} - \left[{}^{C}\hat{u}_j \times \right] \hat{C}_E^C \delta\, {}^{E}\theta_B^E + \frac{1}{\hat{\rho}} \hat{C}_B^C \hat{C}_E^B \delta\, {}^{E}r_j - \frac{1}{\hat{\rho}} \hat{C}_B^C \hat{C}_E^B \delta\, {}^{E}r + v \tag{11.46} $$

which can also be written as

$$ \Delta u = m\,{}^{C}u_j - {}^{C}\hat{u}_j = \begin{bmatrix} \left[{}^{C}\hat{u}_j \times \right] & -\left[{}^{C}\hat{u}_j \times \right] \hat{C}_E^C & -\frac{1}{\hat{\rho}} \hat{C}_B^C \hat{C}_E^B \end{bmatrix} \begin{bmatrix} \delta\theta_{sma} \\ \delta\, {}^{E}\theta_B^E \\ \delta\, {}^{E}r \end{bmatrix} + \frac{1}{\hat{\rho}} \hat{C}_B^C \hat{C}_E^B \delta\, {}^{E}r_j + v \tag{11.47} $$

where $\delta\theta_{sma}$ is the sensor (camera) misalignment error expressed in the camera frame.

Although the unit vector-based equations at which we arrived above are clean, the reality is that the LOS measurement is a 2-degree-of-freedom measurement and it is more conveniently represented by the camera frame h and v measurements that are the angularized image anchor point position coordinates in the camera FOV.

$$h = \frac{x}{d} = \tan\left(\theta_h\right)$$
$$v = \frac{y}{d} = \tan\left(\theta_v\right)$$

The LOS unit vector is therefore given by

$$^{C}u = \frac{1}{\sqrt{x^2 + y^2 + d^2}}\begin{bmatrix} x \\ y \\ d \end{bmatrix} = \frac{1}{\sqrt{h^2 + v^2 + 1^2}}\begin{bmatrix} h \\ v \\ 1 \end{bmatrix}$$

The h and v coordinates of the star position in the star-tracker FOV can be calculated using

$$h = \frac{^{C}u(1)}{^{C}u(3)}, \quad v = \frac{^{C}u(2)}{^{C}u(3)}$$

Conversely, we can also calculate the unit vector from h and v:

$$^{C}u(1) = \frac{h}{\sqrt{1 + h^2 + v^2}}, {}^{C}u(2) = \frac{v}{\sqrt{1 + h^2 + v^2}}, {}^{C}u(3) = \frac{1}{\sqrt{1 + h^2 + v^2}}$$

We will use the above nonlinear equations as the measurement prediction equations. To derive the H-matrices for the Kalman filter, we will need to linearize the above equations:

$$\delta h = \frac{\delta {}^{C}u(1)}{^{C}u(3)} - \frac{^{C}u(1)}{^{SC}u^2(3)}\delta {}^{C}u(3) = \left(\frac{1}{^{C}u(3)}\right)\left(\delta {}^{C}u(1) - h\delta {}^{C}u(3)\right)$$
$$\delta v = \frac{\delta {}^{C}u(2)}{^{C}u(3)} - \frac{^{C}u(2)}{^{C}u^2(3)}\delta {}^{C}u(3) = \left(\frac{1}{^{C}u(3)}\right)\left(\delta {}^{C}u(2) - v\delta {}^{C}u(3)\right)$$

which can be written as

$$\begin{bmatrix} \delta h \\ \delta v \end{bmatrix} = \frac{1}{^{C}u(3)}\begin{bmatrix} 1 & 0 & -h \\ 0 & 1 & -v \end{bmatrix}\delta {}^{C}u = \frac{1}{\sqrt{1 + h^2 + v^2}}\begin{bmatrix} 1 & 0 & -h \\ 0 & 1 & -v \end{bmatrix}\delta {}^{C}u \quad (11.48)$$

Note that, for the above equations to be true, h and v do not need to be small. Combining the results together, we have

$$\begin{bmatrix} \delta h \\ \delta v \end{bmatrix} = \frac{1}{\sqrt{1 + h^2 + v^2}}\begin{bmatrix} 1 & 0 & -h \\ 0 & 1 & -v \end{bmatrix}\delta {}^{C}u_j = H_{sma}\delta\theta_{sma} + H_{att}\theta_{\hat{E}}^{E} + H_{pos}\delta {}^{E}r + H_{ppe}\delta {}^{E}r_j + v \quad (11.49)$$

Each of the H-submatrices is calculated below:
For camera frame misalignment errors, we have

$$H_{stma} = \frac{1}{{}^{C}\hat{u}_j(3)}\begin{bmatrix} 1 & 0 & -h \\ 0 & 1 & -v \end{bmatrix}\left[{}^{C}\hat{u}_j \times\right]$$

or

$$H_{stma} = \frac{1}{{}^{C}\hat{u}_j(3)}\begin{bmatrix} 1 & 0 & -h \\ 0 & 1 & -v \end{bmatrix}\left[{}^{C}\hat{u}_j \times\right] = \frac{1}{{}^{C}\hat{u}_j(3)}\begin{bmatrix} 1 & 0 & -h \\ 0 & 1 & -v \end{bmatrix}\begin{bmatrix} 0 & -{}^{C}u_j(3) & {}^{C}u_j(2) \\ {}^{C}u_j(3) & 0 & -{}^{C}u_j(1) \\ -{}^{C}u_j(2) & {}^{C}u_j(1) & 0 \end{bmatrix}$$

$$= \begin{bmatrix} 1 & 0 & -h \\ 0 & 1 & -v \end{bmatrix}\begin{bmatrix} 0 & -1 & v \\ 1 & 0 & -h \\ -v & h & 0 \end{bmatrix} = \begin{bmatrix} hv & -\left(1+h^2\right) & v \\ 1+v^2 & -hv & -h \end{bmatrix}$$

In summary, we have

$$H_{sma} = \begin{bmatrix} hv & -\left(1+h^2\right) & v \\ 1+v^2 & -hv & -h \end{bmatrix} \tag{11.50}$$

Similarly, we have the H-submatrix for the ECEF frame attitude error states

$$H_{att} = \frac{-1}{{}^{C}\hat{u}_j(3)}\begin{bmatrix} 1 & 0 & -h \\ 0 & 1 & -v \end{bmatrix}\left[{}^{C}\hat{u}_j \times\right]\hat{C}_E^C = -\begin{bmatrix} hv & -\left(1+h^2\right) & v \\ 1+v^2 & -hv & -h \end{bmatrix}\hat{C}_E^C \tag{11.51}$$

For the ECEF position states, we have

$$H_{pos} = -\frac{1}{{}^{C}\hat{u}_j(3)}\begin{bmatrix} 1 & 0 & -h \\ 0 & 1 & -v \end{bmatrix}\frac{1}{\hat{\rho}}\hat{C}_B^C\hat{C}_E^B = \frac{-1}{\hat{\rho}\sqrt{1+h^2+v^2}}\begin{bmatrix} 1 & 0 & -h \\ 0 & 1 & -v \end{bmatrix}\hat{C}_E^C \tag{11.52}$$

For the anchor point knowledge error (database error or Earth-piercing point error) in the ECEF

$$\begin{aligned} H_{rppe} &= \frac{1}{{}^{C}\hat{u}_j(3)}\begin{bmatrix} 1 & 0 & -h \\ 0 & 1 & -v \end{bmatrix}\frac{1}{\hat{\rho}}\hat{C}_B^C\hat{C}_E^B = \frac{1}{\hat{\rho}\sqrt{1+h^2+v^2}}\begin{bmatrix} 1 & 0 & -h \\ 0 & 1 & -v \end{bmatrix}\hat{C}_E^C \\ H_{rppe} &= -H_{pos} \end{aligned} \tag{11.53}$$

Although, in reality, there are various combination of sensors used to achieve specific purposes of a system with various constraints, in the following we will simply show the Kalman filter constructed with the camera included.

We will need to augment the Kalman state vector to include the camera misalignment errors

$$X^T = \left[X_I^T, (\delta\theta_{sma})^T\right]$$
$$\delta X^T = \left[\delta X_I^T, (\delta\delta\theta_{sma})^T\right] \tag{11.54}$$

The linearized propagation error model is then given by

$$\frac{d}{dt}(\delta X) = \begin{bmatrix} A_{II} & 0_{30x3} \\ 0_{3x30} & 0 \end{bmatrix} \delta X + \begin{bmatrix} W_I \\ smarw \end{bmatrix} \tag{11.55}$$

where *smarw* is the white noise process that integrates into the random walk process, which models the camera misalignment.

11.5 SLAM-Based Image and Vision Navigation

11.5.1 The Concept

SLAM was developed in the field of robotics in the 1980s. What it attempts to do is to construct the map and perform localization (i.e., positioning) at the same time. In particular, the methods developed by Smith and Cheeseman [37] use the extended Kalman filter (EKF), the very approach used by traditional PNT system developers.

The idea of EKF-SLAM is actually very simple. It will simply add the position of the landmarks (i.e., the map states) as part of the Kalman filter states and keep track of the uncertainty of the map knowledge and the navigation state knowledge using the Kalman filter covariance matrices. As the result, as new measurements become available, the uncertainty of the entire state vector, which consists of the navigation states and map states, will be reduced as things becomes observable.

The fundamental reason that the positions of the landmarks can become observable is the fact that two (or more) angular LOS measurements of the same landmark from two different location or orientations allow us to triangulate the position of the landmark. In the ideal case where there are no errors, the location of the landmark is simply the intersection of two LOS vectors. This is illustrated in Figure 11.14.

Obviously, in order for the triangulation to work, the position and the attitude of the camera at the two locations needs to be known. The reality is that the position, the attitude, and the landmark locations are maintained as part of the

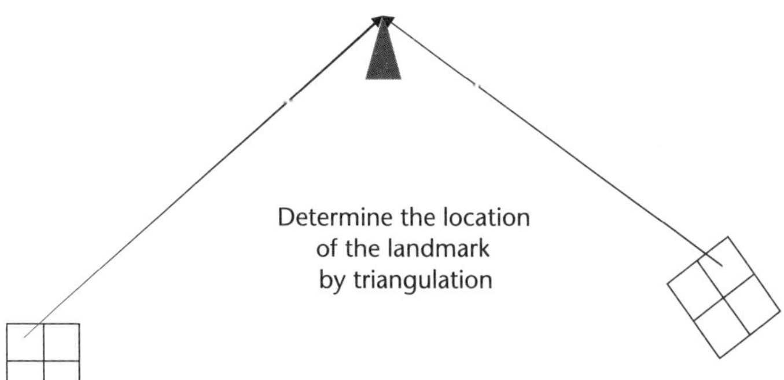

Figure 11.14 Determine the location of the landmark by Line Of Sight (LOS) measurements from two different known locations (and known attitude).

state vector of the Kalman filter and the uncertainty is tracked by the covariance matrices. As the vehicle travels through different locations and attitude, and as the sensor measurements become available (IMU, GNSS, camera, and other sensors), the states become observable over time.

11.5.2 EKF-SLAM

Let the location of the landmarks be given by

$$L_i \quad \left(i = 1,\ 2,\ \cdots,\ n_l\right) \tag{11.56}$$

When we actually build the system, we want to carefully choose landmark states to reflect the knowledge of the constraints and to minimize the number of states. Specifically, each of the landmark location variables is most likely chosen to be a 2-D vector since each of the landmarks is constrained to be on the Earth surface and we typically have good-accuracy terrain models available to allow us to convert 2-D coordinates to 3-D location vectors. (Note that there is nothing fundamental about this assumption; if the landmarks are not constrained for some applications, then the 3-D positions of the landmarks will be included as the states.)

To make this more explicit, let us assume that such a function that converts 2-D landmark states to the 3-D ECEF position vector for the landmark is given by

$$^{E}r_j = g\left(L_i\right) \quad \left(i = 1,\ 2,\ \cdots,\ n_l\right) \tag{11.57}$$

In addition to the IMU, we assume that there are two other types of sensors available. The first type is any sensors that produce measurements as function of position and attitude that are independent of the landmarks. The value of these sensors is to provide the position and attitude knowledge uncertainty reduction when they are available. One example of such a sensor is a GNSS receiver. We will want to use GNSS receiver-provided information as much as we can when GNSS is available and has integrity.

The second type of the sensors is the vision-based navigation sensors, specifically the camera. In this case, the camera will be able to observe the landmarks and produce measurements that can be eventually measured as the LOS of the landmarks as seen from the camera. Obviously, the LOS measurements of the landmarks will produce the observability of the position of the landmarks and the position of the navigator. As the navigator experiences motion, both the position of the navigator and the position of the landmarks become more observable.

Without loss of generality, we assume that there is a single camera and another type 1 sensor mentioned above available. The state of the system is given by

$$X = \begin{bmatrix} X_I \\ \delta\theta_{sma} \\ L_1 \\ L_2 \\ \vdots \\ L_{n_l} \end{bmatrix} \tag{11.58}$$

where X_I are the states associated with INS.

The nonlinear measurement model for the type 1 sensor is assumed to be

$$y_1 = h\left(X_I\right) + v_1 \tag{11.59}$$

and the linearized measurement model is simply

$$\delta y_1 = \frac{\partial h\left(X_I\right)}{\partial X_I}\delta X_I + v_1 \tag{11.60}$$

The nonlinear measurement model for the camera landmark LOS measurements can be described by the following equations.

- Step 1: Convert from a 2-D landmark to a 3-D landmark:

$$^{E}r_j = g\left(L_i\right) \quad \left(i = 1,\ 2,\ \cdots,\ n_l\right) \tag{11.61}$$

- Step 2: Compute the LOS unit vector from the camera to the landmark:

$$^{E}u_j = \frac{1}{\left\| ^{E}r_j - {}^{E}r \right\|}\left(^{E}r_j - {}^{E}r \right) = \frac{1}{\rho_j}\left(^{E}r_j - {}^{E}r \right) \tag{11.62}$$

- Step 3: Transform the ECEF frame LOS vector to the camera frame using attitude knowledge and camera alignment knowledge:

$$^{C}u_j = C_B^C C_E^B \, {}^{E}u_j \tag{11.63}$$

- Step 4: Convert the unit vector to (h, v):

$$h = \frac{^{C}u(1)}{^{C}u(3)}, \quad v = \frac{^{C}u(2)}{^{C}u(3)} \tag{11.64}$$

and the camera output is modeled as

$$\begin{bmatrix} mh \\ mv \end{bmatrix} = \begin{bmatrix} h \\ v \end{bmatrix} + \begin{bmatrix} v_h \\ v_v \end{bmatrix}$$

where $\begin{bmatrix} v_h \\ v_v \end{bmatrix}$ is the measurement noise.

The linearized measurement model can be written as

$$\begin{aligned} \begin{bmatrix} \delta h \\ \delta v \end{bmatrix} &= \frac{1}{\sqrt{1 + h^2 + v^2}} \begin{bmatrix} 1 & 0 & -h \\ 0 & 1 & -v \end{bmatrix} \delta^{C}u \\ &= H_{sma}\delta\theta_{sma} + H_{att}\delta^{E}\theta_B^E + H_{pos}\delta^{E}r + H_{ppe}\frac{\partial g}{\partial L_j}\delta L_j + \begin{bmatrix} v_h \\ v_v \end{bmatrix} \end{aligned} \tag{11.65}$$

where the H-submatrices are the same as what are provided in Section 11.4.

This time, the term due to the landmark is treated as part of the state vector, instead of the measurement error.

The above information is sufficient to allow us to construct the Kalman filter in the following sense:

1. The state propagation:
   a. We will perform the nonlinear propagation of the inertial navigation states by the INS equations.
   b. We will propagate the IMU error parameters as the random walk or first-order Markov process.
   c. We will propagate the landmark states as constant vectors.
2. Type 1 and Type 2 sensor output prediction: The equations are provided in the earlier paragraphs.
3. The covariance propagation: The standard covariance propagation equations are used with the following input:
   a. The A-matrix of the total system is given by

$$A = \begin{bmatrix} A_{II} & 0_{30x3} & 0_{30x2N_L} \\ 0_{3x30} & 0_{3x3} & 0_{3x2N_L} \\ 0_{2N_Lx30} & 0_{2N_Lx3} & 0_{2N_Lx2N_l} \end{bmatrix} \tag{11.66}$$

   b. The Q-matrix of the total system is given by

$$Q = \begin{bmatrix} Q_{II} & 0_{30x3} & 0_{30x2N_L} \\ 0_{3x30} & \mathrm{diag}\left(\sigma^2{}_{sma-x,}\sigma^2{}_{sma-y,}\sigma^2{}_{sma-z,}\right) & 0_{3x2N_L} \\ 0_{2N_Lx30} & 0_{2N_Lx3} & 0_{2N_Lx2N_l} \end{bmatrix} \tag{11.67}$$

4. Gain computation and covariance update: The standard equations are used with the following input:
   a. H-matrix for the type 1 sensor:

$$H = \begin{bmatrix} \dfrac{\partial h(x_I)}{\partial x_I} & 0_{m\times 3} & 0_{m\times(2n_l)} \end{bmatrix} \tag{11.68}$$

   b. H-matrix for the camera for LOS measurements associated with landmark *j*:

$$H_j = \begin{bmatrix} H_{att} & \leftarrow 0 \rightarrow & H_{pos} & H_{sma} & \leftarrow 0 \rightarrow & H_{ppe}\dfrac{\partial g}{\partial L_j} & \leftarrow 0 \rightarrow \end{bmatrix} \tag{11.69}$$

It is important to realize that it is impractical to maintain tens of thousands or even more of the landmarks as part of the Kalman filter states at the same time. In practice, we will need to implement the design in such ways that are similar to the following list:

1. A database of landmarks is maintained that can add new landmarks dynamically.
2. There are efficient searching or retrieving algorithms associated with the database to retrieve the landmarks that are likely to be used for a while. Such actions happen periodically.
3. The landmarks that will be used for a while will be retrieved and will be used as part of the Kalman filter states that are currently in use.
4. In addition, if a new landmark is discovered and considered useful, the states associated with the new landmarks will be added into the states currently in use.
5. As the landmark states are observed and updated, the landmark position, the associated variance of the position error, and other relevant information will be updated and saved into the database.
6. It is often impractical to save the entire covariance matrix associated with the landmarks. One compromise that we will have to do is to only save the variance for each one of the landmarks and ignore the cross-correlations between landmarks and the cross-correlations between the landmarks and other states. We can practically do this if we apply a conservativeness factor to these variances before saving into the database for use in the future.
7. For the currently-in-use landmarks, the variance and covariance matrices are properly maintained by the Kalman filter. The compromise of ignoring the cross-correlation and adding a factor of conservativeness only happens when the variance is saved into the database for future use. When this happens, we will be eliminating some of the landmarks from the currently used landmarks and retrieving from the database the landmarks that will be used.
8. One practice that can greatly simplify the design is to consider the landmarks with small position errors as converged and, consequently, they will no longer need to be included as part of the Kalman filter states when these landmarks become useful again. This will reduce the need to keep a large collection of landmarks as part of the Kalman filter state vector at the same time.

11.6 Visual Odometry

11.6.1 The Concept

As mentioned before, SLAM requires very carefully designed and complicated software to manage the landmarks in the database and the currently-in-use landmarks that are included as part of the Kalman filter state vector. In many cases, we may consider SLAM not profitable for the following reasons:

1. When we will unlikely revisit the same place again.
2. When we desire simplified software design at the expense of some loss of performance.

In these scenarios, visual odometry becomes more practical. As discussed in Section 11.2, visual odometry attempts to determine the change of attitude and the change of position by comparing images over consecutive frames and then accumulating these changes over time as the navigation solution.

Fundamentally, visual odometry is relative. However, if the solution is initialized with respect to an absolute reference, the solution based on visual odometry can be absolutely referenced as well.

Since both the attitude and position solutions based on visual odometry are the results of accumulation, the errors are typically functions of the distance traveled.

As discussed in Section 11.2, visual odometry is often integrated with an inertial-based solution using one of the several possible approaches. Fundamentally, visual odometry derives motion information from consecutive frames of pictures. There are two classes of algorithms for this purpose. The optical flow-based approach directly processes the raw images over consecutive frames and derives information on the motion. A landmark or feature-based approach attempts to match landmarks or features over multiple frames and derives the motion information from them. For the rest of this section, we will focus mainly on the landmark or feature-based approach.

11.6.2 Deriving Motion Information from Measured Landmarks over Consecutive Frames

The problem that we will try to understand and solve is stated below in a factored fashion so that we can be clear on what we are trying to solve by using what knowledge under what constraints.

1. We will want to determine the change of attitude and the change of the position of the camera.
2. The only things that we can use are the measured LOS vectors (fundamentally 2-D information), in the camera coordinate frame at the first location and the second location.
3. The 3-D position knowledge of the landmarks is not available.

Before we introduce the formal mathematical equations for the solutions of the above problem, we will first look at some simple scenarios to understand the essence of the problem and the fundamental reasons why such problem can be solved.

11.6.2.1 Simplified Case 1: No Attitude Change, Unknown Translational Position Change

Figure 11.15 illustrates a simple case when the camera only moved translationally, and we want to determine the distance vector between the two points of observation.

First, with the two LOS vectors already determined, the intersection of the two unit vectors will be the location of the landmark. However, the location of the camera at the first and second locations can be allowed to slide along the two LOS vectors without violating the requirements that the landmarks are observed by the camera from these two directions.

In other words, even if the attitude change is known (or no change at all), a single pair of LOS vectors for one landmark with an unknown position is insufficient to determine the relative position between the two observation locations. The only things that can be determined are that the relative distance vector is located in the plane spanned by the two unit vectors.

However, if we introduce two landmarks that are observed from the two locations, two pairs of unit vectors are available, and the camera location has to be at the intersection of the two LOS vectors from the two landmarks. This is true for both locations.

In this case, other than the scale, the relative distance vector is determined. This is illustrated in Figure 11.16. It can be easily seen that if the two landmarks are well separated (in angle), we will have better accuracy determining the position. If the separation angles are smaller, the geometric dilution of precision will be larger.

Simplified Case 1 Conclusion

1. When attitude is known, the measurements of two landmarks from the two camera locations are required to determine the direction of the relative distance vector.
2. The scale cannot be determined with angular information.
3. In this case, the location of the two landmarks can also be determined (up to a variable scale).
4. Each pair of the unit vectors constrains 1 degree of freedom. Two pairs constrain 2 degrees of freedom. The direction of the distance vector has 3 degrees of freedom. The two pairs leave the scale undefined.

11.6.2.2 Simplified Case 2: Known Translational Motion, Unknown Attitude Change

Figure 11.17 illustrates the simple case where the translational motion is already known and the attitude is not determined. Since we are talking about relative attitude, we can assume that the coordinate frame for the first location is fixed and we will focus on how the coordinate frame for the second location can be defined.

Let us start with two landmarks with unknown locations in order to keep the coordinates of the two unit vectors unchanged in the camera frame, the camera and the two unit vectors can still rotate together around the axis normal to the plane spanned by the two vectors. In other words, the attitude of the camera frame is not yet fully constrained.

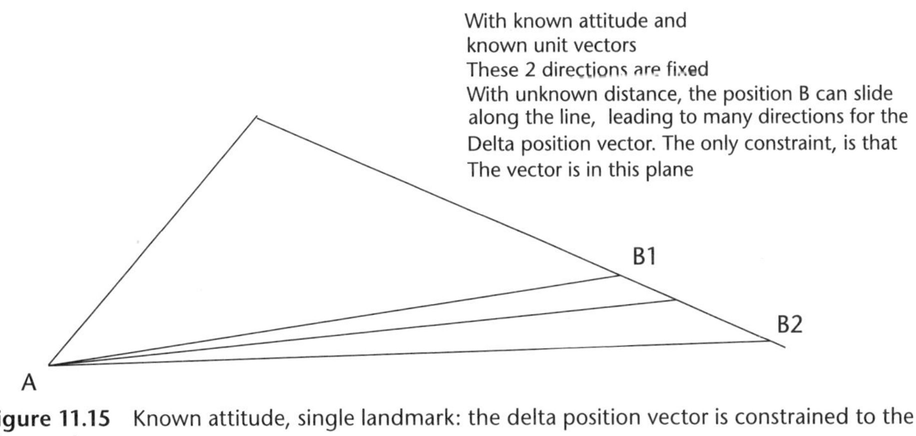

Figure 11.15 Known attitude, single landmark: the delta position vector is constrained to the plane only.

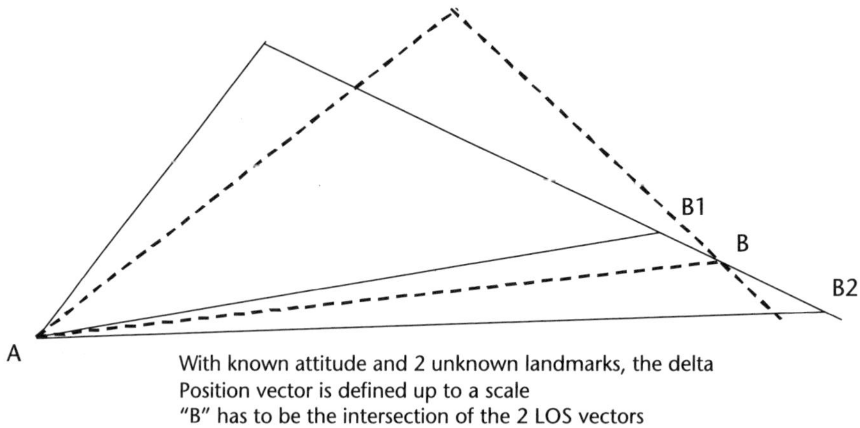

Figure 11.16 Known attitude, two unknown landmarks: the delta position vector is constrained to a fixed direction. Only the scale can change.

If we add a third landmark but in such a way that it is in the plane spanned the first two LOS vectors, in this case, we can still move the camera frame around the axis orthogonal to the plane without violating the requirements that the three unit vectors are in the same plane with the predefined relative angles.

If the third landmark is such that the LOS vector is not in the plane spanned by the first two LOS vectors, in this case, we can no longer rotate the camera frame

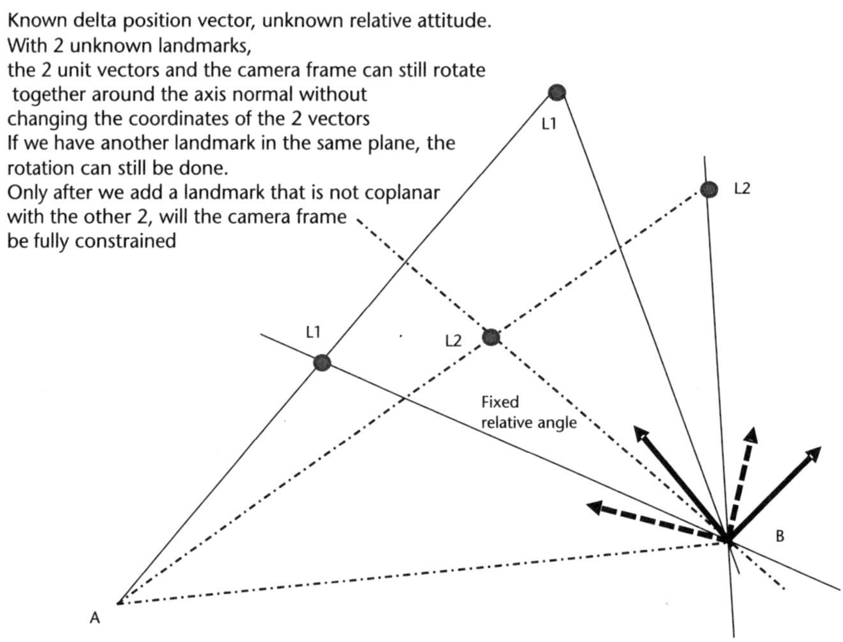

Figure 11.17 Known delta position vector, unknown relative attitude. With two unknown landmarks, the two unit vectors and the camera frame can still rotate together around the axis normal to plane without changing the two coordinates of the two unit vectors. If we have another landmark in the same plane, the rotation can still be done. Only after we add a landmark that is not coplanar with the other two, the camera frame will be fully constrained.

around any axis without violating the relative orientations of the three unit vectors; in other words, the relative attitude is determined if there are three or more landmarks such that the LOS vectors do not fall in the same plane.

Simplified Case 2 Conclusion

1. When translational relative position is known, measurements from three landmarks from the two camera locations are required to determine the relative attitude provided that the three LOS vectors at each location do not fall into the same plane.
2. Each pair of the unit vectors provides constraint in 1 degree of freedom. Three pairs constrain 3 degrees of freedom. The relative orientation is a 3 degree-of-freedom motion and can be constrained by three pairs of unit vectors.

Combining the two cases together, we can expect that it takes at least five pairs of unit vectors to determine both the relative position (to a scale) and the relative attitude. Among the five pairs, at least three of them shall not be coplanar and none of them should be collinear.

11.6.2.3 Epipolar Geometry and Essential Matrix

With the understanding of the simple cases, we can now discuss the generic epipolar geometry and the equations associated with each pair of the unit vector measurements for the same landmark.

Figure 11.18 illustrates the basic epipolar geometry where l is the 3-D location of the landmark, and $r(k)$ and $r(k + 1)$ are the locations of the camera at time k and $k + 1$.

$$\delta r = \delta r(k+1,k) = r(k+1) - r(k)$$

E represents the ECEF frame, and C_k represents the camera frame at time k. $C_k^{k+1} \triangleq C_{C_k}^{C_{k+1}}$ is the DCM from camera frame C_k to camera frame C_{k+1}.

We will also use the symbol $C(k) \triangleq C_E^{C_k}$ to represent the absolute attitude DCM from the Earth frame to the camera frame at time k. Note that $C(k)$ is a DCM and should not be confused with C_k, which is a coordinate frame.

We define the LOS unit vectors of the landmark from location k and $k + 1$ to be

$$u(k) = \frac{(l - r(k))}{\|l - r(k)\|}, \quad u(k+1) = \frac{(l - r(k+1))}{\|l - r(k+1)\|}$$

We will derive the basic equation by using the fact that the three vectors $u(k)$, $u(k + 1)$, $\delta r(k + 1, k)$, are in the same plane.

Since the cross-product of any two vectors is a vector normal to the plane spanned by the two vectors, the third vector left has to be normal to the cross-product vector. This forms our basic equation:

$$\langle u(k+1),\ \delta r(k+1,k) \times u(k) \rangle = 0$$

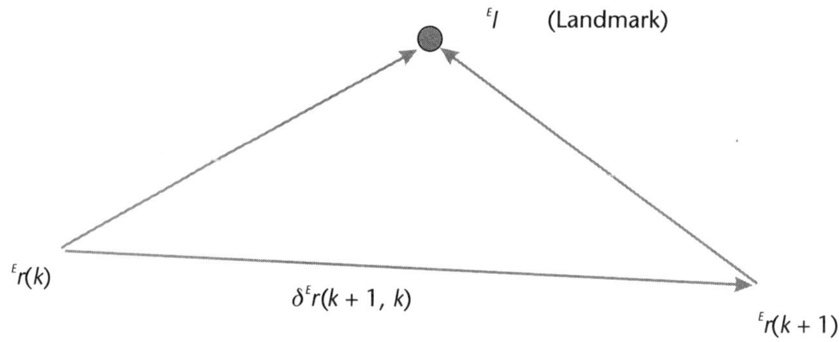

Figure 11.18 Epipolar geometry between two consecutive frames.

Let us write the above equation in the coordinate frame C_{k+1}:

$$\left({}^{C_{k+1}}u(k+1)\right)^T\left(\delta\,{}^{C_{k+1}}r(k+1,k)\right)\times\left(C_k^{k+1}\,{}^{C_k}u(k)\right)=0 \tag{11.70}$$

Using simplified symbols and the fact that the cross-product can be written as the skew-symmetric matrix derived from the vector multiplied by the second vector, we have

$$\left({}^{C_{k+1}}u(k+1)\right)^T\left(\left[\left(\delta\,{}^{C_{k+1}}r\right)\times\right]C_k^{k+1}\right){}^{C_k}u(k)=0$$

Define the matrix E, which is called the essential matrix, to be

$$E=\left[\left(\delta\,{}^{C_{k+1}}r\right)\times\right]C_k^{k+1} \tag{11.71}$$

and the basic equation can be written as

$$\left({}^{C_{k+1}}u(k+1)\right)^T(E)\,{}^{C_k}u(k)=0 \tag{11.72}$$

The essential matrix is a product of a rank = 2 skewed symmetric matrix multiplied by a full rank rotation matrix. The resulting matrix must also have rank = 2.

The essential matrix E fundamentally contains 5 degrees of freedom: 3 attitude degrees of freedom for the rotation matrix as captured by C_k^{k+1} and 2 degrees of freedom for the translation motional (3 minus 1 unobservable scale).

One epipolar geometry constraint equation can be formed by each pair of matching landmarks and it forms a scalar (1-D) constraint. Theoretically speaking, five such equations are needed to solve for the solution. This is consistent with our expectation based on the earlier discussions of simple examples.

There is an algorithm developed by Nister in his 2003 article [57] that describes how five pairs of unit vectors can be used to solve for the essential matrix. However, one can imagine that the algorithm has to be nonlinear since it takes a nonlinear parameterization to describe a DCM in 3 degrees of freedom.

When we have eight or more pairs of the unit vectors available, we can solve the 3×3 essential matrix without using any nonlinear parameterization. This algorithm is described below.

Let

$$u_1 = {}^{C_k}u(k) = \begin{bmatrix} x_1 \\ y_1 \\ z_1 \end{bmatrix}$$

$$u_2 = {}^{C_{k+1}}u(k+1) = \begin{bmatrix} x_2 \\ y_2 \\ z_2 \end{bmatrix}$$

and we have

$$u_2^T E u_1 = 0$$

which can be written as

$$\begin{bmatrix} x_1x_2 & y_1x_2 & z_1x_2 & x_1y_2 & y_1y_2 & z_1y_2 & x_1z_2 & y_1z_2 & z_1z_2 \end{bmatrix} \begin{bmatrix} e_{11} \\ e_{12} \\ e_{13} \\ e_{21} \\ e_{22} \\ e_{23} \\ e_{31} \\ e_{32} \\ e_{33} \end{bmatrix} = w^T vec(E) = 0 \quad (11.73)$$

Stacking the equations formed by m-pairs of unit vectors together, we have

$$\begin{bmatrix} w_1^T \\ w_2^T \\ \vdots \\ w_m^T \end{bmatrix} vec(E) = 0$$

which can be written as

$$W = \begin{bmatrix} w_1^T \\ w_2^T \\ \vdots \\ w_m^T \end{bmatrix} \in R^{m\times 9} \quad e_v = vec(E) \in R^{9\times 1} \quad (11.74)$$

$$We_v = 0$$

This is an mx9 system of linear homogeneous equations. A trivial solution is obviously $e_v = 0$. We would like to solve for a nontrivial solution with its norm, for example, and without a loss of generality, constrained to be equal to 1.

11.6.2.4 The Preliminary Solution

Let $W = UDV^T$ be the singular value decomposition (SVD) of the matrix W where $U \in R^{m\times 9}$ is an $m \times 9$ orthonormal matrix, and $V^T \in R^{9\times 9}$ is a 9×9 orthonormal matrix. $D = \text{diag}([d_1, d_2, \cdots d_9])$ and $(d_1 \geq d_2 \cdots \geq d_9 \geq 0)$

We would like to find a solution $e_v = X$ such that

$$\|e_v\| = \|X\| = 1$$

and $\|We_v\| = \|WX\|$ is minimized.

Note that when there are more than eight equations available and with measurement noises and numerical errors, there is no guarantee that we may be able to make $We_v = WX = 0$, instead we may minimize $\|We_v\| = \|WX\|$.

Since U is an orthonormal matrix, we must have

$$\|WX\| = \|UDV^T X\| = \|DV^T X\| = \|DY\|$$

where

$$Y = V^T X$$

and

$$\|Y\| = \|V^T X\| = \|X\|$$

Since D is positive and diagonal and sorted from maximum to minimum, the solution that satisfies the constraint $\|Y\| = \|V^T X\| = \|X\| = 1$ and minimizes $\|WX\|$ $= \|DY\|$ is simply

$$Y = \begin{bmatrix} 0 \\ \vdots \\ 0 \\ 1 \end{bmatrix}$$

which means

$$e_v = X = VY = V(:,9) \tag{11.75}$$

In other words, the solution is simply the last column of the matrix V. We can write the solution for the essential matrix to be

$$E = mat(e_v) = mat(V(:,9)) \tag{11.76}$$

11.6.2.5 Force into the Essential Matrix

Because of measurement noises and numerical errors, the 3×3 matrix obtained above may not necessarily be a rank = 2 matrix. For that reason, we will need to calculate the SVD of the 3×3 matrix and find the nearest (in terms of distance between two matrices in the Frobenius norm) solution that qualifies to be an essential matrix. This can be done by performing (yet another) SVD:

$$E = USV^T \tag{11.77}$$

A solution that has rank = 2 can be constructed as

$$E = U\begin{bmatrix} 1 & 0 & 0 \\ 0 & 1 & 0 \\ 0 & 0 & 0 \end{bmatrix}V^T \tag{11.78}$$

11.6.2.6 Factor into the Product of the Skew-Symmetric Matrix and Rotation Matrix

There are four different ways in which the matrix E can be factored. They are described next.

Let

$$Q = \begin{bmatrix} 0 & \pm 1 & 0 \\ \mp 1 & 0 & 0 \\ 0 & 0 & 1 \end{bmatrix} \tag{11.79}$$

We have

$$C = U\left(\pm Q^T\right)V^T \tag{11.80}$$

and

$$\left[\delta r \times\right] = U(\pm Q)\begin{bmatrix} 1 & 0 & 0 \\ 0 & 1 & 0 \\ 0 & 0 & 0 \end{bmatrix}U^T \tag{11.81}$$

We can easily verify by directly multiplying the above matrices together that

$$E = U\begin{bmatrix} 1 & 0 & 0 \\ 0 & 1 & 0 \\ 0 & 0 & 0 \end{bmatrix}V^T = \left[\delta r \times\right]C \tag{11.82}$$

Among the four candidate solutions, the proper solution has to be selected by directly checking the consistency of the solution against earlier solutions or the consistency against the location of the landmarks.

11.6.2.7 Rescale of the Delta Distance Vector

As discussed before, just based on two frames of pictures, we do not have a way to determine the absolute scale. The strategy is therefore:

1. We will determine the absolute scale when other information is available.
2. We will try to maintain the same relative scale across the sequence of images when absolute scale is not available.

There are several useful scenarios where the other information can help us to determine the absolute scale. One such other information can be relative attitude change and relative translational motion as determined by the inertial system. By matching the translational motion as determined by inertial navigation and the same solution as determined by visual odometry, we can determine the proper absolute scale to be used. Such other information can be known landmarks or known ranges of the landmarks.

When no other information is available, we should try to maintain the same scale frame to frame. This can be done in the following way:

1. Every time that we solve for the attitude change and position change using the essential matrix solutions with a chosen (arbitrary) scale, we immediately can derive the 3-D position of the landmarks.
2. By choosing two or more well-separated landmarks as reference points, we can calculate their relative distance.
3. If the same landmarks are tracked for at least three consecutive frames, then we can calculate the distances for the second frame and the third frame. The ratio of the second frame and third frame can then be used to adjust the scale used, in such a way that the same scales are used from frame to frame.
4. If we used an inertial system or other information to determine the absolute scale, then the above relative scaling approach will help us to maintain the same scale.

11.6.3 Attitude and Position Propagation Using Visual Odometry

Let us assume that we start with an initial attitude and position knowledge:

$$\begin{aligned} C(0) &= C_E^{C_0} \\ r(0) &= {}^{C_0}r(0) \end{aligned} \tag{11.83}$$

where the frame E is some kind of Earth frame. It could be an absolute frame such as the ECEF frame or an arbitrary Earth-fixed reference frame.

For some applications, an arbitrary reference E is sufficient. For many other applications, E is actually the frame that the system has been using when GPS or other sensors are available, including vision-based navigation with a geo-referenced database.

Assume that we have achieved

$$\begin{aligned} C(k) &= C_E^{C_k} \\ r(k) &= {}^{C_k}r(k) \end{aligned} \tag{11.84}$$

Using the methods discussed in the previous section, we can determine

$$C(k+1,k) = C_{C_k}^{C_{k+1}}, \quad \delta r(k) = {}^{C_{k+1}}\delta r(k) \tag{11.85}$$

We can propagate the solution forward by using

$$C(k+1) = C_E^{C_{k+1}} = C_{C_k}^{C_{k+1}} C_E^{C_k} = C(k+1,k)C(k),$$
$$r(k+1) = {}^{C_{k+1}}r(k+1) = C_{C_k}^{C_{k+1}}\, {}^{C_k}r(k) + {}^{C_{k+1}}\delta r(k) = C(k+1,k)r(k) + \delta r(k) \quad (11.86)$$

As we discussed in Section 11.2, the attitude change and position change information that we calculated through visual odometry can be directly used as described above to propagate the attitude and position forward. It can also be used as measurements that correct the attitude change and position change derived from inertial navigation. A small Kalman filter can be used to generate gains to allow the visual odometry measurements to update the inertial-based odometry solution and calibrate the IMU.

References

[1] Zuo, C., et al., "Display and Detail Enhancement for High-Dynamic-Range Infrared Images," *Opt. Eng.*, Vol. 50, No. 12, December 2011, p. 127401.

[2] https://en.wikipedia.org/wiki/Histogram_equalization.

[3] https://www.flir.com/globalassets/imported-assets/document/boson-camera-adjustments-appnote.pdf (FLIR Camera Adjustments).

[4] Pizer, S. M., et al., "Adaptive Histogram Equalization and Its Variations," *Comput. Vis. Graph. Image Process*, Vol. 39, No. 3, 1987, pp. 355–368.

[5] Zuiderveld, K., "Contrast Limited Adaptive Histogram Equalization," in *Graphics Gems IV*, San Diego, CA: Academic Press, 1994, pp. 474–485.

[6] Vickers, V. E., "Plateau Equalization Algorithm for Real-Time Display of High-Quality Infrared Imagery," *Opt. Eng.*, Vol. 35, 1996, p. 1921.

[7] Branchitta, M., et al., "Dynamic-Range Compression and Contrast Enhancement in Infrared Imaging Systems," *Opt. Eng.*, Vol. 47, No. 7, 2008, p. 076401.

[8] Fitzgibbon, A. W., "Simultaneous Linear Estimation of Multiple View Geometry and Lens Distortion," *Proc. of the 2001 IEEE Computer Society Conf. on Computer Vision and Pattern Recognition (CVPR '01)*, 2001, pp. 125–132.

[9] Claus, D., and A. W. Fitzgibbon, "A Rational Function Lens Distortion Model for General Cameras," *IEEE Computer Society Conf. on Computer Vision and Pattern Recognition (CVPR '05)*, 2005, pp. 213–219.

[10] Wu, F., et al., "Correction of Image Radial Distortion Based on Division Model," *Opt. Eng.*, Vol. 56, No. 1, 2017, p. 013108.

[11] Chowdhary, G., et al., "GPS-Denied Indoor and Outdoor Monocular Vision Aided Navigation and Control of Unmanned Aircraft," *Journal of Field Robotics*, Vol. 30, No. 3, pp. 415–438.

[12] Won, D. H., et al., "Integration of Limited GNSS Signals with Monocular Vision Based Navigation," *12th International Conference on Control, Automation and Systems (ICCAS)*, pp. 1364–1367. October 2012.

[13] George, M., and S. Sukkarieh, "Camera Aided Inertial Navigation in Poor GPS Environments," *IEEE Aerospace Conference*, 2007, pp. 1–12.

[14] Giebner, M. G., "Tightly-Coupled Image-Aided Inertial Navigation System via a Kalman Filter," M.S. thesis, AFIT/GE/ENG/03-10, Graduate School of Engineering and Management, Air Force Institute of Technology (AU), Wright-Patterson AFB OH, ADA415246, March 2003.

[15] Kauffman, K., et al., "Simulation Study of UWB-OFDM SAR for Navigation with INS

Integration," *Proc. of the 2011 International Technical Meeting of the Institute of Navigation*, 2011.

[16] Veth, M. J., "Fusion of Imaging and Inertial Sensors for Navigation," Ph.D. dissertation, Air Force Institute of Technology (AU), Wright-Patterson AFB OH, ADA468382, September 2006.

[17] Chiu, H. -P., et al., "Precise Vision-Aided Aerial Navigation," *IEEE/RSJ International Conference on Intelligent Robots and Systems (IROS)*, 2014.

[18] Chiu, H. -P., et al., "Constrained Optimal Selection for Multi-Sensor Robot Navigation Using Plug-and-Play Factor Graphs," *IEEE International Conference on Robotics and Automation (ICRA)*, 2014.

[19] Chiu, H. -P., et al., "Robust Vision-Aided Navigation Using Sliding-Window Factor Graphs," *IEEE International Conference on Robotics and Automation (ICRA)*, 2013.

[20] Kaess, M., et al., "iSAM2: Incremental Smoothing and Mapping Using the Bayes Tree," *Intl. J. of Robotics Research*, Vol. 31, February 2012, pp. 217–236.

[21] Lindsten, F., et al., "Geo-Referencing for UAV Navigation Using Environment Classification," *Proc. IEEE Intl. Conf. on Robotics and Automation (ICRA)*, 2010.

[22] Patterson, T., et al., "Utilizing Geographic Information System Data for Unmanned Aerial Vehicle Position Estimation," *Proc. of Canadian Conf. on Computer and Robot Vision*, 2011.

[23] Sim, D., et al., "Integrated Position Estimation Using Aerial Image Sequences," *IEEE Transactions on Pattern Analysis and Machine Intelligence*, Vol. 24, No. 1, 2002.

[24] Conte, G., and P. Doherty, "Vision-Based Unmanned Aerial Vehicle Navigation Using Geo-Referenced Information," *EURASIP Journal of Advances in Signal Processing*, 2009.

[25] Kschischang, F., et al., "Factor Graphs and the Sum Product Algorithm," *IEEE Transactions on Information Theory*, Vol. 47, No. 2, February 2001.

[26] Mourikis, A., et al., "Vision-Aided Inertial Navigation for Spacecraft Entry, Descent, and Landing," *IEEE Transactions on Robotics*, Vol. 25, No. 2, April 2009.

[27] Chiu, H., et al., "Robust Vision-Aided Navigation Using Sliding-Window Factor Graphs," *Proc. IEEE Intl. Conf. on Robotics and Automation (ICRA)*, 2013.

[28] Kumar, R., et al., "Registration of Video to Geo-Referenced Imagery," *Proc. IEEE Intl. Conf. on Pattern Recognition (ICPR)*, 1998.

[29] Mourikis, A., and S. Roumeliotis, "A Multi-State Constraint Kalman Filter for Vision-Aided Inertial Navigation," *Proc. IEEE Intl. Conf. on Robotics and Automation (ICRA)*, 2007.

[30] Cranos, W., "Rigid-Body Image Registration Using Mutual Information," https://pdfs.semanticscholar.org/4aa2/8f76699a59ea924858c2fb461193202af288.pdf.

[31] Pluim, J. P., et al., "Mutual Information Matching and Interpolation Artifacts," *Proc. of SPIE Medical Imaging: Image Processing*, 1999, Vol. 3, p. 661.

[32] Viola, P. and W. M. Wells III, "Alignment by Maximization of Mutual Information," *International Conference on Computer Vision*, 1995.

[33] Pluim, J. P., et al., "Image Registration by Maximization of Combined Mutual Information and Gradient Information," *IEEE Transactions on Medical Imaging*, Vol. 19, No. 8, 2000, pp. 809–814.

[34] Chen, H., and P. K. Varshney, "Registration of Multimodal Brain Images: Some Experimental Results," *Proc. of SPIE Conference on Sensor Fusion: Architectures, Algorithms, and Applications VI*, Orlando, FL, 2002.

[35] Freire, L., and F. Godinho, "Registration by Maximization of Mutual Information—A Cross Validation Study," *IEEE International Symposium on Bio-Informatics and Biomedical Engineering*, 2000, pp. 322–329.

[36] Roussillon, C., et al., "RT-SLAM: A Generic and Real-Time Visual SLAM Implementation," *Int. Conf. on Computer Vision Systems (ICVS)*, Sophia Antipolis, France, September 2011.

[37] Smith, R., and P. Cheeseman, "On the Representation and Estimation of Spatial Uncertainty," *Int. Journal of Robotics Research*, Vol. 5, No. 4, 1987, pp. 56–68.

[38] Sola, J., "Impact of Landmark Parameterization on Monocular EKF-SLAM with Points and Lines," *Int. Journal of Computer Vision*, Vol. 97, No. 3, September 2012, pp. 339–368.

[39] Leutenegger, S., et al., "Key Frame-Based Visual-Inertial SLAM Using Nonlinear Optimization," *Int. J. Robot. Research*, 2015.

[40] Mur-Artal, R., et al., "ORB-SLAM: A Versatile and Accurate Monocular SLAM System," *IEEE Transactions on Robotics*, Vol. 31, No. 5, 2015, pp. 1147–1163.

[41] Forster, C., et al., "On-Manifold Preintegration for Real-Time Visual-Inertial Odometry," *IEEE Transactions on Robotics*, Vol. 33, No. 1, 2017, pp. 1–21.

[42] Lupton, L., and S. Sukkarieh, "Visual-Inertial-Aided Navigation for High Dynamic Motion in Built Environments Without Initial Conditions," *IEEE Transactions on Robotics*, 2012.

[43] Hartley, R., and A. Zisserman, *Multiple View Geometry in Computer Vision*, 2nd ed., New York: Cambridge University Press.

[44] Amado, A., "Pre-Integrated Dynamics Factors and a Dynamical Agile Visual-Inertial Dataset for UAV Perception," Master's thesis, Massachusetts Institute of Technology, 2018.

[45] Koliraov, M., et al., "Differential Dynamic Programming for Optimal Estimation," *IEEE Int. Conf. Robot. Autom (ICRA)*, 2015.

[46] Zhang, Z., and D. Scaramuzza, "A Tutorial on Quantitative Trajectory Evaluation for Visual-Inertial Odometry," *IEEE/RSJ Int. Conf. Intell. Robot. Syst. (IROS)*, 2018.

[47] Hesch, J. A., et al., "Camera-IMU-Based Localization: Observability Analysis and Consistency Improvement," *Int. J. Robot. Research*, Vol. 33, No. 1, 2014, pp. 182–201.

[48] Burri, M., et al., "The EuRoC Micro Aerial Vehicle Datasets," *Int. J. Robot. Research*, Vol. 35, 2015, pp. 1157–1163.

[49] Nyul, L. G., and Udupa J. K., "Incorporating a Measure of Local Scale in Voxel-Based 3-D Image Registration," *IEEE Transactions on Medical Imaging*, Vol. 22, No. 2, 2003, pp. 228–237.

[50] Khalid, Y., et al., "An Overview to Visual Odometry and Visual SLAM: Applications to Mobile Robotics," *Intell. Ind. Syst.*, Vol. 1, 2015, pp. 289–311.

[51] Shaharyar, A., and T. Khan, "A Comparative Analysis of SIFT, SURF, KAZE, AKAZE, ORB, and BRISK," *2018 International Conference on Computing, Mathematics and Engineering Technologies (iCoMET 20)*, 2018.

[52] Scaramuzza D., and F. Fraundorfer, "Visual Odometry: Part I—The First 30 Years and Fundamentals," *IEEE Robotics and Automation Magazine*, Vol. 18, No. 4, 2011.

[53] Fraundorfer, D., and D. Scaramuzza, "Visual Odometry: Part II—Matching, Robustness, Optimization, and Applications," *IEEE Robotics and Automation Magazine*, Vol. 19, No. 2, 2012.

[54] Scaramuzza, D., "1-Point-RANSAC Structure from Motion for Vehicle-Mounted Cameras by Exploiting Non-Holonomic Constraints," *International Journal of Computer Vision*, Vol. 95, No. 1, 2011.

[55] Kneip, L., D. Scaramuzza, and R. Siegwart, "A Novel Parameterization of the Perspective-Three-Point Problem for a Direct Computation of Absolute Camera Position and Orientation," *IEEE Conference on Computer Vision and Pattern Recognition (CVPR)*, Colorado Springs, CO, 2011.

[56] Fraundorfer, F., D. Scaramuzza, and M. Pollefeys, "A Constricted Bundle Adjustment Parameterization for Relative Scale Estimation in Visual Odometry," *IEEE International Conference on Robotics and Automation (ICRA 2010)*, Anchorage, AK, May 2010.

[57] Nister, D., "An Efficient Solution to the Five-Point Relative Pose Problem," *Proc. Computer Vision and Pattern Recognition (CVPR '03)*, 2003, pp. II: 195–202.

[58] Forster, C., et al., "SVO: Semi-Direct Visual Odometry for Monocular and Multi-Camera Systems," *IEEE Transactions on Robotics*, Vol. 33, No. 2, 2017, pp. 249–265.

CHAPTER 12

Gravity Anomaly and Magnetic Anomaly Referenced Navigation

12.1 Introduction

Gravity anomaly referenced navigation (GARN) and magnetic anomaly referenced navigation (MARN) can be viewed as extensions or even types of the wide-sense TRN (WSTRN) [1–37] discussed in Chapter 9.

Compared with the classic TRN, GARN and MARN have a number of advantages:

1. Gravity anomaly and magnetic anomaly aiding are self-contained, that is, they do not need any man-made infrastructure (such as GPS). One exception to this claim is that both GARN and MARN require good-quality Earth gravity anomaly and magnetic anomaly model, map, or database.
2. Both the gravity field and the magnetic field of the Earth are available everywhere and the variations of anomaly are practically also available everywhere. Consequently, the approach is applicable to air vehicles, ground vehicles, and marine and submarine vehicles as well. As a comparison, vision-based navigation, for example, will suffer from a lack of reference when the camera is looking at featureless waters. Classic TRN will not be effective over flat ocean or flat land. This does not happen for gravity or magnetic field anomaly referenced navigation.
3. Only passive sensors are required. Neither gravity sensor nor magnetic field sensor generates any emissions and/or detectable signals.
4. Spoofing and jamming are very difficult. They require a very large mass to change the gravity field. The gravity field and magnetic field go down by $1/r^2$ (gravity), $1/r^3$ (gravity gradient), and $1/r^5$ (magnetic field) with respect to the field generating sources. Consequently, it will be extremely expensive, if not infeasible at all, to jam or spoof the Earth gravity or magnetic field.

The main advantage of the gravity anomaly-based approach over the magnetic anomaly-based approach is that the gravity distortions by masses on the vehicle are typically very small and are rarely a concern. However, the magnetic field generated by other devices on the vehicle can easily be overwhelming and can be much bigger than the Earth magnetic field anomaly.

However, MARN has the following advantages over GARN:

1. The sensors are much cheaper and typically have low SWAP. High-accuracy, reasonable cost, low SWAP magnetometers, such as optically pumped scalar magnetometers, exist today that are sufficient to support magnetic field referenced navigation. However, the sensors for gravity anomaly and gravity gradient anomaly are still some distances away from being good enough to support affordable navigation purposes.
2. Unlike gravity sensors, the magnetometer measurement is not mixed with vehicle motions.

Both GARN and MARN suffer today from the lack of a high-quality model, map, or database for the Earth gravity and magnetic field.

There are a number of variations of the various types of GARN and MARN systems that will be discussed in this chapter.

Strapdown gravity gradient anomaly referenced navigation (GGARN) uses the gravity gradient instrument (GGI) output to assist the IMU-based inertial navigation. Both the IMU and the GGIs are strapped down to the vehicle. The strapdown inertial navigation provides the primary solution. The GGI outputs are used to match against the predicted GGI outputs that are based on the gravity gradient map or database, the current position knowledge, the current attitude knowledge, and other information. The differences are used to generate corrections to the navigation solutions as well as sensor parameters. The gravity gradient forms a 3×3 matrix called the gravity gradient tensor. Because of the symmetry and other constraints, five of the variables are independent. In a strapdown system, the GGI must be able to provide measurements of all five variables, as the minimum, in order to provide the measurements that can be transformed to various frames to support the navigation equations.

When the GGI is placed on, for example, a local-level platform, instead of being strapped down to the vehicle body, there is an opportunity for a single-axis GGI, typically for the local vertical axis, to be used. This is because there is no longer the need to mathematically transform the complete gravity gradient tensor matrix from the body frame to the local-level frame. Depending on the specific technology involved, a single-axis GGI can be significantly cheaper and requires much less SWAP.

Gravity gradient anomaly is typically measured by differencing high-precision accelerometers with some sophisticated motion modulation to eliminate various error sources. One of the most important error sources that is canceled out in this process is the accelerations due to vehicle motion as well as the nominal part of the Earth's gravity. For this very reason, the gravity gradient referenced navigation system does not have the following two disadvantages when only gravity anomaly, instead of gravity gradient anomaly, is used for navigation purposes:

1. Gravity anomaly is hard to distinguish from accelerations due to vehicle motion. The accelerations due to motion can be orders of magnitude bigger than the anomaly.
2. The projection error of gravity due to attitude and misalignment errors can be orders of magnitude bigger than the anomaly.

As a result, for GARN, when the gravity gradient is not used, it is fair to suggest the following:

1. Long-term averaging and filtering of gravity measurement are required in order to average out or filter out the vehicle motion contributions.
2. The vehicle that can use this type of navigation system must not have frequent high-dynamic motions; as a result, only low-dynamic marine and submarine vehicles seem to be the proper platform to host such navigation systems.
3. A strapdown gravimeter does not make much sense due to the projection errors of gravity introduced by even very small attitude and misalignment knowledge errors.
4. Even for a gravimeter on a platform, a 3-axis gravimeter does not provide much advantage over a single-axis gravimeter along the local vertical axis. This is because the locally level axes will suffer from the same platform control and misalignment knowledge error that is often much bigger than the local-level axis gravity anomaly. This is not a problem for the local vertical axis since the contribution of attitude errors is through cosine effect.

As the result, the only gravimeter referenced navigation system that makes sense is based on a single-axis gravimeter mounted on a local-level platform. The local-level platform does not need to be very precise due to the second-order effects of the attitude and misalignment errors on the local vertical axis.

For the MARN system, because of the availability of high-accuracy, low-SWAP scalar magnetometers, a strapdown, scalar sensor-based system makes the most economical sense. Vector magnetometers such as fluxgate-based magnetometers are much less feasible due to three reasons:

1. Vector magnetometers such as fluxgate magnetometers are 10 to 100 times less accurate.
2. Vector field measurements, when transformed from the sensor frame to the Earth frame, will suffer from large errors due to even a small error in attitude knowledge (the same reason why a 3-axis gravimeter is not profitable). These errors will typically overwhelm the magnetic field anomaly.
3. The vector Earth magnetic field anomaly map or database is much less available than the scalar version.

As mentioned before, unlike GGIs, the state-of-the-art, high-accuracy magnetometers are very much ready to support magnetic field anomaly referenced navigation. The performance is sufficient, the SWAP is low, and the cost is very reasonable. The hurdles are the availability of high-quality Earth magnetic field reference maps and the ability to calibrate the magnetic field added by the devices on the vehicle, which, in many cases, will be much bigger than the Earth magnetic field anomaly.

To summarize, Table 12.1 compares the GARN and MARN against the classic TRN and lists some of the key points that we discussed in this section.

Table 12.1 Comparison of Wide-Sense TRN Methods

	Type	*Advantage*	*Disadvantage*	*State of the Art and Issues*
1	TRN (Classic Terrain Altitude Referenced Navigation)	Good accuracy, mature technology	Will not work over flat land and water; active signal, detectable	Mature; being improved for better performance, lower cost, and lower SWAP
2	GARN	Passive; hard to jam or spoof; available everywhere	Need to average for a long time to remove real motion; lower performance	Lack of low-cost sensor; lack of high-quality map or database; 3-axis hard to be helpful; single-axis low-accuracy platform makes most sense
3	GGARN	Passive; hard to jam or spoof; available everywhere; motion is no longer a problem	Need high-accuracy expensive sensors to achieve reasonable performance	Lack of low-cost and high-accuracy sensor is the primary concern; lack of high-quality database is also a concern; strapdown requires all axes (at least 5 sensors to form GG (Gravity Gradient) tensor); may use less (e.g., single axis on a platform)
4	MARN	Passive; hard to jam or spoof; available everywhere; motion not a problem; good quality, low-cost, low SWAP sensor exists today	Need high-quality database to achieve good performance; need to calibrate field induced by other devices on the vehicle	Biggest concern is that the existing maps or databases are not good enough to support good performance; high-quality modeling and calibration of the field in the vehicle environment are also challenges; scalar sensor-based approach makes most sense; vector measurement and gradient measurement much less feasible

12.2 Generic TRN: Fundamental Performance Drivers

In this section, we will discuss the generic TRN in general, which at least includes the classic radar or laser altimeter-based TRN, GARN, GGARN, and MARN.

Although the performance of the navigation system can be easily predicted when we build the mathematical model and run the simulation, it is still important to develop an intuitive understanding on the most important drivers to the performance and how the performance are impacted by these most important drivers. This is the purpose of this section. We will discuss the performance conceptually, which will help us to understand the fundamental issues and hopefully these concepts can help the readers to drive their own designs into the right direction.

Let a generalized terrain model be given by

$$y = f(x) + v \tag{12.1}$$

where x is the position and y is the generalized terrain variable such as altitude, magnetic field, gravity, or gravity gradient and v is a combination of the measurement noise and model error including database errors.

We will linearize this model around a known point

$$y_0 = f(x_0) \tag{12.2}$$

The linearized equation is then given by

$$\begin{aligned} y - y_0 &= f(x) - f(x_0) + v \\ &\approx \left(\left. \frac{\partial f(x)}{\partial x} \right|_{x=x_0} \right)(x - x_0) + v = \left(grad\left(f(x_0)\right)\right)^T (x - x_0) + v \end{aligned} \tag{12.3}$$

In order to provide 3-D position observability, we will need to generate measurements at a number of locations where the gradient is along different directions. In addition, we often simply need to allow time for the averaging of the measurement noise. Between these different locations or during the averaging period, we will need to count on inertial navigation to bridge the solution.

Let

$$x_j = x + \delta x_j + w_j \tag{12.4}$$

where δx_j is the position change as predicted by inertial navigation and w_j is the error of the inertial navigation solution.

We will then have

$$\bar{y}_j - y_0 = \left(grad\left(f\left(x_0 + \delta x_j\right)\right)\right)^T \left(x + \delta x_j + w_j - x_0\right) + \bar{v} \tag{12.5}$$

Stacking them together, we have

$$\begin{bmatrix} \bar{y}_1 - y_0 \\ \vdots \\ \bar{y}_m - y_0 \end{bmatrix} = \begin{bmatrix} \left(grad\left(f\left(x_0 + \delta x_1\right)\right)\right)^T \\ \vdots \\ \left(grad\left(f\left(x_0 + \delta x_m\right)\right)\right)^T \end{bmatrix} (x - x_0) + \begin{bmatrix} \left(grad\left(f\left(x_0 + \delta x_1\right)\right)\right)^T \left(\delta x_1 + w_1\right) \\ \vdots \\ \left(grad\left(f\left(x_0 + \delta x_m\right)\right)\right)^T \left(\delta x_m + w_m\right) \end{bmatrix} + \bar{v} \tag{12.6}$$

which can be written as

$$\delta Y = G(x - x_0) + \delta y(\delta x_1, \delta x_2, \cdots, \delta x_m) + \delta y(w_1, w_2, \cdots, w_m) + V \tag{12.7}$$

Solving for x, we have the following equation:

$$x = x_0 + pinv(G)\left(\delta Y - \delta y(\delta x_1, \delta x_2, \cdots, \delta x_m) - \delta y(w_1, w_2, \cdots, w_m)\right) + pinv(G)(V) \tag{12.8}$$

where pinv( ) is the pseudo inverse of a matrix.

An estimate of the position is therefore

$$\hat{x} = x_0 + pinv(G)\left(\delta Y - \delta y(\delta x_1, \delta x_2, \cdots, \delta x_m)\right) \tag{12.9}$$

We see three important contributors:

1. Measurement noise (combined database and sensor errors) V;
2. The inertial navigation error $(w_1, w_2, \cdots, w_m)$;
3. The dilution of the precision, which is the inverse of the gradient matrix.

To help to develop an even clearer intuitive sense, let us look at a 1-D navigation problem. In this case, we have the following simplified scalar equations:

$$\begin{aligned} x + \delta x_j + w_j &= x_0 + \left(grad\left(f\left(x_0 + \delta x_j\right)\right)\right)^{-1}\left(\bar{y} - y_0 - \bar{v}\right) \\ \hat{x} &= x_0 - \delta x_j + \left(grad\left(f\left(x_0\right)\right)\right)^{-1}\left(\bar{y} - y_0\right) \end{aligned} \tag{12.10}$$

and

$$\sigma^2(\hat{x}) = \sigma^2\left(w_j\right) + \left(grad\left(f\left(x_0\right)\right)\right)^{-2}\sigma^2(\bar{v}) \tag{12.11}$$

which says that:

1. The inertial navigation error comes in without amplification or reduction during the time that we are seeking observability.
2. The measurement errors are averaged during the time of seeking observability.
3. The measurement errors are amplified by the inverse of the gradient.

The following term, in most cases, will be the dominant contributor to the total performance:

$$\left(grad\left(f\left(x_0\right)\right)\right)^{-1}\sigma(\bar{v}) = \frac{\sigma(\bar{v})}{grad\left(f\left(x_0\right)\right)} \tag{12.12}$$

In other words, the performance is mostly determined by the sensor or database error after averaging divided by the gradient of the terrain reference signal. How much we are allowed to average out the sensor or database error depends on the quality of the inertial navigation sensors.

Table 12.2 lists some typical terrain-dependent variable sensors and the gradient of the field on the Earth and examples of positioning performance.

To develop an understanding of why the averaged noise over the gradient is the dominant term, we can look at the budget shown in Table 12.3, which includes both inertial navigation and a gravimeter-based aiding mechanism (assuming that we have magically taken care of the contributions of the acceleration).

In the simple budget, we used a low-quality, IMU-based inertial navigation aided by a gravimeter. We also assumed that it took the same amount of time to create observability, wait for noises to be averaged down, and experience enough changes of the measurements. What is important to recognize is that even for a low-quality IMU, the inertial navigation error during the seeking observability and averaging

Table 12.2 ROM (Rough Order of Magnitude) Values of Terrain Sensor Errors, Terrain Gradients, and Navigation Performance

	Sensor Type	*Generalized Terrain Sensor and Database Error After Averaging (Rough Order of Magnitude)*	*Unit of the Generalized Terrain Variable*	*Example Gradient Values (Generalized Terrain Variable Unit/km)*	*Example Performance (meter) (Sensor and Database Errors After Averaging)/ (Gradient)*	*Main Challenges*
Traditional TRN	Radar altimeter	5	Meter	100 (not required all the time, since we have inertial in between)	50	RF signals are detectable; terrain changes not available everywhere
GARN	Gravimeter	2	mGal (= 1e-5 m/sec/sec ~= 1 μG)	5 (not required all the time, since we have inertial in between)	400	Long-term average needed to separate gravity from motion; expensive sensors
Gravity Gradient Referenced Navigation	GGI	0.1 (could be the result of averaging 1 Eo/sqrt(Hz) for 100 seconds)	Eotovos (= 1e-9/ s^2 = 1e-4 mGal/m)	0.1 (smooth terrain)	500 (assuming average of 4 axes)	Expensive sensors and insufficient accuracy of database
Gravity Gradient Referenced Navigation	GGI	0.1 (could be the result of averaging 1 Eo/sqrt(Hz) for 100 seconds)	Eotovos (= 1e-9/ s^2 = 1e-4 mGal/m)	5 (rough terrain, not required all time)	10 (assuming average of 4 axes)	Expensive sensors and insufficient accuracy of database
Magnetic Field Referenced Navigation	Magnetometer	30 (typical database quality today)	nTesla	10 (at higher altitude, lower gradient is expected)	3,000	Insufficient accuracy of database or model; nearby field disturbance
Magnetic Field Referenced Navigation	Magnetometer	2 (specially built database for a local area)	nTesla	20 (Lower altitude, higher gradient is expected)	100	Database error and calibration residual of environment is typically much larger than sensor errors

Table 12.3 Example ROM Error Budget

Contributors	*Value*	*Unit*
Sensor noise after averaging	2	milliGal
Database error after averaging	2	milliGal
Combined	2.83	milliGal
Gradient	5	milliGal/km
After update error	566	meter
Distance to cause enough change	0.566	km
Velocity	20	km/hour
Time to cause enough gravity change	305.5	Second
INS velocity residual	0.1	m/sec
INS steady state gyro bias calibration residual	0.1	°/hour
INS attitude error and acceleration bias residual	0.3	mg
INS drift due to gyro bias	22.6	meter
INS drift due attitude and acceleration bias	137.2	meter
INS drift due to velocity error	30.5	meter
Performance before update	583	meter

time period is typically much smaller than the error calculated from the averaged-down sensor or map error divided by the gradient.

12.3 Gravimeter, GGI, and Magnetometers

In this section, we will only briefly discuss the sensors associated with GARN and MARN. The readers are referred to a large number of references that provide comprehensive coverage of a broad spectrum of sensors.

12.3.1 Gravimeters

There is no fundamental difference between a gravimeter and an accelerometer. In other words, a gravimeter is simply a high-precision accelerometer. In order to measure gravity to the level of Earth gravity anomaly, accuracy down to milliGal or microG level or sub-milliGal/sub-micro-G level will be required.

The classic navigation-grade accelerometers are getting close to this accuracy with careful modeling and calibration of the errors. New technologies such as an atom interferometry-based gravimeter can achieve performance levels down to microGal or nanoG level, which will be sufficient to support GARN systems.

As mentioned before, the biggest challenge for GARN is to be able to remove the impact of the real motion from the measurement. For this reason, GARN only makes sense for vehicles that will travel long and stable low-dynamic journeys such as submarines and low-dynamic marine vehicles.

One important thing is that it is often not very profitable to use a 3-axis gravimeter. When local vertical axis gravity is projected to other axes, a small attitude error will typically introduce an error that is orders of magnitude larger than the gravity anomaly. For example, a 1-mrad attitude error will introduce 1,000 μG of error in the local-level frame. In order to achieve accuracy down to 1 μG, the attitude knowledge error needs to be accurate down to 1 μrad, which is typically very difficult due to thermal deformations and many other error sources.

However, the small attitude errors do not contribute much to the local vertical axis due to the second-order nature of cosine function of small angles. For example, an attitude error of 1 mrad will only cause the local vertical axis gravity measurement to have an error of 0.5 μG.

Consequently, it makes sense that:

1. A single-axis local vertical gravimeter is a good choice for a practical design.
2. The platform that maintains the local-level attitude does not need to be very accurate; attitude errors better than 1 mrad should be acceptable to most applications.

12.3.2 GGI

GGIs attempt to measure the gravity gradient. The process of deriving the gradient removes some of the key error contributors for gravimeters:

1. The vehicle linear acceleration is canceled.
2. The large value of the gravity is canceled.

If we use a strapdown GGI, we must have all the axes (at least 5 axes) available to be able to compute the full 3×3 gravity gradient tensor matrix in order for the matrix to be transformed to other coordinate frames.

There have been a number of GGI development projects with prototype system built in the last 50 years or so. We will not describe each of these development efforts; the readers are referred to [1, 2] for a more comprehensive coverage of this history. Instead, we use Table 12.4 to cover some of the most significant development efforts that, in our opinion, are important for the future.

In Table 12.4 there are three classes of technologies included.

The Bell Aerospace design, which has been used as a starting point for several other developments, uses a number of classic accelerometers mounted on a rotating disk. The separation of accelerometers allows us to measure the change of acceleration over distance, hence the gradient. The rotation modulates some of the constant or slow-changing errors of the classic accelerometers into high-frequency, zero-mean errors, consequently achieving performance levels that would not be possible to achieve if the accelerometers were not on a rotating disk.

The University of Maryland and the European Space Agency use the Superconducting Quantum Interference Device (SQID) and the wire Electric Discharge Machining (EDM) technology to achieve very high levels of performance. AOSense and some other companies use atom interferometry to achieve very high-performance levels for both gyros and accelerometers.

Table 12.4 Important GGI Development Projects

Developer	*Current Status*	*Technology*	*Noise (Rough Order of Magnitude)*
Bell Aerospace, Textron, Lockheed-Martin	Multiple generations of designs and prototypes since 1960s	Rotating Accelerometers	1 to 1 Eo/sqrt(Hz)
University of Maryland	R&D, low maturity level	Superconducting Quantum Interference Device (SQID)	0.02 to 1 Eo/sqrt(Hz)
European Space Agency Electrostatic Gravity Gradiometer (EGG)	Science mission, R&D	and wire Electric Discharge Machining (EDM)	3e-3 EO/sqrt(Hz)
AOSense; Cold Atom Gravity Gradiometer	R&D, low maturity level	Atom interferometry	Not disclosed

Because we integrate the GGI with INS, it is important to mention that the noises mentioned in Table 12.4 can be significantly reduced by fundamentally averaging these noises down while we are maintaining the navigation accuracy using INS. In this sense, what appears in the table as higher values may not be as bad as the values indicate. The more important performance parameters, which are unfortunately missing from this table due to lack of publicly available information, are the errors that cannot be averaged down such as unknown but slow-changing sensor biases. Some of the sensors, for example, the rotating disk of accelerometers, modulate the accelerometer biases into sinusoidal signals and, consequently, can be eliminated. In this case, the noise is indeed the performance driver.

12.3.3 Magnetometers

Magnetometer technology today has probably already surpassed the need for Earth magnetic field referenced navigation. As the result, the dominant error contributors to magnetic field referenced navigation are the magnetic field map or database errors and the calibration residuals of the distortions introduced by the nearby environments, typically due to irons, power supplies, and other electronic devices.

Again, it is not our plan to provide a comprehensive discussion of the magnetometers. The readers are referred to [38–43] if interested. Instead, we will briefly discuss the following two types of magnetometers that are most practical for our applications: (1) optically pumped scalar magnetometers (OPMs), and (2) fluxgate vector magnetometers.

12.3.3.1 The OPMs

The OPM measures the absolute magnetic field intensity and it measures the magnitude of the field not the vector of the field. It is the de facto standard sensor used for magnetic field survey.

OPMs belong to a larger class of nuclear resonance magnetometers that use atomic properties of gases to measure the magnetic field. Within this larger class, there are at least proton precession magnetometers, alkali vapor optically pumped magnetometers, and Overhauser effect magnetometers.

OPM is a mature technology. It features high sensitivity, a high sampling rate, and low SWAP. The typical performance level is at 0.01% or better. For a dynamic range up to 10,000 nT, that means that the sensor errors are typically at the level of 1 nT. Because of all of these advantages, OPM is a natural choice for MARN.

OPM is a scalar magnetometer. It measures the magnitude of the field regardless of the direction of the magnetic field vector. However, it does have dead zones. More appropriately, it has limited FOV. It can only provide an output when the magnetic field vector falls into the FOV of the sensor. For that reason, it is often necessary to have two sensors in order to avoid dead zones.

OPM also has attitude-dependent errors (called heading errors), despite being a scalar sensor, which need to be properly modeled when OPM is used for navigation.

OPMs use alkali metals such as cesium and potassium. When a cell containing the gaseous metal is polarized by exposure to light of a specific wavelength, the light will depopulate one electron energy level in the gas by pumping the electrons to a higher energy level. These electrons will spontaneously decay to both energy levels, and in the end, a lower energy level will be fully populated.

When a lower wavelength RF energy is applied to the cell, the cell will be depolarized by shifting the electrons in the lower energy level back to their original position.

It turns out that the energy required to repopulate the original energy level is a function of the ambient magnetic field, following the Zeeman effect. As the result, measuring the frequency of the depolarizing RF power means that we are also measuring the magnetic field.

The intensity of the Zeeman effect is dependent on the direction of the ambient magnetic field with respect to the direction of the applied light and the RF power. When the relative orientation is not within the proper range, there will be a loss of signal, which leads to the dead zone mentioned earlier. Fortunately, there is a high tolerance to this effect; the sensor will stay operational when the field vector is within (typically) 10° to 80° of sensor orientation.

12.3.3.2 The Flux Gate

The classic fluxgate magnetometers are different from OPMs in two fundamentally important aspects:

1. The flux gate magnetometers are relative sensors. They will need to be calibrated against a reference field.
2. The flux gate magnetometers measure the projection of the field vector on the sensor axis of sensitivity. Consequently, three or more sensors can be used to measure the magnetic field vector.

The performance of most of the flux-gate magnetometers ranges from 0.1% to 1%. For a dynamic range of 10,000 nT, 0.1% means 10 nT and 1% means 100 nT. In other words, the flux gate is typically 10 to 100 times less accurate than OPMs.

The advance of sensor technology leads to continuous improvement of performance. OPM with accuracy down to picoTesla or lower have already been reported. Flux-gate magnetometers with 0.01% accuracy have also been reported in the literature. Due to the limitation of human capability to survey and build an accurate Earth magnetic map or database, very high-accuracy sensors will not be useful since the errors in the map or database will dominate anyway. However, the 0.01% fluxgate performance means that we can also use the flux-gate magnetometer as the primary sensor for the MARN.

As the result, MARN can be done in many different ways depending on what type of sensor measurements will be used:

1. Scalar magnetic intensity measurement (1 axis);
2. Some or all of the components of a vector field measurements (1 to 3 axes);
3. The gradient tensor matrix (1 to 6 axes) (3 × 3 matrix).

Realistically, the scalar measurement approach requires the least amount of magnetic field anomaly data and has the best chance to be used in practice. The state of the art is such that the map or database support for the vector field measurement-based approach or the gradient tensor matrix-based approach is far from reality.

In addition to the problems associated with the magnetic anomaly map or database, vector-based measurement also suffers from the same problem as gravity anomaly measurement. That is, a small attitude knowledge error, when used to project the magnetic field vector from one frame to another frame, can introduce an error that is large enough to significantly corrupt the anomaly measurement because one has to transform the total field in that computation.

Similar to the gravity gradient tensor, the magnetic gradient tensor will remove the large value transformation error problem mentioned above, making the situation better. However, the lack of a high-accuracy map or database is still a big hurdle.

12.4 Earth Gravity Anomaly and Magnetic Anomaly Model and Database

12.4.1 Earth Gravity Field and Gravity Gradient

The gravitational potential at location r due to a volume of mass V is defined to be

$$V_g(r) = G\iiint_V \frac{\rho(\eta)}{\|r - \eta\|} dV \tag{12.13}$$

where $\rho(\eta)$ is the density of the mass at location η and G is Newton's gravitational constant, which equals 6.6742e-11 m^3/kg/sec^2.

The first-order derivative against the position vector r, or the gradient, is the gravitational acceleration:

$$g = \frac{\partial V_g(r)}{\partial r} = \nabla V_g(r) \tag{12.14}$$

and the second-order derivative is the gravity gradient matrix or the gravity gradient tensor

$$\Gamma = \frac{\partial \partial V_g(r)}{\partial r\, \partial r^T} = \nabla \nabla^T V_g(r) = \begin{bmatrix} \Gamma_{xx} & \Gamma_{xy} & \Gamma_{xz} \\ \Gamma_{xy} & \Gamma_{yy} & \Gamma_{yz} \\ \Gamma_{xz} & \Gamma_{yz} & \Gamma_{zz} \end{bmatrix} \tag{12.15}$$

which is a symmetric matrix. In addition, the trace of the matrix satisfies the following Fourier equation

$$\Delta V_g = \mathrm{trace}(\Gamma) = \Gamma_{xx} + \Gamma_{yy} + \Gamma_{zz} = -4\pi G \rho(r) \tag{12.16}$$

If we ignore the free-air gravitational potential, we have the Laplace equation:

$$\Delta V_g = \mathrm{trace}(\Gamma) = \Gamma_{xx} + \Gamma_{yy} + \Gamma_{zz} = 0 \tag{12.17}$$

Regardless of the Laplace or Fourier equation, the 3×3 matrix only has $9 - 3 - 1 = 5$ independent variables because of the symmetry and the constraint on the trace.

Spherical harmonics are used to model the gravity potential of the Earth. The scalar gravitational potential V_g is a function of the position in the ECEF frame and is modeled as

$$V_g\left({}^E R_E^B\right) = -\frac{\mu}{r}\left(1 + \sum_{n=2}^{\infty}\left(\frac{a}{r}\right)^n \sum_{m=0}^{n} P_{n,m}(\sin\lambda)\left(C_{n,m}\cos m\phi + S_{n,m}\sin m\phi\right)\right) \tag{12.18}$$

where the ECEF position vector is converted to the geocentric (not geodetic) spherical coordinates in terms of distance from the center of the Earth r, geocentric latitude λ, and longitude ϕ, which satisfy the following equation:

$${}^E R_E^B = \begin{bmatrix} r\cos(\lambda)\cos(\phi) \\ r\cos(\lambda)\sin(\phi) \\ r\sin(\lambda) \end{bmatrix} \tag{12.19}$$

where a is the semi-major axis of the Earth, which is used to normalize the coefficients to unitless; and $P_{n,0}$ are the Legendre polynomials and $P_{n,m}$ for $1 \le m \le n$ are the associated Legendre functions that can be recursively derived from $P_{n,0}$.

The EGM96 model has $n = 360$ with 130,317 coefficients.

The EGM2008 model that represents the state of art of the comprehensive Earth gravity model development is complete to a spherical harmonic degree and an order of 2,159 ($n = m = 2{,}159$) and contains additional coefficients extending to degree 2,190 and order 2,159. EGM2008 represents a factor of 6 improvements in resolution and factors of 3 to 6 in accuracy over EGM96.

The resolution of a particular model is often calculated as $\pi r/n$ where r is the average Earth radius and n is the harmonic degree of the model. Using $r = 6{,}371$ km, the spatial resolution of the EGM96 model is calculated to be approximately 56 km. In other words, only long-wave effects are captured by EGM96.

For EGM2008, the resolution is 9.3 km (i.e., about a factor of 6 improvement). It captures more details of the gravity variation.

Neither EGM 96 nor EGM 2008 has captured the detailed terrain effects. Consequently, for navigation purposes, the gravity model needs to be a combination of the EGM model and the terrain effect model. The latter can be approximately derived from the Earth terrain profile.

Since the impact of a piece of mass to gravity goes down by ($1/r^2$) and the gravity gradient goes down by ($1/r^3$), the terrain effect on the gravity gradient goes down quickly as altitude increases. However, at low altitude, the gravity gradient due to the terrain effects is much larger than that modeled by the EGM model. As the altitude increases, the terrain effects on the gravity gradient become less and eventually the gradient will be dominated by what is modeled by the EGM model.

In order to achieve a high performance using gravity anomaly and gravity gradient anomaly, the terrain effect data ultimately has to be created by an aerial survey. Before that happens, Parker [6] and Jekeli and Zhu [7] demonstrated how we can calculate reasonably accurate gravity potential and gravity gradient from terrain data.

12.4.2 Earth Magnetic Field

The Earth magnetic field has the following key components:

1. Earth main field due to the Earth core (core field);
2. Earth field due to the Earth crust (crustal field);
3. Induced field by the external field;
4. The external field.

The Earth main magnetic field due to the Earth core accounts for more than 95% of the field and its intensity ranges from 30e3 nT to 70e3 nT. The core field exhibits secular variations and for this reason, it is remodeled every 5 years.

The wavelength associated with the Earth core field is long, and it is captured by the first to thirteenth waves of the spherical harmonic model. The shortest wavelength is around 40,000 km. In other words, the core field is not very useful for magnetic field anomaly-based navigation (although it is useful for the traditional use of the magnetic field such as heading and attitude determination).

The crustal field accounts for 1% to 5% of the total field and the magnitude is on the order of hundreds of nano-Tesla. It changes slowly and there exist wavelengths as short as 1 km that are associated with the crustal field. The variations over distance become less as the altitude becomes higher. The crustal field is the component that can really help navigation.

The induced field is the field induced by the external field. External sources will induce currents in the conductive mantle of the Earth. These currents will create their own magnetic field. The induced field is typically time-varying.

The external fields are mainly the fields generated by the ionosphere and the magnetosphere.

There exist many geomagnetic field models. The International Geomagnetic Reference Field (IGRF) and the World Magnetic Model (WMM) are among the most popular ones. The WMM is a predictive-only model and is valid for the current epoch (2020 to 2025). The IGRF is retrospectively updated. Both are updated every 5 years. The reader is referred to [5] for a comprehensive comparison of various global geomagnetic field models.

All these models are spherical harmonics-based. Specifically, let V_m be the scalar magnetic potential that is a function of the position in the ECEF frame:

$$V_m = V_m\left({}^{E}R_{E}^{B}\right) \tag{12.20}$$

The Earth magnetic field B is the gradient of the magnetic potential function:

$${}^{E}B = -\frac{\partial V_m\left({}^{E}R_{E}^{B}\right)}{\partial {}^{E}R_{E}^{B}} = -\nabla V_m\left({}^{E}R_{E}^{B}\right) \tag{12.21}$$

V_m is modeled by spherical harmonics as

$$V_m\left({}^{E}R_{E}^{B}\right) = a\left(\sum_{n=1}^{\infty}\left(\frac{a}{r}\right)^{n+1}\sum_{m=0}^{n}P_{n,m}(\sin\lambda)\left(g_{n,m}\cos m\phi + h_{n,m}\sin m\phi\right)\right) \tag{12.22}$$

where the ECEF position vector is converted to the geocentric spherical coordinates in terms of distance from the center of the Earth r, geocentric latitude λ, and longitude ϕ.

a is the semi-major axis of the Earth, which is used to normalize the coefficients to unitless, $P_{n,0}$ are the Legendre polynomials, and $P_{n,m}$ for $1 \le m \le n$ are the associated Legendre functions that are identical to the ones used for the gravity model.

Wave numbers 1 to 12 are primarily from the core. Wave numbers 13 to 15 come from both the core and the crust. Wave numbers 16 and higher are from the crust.

The IGRF and WMM primarily capture the core field. One important thing to know is that the model can be continued or extended for the higher altitude using the data from the lower altitude. The data currently available for IGRF and WMM are such that they do not provide sufficient details or sufficient accuracy to support MARN.

There are additional maps or databases available that are specifically to capture information about the magnetic field anomaly. In particular, North American Magnetic Anomaly Database (NAMAD) [8–10] was created as a joint effort between the United States Geological Survey, the Geological Survey of Canada, and the Consejo de Recursos Minerales of Mexico. It provides data of variable quality over 2–8-km spacing over North America and has the potential to provide reasonable performance for magnetic field referenced navigation.

The World Digital Magnetic Anomaly Map (WDMAM) [11–13] is the result of an international effort to create a world magnetic anomaly map. The map or database models the Earth's magnetic field anomaly at a 5-km altitude with a 3-arc-minute resolution (~5.5 km at the equator). The World Digital Magnetic Anomaly was also created from data with varying qualities. It does have the potential to provide some level of accuracy to support MARN.

12.5 Architecture

The architecture to allow Earth gravity anomaly and Earth magnetic anomaly to be used as reference for navigation is essentially the same as the TRN. In this section, we will describe the architecture for three most appropriate variations:

1. The strapdown GGI-aided INS;
2. Single-axis gravimeter-aided INS;
3. Scalar magnetometer-aided INS.

Many other variations can be created, but the fundamental principles are the same.

Figure 12.1 illustrates the architecture for the GGI-aided GGARN system. Obviously, this architecture is consistent with the generic All Source PNT architecture discussed in earlier chapters and the GGI update lane can be easily incorporated into the All Source PNT architecture.

Figure 12.2 illustrates the architecture for a single-axis gravimeter-based GARN system. As we discussed before, because of the large gravity projection problem, it is not very profitable to use a 3-axis gravimeter system. Only a single-axis sensor along the local vertical makes the most sense. A low-cost platform that is controlled using the INS output is assumed to control the platform to be locally level. Both the IMU and the gravimeter are assumed to be on the low-cost platform.

As can be seen, this architecture uses a low-cost platform to keep the gravimeter locally vertical. As mentioned before, because of the cosine effect, the platform

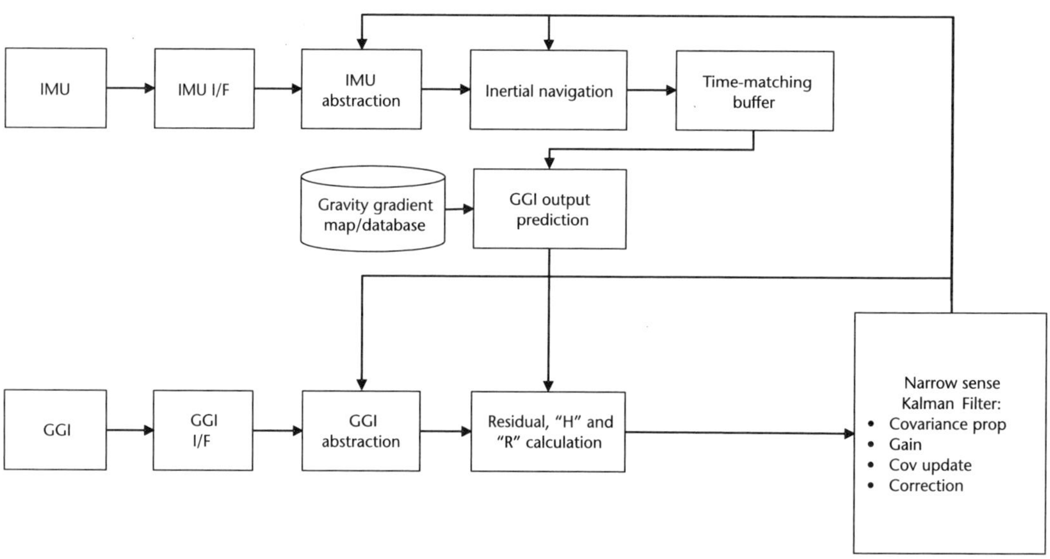

Figure 12.1 GGARN.

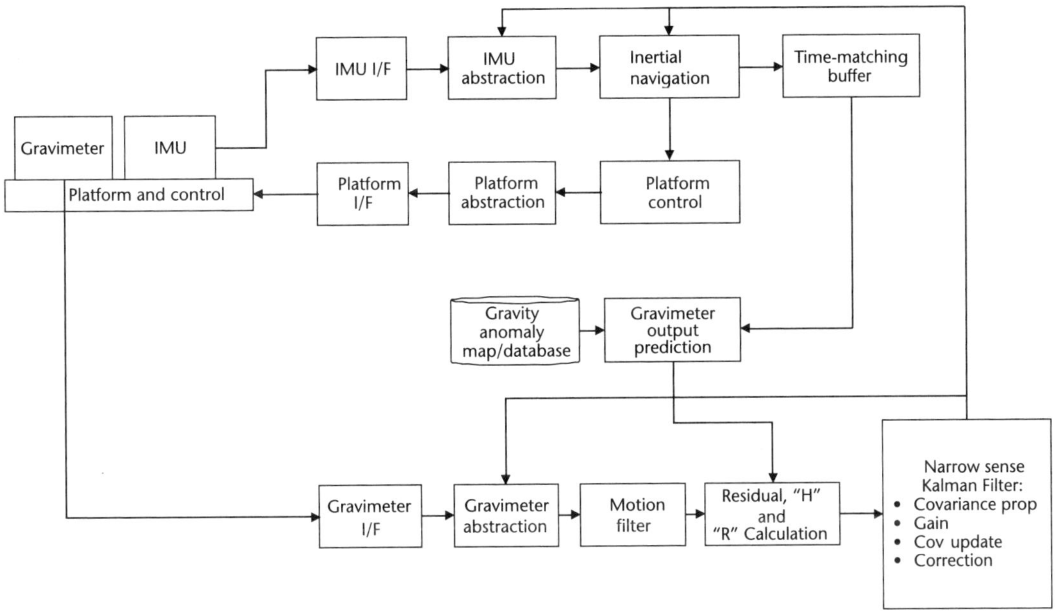

Figure 12.2 Single-axis gravimeter on a low-cost, platform-based GARN system.

does not need to be precise. A low-cost platform with error on the order of 1 mrad should be sufficient.

In addition, a motion filter has to be used to filter out the accelerations due to motion. Typically, this means an extremely low bandwidth lowpass filter with additional notches for vehicle-specific frequencies.

Figure 12.3 illustrates the magnetometer-based MARN. As can be seen, it is architecturally identical to both the classic TRN and GARN. Although the diagram is generic, as mentioned before, the precision scalar magnetometer such as OPM is considered the correct choice for this type of system.

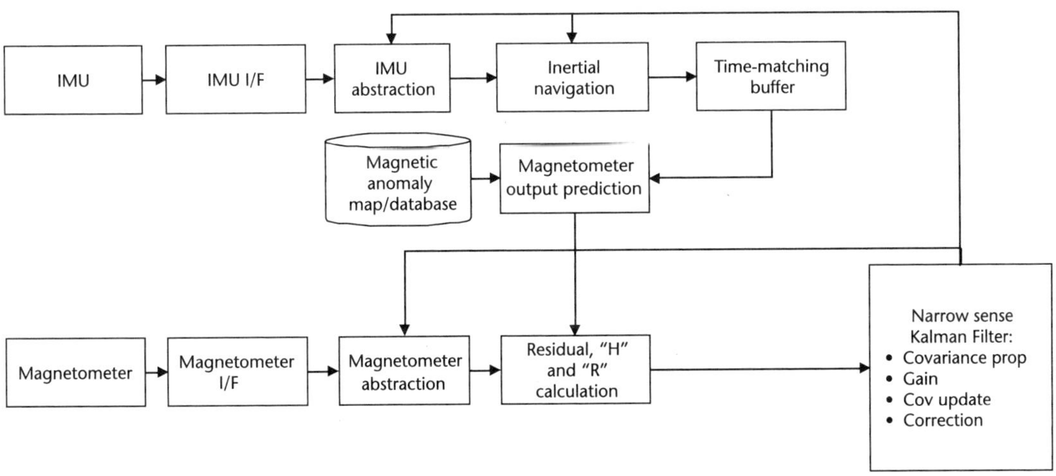

Figure 12.3 MARN.

12.6 Propagation Algorithms and Error Models

The inertial navigation algorithms are discussed in Chapter 3, and there are several types of the algorithm depending in which frame the equations are written. For the purpose of this chapter, we use the INS algorithms presented in the ECEF frame, which are summarized here:

1. The attitude propagation equation in the quaternion form is simply

$$\dot{q}_B^E = \frac{1}{2}\left(q_B^E \otimes {}^B\omega_I^B - {}^E\omega_I^E \otimes q_B^E\right) \tag{12.23}$$

2. The DCM can be derived from the quaternion

$$C_B^E = q2DCM\left(q_B^E\right) \tag{12.24}$$

3. The velocity propagation equation is given by

$$\begin{aligned} \frac{d}{dt}{}^E v_E^B &= C_B^E {}^B f_I^B + {}^E g - 2{}^E\omega_I^E \times {}^E v_E^B - \left[{}^E\omega_I^E \times\right]\left[{}^E\omega_I^E \times\right]{}^E r_E^B \\ &= C_B^E {}^B f_I^B + {}^E g_c - 2{}^E\omega_I^E \times {}^E v_E^B \end{aligned} \tag{12.25}$$

where

$${}^E g_c = {}^E g - \left[{}^E\omega_I^E \times\right]\left[{}^E\omega_I^E \times\right]{}^E r_E^B$$

4. Finally, the position propagation equation is simply

$$\frac{d}{dt}{}^E r_E^B = {}^E v_E^B \tag{12.26}$$

For the covariance propagation of the Kalman filter, we will need the linearized error models for attitude, velocity, and position, which can be repeated from Chapter 3:

1. Attitude error propagation

$$\delta {}^E\dot{\theta}_B^E = -\left[{}^E\omega_I^E \times\right]\delta {}^E\theta_B^E - C_B^E \delta {}^B\omega_I^B \tag{12.27}$$

2. Velocity error propagation

$$\frac{d}{dt}\delta {}^E v_E^B = \left[\left(C_B^E m {}^B f_I^B\right)\times\right]\delta {}^E\theta_B^E + C_B^E \delta {}^B f_I^B + \left[\frac{\partial {}^E g}{\partial {}^E r_E^B}\right]\delta {}^E r_I^B - 2\left[\left({}^E\omega_I^E\right)\times\right]\delta {}^E v_E^B \tag{12.28}$$

3. Position error propagation

$$\frac{d}{dt}\delta {}^E r_E^B = \delta {}^E v_E^B \tag{12.29}$$

We will use the following gyro error models

$$\delta\,{}^{B}\omega_{I}^{B} = m\,{}^{B}\omega_{I}^{B} - {}^{B}\omega_{I}^{B} = C_{G}^{B}\mathrm{diag}\left({}^{G}s\right){}^{G}\omega_{I}^{B} + \left(\left[{}^{B}\theta_{G}\times\right] + L\left({}^{B}\beta_{G}\right)\right){}^{B}\omega_{I}^{B} + {}^{B}b_{G} + C_{G}^{B}ARW \tag{12.30}$$

where

$$\left[{}^{B}\theta\times\right] = \begin{bmatrix} 0 & -\theta_{z} & \theta_{y} \\ \theta_{z} & 0 & -\theta_{x} \\ -\theta_{y} & \theta_{x} & 0 \end{bmatrix},\quad L\left({}^{B}\beta\right) = \begin{bmatrix} 0 & 0 & 0 \\ \beta_{z} & 0 & 0 \\ -\beta_{x} & \beta_{y} & 0 \end{bmatrix} \tag{12.31}$$

which can also be written in a more Kalman filter-friendly way as

$$\delta\,{}^{B}\omega_{I}^{B} = C_{G}^{B}\mathrm{diag}\left({}^{G}\omega_{I}^{B}\right){}^{G}s - \left[{}^{B}\omega_{I}^{B}\times\right]{}^{B}\theta_{G} + J\left({}^{B}\omega_{I}^{B}\right){}^{B}\beta_{G} + {}^{B}b_{G} + C_{G}^{B}ARW \tag{12.32}$$

where

$$J(\omega) = \begin{bmatrix} 0 & 0 & 0 \\ 0 & 0 & \omega_{x} \\ \omega_{y} & -\omega_{x} & 0 \end{bmatrix} \tag{12.33}$$

We use the following accelerometer error model:

$$\delta\,{}^{B}f_{I}^{B} = m\,{}^{B}f_{I}^{B} - {}^{B}f_{I}^{B} = C_{A}^{B}\mathrm{diag}\left({}^{A}s\right){}^{A}f_{I}^{B} + L\left({}^{B}\beta_{A}\right){}^{B}f_{I}^{B} + {}^{B}b_{A} + C_{A}^{B}VRW \tag{12.34}$$

which can also be written in a more Kalman filter-friendly way as

$$\delta\,{}^{B}f_{I}^{B} = C_{A}^{B}\mathrm{diag}\left({}^{A}f_{I}^{B}\right){}^{A}s + J\left({}^{B}f_{I}^{B}\right){}^{B}\beta_{A} + {}^{B}b_{A} + C_{A}^{B}VRW \tag{12.35}$$

$$J\left({}^{B}f_{I}^{B}\right) = \begin{bmatrix} 0 & 0 & 0 \\ 0 & 0 & {}^{B}f_{I\,x}^{B} \\ {}^{B}f_{I\,y}^{B} & -{}^{B}f_{I\,x}^{B} & 0 \end{bmatrix} \tag{12.36}$$

where VRW is the accelerometer velocity random walk, ${}^{A}s$ is the scale factor error in the accelerometer sensor frame, ${}^{B}\beta_{A}$ is the accelerometer nonorthogonality misalignments, and ${}^{B}b_{A}$ is the accelerometer bias vector in the body frame.

If we model each one of the IMU error parameters as a random walk process and let

$$\begin{aligned} \delta X_{I}^{T} &= \left[\delta\,{}^{E}\dot{\theta}_{B}^{E}, \delta\,{}^{B}b_{G}^{T}, \delta\,{}^{G}s^{T}, \delta\,{}^{B}\beta_{G}^{T}, \delta\,{}^{B}\theta_{G}^{T}, \left(\delta\,{}^{E}v_{E}^{B}\right)^{T}, \delta\,{}^{B}b_{A}^{T}, \delta\,{}^{A}s^{T}, \delta\,{}^{B}\beta_{A}^{T}, \left(\delta\,{}^{E}r_{E}^{B}\right)^{T}\right] \\ W_{I}^{T} &= \left[-\left(C_{G}^{E}ARW\right)^{T}, RRW^{T}, SRW_{G}^{T}, \beta RW_{G}^{T}, \theta RW_{G}^{T}, \left(C_{A}^{E}VRW\right)^{T}, ABRW^{T}, SRW_{A}^{T}, \beta RW_{A}{}^{T}\, 0\right] \end{aligned} \tag{12.37}$$

We then have

$$\delta\dot{X}_I = A_{II}\delta X_I + W_I$$

$$A_{II} = \begin{bmatrix} -[{}^E\omega_I^E\times] & -C_B^E & -C_G^E\mathrm{diag}({}^G\omega_I^B) & -C_B^E J({}^B\omega_I^B) & C_B^E[{}^B\omega_I^B\times] & 0_{3x3} & 0_{3x3} & 0_{3x3} & 0_{3x3} & 0_{3x3} \\ 0_{3x3} & 0_{3x3} & 0_{3x3} & 0_{3x3} & 0_{3x3} & 0_{3x3} & 0_{3x3} & 0_{3x3} & 0_{3x3} & 0_{3x3} \\ 0_{3x3} & 0_{3x3} & 0_{3x3} & 0_{3x3} & 0_{3x3} & 0_{3x3} & 0_{3x3} & 0_{3x3} & 0_{3x3} & 0_{3x3} \\ 0_{3x3} & 0_{3x3} & 0_{3x3} & 0_{3x3} & 0_{3x3} & 0_{3x3} & 0_{3x3} & 0_{3x3} & 0_{3x3} & 0_{3x3} \\ 0_{3x3} & 0_{3x3} & 0_{3x3} & 0_{3x3} & 0_{3x3} & 0_{3x3} & 0_{3x3} & 0_{3x3} & 0_{3x3} & 0_{3x3} \\ [(C_B^E m\,{}^B f_I^B)\times] & 0_{3x3} & 0_{3x3} & 0_{3x3} & 0_{3x3} & -2[({}^E\omega_I^E)\times] & C_B^E & C_A^E\mathrm{diag}({}^A f_I^B) & C_B^E J({}^B f_I^B) & \left[\dfrac{\partial\,{}^E g}{\partial\,{}^E r_E^B}\right] \\ 0_{3x3} & 0_{3x3} & 0_{3x3} & 0_{3x3} & 0_{3x3} & 0_{3x3} & 0_{3x3} & 0_{3x3} & 0_{3x3} & 0_{3x3} \\ 0_{3x3} & 0_{3x3} & 0_{3x3} & 0_{3x3} & 0_{3x3} & 0_{3x3} & 0_{3x3} & 0_{3x3} & 0_{3x3} & 0_{3x3} \\ 0_{3x3} & 0_{3x3} & 0_{3x3} & 0_{3x3} & 0_{3x3} & 0_{3x3} & 0_{3x3} & 0_{3x3} & 0_{3x3} & 0_{3x3} \\ 0_{3x3} & 0_{3x3} & 0_{3x3} & 0_{3x3} & 0_{3x3} & I_{3x3} & 0_{3x3} & 0_{3x3} & 0_{3x3} & 0_{3x3} \end{bmatrix} \tag{12.38}$$

12.7 Measurement Equations and Error Models

12.7.1 Strapdown GGI Update

Assuming that that we have a sufficient number of gravity gradient sensors to form a GGI that measures the gravity gradient tensor:

$$m\,{}^S\Gamma = C_B^S\,{}^B\Gamma\left(C_B^S\right)^T + V \tag{12.39}$$

where S stands for the sensor frame, C_B^S is assumed to be the known sensor to body frame DCM, ${}^B\Gamma$ is the gravity gradient matrix expressed in the vehicle body frame, and V is the measurement error.

We can predict the measurement using the current navigation states and the map database, which can be used to derive the predicted gravity gradient matrix in the ECEF frame.

The transformation of the gravity gradient matrix expressed in the ECEF frame to that in the body frame can be done using the following equation:

$${}^B\Gamma = C_E^B\,{}^E\Gamma\left(C_E^B\right)^T \tag{12.40}$$

This matrix transformation can be alternatively written in vector form. Since the matrices ${}^B\Gamma$ and ${}^E\Gamma$ are symmetric, we will represent these matrices using a vector consisting of six elements.

Let us define the following two vectors derived from the matrices

$${}^B v\Gamma = \begin{bmatrix} {}^B\Gamma_{11} \\ {}^B\Gamma_{12} \\ {}^B\Gamma_{13} \\ {}^B\Gamma_{22} \\ {}^B\Gamma_{23} \\ {}^B\Gamma_{33} \end{bmatrix}, \quad {}^E v\Gamma = \begin{bmatrix} {}^E\Gamma_{11} \\ {}^E\Gamma_{12} \\ {}^E\Gamma_{13} \\ {}^E\Gamma_{22} \\ {}^E\Gamma_{23} \\ {}^E\Gamma_{33} \end{bmatrix} \tag{12.41}$$

If we symbolically expand (12.40) element by element, we will be able to show that

$$ {}^{B}v\Gamma = T_E^B \, {}^{E}v\Gamma \tag{12.42} $$

where the 6×6 matrix T_E^B is given by (assuming that $C_E^B = [c_{ij}]$)

$$ T_E^B = \begin{bmatrix} c_{11}^2 & 2c_{11}c_{12} & 2c_{11}c_{13} & c_{12}^2 & 2c_{11}c_{13} & c_{13}^2 \\ c_{11}c_{21} & c_{11}c_{22} + c_{12}c_{21} & c_{11}c_{23} + c_{13}c_{21} & c_{12}c_{22} & c_{12}c_{23} + c_{13}c_{22} & c_{13}c_{23} \\ c_{11}c_{31} & c_{11}c_{32} + c_{12}c_{31} & c_{11}c_{33} + c_{13}c_{31} & c_{12}c_{32} & c_{12}c_{33} + c_{13}c_{32} & c_{13}c_{33} \\ c_{21}^2 & 2c_{21}c_{22} & 2c_{21}c_{23} & c_{22}^2 & 2c_{22}c_{23} & c_{23}^2 \\ c_{21}c_{31} & c_{21}c_{32} + c_{22}c_{31} & c_{21}c_{33} + c_{23}c_{31} & c_{22}c_{32} & c_{22}c_{33} + c_{23}c_{32} & c_{23}c_{33} \\ c_{31}^2 & 2c_{31}c_{32} & 2c_{31}c_{33} & c_{32}^2 & 2c_{32}c_{33} & c_{33}^2 \end{bmatrix} \tag{12.43} $$

Now let us derive the linearized equation for the sensitivity of the gravity gradient tensor error with respect to platform attitude error and position error.

Given

$$ \begin{aligned} {}^{B}\Gamma &= C_E^B \, {}^{E}\Gamma \left(C_E^B\right)^T \\ {}^{B}\hat{\Gamma} &= \hat{C}_E^B \, {}^{E}\Gamma \left(\hat{C}_E^B\right)^T \end{aligned} \tag{12.44} $$

we have

$$ \begin{aligned} {}^{B}\Gamma - {}^{B}\hat{\Gamma} &= C_E^B \, {}^{E}\Gamma \left(C_E^B\right)^T - \hat{C}_E^B \, {}^{E}\hat{\Gamma} \left(\hat{C}_E^B\right)^T \\ &\approx \left(C_E^B - \hat{C}_E^B\right) \Gamma \left(C_E^B\right)^T + C_E^B \, {}^{E}\Gamma \left(C_E^B - \hat{C}_E^B\right)^T + C_E^B \left({}^{E}\Gamma - {}^{E}\hat{\Gamma}\right) C_E^B \end{aligned} \tag{12.45} $$

The first two terms are the errors due to attitude knowledge error. The last term is the error due to the knowledge of the gravity gradient tensor error. Assuming that the database errors are modeled as random noise and are lumped into the sensor noises, the primary contributor of the last term is the position knowledge error, which is exactly the term through which the position becomes observable.

Note that

$$ \left(C_E^B \quad \hat{C}_E^B\right) = C_E^B \left[\theta_{\hat{E}}^E \times\right] \tag{12.46} $$

and we have

$$ \begin{aligned} &\left(C_E^B - \hat{C}_E^B\right) \Gamma (C_E^B)^T + \hat{C}_E^B \, {}^{E}\Gamma (C_E^B - \hat{C}_E^B)^T \\ &= C_E^B \left[\theta_{\hat{E}}^E \times\right] \Gamma \left(C_E^B\right)^T - \hat{C}_E^B \, {}^{E}\Gamma \left[\theta_{\hat{E}}^E \times\right] \left(C_E^B\right)^T \\ &\approx C_E^B \left[\theta_{\hat{E}}^E \times\right] \Gamma \left(C_E^B\right)^T - C_E^B \, {}^{E}\Gamma \left[\theta_{\hat{E}}^E \times\right] \left(C_E^B\right)^T \\ &= C_E^B \left(\left[\theta_{\hat{E}}^E \times\right] \Gamma - {}^{E}\Gamma \left[\theta_{\hat{E}}^E \times\right]\right) \left(C_E^B\right)^T \end{aligned} \tag{12.47} $$

Define

$$L_\theta^E = \left(\left[\theta_{\hat{E}}^E \times\right]{}^E\Gamma - {}^E\Gamma\left[\theta_{\hat{E}}^E \times\right]\right) = \left(\left[\delta\,{}^E\theta_B^E \times\right]{}^E\Gamma - {}^E\Gamma\left[\delta\,{}^E\theta_B^E \times\right]\right) \tag{12.48}$$

Here we used the preferred symbol $\delta\,{}^E\theta_B^E = \theta_{\hat{E}}^E$ and we have

$$\begin{aligned}\delta\,{}^B\Gamma &= {}^B\Gamma - {}^B\hat{\Gamma} = C_E^B\left(\left[\delta\,{}^E\theta_B^E \times\right]{}^E\Gamma - {}^E\Gamma\left[\delta\,{}^E\theta_B^E \times\right]\right)\left(C_E^B\right)^T + C_E^B\delta\,{}^E\Gamma\left(C_E^B\right)^T \\ &= C_E^B L_\theta^E\left(C_E^B\right)^T + C_E^B\delta\,{}^E\Gamma\left(C_E^B\right)^T\end{aligned} \tag{12.49}$$

Expanding (12.48), we have

$$L_\theta^E = \left(\left[\delta\,{}^E\theta_B^E \times\right]{}^E\Gamma - {}^E\Gamma\left[\delta\,{}^E\theta_B^E \times\right]\right)$$

$$= \begin{bmatrix} 2\left({}^E\Gamma_{13}\delta\,{}^E\theta_B^E(2) - {}^E\Gamma_{12}\delta\,{}^E\theta_B^E(3)\right) & \begin{matrix}-{}^E\Gamma_{13}\delta\,{}^E\theta_B^E(1) + {}^E\Gamma_{23}\delta\,{}^E\theta_B^E(2) \\ +\left({}^E\Gamma_{11} - {}^E\Gamma_{11}\right)\delta\,{}^E\theta_B^E(3)\end{matrix} & \begin{matrix}{}^E\Gamma_{12}\delta\,{}^E\theta_B^E(1) + \left({}^E\Gamma_{33} - {}^E\Gamma_{11}\right)\delta\,{}^E\theta_B^E(2) \\ + {}^E\Gamma_{23}\delta\,{}^E\theta_B^E(3)\end{matrix} \\ \text{symmetric} & -2\left({}^E\Gamma_{23}\delta\,{}^E\theta_B^E(1) - {}^E\Gamma_{12}\delta\,{}^E\theta_B^E(3)\right) & \begin{matrix}\left({}^E\Gamma_{22} - {}^E\Gamma_{11}\right)\delta\,{}^E\theta_B^E(1) - {}^E\Gamma_{12}\delta\,{}^E\theta_B^E(2) \\ + {}^E\Gamma_{13}\delta\,{}^E\theta_B^E(3)\end{matrix} \\ \text{symmetric} & \text{symmetric} & 2\left({}^E\Gamma_{23}\delta\,{}^E\theta_B^E(1) - {}^E\Gamma_{13}\delta\,{}^E\theta_B^E(2)\right) \end{bmatrix} \tag{12.50}$$

Taking the six elements of the symmetric matrix and forming a vector, we have the following equation:

$$\begin{aligned} vL_\theta^E &= \begin{bmatrix} L_\theta^E(1,1) \\ L_\theta^E(1,2) \\ L_\theta^E(1,3) \\ L_\theta^E(2,2) \\ L_\theta^E(2,3) \\ L_\theta^E(3,3) \end{bmatrix} = \begin{bmatrix} 2\left({}^E\Gamma_{13}\delta\,{}^E\theta_B^E(2) - {}^E\Gamma_{12}\delta\,{}^E\theta_B^E(3)\right) \\ -{}^E\Gamma_{13}\delta\,{}^E\theta_B^E(1) + {}^E\Gamma_{23}\delta\,{}^E\theta_B^E(2) + \left({}^E\Gamma_{11} - {}^E\Gamma_{11}\right)\delta\,{}^E\theta_B^E(3) \\ {}^E\Gamma_{12}\delta\,{}^E\theta_B^E(1) + \left({}^E\Gamma_{33} - {}^E\Gamma_{11}\right)\delta\,{}^E\theta_B^E(2) \\ -2\left({}^E\Gamma_{23}\delta\,{}^E\theta_B^E(1) - {}^E\Gamma_{12}\delta\,{}^E\theta_B^E(3)\right) \\ \left({}^E\Gamma_{22} - {}^E\Gamma_{11}\right)\delta\,{}^E\theta_B^E(1) - {}^E\Gamma_{12}\delta\,{}^E\theta_B^E(2) + {}^E\Gamma_{13}\delta\,{}^E\theta_B^E(3) \\ 2\left({}^E\Gamma_{23}\delta\,{}^E\theta_B^E(1) - {}^E\Gamma_{13}\delta\,{}^E\theta_B^E(2)\right) \end{bmatrix} \\ &= \begin{bmatrix} 0 & 2\,{}^E\Gamma_{13} & -2\,{}^E\Gamma_{12} \\ -{}^E\Gamma_{13} & {}^E\Gamma_{23} & {}^E\Gamma_{11} - {}^E\Gamma_{22} \\ {}^E\Gamma_{12} & {}^E\Gamma_{33} - {}^E\Gamma_{11} & -{}^E\Gamma_{23} \\ -2\,{}^E\Gamma_{23} & 0 & 2\,{}^E\Gamma_{12} \\ {}^E\Gamma_{22} - {}^E\Gamma_{33} & -{}^E\Gamma_{12} & {}^E\Gamma_{13} \\ 2\,{}^E\Gamma_{23} & -2\,{}^E\Gamma_{13} & 0 \end{bmatrix} \delta\,{}^E\theta_B^E \end{aligned} \tag{12.51}$$

We can use the symbol

$$\frac{\partial vL_\theta^E}{\partial \delta^E\theta_B^E} = \begin{bmatrix} 0 & 2^E\Gamma_{13} & -2^E\Gamma_{12} \\ -^E\Gamma_{13} & ^E\Gamma_{23} & ^E\Gamma_{11} - ^E\Gamma_{22} \\ ^E\Gamma_{12} & ^E\Gamma_{33} - ^E\Gamma_{11} & -^E\Gamma_{23} \\ -2^E\Gamma_{23} & 0 & 2^E\Gamma_{12} \\ ^E\Gamma_{22} - ^E\Gamma_{33} & -^E\Gamma_{12} & ^E\Gamma_{13} \\ 2^E\Gamma_{23} & -2^E\Gamma_{13} & 0 \end{bmatrix} \tag{12.52}$$

Then the equation can be written simply as

$$vL_\theta^E = \left(\frac{\partial vL_\theta^E}{\partial \delta^E\theta_B^E}\right)\delta^E\theta_B^E \tag{12.53}$$

Now looking at the overall derivation from the current knowledge, we have

$$\delta^B\Gamma = {}^B\Gamma - {}^B\hat{\Gamma} = C_E^B L_\theta^E \left(C_E^B\right)^T + C_E^B \delta^E\Gamma \left(C_E^B\right)^T \tag{12.54}$$

which can be written in the six-element vector form as

$$\delta^B v\Gamma = {}^B v\Gamma - {}^B v\hat{\Gamma} = T_E^B vL_\theta^E + T_E^B \delta^E v\Gamma \tag{12.55}$$

As discussed before, the primary contributor to $\delta^E v\Gamma$ is the position error, which can be derived from the database values and we have

$$\delta^E v\Gamma = \left(\frac{\partial^E v\Gamma}{\partial^E r}\right)\delta^E r \tag{12.56}$$

Combining everything together, we have the following linearized error sensitivity equation

$$\delta^B v\Gamma = {}^B v\Gamma - {}^B v\hat{\Gamma} = T_E^B \left(\frac{\partial vL_\theta^E}{\partial \delta^E\theta_B^E}\right)\delta^E\theta_B^E + T_E^B \left(\frac{\partial^E v\Gamma}{\partial^E r}\right)\delta^E r \tag{12.57}$$

Now going back to the measurement model

$$m^S\Gamma = C_B^S {}^B\Gamma \left(C_B^S\right)^T + V$$

the corresponding six-element vector form can be written as

$$m^S v\Gamma = T_B^S {}^B v\Gamma + vV \tag{12.58}$$

The linearized measurement model (or the Kalman filter residual equation) can then be written as

$$m^S v\Gamma - {}^S v\hat{\Gamma} = T_B^S T_E^B \left(\left(\frac{\partial vL_\theta^E}{\partial \delta^E\theta_B^E}\right)\delta^E\theta_B^E + \left(\frac{\partial^E v\Gamma}{\partial^E r}\right)\delta^E r\right) + vV \tag{12.59}$$

Consequently, the H-matrix can be written as

$$H_{GGI} = \left[T_B^S T_E^B \left(\frac{\partial v L_\theta^E}{\partial \delta^E \theta_B^E} \right), 0, \ldots, 0, T_B^S T_E^B \left(\frac{\partial^E v \Gamma}{\partial^E r} \right) \right] \tag{12.60}$$

Note that the above model is much simplified. For the real world, the following additional complexities can be introduced:

1. The sensor alignment knowledge errors can be included and therefore possibly estimated as part of the Kalman filter state. An approach similar to the attitude error discussed above can be taken to include this error term.
2. The additional error terms based on physics of the sensor may be included in the model (such as bias and scale factor errors).
3. Depending on the physics of the specific GGI, there might be other terms such as the rotational rate of the body frame that can be included in the measurement model.

12.7.2 Single-Axis Gravity Anomaly Update

Gravity gradient is very small and it requires high precision and typically expensive GGIs to measure. However, gravity anomaly is on the order of 100 μG, which typically requires less precise sensors. The problem with the gravity anomaly is that it is indistinguishable from motion-induced accelerations. Consequently, gravity anomaly (instead of gravity gradient anomaly) aided navigation requires heavy averaging or filtering to remove the impact of the motion in the gravity measurement.

Another important issue for gravity anomaly is that the gravity anomaly along local-level axes are often difficult to measure because the projections of the gravity through small attitude or misalignment errors on the local-level axes are often orders of magnitude bigger than the gravity anomaly. For example, in order to maintain 1-μG level accuracy for the gravity measurement along the local-level axes, the attitude knowledge error needs to be better than 1 μrad. Even if we have sensors that can achieve that level of accuracy, the thermal deformations and other contributors can easily go above that.

However, small attitude errors do not generate significant gravity measurement errors along the local vertical axis due to the cosine effect. In other words, first-order attitude errors only contribute second-order errors to the local vertical axis gravity measurement. For example, 1-mrad roll or pitch error only generates 0.5 μG of gravity error in the local vertical axis.

Consequently, we conclude that:

1. Strapdown sensor-based, gravity anomaly-aided navigation is difficult to implement due to the overwhelming errors introduced when the vertical axis gravity is projected through attitude and misalignment errors onto the horizontal axes.
2. Single vertical axis gravity anomaly on a not-very-precise platform is a good compromise between performance and cost.
3. The long-term average is required to average or filter out the vehicle motion.

The measured local vertical axis gravity can be modeled as

$$ma_z = \eta(lat, lon, h) + v \tag{12.61}$$

where $\eta(lat, lon, h)$ is the local vertical gravity predicted by the gravity map for the location in geocentric coordinate (lat, lon, h) (note that we typically use the geocentric coordinate for gravity models).

To predict the measurement, we first convert the ECEF position to (Lat, Lon, and Altitude)

$$L\hat{L}A = \begin{bmatrix} l\hat{a}t \\ l\hat{o}n \\ \hat{h} \end{bmatrix} = ecef2lla\_geocentric\left({}^{E}\hat{r}_{E}^{B}\right)$$

then calculate

$$m\hat{a}_z = \eta\left(l\hat{a}t, l\hat{o}n, \hat{h}\right)$$

We will derive the linearized measurement model for the purpose of the Kalman filter H-matrix:

$$\begin{aligned} \delta ma_z &= \frac{\partial\eta(lat,lon,h)}{\partial lat}\frac{\partial lat}{\partial {}^{E}r_{E}^{B}}\delta\,{}^{E}r_{E}^{B} + \frac{\partial\eta(lat,lon,h)}{\partial lon}\frac{\partial lon}{\partial {}^{E}r_{E}^{B}}\delta\,{}^{E}r_{E}^{B} + \frac{\partial\eta(lat,lon,h)}{\partial h}\frac{\partial h}{\partial {}^{E}r_{E}^{B}}\delta\,{}^{E}r_{E}^{B} + v \\ &= \left(\frac{\partial\eta(lat,lon)}{\partial lat}\frac{\partial lat}{\partial {}^{E}r_{E}^{B}} + \frac{\partial\eta(lat,lon)}{\partial lon}\frac{\partial lon}{\partial {}^{E}r_{E}^{B}} + \frac{\partial\eta(lat,lon,h)}{\partial h}\frac{\partial h}{\partial {}^{E}r_{E}^{B}}\right)\delta\,{}^{E}r_{E}^{B} + v \end{aligned} \tag{12.62}$$

and the corresponding H-matrix is simply

$$H_{GA} = \left[0,0,\ldots,0,\left(\frac{\partial\eta(lat,lon)}{\partial lat}\frac{\partial lat}{\partial {}^{E}r_{E}^{B}} + \frac{\partial\eta(lat,lon)}{\partial lon}\frac{\partial lon}{\partial {}^{E}r_{E}^{B}} + \frac{\partial\eta(lat,lon,h)}{\partial h}\frac{\partial h}{\partial {}^{E}r_{E}^{B}}\right)\right] \tag{12.63}$$

12.7.3 Scalar Magnetometer Update

The magnetometer needs to be corrected for a number of errors before it can be used for MARN. Specifically, we want to correct the following errors based on the best knowledge:

1. Correction of the heading error: We want to correct the attitude-dependent errors using our current attitude knowledge. Typically, the sensitivity of the heading errors is not such that it can justify modeling the measurement errors as a function of attitude knowledge error (consequently using a nonzero term in the H-matrix with respect to the attitude error states).
2. Correction of the vehicle environment-induced errors: This step is vital to the success of magnetic referenced navigation. In order to achieve a reasonable performance, we must ensure that the magnetic field induced by the vehicle environment is either calibrated by estimating the parameters of a model in a calibration process, has a high enough frequency, or is randomized so

that it is filtered out. The correction and calibration model is most likely a function of the vehicle attitude as well. Again, we do not anticipate that the attitude-dependent corrections are such that they justify the inclusion of the attitude errors in the measurement model.

3. Correction of other known contents of the magnetic field (such as the baseline IGRF or WMM field model) before the magnetic anomaly model can be used.

In other words, we will generate the following magnetic anomaly measurement from the raw total magnetic field measurement

$$y_m = z_m - IGRF(lat,lon,h) - \text{Heading Correction}\left(C_E^B\right) - \text{Vehicle Field Correction}\left(C_E^B\right) \tag{12.64}$$

where z_m is the raw sensor output that measures the total magnetic intensity as seen by the sensor plus the measurement noise and y_m is the derived magnetic anomaly measurement plus measurement noise.

Using the magnetic anomaly model such as NAMAD, we can model the derived magnetic anomaly measurement as

$$y_m = \eta(lat,lon,h) + v \tag{12.65}$$

where $\eta(lat, lon, h)$ is the function representing the magnetic anomaly model such as NAMAD using the location in the geocentric coordinate (lat, lon, h) as inputs.

To predict the measurement, we first convert the ECEF position to (Lat, Lon, and Altitude) (again note that we are using geocentric coordinates, not geodetic coordinates, here).

$$L\hat{L}A = \begin{bmatrix} l\hat{a}t \\ l\hat{o}n \\ \hat{h} \end{bmatrix} = ecef2lla\_geocentric\left({}^E\hat{r}_E^B\right)$$

then calculate

$$\hat{y}_m = \eta\left(l\hat{a}t, l\hat{o}n, \hat{h}\right) \tag{12.66}$$

We will derive the linearized measurement model for the purpose of Kalman filter H-matrix:

$$\begin{aligned} \delta y_m &= \frac{\partial\eta(lat,lon,h)}{\partial lat}\frac{\partial lat}{\partial {}^E r_E^B}\delta {}^E r_E^B + \frac{\partial\eta(lat,lon,h)}{\partial lon}\frac{\partial lon}{\partial {}^E r_E^B}\delta {}^E r_E^B + \frac{\partial\eta(lat,lon,h)}{\partial h}\frac{\partial h}{\partial {}^E r_E^B}\delta {}^E r_E^B + v \\ &= \left(\frac{\partial\eta(lat,lon)}{\partial lat}\frac{\partial lat}{\partial {}^E r_E^B} + \frac{\partial\eta(lat,lon)}{\partial lon}\frac{\partial lon}{\partial {}^E r_E^B} + \frac{\partial\eta(lat,lon,h)}{\partial h}\frac{\partial h}{\partial {}^E r_E^B}\right)\delta {}^E r_E^B + v \end{aligned} \tag{12.67}$$

and the corresponding H-matrix is simply

$$H_m = \left[0,0,\ldots,0,\left(\frac{\partial\eta(lat,lon)}{\partial lat}\frac{\partial lat}{\partial {}^E r_E^B} + \frac{\partial\eta(lat,lon)}{\partial lon}\frac{\partial lon}{\partial {}^E r_E^B} + \frac{\partial\eta(lat,lon,h)}{\partial h}\frac{\partial h}{\partial {}^E r_E^B}\right)\right] \tag{12.68}$$

12.8 Simulation Examples

We will show simple examples of a single-axis gravimeter-aided navigation system and its performance for a trajectory in the Pacific Ocean and over the Rocky Mountains. In this example, we will assume that we magically filtered out the acceleration due to the real motion.

Figure 12.4 illustrates the variation of gravity field in the Pacific Ocean where the RMS (Root Mean Square) value of the gravity anomaly variations is at 9.5 mGal. In comparison, much more variations can be seen over the Rocky Mountains, which is illustrated in Figure 12.5.

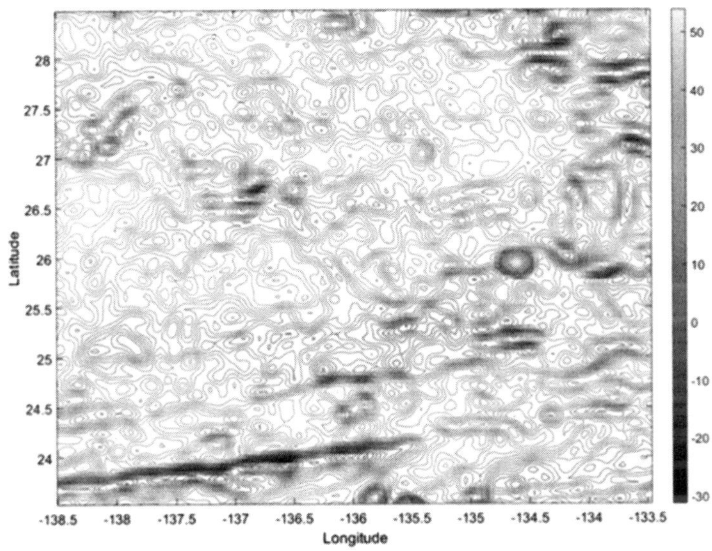

Figure 12.4 Gravity anomaly over Pacific Ocean, RMS value at 9.5 mGal.

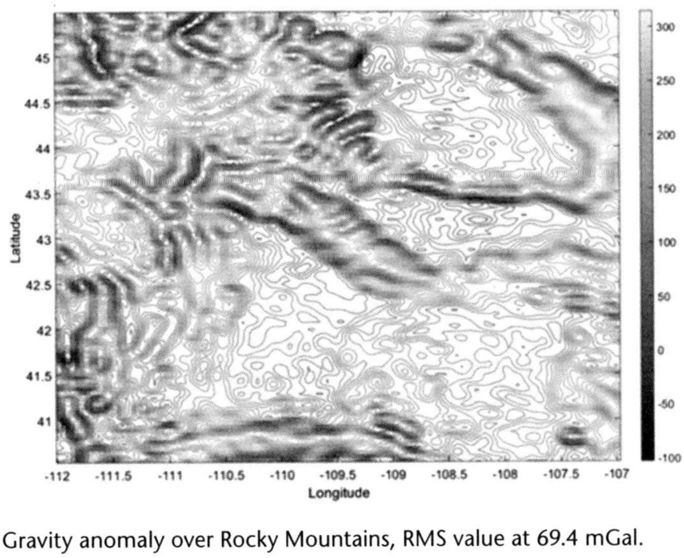

Figure 12.5 Gravity anomaly over Rocky Mountains, RMS value at 69.4 mGal.

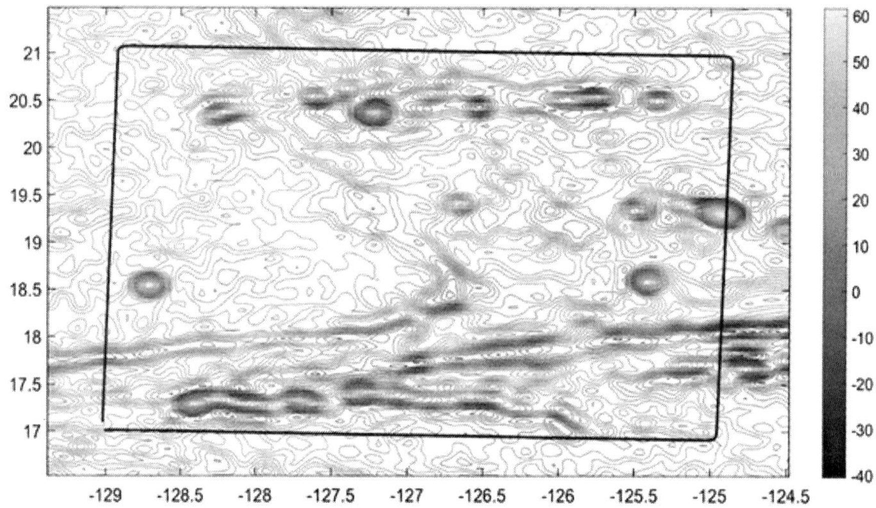

Figure 12.6 Hypothetical trajectory in the Pacific Ocean.

Figure 12.6 shows the trajectory that we simulated in the Pacific Ocean. Figure 12.7 shows the navigation performance using a gravimeter-aided, low-cost INS for the Pacific Ocean trajectory.

In the following, we will present a totally hypothetical simulation case just for the purposes of comparison. Assume again that we magically averaged or filtered out

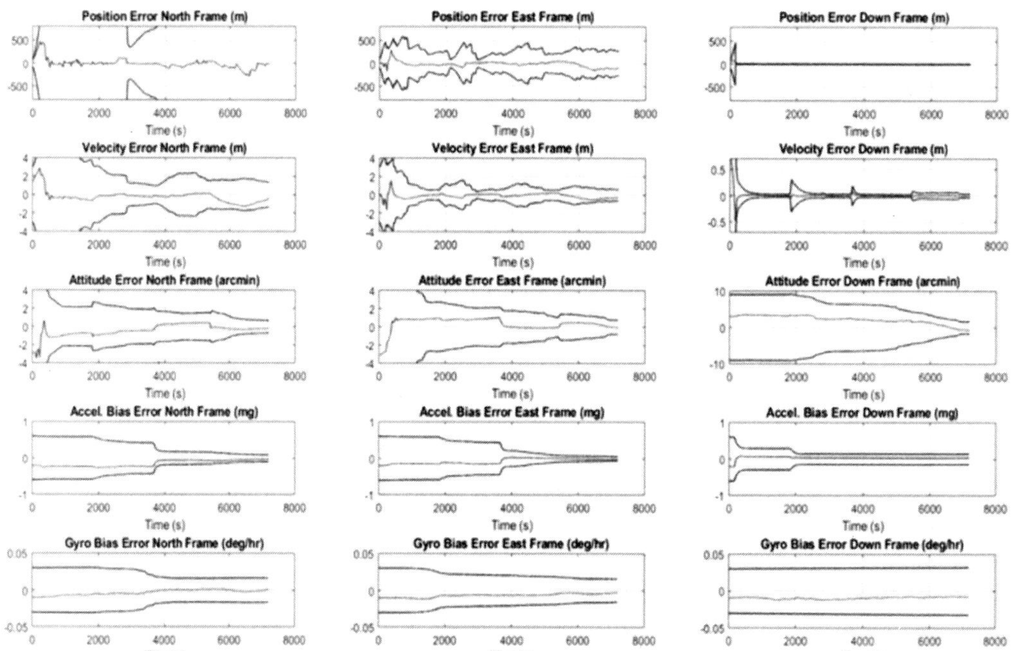

Figure 12.7 Hypothetical trajectory in the Pacific Ocean and the gravimeter-aided INS performance.

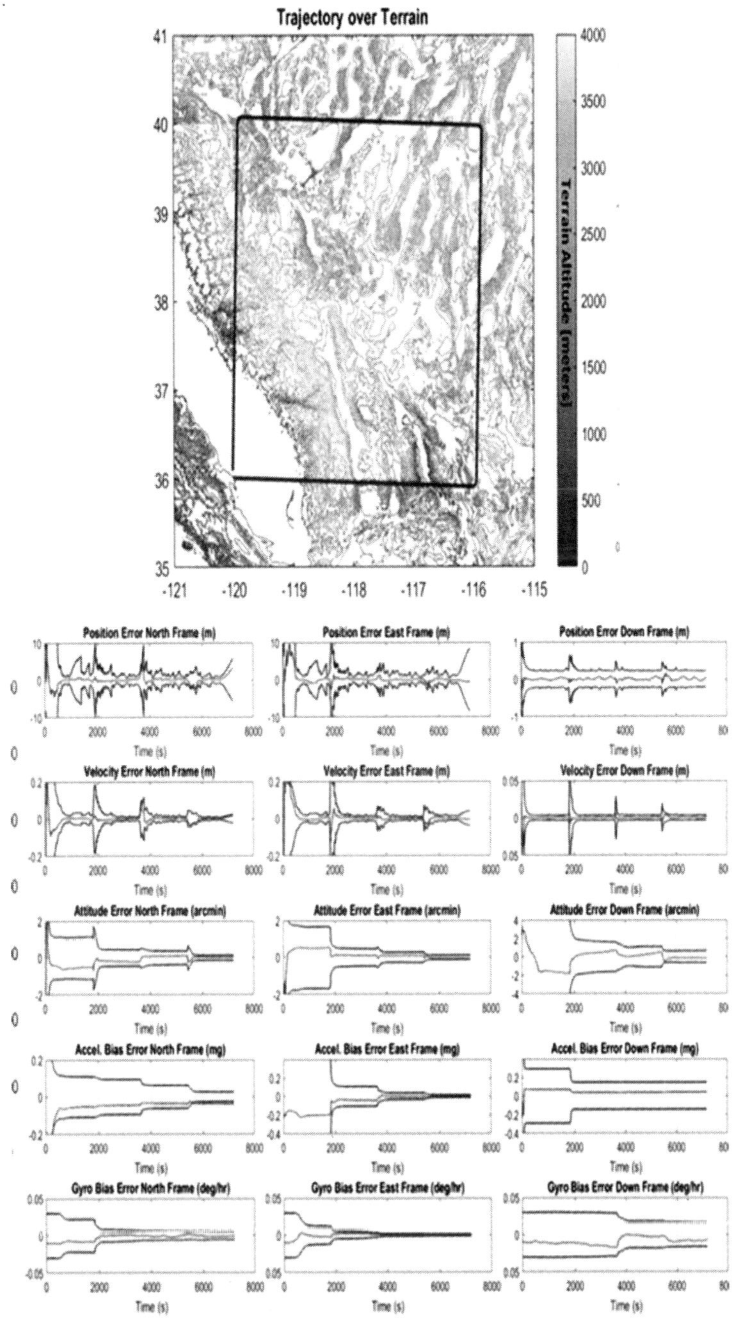

Figure 12.8 Hypothetical trajectory over rough terrains and the gravimeter-aided INS performance.

the acceleration and we fly at a low altitude over rough terrains of California. The predicted performance becomes much better due to the high variations of gravity anomaly and our high speed, which causes us to experience the variations quickly. Figure 12.8 shows the trajectory and the simulation results.

References

[1] Richeson, J. A., "Gravity Gradiometer Aided Inertial Navigation Within Non-GNSS Environments," Ph.D. Dissertation, University of Maryland, College Park, MD, 2008.

[2] Rogers, M. M., "An Investigation into the Feasibility of Using a Modern Gravity Gradient Instrument for Passive Aircraft Navigation and Terrain Avoidance," Master's Thesis, Air Force Institute of Technology, 2009.

[3] Canciani, A. J., "Absolute Positioning Using the Earth's Magnetic Anomaly Field," Ph.D. Dissertation, Air Force Institute of Technology, 2016.

[4] Canciani, A., and J. Raquet, "Airborne Magnetic Anomaly Navigation," *IEEE Transactions on Aerospace and Electronic Systems*, Vol. 53, No. 1, February 2017, pp. 67–80.

[5] Oehler, J., D. Rouxel, and M. Lequentrec-Lalancette, "Comparison of Global Geomagnetic Field Models and Evaluation Using Marine Datasets in the North-Eastern Atlantic Ocean and Western Mediterranean Sea," *Earth Planets Space*, Vol. 70, No. 99, 2018.

[6] Parker, R. L., "The Rapid Calculation of Potential Anomalies," *The Geophysical Journal of the Royal Astronomical Society*, Vol. 31, No. 4, 1972, pp. 447–455.

[7] Jekeli, C., and L. Zhu, "Comparison of Methods to Model the Gravitational Gradients from Topographic Data Bases," *Geophys. J. Int.*, Vol. 166, 2006, pp. 999–1014.

[8] https://mrdata.usgs.gov/magnetic/, USGS. North American Magnetic Anomaly Database.

[9] https://catalog.data.gov/dataset/digital-data-grids-for-the-magnetic-anomaly-map-of-north-america, North American Magnetic Anomaly Database.

[10] Bankey, V., et al., "Digital Data Grids for the Magnetic Anomaly Map of North America," U.S. Geological survey open file report 02-414, U.S. Geological Survey, 2002.

[11] http://wdmam.org/, The World Digital Magnetic Anomaly Map.

[12] http://geomag.org/models/wdmam.html, The World Digital Magnetic Anomaly Map.

[13] Maus, S., "Emag2: Earth Magnetic Anomaly Grid (2-Arc-Minute Resolution)," National Geophysical Data Center, NOAA, 2009.

[14] Metzger, J., et al., "Sigma-Point Filter for Terrain Referenced Navigation," *AIAA Guidance, Navigation, and Control Conference and Exhibit*, San Francisco, CA, August 15–18, 2005.

[15] Wu, X., "Gravity Gradient Survey with a Mobile Atom Interferometer," Ph.D. Dissertation, Stanford University, 2009.

[16] Metzger, E. H., and A. Jircitano, "Inertial Navigation Performance Improvement Using Gravity Gradient Matching Techniques," *J. Spacecraft*, June 1976.

[17] Moryl, J., H. Rice, and S. Shinners, "The Universal Gravity Module for Enhanced Submarine Navigation," *IEEE 1998 Position Location and Navigation Symposium*, Palm Springs, CA, 1998.

[18] Wu, L., et al., "Performance Evaluation and Analysis for Gravity Matching Aided Navigation," *Sensors*, April 2017.

[19] Wang, H., et al., "Location Accuracy of INS/Gravity-Integrated Navigation System on the Basis of Ocean Experiment and Simulation," *Sensors*, December 2017.

[20] Wang, B., et al., "A Particle Filter-Based Matching Algorithm with Gravity Sample Vector for Underwater Gravity Aided Navigation," *IEEE/ASME Transactions on Mechatronics*, Vol. 21, No. 3, June 2016.

[21] Morales, G., "Magnetometer Aided Navigation Filters for Improved Observability and Estimation on Ground Vehicles," Master's Thesis, Auburn University, 2016.

[22] Wang, H., et al., "Characteristics of Marine Gravity Anomaly Reference Maps and Accuracy Analysis of Gravity Matching-Aided Navigation," *Sensors (Basel)*, Vol. 17, No. 8, August 2017, p. 1851.

[23] Wang, H., et al., "Location Accuracy of INS/Gravity-Integrated Navigation System on the Basis of Ocean Experiment and Simulation," *Sensors*, Vol. 17, 2017, p. 2961.

[24] Dai, T., et al., "A Robust Underwater Navigation Method Fusing Data of Gravity Anomaly and Magnetic Anomaly," *International Journal of Systems Science*, Vol. 50, No. 4, 2019, pp. 679–693.

[25] Dai, T., et al., "Solving Gravity Anomaly Matching Problem Under Large Initial Errors in Gravity Aided Navigation by Using an Affine Transformation Based Artificial Bee Colony Algorithm," *Frontiers in Neurorobotics*, Vol. 13, 2019.

[26] Bishop, G. C., "Gravitational Field Maps and Navigational Errors Unmanned Underwater Vehicles," *IEEE J. Ocean. Eng.*, Vol. 27, 2002, pp. 726–737.

[27] Han, Y., et al., "An Improved TERCOM Based Algorithm for Gravity-Aided Navigation," *IEEE Sens. J.*, Vol. 16, 2016, pp. 2537–2544.

[28] Wu, L., et al., "Modelling Full-Tensor Gravity Gradient Maps for Gravity Matching Navigation," Vol. 21, pp. 7037–7041, https://www.researchgate.net/publication/291987927_Modelling_full-tensor_gravity_gradient_maps_for_gravity_matching_navigation.

[29] Claus, B., and R. Bachmayer, "Terrain-Aided Navigation for an Underwater Glider," *Journal of Field Robotics*, Vol. 32, 2015, pp. 935–951.

[30] Hollowell, J., "Heli/SITAN: A Terrain Referenced Navigation Algorithm for Helicopters," *IEEE Symposium on Position Location and Navigation. A Decade of Excellence in the Navigation Sciences*, March 1990.

[31] McGuffin, J. T., "Terrain Referenced Navigation Electromagnetic-Gravitational Correlation," U.S. Patent 5272639A, 1992 (2019-11-14 Application status is Expired—Lifetime).

[32] Fellerhoff, J. R., "SITAN Implementation in the Saint System," *IEEE Journal*, 1986, pp. 89–95.

[33] Baird, C. A., and M. R. Abramson, "A Comparison of Several Digital Map-Aided Navigation Techniques," *IEEE Journal*, 1984, pp. 286–293.

[34] Hostetler, L. D., and R. D. Andreas, "Nonlinear Kalman Filtering Techniques for Terrain-Aided Navigation," *IEEE Transactions on Automatic Control*, Vol. AC-28, No. 3, March 1983, pp. 315–323.

[35] Priestley, N., "Terrain Referenced Navigation," *IEEE Symposium on Position Location and Navigation. A Decade of Excellence in the Navigation Sciences*, March 20, 1990.

[36] Henley, A. J., "Terrain Aided Navigation: Current Status, Techniques for Flat Terrain and Reference Data Requirements," *IEEE Symposium on Position Location and Navigation. A Decade of Excellence in the Navigation Sciences*, March 20, 1990.

[37] Jalving, B., et al., "Terrain Referenced Navigation of AUVs and Submarines Using Multi-Beam Echo Sounders," Norwegian Defense Research, 2004, http://citeseerx.ist.psu.edu/viewdoc/download?doi=10.1.1.459.3521&rep=rep1&type=pdf.

[38] https://www.sciencedirect.com/topics/earth-and-planetary-sciences/magnetometer.

[39] https://en.wikipedia.org/wiki/Magnetometer.

[40] http://www.gemsys.ca/pdf/GEM_Scalar_Magnetometers_Optically_Pumped.pdf.

[41] https://gmw.com/product/mag-03-mag-13/.

[42] Tierney, T., et al., "Optically Pumped Magnetometers: From Quantum Origins to Multi-Channel Magnetoencephalography," *NeuroImage*, Vol. 199, 2019, pp. 598–608.

[43] Ripka, P., "Review of Fluxgate Sensors," *Sensors and Actuators A: Physical*, Vol. 33, No. 3, June 1992, pp. 129–141.

CHAPTER 13

Collaborative PNT

13.1 Introduction

In this chapter, we will deal with the PNT problem when a group of navigators, users, and platforms, each of them called a node, is operating together in a networked environment. Each of the nodes is located at a different geographical location and is exposed to different GNSS environment. Some of the nodes may be located where the GNSS signals are widely available and are of high integrity. Some of the nodes may be located in an urban canyon where high-rise buildings will block some of the GNSS signals from the user. Some of nodes may be located in an area where there is a significant amount of RF interferences or jamming, which leads to the unavailability of GNSS signals.

Each of the nodes may be equipped with different types of PNT devices and/or sensors. Some of them may just have a GNSS receiver. Some may have GNSS or INS available. Some may have vision-based navigation and/or celestial navigation capabilities that can provide PNT solutions even when GNSS is completely denied.

Collaborative PNT, also known as cooperative PNT, is to leverage both the information available locally and the information available through network communications to achieve better PNT performance, availability, and integrity for each of the operating nodes. Unlike other complementary PNT topics, collaborative or cooperative PNT is not an area that has received sufficient attention by the researchers, in the author's opinion, although there has been some limited coverage on this topic in the literature [1–19].

The communication network plays two important roles:

1. Allows the operating nodes to share information.
2. Allows new collaborating measurements to be created in addition to the PNT sensor information available locally at each node.

There are at least three important types of collaboration and collaborative measurements that can be achieved in a network environment:

1. Time transfer and time synchronization between the operating nodes (as an example, the popular NTP and IEEE 1588 Precision Timing Protocol (PTP) [20–22] are open standards that are probably running on every computer today);
2. Range measurements between collaborative partners;

3. LOS, angle of arrival (AOA), or direction of arrival (DOA) measurements between collaborative partners.

Collaborative PNT achieves better PNT solutions than each of the nodes can achieve, in at least two different ways:

1. Each node can improve their solution by leveraging its own information and its neighbor's information (local with immediate-neighbor (LWIN) solution).
2. A network or subnetwork can achieve better solutions for each of the nodes in the network or subnetwork by leveraging all the information available for the entire network or subnetwork (integrated total network solution (ITNS)).

Figure 13.1 illustrates the concept of collaborative PNT. In this figure, we illustrate three types of interconnections or collaborations:

1. The dashed line shows the communication-only connections.
2. The thin solid line shows the communication, time-transfer, or range measurement collaboration connections.
3. The thick solid line shows the communication, time-transfer, range, or LOS direction collaboration.

It is important to point out that a network can often be decomposed into a number of subnetworks, and, between these subnetworks, there are no collaborative measurements available. In other words, we can reduce the complexity and the size of the problem by dealing with simpler and smaller subnetworks. Without the collaborative measurements, there is nothing better we can achieve by treating the network as a whole.

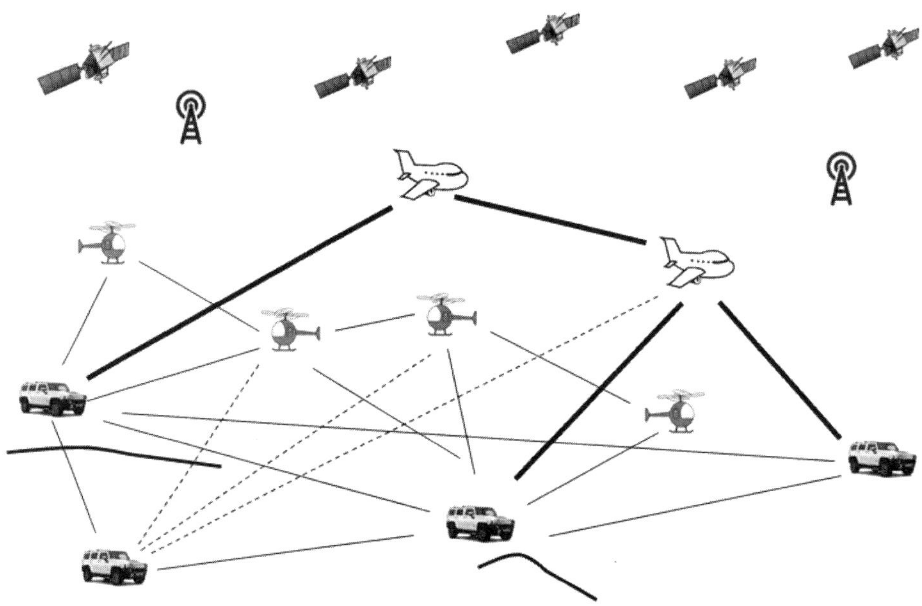

Figure 13.1 Network collaborative PNT.

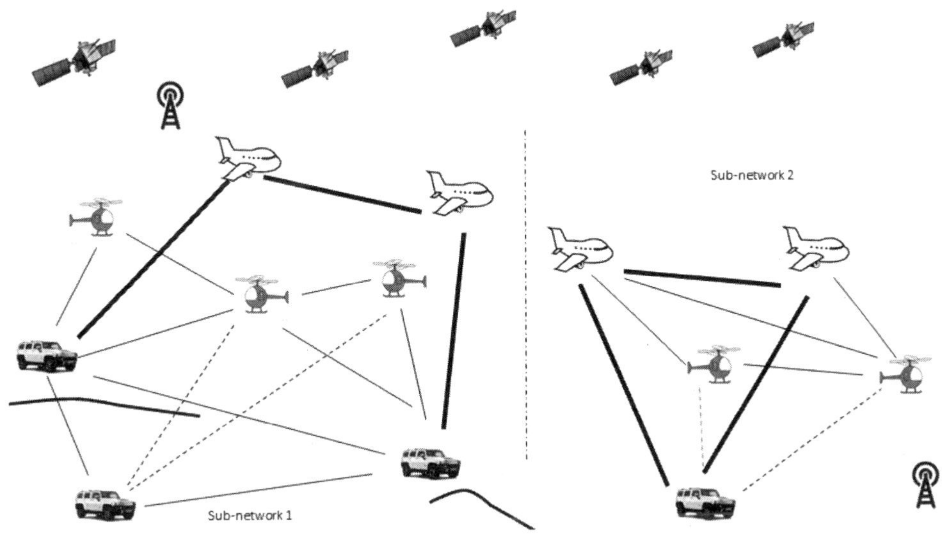

Figure 13.2 A network is decomposed into two independent subnetworks.

Figure 13.2 illustrates a network that is decomposed into two independent subnetworks.

13.2 Collaborative PNT Architecture

13.2.1 Generic Kalman Filter Architecture for a Network Collaborative Estimator

We will first lay the foundation for this chapter by discussing the generic Kalman filter for a collaborative network. The contents in this chapter are generally applicable to many other types of applications, not just PNT.

Let there be n nodes in a network or a subnetwork, and each of the nodes has the following equations as the model of states and measurements:

1. The local state model for node i:

$$x_i(k+1) = f_i\left(x_i(k), u_i(k), k\right) + w_i(k) \tag{13.1}$$

2. The local measurement model for node i:

$$z_i(k) = g_i\left(x_i(k), u_i(k), k\right) + v_i(k) \tag{13.2}$$

3. The collaboration measurement model between node i and node j as measured by node i:

$$y_{ij}(k) = h_{ij}\left(x_i(k), x_j(k), k\right) + s_{ij}(k) \tag{13.3}$$

where $w_i(k)$, $v_i(k)$, and $s_{ij}(k)$ are independent white noise sequences with

$$
\begin{aligned}
E\left(w_i(k)w_i^T(l)\right) &= Q_d(k)\delta_{kl} \\
E\left(v_i(k)v_i^T(l)\right) &= R_{vdi}(k)\delta_{kl} \\
E\left(s_{ij}(k)s_{ij}^T(l)\right) &= R_{sdij}(k)\delta_{kl}
\end{aligned} \tag{13.4}
$$

Note that there could be as many as $n(n - 1)$ collaborative measurements if each node attempts to generate a collaborative measurement by collaborating with all the other nodes.

There are two ways that we can build the corresponding estimator for the above system:

1. *The ITNS approach:* In this approach, all the models and measurements are combined together to form the model and measurements for the integrated system. A centralized estimator is built and the results can be distributed to each node. Since this approach uses all the information available in an integrated manner, we can imagine that the performance of this approach will be the best. The disadvantages of this approach are that it requires the most amount of communication and computational resources. However, it is important to note that even for this integrated approach, a significant amount of computations, for example, the state propagation, can be done locally and only the propagated states have to be communicated instead of all the models and input data, reducing the need for communication and centralized computation.
2. *The LWIN approach:* In this approach, an estimator is developed for each node using the local model and local measurements together with the collaborative measurements with its immediate neighbors. This approach is much more practical in real implementations and practices and typically the performance will not be significantly worse. With that said, it is always possible to construct a pathological example demonstrating that the ITNS approach can provide a much better performance than the LWIN approach.

13.2.2 Algorithm and Equations for ITNS Extended Kalman Filter

As we discussed before, the ITNS is a centralized approach. The Kalman filter built using this approach also requires a significant amount of centralized computation. Even so, there are still some parts of the computation, as will be seen below, that can be done in a distributed manner:

1. State propagation can be done completely locally for each node i ($i = 1, 2, \ldots, n$)

$$
\hat{x}_i(k|k-1) = f_i\left(\hat{x}_i(k-1|k-1), u_i(k-1), k-1\right) \tag{13.5}
$$

2. The state transition matrix for the entire system is block-diagonal

$$\Phi(k-1) = \begin{bmatrix} \Phi_{11}(k-1) & 0 & 0 & 0 \\ 0 & \Phi_{22}(k-1) & 0 & 0 \\ 0 & 0 & \ddots & 0 \\ 0 & 0 & 0 & \Phi_{nn}(k-1) \end{bmatrix} \quad (13.6)$$

where

$$\Phi_{ii}(k-1) \triangleq \frac{\partial f_i}{\partial x_i}_{|x=\hat{x}_i(k-1|k-1)} \triangleq f_{i,x_i|x_i=\hat{x}_i(k-1|k-1)} \quad (13.7)$$

The system noise covariance matrix will be block-diagonal as well

$$Q(k) = \begin{bmatrix} Q_1(k) & 0 & \cdots & 0 \\ 0 & Q_2(k) & \cdots & 0 \\ \vdots & \vdots & \ddots & \vdots \\ 0 & 0 & \cdots & Q_n(k) \end{bmatrix} \quad (13.8)$$

However, the covariance matrix will not be block-diagonal; consequently, we have the total covariance propagation equation for the system.

$$P(k|k-1) = \Phi(k-1)P(k-1|k-1)\Phi^T(k-1) + Q(k) \quad (13.9)$$

Obviously, if we expand this equation block by block, we will be able to save a significant amount of computations because the matrices Φ and Q are block diagonal.

3. The prediction of local measurement and the local residual can be done locally

$$\hat{z}_i(k|k-1) = g_i\left(\hat{x}_i(k|k-1), u_i(k-1), k\right) \quad (13.10)$$

and the local residual can be computed as

$$\begin{aligned} \mu_i &= \left(z_i(k) - \hat{z}_i(k|k-1)\right) \\ \mu &= \begin{bmatrix} \mu_1^T & \mu_2^T & \cdots & \mu_n^T \end{bmatrix}^T \end{aligned} \quad (13.11)$$

4. The H-matrix for the all the local measurements stacked together is block-diagonal:

$$H_z(k) \triangleq \frac{\partial g}{\partial x}_{|x=\hat{x}(k|k-1)} = \begin{bmatrix} \frac{\partial g_1}{\partial x_1} & 0 & \cdots & 0 \\ 0 & \frac{\partial g_2}{\partial x_2} & \cdots & 0 \\ \vdots & \vdots & \ddots & \vdots \\ 0 & 0 & 0 & \frac{\partial g_n}{\partial x_n} \end{bmatrix} \quad (13.12)$$

5. The R-matrix for local measurements $z(k)$ is also block-diagonal:

$$R_z(k) = \begin{bmatrix} R_{z1}(k) & 0 & \cdots & 0 \\ 0 & R_{z2}(k) & \cdots & 0 \\ \vdots & \vdots & \ddots & 0 \\ 0 & 0 & 0 & R_{zn}(k) \end{bmatrix} \tag{13.13}$$

The covariance matrix of the local residual vector (all the local residuals stacked together) can be calculated as

$$P_\mu = \begin{bmatrix} \left(\frac{\partial g_1}{\partial x_1}\right) P_{11} \left(\frac{\partial g_1}{\partial x_1}\right)^T + R_{z11} & \left(\frac{\partial g_1}{\partial x_1}\right) P_{12} \left(\frac{\partial g_2}{\partial x_2}\right)^T & \cdots & \left(\frac{\partial g_1}{\partial x_1}\right) P_{1n} \left(\frac{\partial g_n}{\partial x_n}\right)^T \\ \left(\frac{\partial g_2}{\partial x_2}\right) P_{21} \left(\frac{\partial g_1}{\partial x_1}\right)^T & \left(\frac{\partial g_2}{\partial x_2}\right) P_{22} \left(\frac{\partial g_2}{\partial x_2}\right)^T + R_{z22} & \cdots & \left(\frac{\partial g_2}{\partial x_2}\right) P_{2n} \left(\frac{\partial g_n}{\partial x_n}\right)^T \\ \vdots & \vdots & \ddots & \vdots \\ \left(\frac{\partial g_n}{\partial x_n}\right) P_{n1} \left(\frac{\partial g_1}{\partial x_1}\right)^T & \left(\frac{\partial g_n}{\partial x_n} P_{n2}\right) \left(\frac{\partial g_2}{\partial x_2}\right)^T & \cdots & \left(\frac{\partial g_n}{\partial x_n}\right) P_{nn} \left(\frac{\partial g_n}{\partial x_n}\right)^T + R_{znn} \end{bmatrix} \tag{13.14}$$

which, unfortunately, is not block-diagonal. The inverse of the above matrix is given by

$$B_\mu = P_\mu^{-1} \tag{13.15}$$

and the gain matrix will be

$$K_z(k) = P(k|k-1)H_z^T(k)B_\mu \tag{13.16}$$

which is normally a full matrix, despite the fact that the H-matrix is block diagonal.

6. The entire state vector is updated using

$$\hat{x}(k|k) = \hat{x}(k|k-1) + K_z\mu(k) \tag{13.17}$$

and the entire covariance matrix needs to be updated using

$$P(k|k) = \left(I - K_z(k)H_z(k)\right)P(k|k-1) \tag{13.18}$$

The collaborative measurement can be used at a different time moment, preferably right after a covariance propagation step. If so, we do not need to do anything in this step.

The collaborative measurements can also be used right after the local measurement update is performed. If so, we will need to copy the updated covariance matrix into the pre-update covariance matrix before we can use the new measurements in a sequential manner.

It is important to realize that for each of the node (i), there could be a number of collaborative measurements associated with different partners. Consequently, we can have up to $n(n-1)$ number of collaborative updates (compared with only n local updates). To avoid making the equations very complicated, we will loop through up to $n(n-1)$ measurements and in each iteration we only deal with one measurement equation, which is associated with one pair of nodes. The equation will be written for node i as measured by node i when collaborating with node j. It is important to remember, between each of the iterations, that if no covariance propagation is done, we always have to copy the current updated covariance matrix into the pre-update covariance matrix:

$$P(k|k-1) = P(k|k) \tag{13.19}$$

Steps 8 through 11 will need to be repeated for each of the nodes. For each of the nodes, these steps will be repeated for every one of the collaborative measurements. The two loops combined can lead to the repeated execution of these steps for up to $n(n-1)$ times.

7. Prediction of collaborative measurement can be calculated as

$$\hat{y}_{ij}(k|k-1) = h_{ij}\left(\hat{x}_i(k|k-1), \hat{x}_j(k|k-1), k\right) \tag{13.20}$$

and the residual can be computed as

$$\upsilon_{ij} = \left(y_{ij}(k) - \hat{y}_{ij}(k|k-1)\right) \tag{13.21}$$

8. The H-matrix for the collaborative measurement as measured by node i when collaborating with node j can be written as

$$\begin{aligned} H_{yij}(k) &= \left[\begin{array}{cccccccc} \cdots & 0 & \dfrac{\partial h_{ij}}{\partial x_i} & 0 & \cdots & 0 & \dfrac{\partial h_{ij}}{\partial x_j} & 0 & \cdots \end{array}\right] \\ &= \left[\begin{array}{cccccccc} \cdots & 0 & h_{yiji} & 0 & \cdots & 0 & h_{yijj} & 0 & \cdots \end{array}\right] \end{aligned} \tag{13.22}$$

9. The measurement variance matrix is assumed to be given by $R_{yij}(k)$. The covariance matrix of the residual can be calculated as

$$P_{\upsilon ij} = \left(h_{yiji}(k)P_{ii}(k|k-1)h^T_{yiii}(k) + h_{yijj}(k)P_{jj}(k|k-1)h^T_{yijj}(k) + R_{yij}(k)\right) \tag{13.23}$$

The inverse of the above matrix is calculated as

$$B_{\upsilon ij} = P_{\upsilon ij}^{-1} \tag{13.24}$$

and the gain matrix for the entire state vector using measurement y_{ij} will be

$$K_{yij}(k) = P(k|k-1)H^T_{yij}(k)B_{\upsilon ij} \tag{13.25}$$

10. The state and covariance matrix update are done using the following equations

$$\hat{x}(k|k) = \hat{x}(k|k-1) + K_{yij}\upsilon \tag{13.26}$$

$$P(k|k) = \left(I - K_{yij}(k)H_{yij}(k)\right)P(k|k-1) \tag{13.27}$$

Figure 13.3 illustrates the architecture of the ITNS approach.

13.2.3 Algorithm and Equations for the LWIN EKF

The LWIN approach is decentralized. The estimator consists of distributed local estimators. Each of them maintains its own state vector and covariance matrix. The local updates, in this case, are done completely locally. The collaborative measurement based updates are also done locally by using the best estimates of the collaborating partner's state in the collaborative measurement equation as a measurement with noises, instead of as part of the total system states. The estimates are typically received from the collaborating partners.

As will be seen in the rest of the descriptions, the EKF for the LWIN approach is basically n decoupled classical EKF. The following is done for each node i ($i = 1, 2, \ldots, n$) independently:

1. State propagation:

$$\hat{x}_i(k \mid k-1) = f_i\left(\hat{x}_i(k-1|k-1), u_i(k-1), k-1\right) \tag{13.28}$$

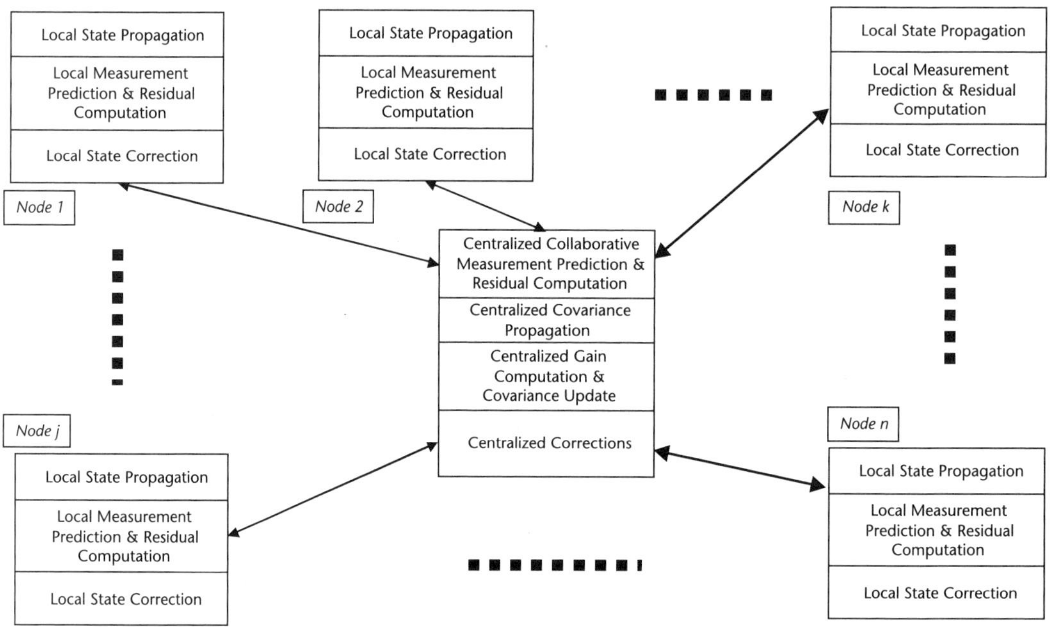

Figure 13.3 The ITNS approach.

2. The covariance propagation:
   The state transition matrix for node i is computed as

$$\Phi_{ii}(k-1) \triangleq \frac{\partial f_i}{\partial x_i}_{|x=\hat{x}_i(k-1|k-1)} \triangleq f_{i,x_i|x_i=\hat{x}_i(k-1|k-1)} \tag{13.29}$$

and the system noise covariance matrix is given as $Q_i(k)$. The covariance propagation for node i is done using:

$$P_i(k|k-1) = \Phi_i(k-1)P_i(k-1|k-1)\Phi_i^T(k-1) + Q_i(k) \tag{13.30}$$

3. The prediction of local measurements and the local residuals can also be done locally

$$\hat{z}_i(k|k-1) = g_i\left(\hat{x}_i(k|k-1), u_i(k-1), k\right) \tag{13.31}$$

and the local residual can be computed as

$$\mu_i = \left(z_i(k) - \hat{z}_i(k|k-1)\right) \tag{13.32}$$

4. The H-matrix for the local measurements is

$$H_{zi}(k) \triangleq \frac{\partial g_i}{\partial x_i}_{|x_i=\hat{x}_i(k|k-1)} \tag{13.33}$$

5. The R-matrix for the local measurements z_i is given by $R_{zi}(k)$. The covariance matrix of the local residual can be calculated as

$$P_{\mu_i} = \left(\frac{\partial g_i}{\partial x_i}\right)P_i\left(\frac{\partial g_i}{\partial x_i}\right)^T + R_{zi} \tag{13.34}$$

   The inverse of the above matrix is given by

$$B_{\mu i} = P_{\mu i}^{-1} \tag{13.35}$$

and the gain matrix will be

$$K_{zi}(k) = P_i(k|k-1)H_{zi}^T(k)B_{\mu i} \tag{13.36}$$

6. The local state vector is updated using

$$\hat{x}_i(k|k) = \hat{x}_i(k|k-1) + K_{zi}\mu_i(k) \tag{13.37}$$

and the local covariance matrix is updated using

$$P_i(k|k) = \left(I - K_{zi}(k)H_{zi}(k)\right)P_i(k|k-1) \tag{13.38}$$

7. The collaborative measurement can be used at a different time moment, preferably right after a covariance propagation step. If so, we do not need to do anything in this step.

The collaborative measurement can also be used right after the local measurement update. If so, we will need to copy the updated covariance matrix into the pre-update covariance matrix before we can use the new measurements in a sequential manner.

It is important to realize that, for each node (i), there could be 0 to (n – 1) of collaborative measurements associated with different partners. To avoid making the equations very complicated, we will loop through up to (n – 1) measurements for each node and in each iteration we only deal with one measurement equation. It is important to remember, between each of the iterations, that if no covariance propagation is done, we always have to copy the current updated covariance matrix into the pre-update covariance matrix:

$$P_i(k|k-1) = P_i(k|k) \tag{13.39}$$

8. The prediction of collaborative measurement can be calculated as

$$\hat{y}_{ij}(k|k-1) = h_{ij}\left(\hat{x}_i(k|k-1), \hat{x}_j(k|k-1), k\right) \tag{13.40}$$

and the residual can be computed as

$$\upsilon_{ij} = \left(y_{ij}(k) - \hat{y}_{ij}(k|k-1)\right) \tag{13.41}$$

9. The H-matrix for the collaboration measurement as measured by node i can be written as

$$H_{yij}(k) = \frac{\partial h_{ij}}{\partial x_i} \tag{13.42}$$

10. The measurement variance matrix is assumed to be given by $R_{yij}(k)$. Since we are now using the collaborating partner's state as measurements, the compound covariance matrix for the measurement will be given by

$$\bar{R}_{yij}(k) = R_{yij}(k) + \left(\frac{\partial h_{ij}}{\partial x_j}\right) P_j(k|k-1) \left(\frac{\partial h_{ij}}{\partial x_j}\right)^T \tag{13.43}$$

The covariance matrix of the residual can be calculated as

$$P_{\upsilon ij} = \left(H_{yij}(k) P_i(k|k-1) H_{yij}^T(k) + \left(\frac{\partial h_{ij}}{\partial x_j}\right) P_j(k|k-1) \left(\frac{\partial h_{ij}}{\partial x_j}\right)^T + R_{yij}(k)\right) \tag{13.44}$$

The inverse of the above matrix is calculated as

$$B_{\upsilon ij} = P_{\upsilon ij}^{-1} \tag{13.45}$$

and the gain matrix for the local estimator using measurement y_{ij} will be

$$K_{yij}(k) = P_i(k|k-1) H_{yij}^T(k) B_{\upsilon ij} \tag{13.46}$$

11. The state and covariance matrix update are done using the following equations:

$$\hat{x}_i(k|k) = \hat{x}_i(k|k-1) + K_{yij}\upsilon \tag{13.47}$$

$$P_i(k|k) = \left(I - K_{yij}(k)H_{yij}(k)\right)P_i(k|k-1) \tag{13.48}$$

As can be seen above, the key step that makes all the local estimators decoupled from each other is the use of the state of the collaborator as measurement, not as part of the state, and the use of its covariance matrix to create the compound measurement covariance matrix $\bar{R}_{yij}(k) = R_{yij}(k) + \left(\frac{\partial h_{ij}}{\partial x_j}\right)P_j(k|k-1)\left(\frac{\partial h_{ij}}{\partial x_j}\right)^T$ in step 10.

Figure 13.4 illustrates the LWIN approach.

13.2.4 Two Types of Collaborative Measurements

Wireless communication between collaborating partners can be achieved through an RF or an optical communication link. The communication links can help to achieve the following purposes:

1. Communication of useful information. From PNT point of view, a communication link will allow the collaborating partners to exchange information about the position, velocity, attitude, and time of each of the collaborating partners.

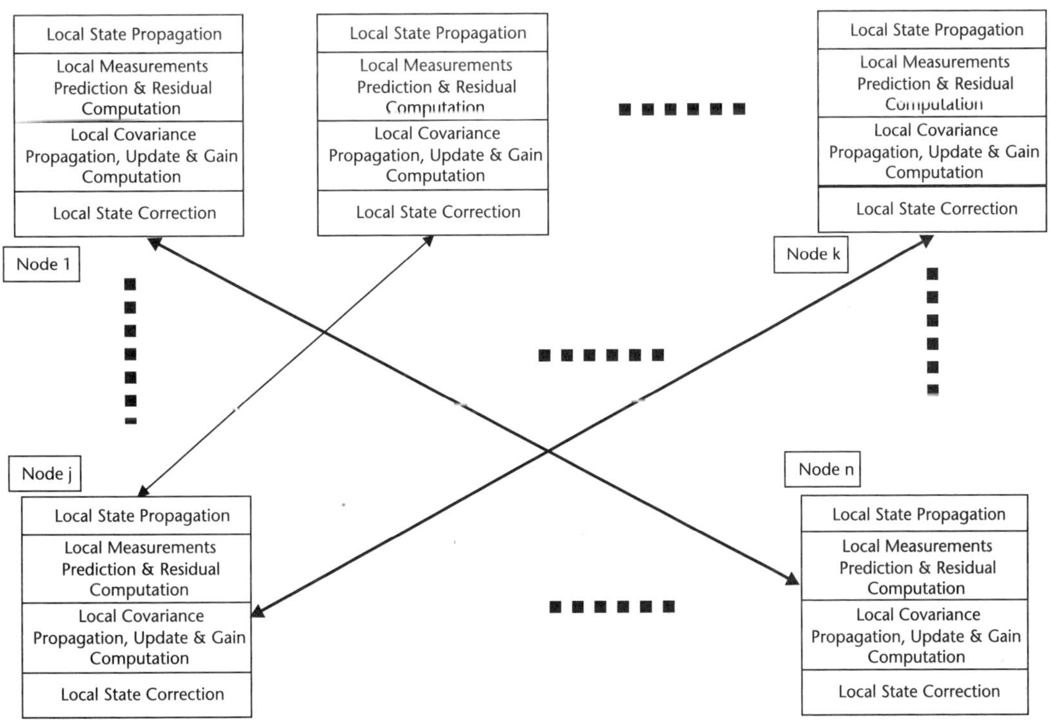

Figure 13.4 LWIN approach.

2. To create measurements of variables that can be used for PNT. Specifically, the following two types of measurements can be produced by many of the intervehicle communication links:
   a. TOA of signals that will lead to time transfer between the partners and measurement of the range between the two partners;
   b. AOA or DOA that are measured using phased-array antennas or using a precision gimbal system.

As discussed in Chapter 4, when the partners send each other a time mark together with the time stamp of transmission according to their own clock, through an RF link or an optical link, both the range between the two partners and the clock bias between the two clocks can be estimated using the two-way TOA measurements. This is the two-way time transfer.

Range measurements of a node with respect to three or more non-coplanar collaborators with known positions will allow the position of the node to be determined.

In general, the range measurements of a node with respect to collaborators with partial or full position knowledge will help the node to reduce the position knowledge uncertainty by taking advantage of the collaborating relationship.

The AOA and DOA combined with range measurements will allow a node to determine the position of another node directly without triangulation measurements with other nodes.

13.3 Collaborative Time Synchronization and NTP and PTP

One of the most important collaborations that can be achieved by a network of collaborators is the collaboration in time synchronization. Thanks to the large amount of developments and practices in computer network engineering, this is a well-developed field with a significant amount of existing technologies that can be leveraged. Specifically, NTP and PTP (IEEE1588) are open standard time synchronization protocols and methods that are widely adopted. Today, almost every computer runs NTP and synchronizes its time with time servers on the internet.

NTP typically achieves time synchronization accuracy from a few microseconds to a few milliseconds. PTP, which is similar to NTP, is designed to achieve higher accuracy up to nanosecond or even picosecond levels. The actual performance depends on network topology, network bandwidth, traffic, and the hardware support of the network interface card. NTP and PTP are very similar. The primary differences are:

1. PTP uses hardware time-stamping to achieve higher accuracy of time stamps. In comparison, NTP has to wait for the operating system to provide a software-based time stamp and consequently suffers from more latency and less accuracy.
2. PTP attempts to compensate for delays introduced by the network interface hardware.

Regardless of the actual network topology, both NTP and PTP treat time resources in the network in a hierarchy and each level of the hierarchy is called a stratum. Stratum 0 devices serve as the master source of the time reference. They are typically high-precision timekeeping devices such as atomic clocks or GPS receivers. Stratum 1 devices are synchronized to stratum 0 devices typically through a direct-network connection that exhibits stable delays caused by network traffic. A time server at stratum 1 will provide services to stratum 2 devices. A stratum n device will provide services to stratum $n + 1$ devices.

Figure 13.5 illustrates the strata of NTP and PTP.

Devices on the network can talk to multiple time servers at a stratum above to achieve performance improvements by statistical averaging. It can also talk to its peers in the same stratum to perform cross-checking to achieve more stable and robust time synchronization.

NTP and PTP use the same algorithm as the two-way time transfer described in Chapter 4. The only difference is that the delay due to the distance traveled by the RF or optical signals will be replaced by multihop network delays between the master and the slave.

Figure 13.6 illustrates the two messages and the four time stamps involved in a two-way time transfer process.

Let the offset between the two time references be given by Δt, that is,

$$^{R2}t = {}^{R1}t + \Delta \mathrm{t}$$

Let $^{R1}t_{11}$ be the time when the master sends out a time stamp message according to its own clock, and $^{R2}t_{12}$ be the time when the slave receives the time stamp message according to the slave's clock.

Clearly, we have

$$^{R2}t_{12} = {}^{R1}t_{11} + \Delta t + d$$

where d is the time delay to transfer the message from the master to slave and vice versa.

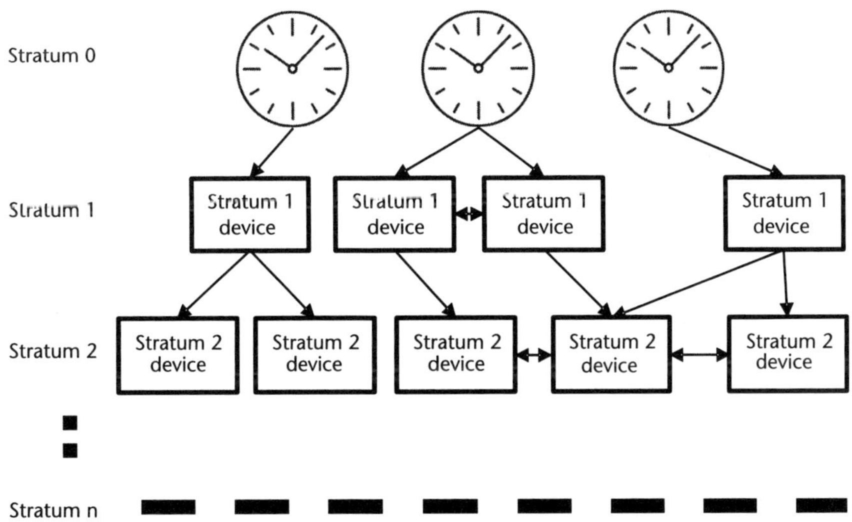

Figure 13.5 The strata of NTP and PTP.

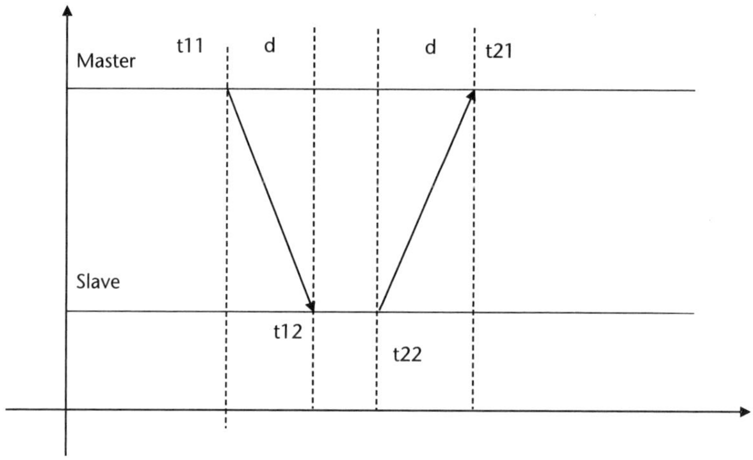

Figure 13.6 Two messages and four time stamps involved in a two-way time transfer process.

We can write similar equations for the time stamp message sent from the slave to the master.

Let ${}^{R2}t_{22}$ be the time when the slave sends out a time stamp message according to its own clock and ${}^{R1}t_{21}$ be the time when the master receives the time stamp message according to its own clock. We have

$$ {}^{R1}t_{21} = {}^{R2}t_{22} - \Delta t + d $$

The above two equations can be written as

$$ \begin{aligned} \Delta t + d &= \left({}^{R2}t_{12} - {}^{R1}t_{11}\right) \\ \Delta t - d &= \left({}^{R2}t_{22} - {}^{R1}t_{21}\right) \end{aligned} $$

The right side of the above two equations can be called pseudo time delays. They are the time delays of the master to salve message and the slave to master message, respectively, with the exception that the time stamps are not using the same reference.

Solving the above equations, we have

$$ \begin{aligned} \Delta t &= \frac{\left({}^{R2}t_{12} - {}^{R1}t_{11}\right) + \left({}^{R2}t_{22} - {}^{R1}t_{21}\right)}{2} \\ d &= \frac{\left({}^{R2}t_{12} - {}^{R1}t_{11}\right) - \left({}^{R2}t_{22} - {}^{R1}t_{21}\right)}{2} \end{aligned} \tag{13.49} $$

and the errors are characterized by

$$ \begin{aligned} \sigma(\Delta t) &= \frac{\sqrt{\sigma^2\left({}^{R2}t_{12}\right) + \sigma^2\left({}^{R1}t_{11}\right) + \sigma^2\left({}^{R2}t_{22}\right) + \sigma^2\left({}^{R1}t_{21}\right)}}{2} \\ \sigma(d) &= \sigma(\Delta t) \end{aligned} \tag{13.50} $$

Over a short period of time, the time offset between the master and slave can be considered as constant and if we send multiple pairs of messages and derive a measurement of the time offset from each pair of messages, we will be able to average these measurements to achieve improved performance.

We obviously can be even more sophisticated by modeling the time offset, its changing rate, and the other errors and using a Kalman filter or a fixed gain filter to estimate the states to achieve even better performance.

In addition, as mentioned before, a node in the network can be talking to multiple time servers and achieve better performance, availability, and integrity. Figure 13.7 illustrates this paradigm.

It is very important that the two-way communication between the master and the slave is symmetric (i.e., the paths and traffic delays from the master to slave and from the slave to master are such that the delays are the same). As one can imagine, the communication path and traffic asymmetry will be the primary error sources for both NTP and PTP-based time synchronization.

When the communication path becomes a direct RF or optical link between the two nodes, we have the best chance for symmetry and, in this case, the delay and the range are related by

$$r = c \cdot d$$

We can derive the range measurement through this process.

13.4 Collaboration with LOS, AOA, and DOA and Range Measurements

This is the highest level of collaboration in terms of richness of information. With range and LOS direction information, we can derive the position of a collaboration partner directly.

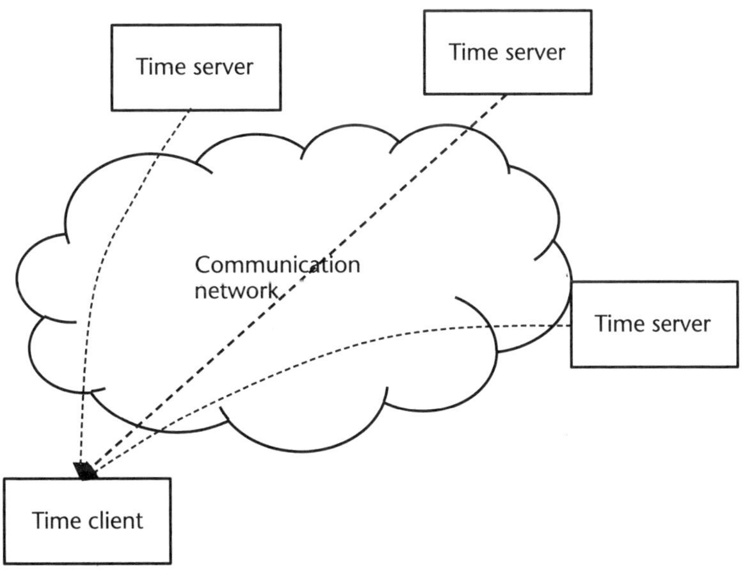

Figure 13.7 Using multiple time servers on the network to achieve better performance, availability, and integrity.

Given node i and node j with their positions in the ECEF frame given by ${}^{E}R_i$, ${}^{E}R_j$, let ${}^{B_i}u_{ji}$ be the unit vector pointing from node i to node j, measured by node i in its body frame B_i and let ${}^{B_j}u_{ij}$ be the unit vector pointing from node j to node i measured by node j in its body frame B_j.

We can write the following two equations

$$ {}^{E}R_j - {}^{E}R_i = C_{B_i}^{E}\left(r_{ji}\,{}^{B_i}u_{ji}\right) \tag{13.51} $$

$$ {}^{E}R_i - {}^{E}R_j = C_{B_j}^{E}\left(r_{ij}\,{}^{B_j}u_{ij}\right) \tag{13.52} $$

The above equations can be used in several different ways as shown next.

13.4.1 LWIN Equations

When taking an LWIN approach, we will either be node i-centric or node j-centric.

Specifically, if the position of node j is unknown, we can immediately derive its position using

$$ {}^{E}R_j = {}^{E}R_i + C_{B_i}^{E}\left(r_{ji}\,{}^{B_i}u_{ji}\right) \tag{13.53} $$

If we are node i-centric, and node j is a collaborator that has communicated to node i its position, we can use the measurement generated by node i to estimate the attitude and position of node i:

$$ z_{iij} = r^{B_i}u_{ji} = C_{E}^{B_i}\left({}^{E}R_j - {}^{E}R_i\right) + v_{iij} \tag{13.54} $$

which can be viewed as measurements of the attitude, as represented by $C_E^{B_i}$ (Earth to body frame DCM), and the position, as represented by ${}^{E}R_i$ (position expressed in the Earth frame) of node i. The noise term v_{iij} lumps together all the measurement errors.

The subscript of the measurement z_{iij} can be read as: the measurement for node i, using measurement generated by node i, through the collaboration of node i with node j.

Similarly, node j can send its measurement, position, and attitude to node i and allow it to form the following measurement equation based on (13.52):

$$ z_{iji} = r^{B_j}u_{ij} + C_{E}^{B_j}\,{}^{E}R_j = C_{E}^{B_j}\,{}^{E}R_i + v_{iji} \tag{13.55} $$

which can be used as the measurement of the position vector of node i as represented by ${}^{E}R_i$.

The noise term v_{iji} lumps together all measurement errors.

The subscript of the measurement z_{iji} can be read as: the measurement for node i, using measurement generated by node j, through collaboration of node j with node i.

Similarly, we can write the same type of two equations symmetrically for node j, which will produce the equations for $z_{jji}$and z_{jij}.

The above views of the collaboration measurement are for the LWIN approach.

13.4.2 ITNS Equations

For the ITNS approach, for each pair of collaborative partners, we can write two equations:

$$z_{ij} = r^{B_i} u_{ji} = C_E^{Bi}\left({}^E R_j - {}^E R_i\right) + v_{ij} \tag{13.56}$$

and

$$z_{ji} = r^{B_j} u_{ij} = C_E^{B_j}\left({}^E R_i - {}^E R_j\right) + v_{ji} \tag{13.57}$$

Equation (13.56) provides observations for node i attitude, node i position, and node j position. Equation (13.57) provides observations for node j attitude, node j position, and node i position.

13.4.3 Linearized Equations

For EKF, and for error analysis, we will want to derive the linearized models for the measurement equations.

First, we can write

$$\delta z_{ij} = \delta\left(r^{B_i} u_{ji}\right) = \delta C_E^{Bi}\left({}^E \hat{R}_j - {}^E \hat{R}_i\right) + \hat{C}_E^{Bi}\left(\delta\, {}^E R_j - \delta\, {}^E R_i\right) + v_{ij}$$

Note that

$$\delta C_E^{Bi} = C_E^{Bi} - C_{\hat{E}}^{Bi} = C_{\hat{E}}^{Bi} C_E^{\hat{E}} - C_{\hat{E}}^{Bi} = C_{\hat{E}}^{Bi}\left(C_E^{\hat{E}} - I\right)$$

and

$$C_E^{\hat{E}} = \left(I - \left[{}^E\theta_{iE}^{\hat{E}} \times\right]\right) = \left(I + \left[{}^E\theta_{i\hat{E}}^{E} \times\right]\right)$$

The precise meaning of symbols used above is

$${}^E\theta_{i\hat{E}}^{E} = \left({}^E\theta_{B_i}^{E} - {}^E\hat{\theta}_{B_i}^{E}\right) = \delta\, {}^E\theta_{B_i}^{E}$$

We then have

$$\delta C_E^{Bi} = C_{\hat{E}}^{Bi}\left(C_E^{\hat{E}} - I\right) = C_{\hat{E}}^{Bi}\left[{}^E\theta_{i\hat{E}}^{E} \times\right] = \hat{C}_E^{Bi}\left[\delta\, {}^E\theta_{B_i}^{E} \times\right]$$

and

$$\delta C_E^{Bi}\left({}^E \hat{R}_j - {}^E \hat{R}_i\right) = C_{\hat{E}}^{Bi}\left[\delta\, {}^E\theta_{B_i}^{E} \times\right]\left({}^E \hat{R}_j - {}^E \hat{R}_i\right) = -\hat{C}_E^{Bi}\left[\left({}^E \hat{R}_j - {}^E \hat{R}_i\right) \times\right]\delta\, {}^E\theta_{B_i}^{E}$$

So we end up with

$$\delta z_{ij} = z_{ij} - \hat{z}_{ij} = \hat{C}_E^{B_i}\delta^E R_j - \hat{C}_E^{B_i}\delta^E R_i - \hat{C}_{\hat{E}}^{Bi}\left[\left({}^E\hat{R}_j - {}^E\hat{R}_i\right)\times\right]\delta^E\theta_{B_i}^E + \upsilon_{ij} \tag{13.58}$$

Similarly, we have

$$\delta z_{ji} = z_{ji} - \hat{z}_{ji} = \hat{C}_E^{B_j}\delta^E R_i - \hat{C}_E^{B_j}\delta^E R_j - \hat{C}_{\hat{E}}^{B_j}\left[\left({}^E\hat{R}_i - {}^E\hat{R}_j\right)\times\right]\delta^E\theta_{B_j}^E + \upsilon_{ji} \tag{13.59}$$

For the ITNS approach, the above equations are directly used.
For the LWIN approach, the equations are viewed for node i as

Measurement by node i and for node i:

$$\delta z_{iij} \equiv \delta z_{ij} - \hat{C}_E^{B_i}\delta^E R_j = -\hat{C}_E^{B_i}\delta^E R_i - \hat{C}_{\hat{E}}^{Bi}\left[\left({}^E\hat{R}_j - {}^E\hat{R}_i\right)\times\right]\delta^E\theta_{B_i}^E + \upsilon_{iij} \tag{13.60}$$

where $\upsilon_{iij} = \upsilon_{ij} - \hat{C}_E^{B_i}\delta^E R_j$

υ_{iij} is the combination of measurement error and the knowledge error of the position of node j.

Measurement by node j and for node i

$$\delta z_{iji} \equiv \delta z_{ji} + \hat{C}_E^{B_j}\delta^E R_j + \hat{C}_{\hat{E}}^{B_j}\left[\left({}^E\hat{R}_i - {}^E\hat{R}_j\right)\times\right]\delta^E\theta_{B_j}^E = \hat{C}_E^{B_j}\delta^E R_i + \upsilon_{iji} \tag{13.61}$$

where

$$\upsilon_{iji} = \upsilon_{ji} + \hat{C}_E^{B_j}\delta^E R_j + \hat{C}_{\hat{E}}^{B_j}\left[\left({}^E\hat{R}_i - {}^E\hat{R}_j\right)\times\right]\delta^E\theta_{B_j}^E$$

υ_{iji} is the combination of measurement error and the knowledge error of the position and attitude of node j.

13.4.4 Integration with Inertial Navigation: The ITNS Approach

We will illustrate how this is done for a simple network that consists of only two nodes, node i and node j, and this can be easily extended to bigger systems.

Without repeating the same equations from Chapter 3, let us abstract the INS's nonlinear algorithms as

$$\dot{X}_I = f\left(X_I\right) + W_I$$

where the state vector of inertial navigation is defined by

$$X_I^T = \left[\left(q_B^E\right)^T, \left({}^Bb_G\right)^T, \left({}^Gs\right)^T, \left({}^B\beta_G\right)^T, \left({}^B\theta_G\right)^T, \left({}^E\upsilon_E^B\right)^T, \left({}^Bb_A\right)^T, \left({}^As\right)^T, \left({}^B\beta_A\right)^T, \left({}^Er_E^B\right)^T\right] \tag{13.62}$$

and the noise process that drives the nonlinear system is defined by

$$W_I^T = \left[-\left(C_G^E ARW\right)^T, RRW^T, SRW_G^T, \beta RW_G^T, \theta RW_G^T, \left(C_A^E VRW\right)^T, ABRW^T, SRW_A^T, \beta RW_A,^T 0 \right] \tag{13.63}$$

All symbols are defined in Chapter 3.

The linearized error model is given by

$$\delta \dot{X}_I = A_{II} \delta X_I + W_I \tag{13.64}$$

where

$$\delta X_I^T = \left[\delta^E \dot{\theta}_B^E, \delta^B b_G^T, \delta^G s^T, \delta^B \beta_G^T, \delta^B \theta_G^T, \left(\delta^E v_E^B\right)^T, \delta^B b_A^T, \delta^A s^T, \delta^B \beta_A^T, \left(\delta^E r_E^B\right)^T \right] \tag{13.65}$$

and

$$A_{II} = \begin{bmatrix} -[^E\omega_I^E \times] & -C_B^E & -C_G^E \mathrm{diag}(^G\omega_I^B) & -C_B^E J(^B\omega_I^B) & C_B^E[^B\omega_I^B \times] & 0_{3x3} & 0_{3x3} & 0_{3x3} & 0_{3x3} & 0_{3x3} \\ 0_{3x3} & 0_{3x3} & 0_{3x3} & 0_{3x3} & 0_{3x3} & 0_{3x3} & 0_{3x3} & 0_{3x3} & 0_{3x3} & 0_{3x3} \\ 0_{3x3} & 0_{3x3} & 0_{3x3} & 0_{3x3} & 0_{3x3} & 0_{3x3} & 0_{3x3} & 0_{3x3} & 0_{3x3} & 0_{3x3} \\ 0_{3x3} & 0_{3x3} & 0_{3x3} & 0_{3x3} & 0_{3x3} & 0_{3x3} & 0_{3x3} & 0_{3x3} & 0_{3x3} & 0_{3x3} \\ 0_{3x3} & 0_{3x3} & 0_{3x3} & 0_{3x3} & 0_{3x3} & 0_{3x3} & 0_{3x3} & 0_{3x3} & 0_{3x3} & 0_{3x3} \\ [(C_B^E m^B f_I^B)\times] & 0_{3x3} & 0_{3x3} & 0_{3x3} & 0_{3x3} & -2[(^E\omega_I^E)\times] & C_B^E & C_A^E \mathrm{diag}(^A f_I^B) & C_B^E J(^B f_I^B) & \left[\dfrac{\partial^E g}{\partial^E r_E^B}\right] \\ 0_{3x3} & 0_{3x3} & 0_{3x3} & 0_{3x3} & 0_{3x3} & 0_{3x3} & 0_{3x3} & 0_{3x3} & 0_{3x3} & 0_{3x3} \\ 0_{3x3} & 0_{3x3} & 0_{3x3} & 0_{3x3} & 0_{3x3} & 0_{3x3} & 0_{3x3} & 0_{3x3} & 0_{3x3} & 0_{3x3} \\ 0_{3x3} & 0_{3x3} & 0_{3x3} & 0_{3x3} & 0_{3x3} & 0_{3x3} & 0_{3x3} & 0_{3x3} & 0_{3x3} & 0_{3x3} \\ 0_{3x3} & 0_{3x3} & 0_{3x3} & 0_{3x3} & 0_{3x3} & I_{3x3} & 0_{3x3} & 0_{3x3} & 0_{3x3} & 0_{3x3} \end{bmatrix} \tag{13.66}$$

In the ITNS approach, there will be two instances of inertial navigation algorithm running:

$$\dot{X}_{Ii} = f\left(X_{Ii}\right) + W_{Ii} \text{ and } \dot{X}_{Ij} = f\left(X_{Ij}\right) + W_{Ij}$$

The state vector of the system is the combination of the two state vectors and is given by

$$X_I = \begin{bmatrix} X_{Ii} \\ X_{Ij} \end{bmatrix} \tag{13.67}$$

The linearized model state vector is then given by

$$\delta X_I = \begin{bmatrix} \delta X_{Ii} \\ \delta X_{Ij} \end{bmatrix} \tag{13.68}$$

The linearized error dynamic model is provided by

$$\delta \dot{X}_I = \begin{bmatrix} \delta \dot{X}_{Ii} \\ \delta \dot{X}_{Ij} \end{bmatrix} = \begin{bmatrix} A_{IIi} & 0 \\ 0 & A_{IIj} \end{bmatrix} \begin{bmatrix} \delta \dot{X}_{Ii} \\ \delta \dot{X}_{Ij} \end{bmatrix} + \begin{bmatrix} W_{Ii} \\ W_{Ij} \end{bmatrix} \tag{13.69}$$

The discrete version is given by

$$\begin{bmatrix} \delta X_{Ii}(k+1) \\ \delta X_{Ij}(k+1) \end{bmatrix} = \begin{bmatrix} \Phi_{Ii}(k+1,k) & 0 \\ 0 & \Phi_{Ii}(k+1,k)_{Ij} \end{bmatrix} \begin{bmatrix} \delta \dot{X}_{Ii} \\ \delta \dot{X}_{Ij} \end{bmatrix} + \begin{bmatrix} W_{dIi} \\ W_{dIj} \end{bmatrix} \tag{13.70}$$

where $\Phi_{Ii}(k+1,k)$ and $\Phi_{Ij}(k+1,k)$ are state transition matrices derived from A_{IIi} and A_{IIj} and W_{dIi} and W_{dIj} are the discrete noise processes derived from W_{Ii} and W_{Ij}.

This equation is the basis for the Kalman filter covariance propagation.

Although both the Φ-matrix and the Q-matrix are block-diagonal, the covariance matrix P is not, and the covariance propagation will need to be done for the whole system.

The nonlinear measurement equations for the two types of measurements will be

$$z_{ij} = r^{B_i} u_{ij} = C_E^{Bi}\left({}^E R_j - {}^E R_i\right) + v_{ij} \tag{13.71}$$

and

$$z_{ji} = r^{B_j} u_{ji} = C_E^{B_j}\left({}^E R_i - {}^E R_j\right) + v_{ji} \tag{13.72}$$

The linearized of measurement error equation will be

$$\delta z_{ij} = z_{ij} - \hat{z}_{ij} = \hat{C}_E^{B_i} \delta\, {}^E R_j - \hat{C}_E^{B_i} \delta\, {}^E R_i - \hat{C}_{\hat{E}}^{Bi}\left[\left({}^E\hat{R}_j - {}^E\hat{R}_i\right)\times\right]\delta\, {}^E\theta_{B_i}^E + v_{ij} \tag{13.73}$$

and

$$\delta z_{ji} = z_{ji} - \hat{z}_{ji} = \hat{C}_E^{B_j} \delta\, {}^E R_i - \hat{C}_E^{B_j} \delta\, {}^E R_j - \hat{C}_{\hat{E}}^{B_j}\left(\left({}^E\hat{R}_i - {}^E\hat{R}_j\right)\times\right)\delta\, {}^E\theta_{B_j}^E + v_{ji} \tag{13.74}$$

Consequently, we have the following H-matrices corresponding to each of the measurement equations above.

$$H_{zij} = \begin{bmatrix} -\hat{C}_{\hat{E}}^{Bi}\left[\left({}^E\hat{R}_j - {}^E\hat{R}_i\right)\times\right] & 0 & \cdots & 0 & -\hat{C}_E^{B_i} & 0 & \cdots & 0 & \hat{C}_E^{B_i} \end{bmatrix} \tag{13.75}$$

The nonzero elements are corresponding to node i attitude error, node i position error, and node j position error.

$$H_{zji} = \begin{bmatrix} 0 & \cdots & 0 & \hat{C}_E^{B_j} & -\hat{C}_{\hat{E}}^{Bi}\left[\left({}^E\hat{R}_j - {}^E\hat{R}_i\right)\times\right] & 0 & \cdots & 0 & -\hat{C}_E^{B_i} \end{bmatrix} \tag{13.76}$$

The nonzero elements are corresponding to node i position error, node j attitude error, and node j position error.

13.4.5 Integration with Inertial Navigation: The LWIN Approach

In this case, we only need to describe one of the local Kalman filters.

The inertial navigation algorithm will be given by

$$\dot{X}_{Ii} = f\left(X_{Ii}\right) + W_{Ii} \tag{13.77}$$

and the linearized error equation is given by

$$\delta\dot{X}_{Ii} = A_{IIi}\delta X_{Ii} + W_{Ii} \tag{13.78}$$

The measurement equations are given by

$$z_{iij} = r^{B_i}u_{ji} = C_E^{B_i}\left({}^E R_j - {}^E R_i\right) + v_{iij} \tag{13.79}$$

and

$$z_{iji} = r^{B_i}u_{ij} + C_E^{B_i}\,{}^E R_j = C_E^{B_j}\,{}^E R_i + v_{iji} \tag{13.80}$$

The linearized equations are given by

$$\delta z_{iij} = z_{iij} - \hat{z}_{iij} = \delta z_{ij} - \hat{C}_E^{B_i}\delta\,{}^E R_j = -\hat{C}_E^{B_i}\delta\,{}^E R_i - \hat{C}_{\hat{E}}^{Bi}\left[\left({}^E\hat{R}_j - {}^E\hat{R}_i\right)\times\right]\delta\,{}^E\theta_{B_i}^E + v_{iij} \tag{13.81}$$

and

$$\delta z_{iji} = z_{iji} - \hat{z}_{iji} + \hat{C}_E^{B_i}\delta\,{}^E R_j + \hat{C}_{\hat{E}}^{B_i}\left[\left({}^E\hat{R}_i - {}^E\hat{R}_j\right)\times\right]\delta\,{}^E\theta_{B_j}^E = \hat{C}_E^{B_i}\delta\,{}^E R_i + v_{iji} \tag{13.82}$$

The corresponding H-matrices are

$$H_{ziij} = \left[\begin{array}{cccc} -\hat{C}_{\hat{E}}^{Bi}\left[\left({}^E\hat{R}_j - {}^E\hat{R}_i\right)\times\right] & 0 \ \cdots \ 0 & -\hat{C}_E^{B_i} \end{array}\right] \tag{13.83}$$

and

$$H_{ziji} = \left[\begin{array}{ccc} 0 & \cdots & 0 \ \ \hat{C}_E^{B_i} \end{array}\right] \tag{13.84}$$

13.4.6 Simulation Example

As shown in Figure 13.8, two vehicles are equipped with the GPS and INS together with cross-communication links that are able to measure the LOS vector and the range. Both vehicles are flying through areas where GPS is available and not available. Without collaboration, each vehicle will suffer a period of bad navigation performance. With collaboration, good performance can be maintained throughout the process. We will look at three types of performances:

1. GPS and INS with no collaboration;
2. GPS and INS with LWIN collaboration;
3. GPS and INS with ITNS collaboration.

Figures 13.9 and 13.10 illustrate the navigation performance for the two vehicles with GPS and INS only. Navigation errors for an instance of Monte Carlo simulation

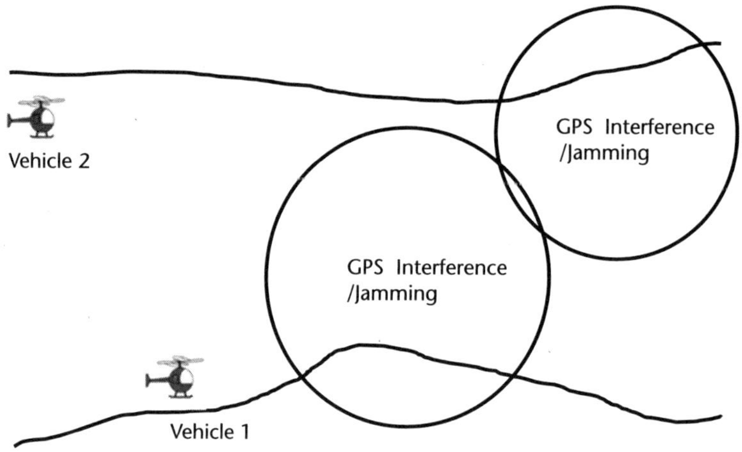

Figure 13.8 Two vehicles traveling through areas with GPS unavailability.

are plotted with the 3 sigma upper and lower bounds as predicted by the Kalman filter covariance matrix. For vehicle 1, the GPS is denied 5 minutes from the start of the run and the GPS becomes available again 10 minutes after. For vehicle 2, the GPS is denied starting at 15 minutes from the start of the run and the GPS becomes available again 10 minutes after.

Figures 13.11 and 13.12 show the performance of vehicle 1 and vehicle 2 with GPS and INS and the LWIN collaboration (LOS and range) updates.

Figures 13.13 and 13.14 show the performance of vehicle 1 and vehicle 2 when we use the ITNS approach. A single Kalman filter with twice as many states is used. As can be seen, the performance is only slightly better than that of the LWIN approach.

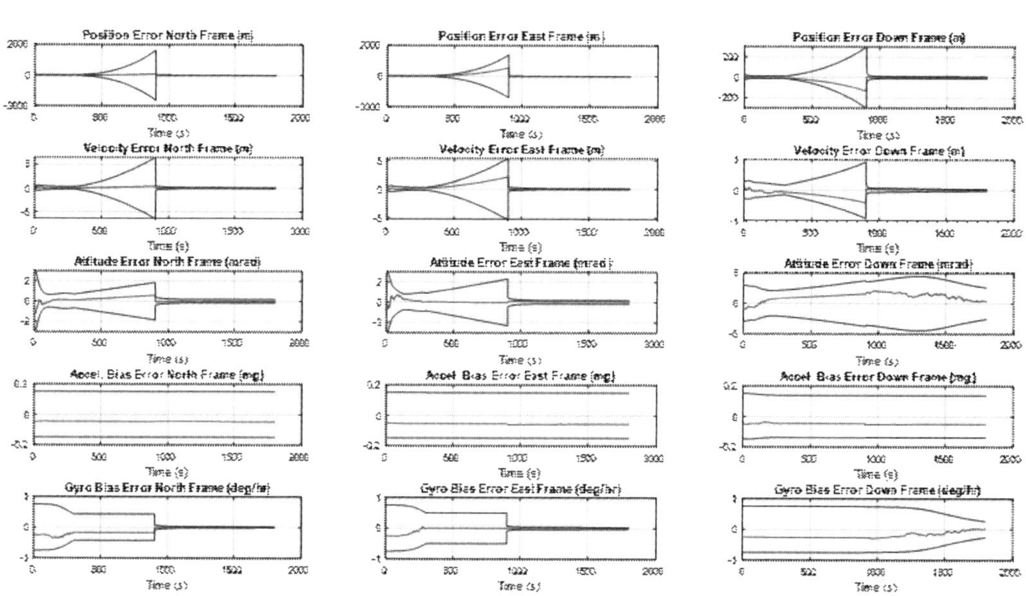

Figure 13.9 Vehicle 1 performance with GPS and INS only.

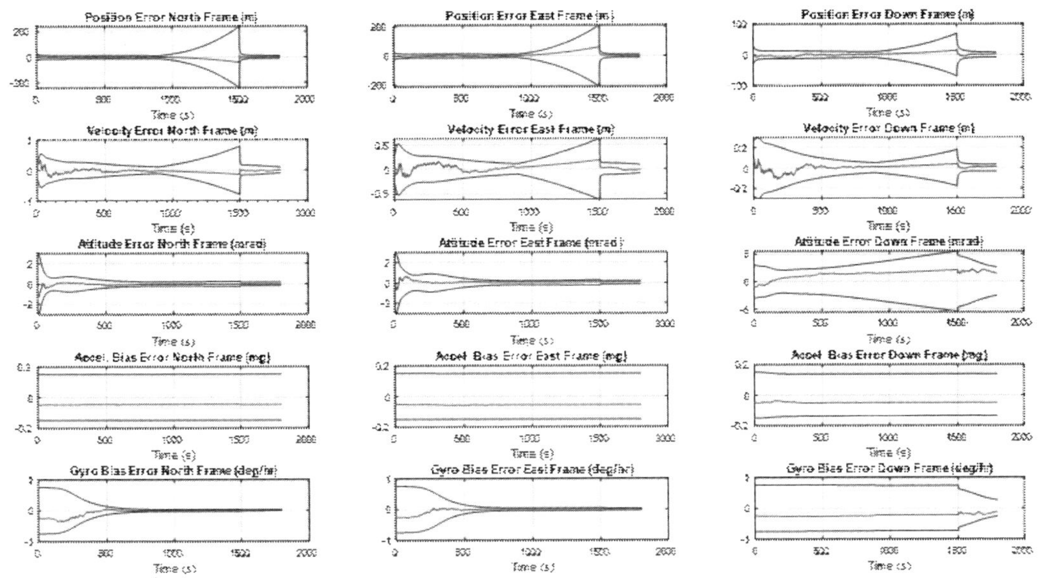

Figure 13.10 Vehicle 2 performance with GPS and INS only.

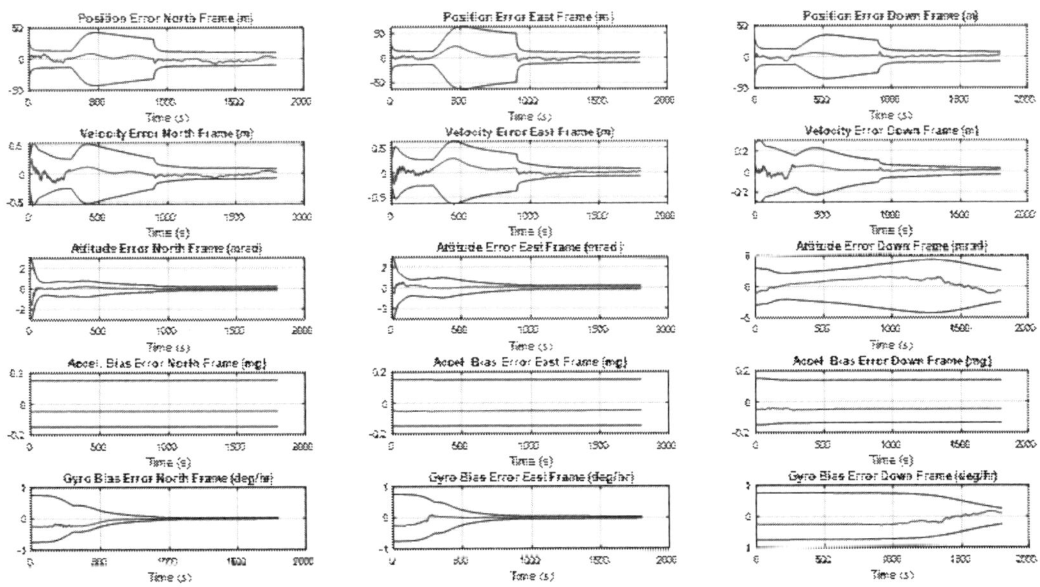

Figure 13.11 Vehicle 1 performance with the GPS and INS and the LWIN collaboration.

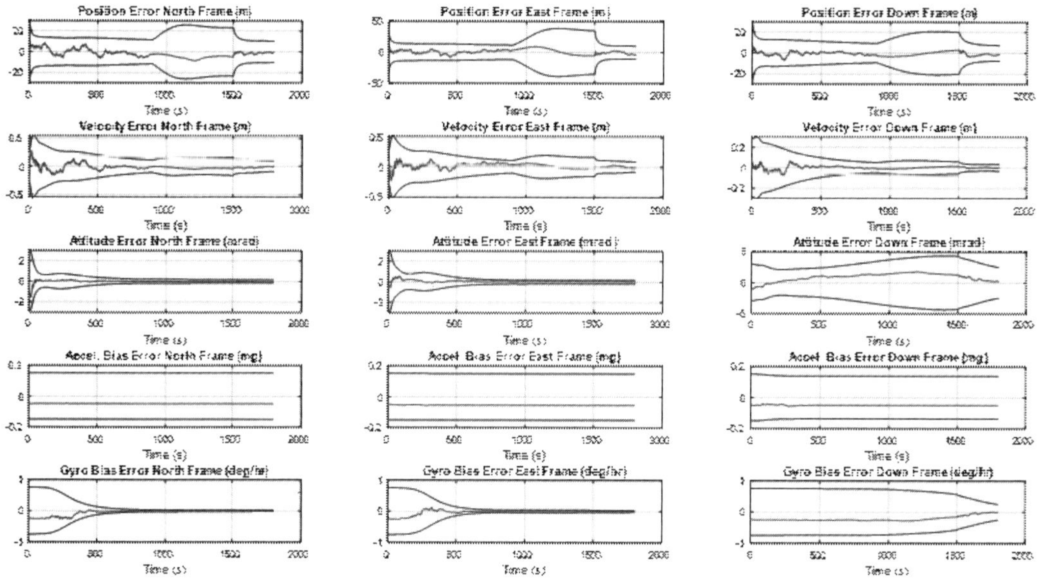

Figure 13.12 Vehicle 2 performance with the GPS and INS and the LWIN collaboration.

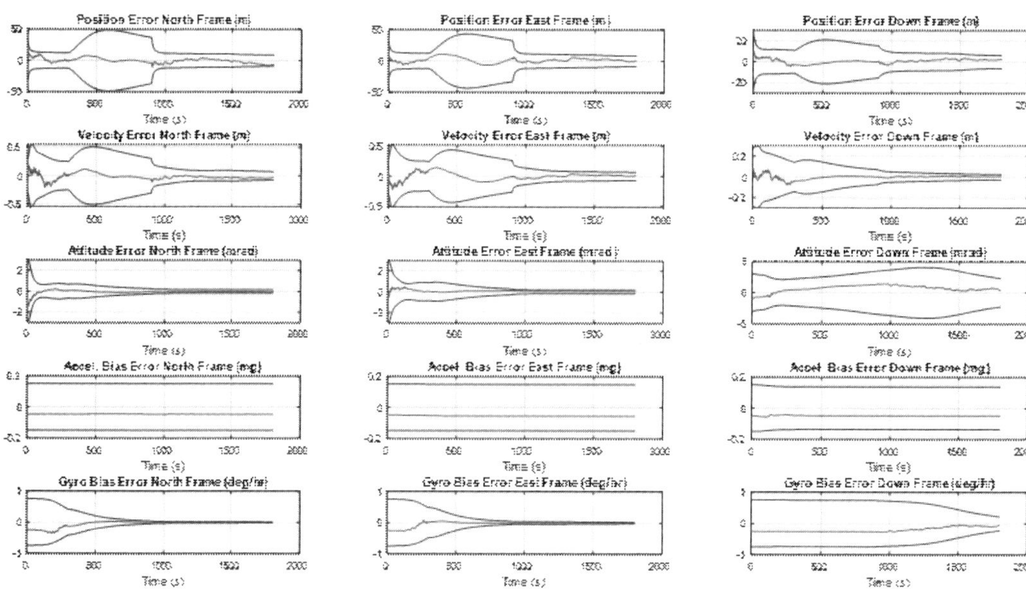

Figure 13.13 Vehicle 1 performance with the GPS and INS and the ITNS collaboration.

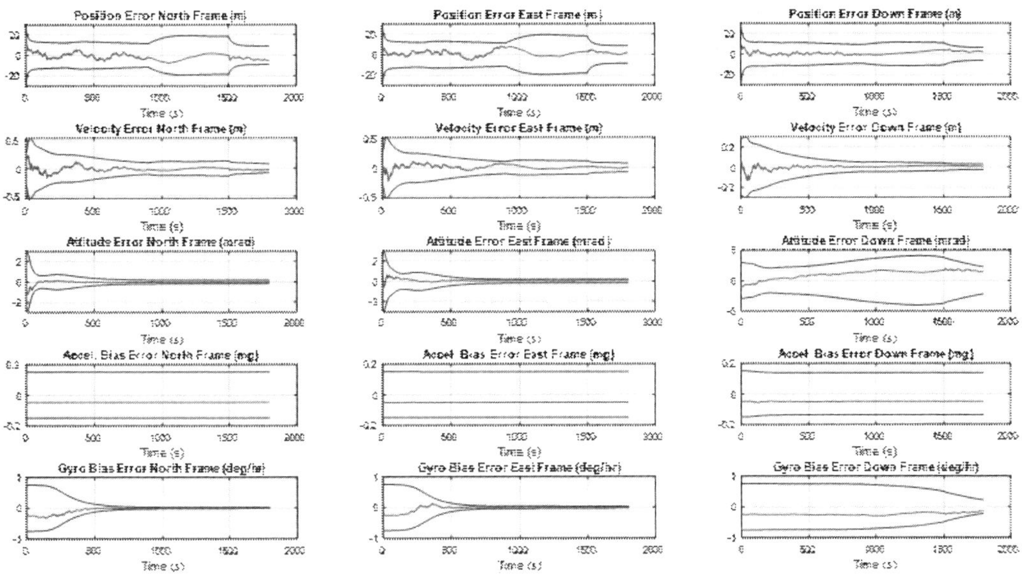

Figure 13.14 Vehicle 2 performance with the GPS and INS and the ITNS collaboration.

13.5 Collaboration with Range Measurements Only

13.5.1 Nonlinear and Linearized Measurement Equations

Given node i and node j with their positions in the ECEF frame given by ${}^{E}R_i$, ${}^{E}R_j$, the range measurement is simply given by

$$d_{ij} = \left\| {}^{E}R_i - {}^{E}R_j \right\| + v_{ij} \tag{13.85}$$

This simple equations can be used in three different ways:

1. If ${}^{E}R_j$ is assumed to be known, then the above equation provides some information about ${}^{E}R_i$. When this equation is viewed this way, we are taking an LWIN approach for node i.
2. Similarly, if ${}^{E}R_i$ is assumed to be known, then the above equation provides some information about ${}^{E}R_j$. When this equation is viewed this way, we are taking an LWIN approach for node j.
3. If we take an ITNS approach, both ${}^{E}R_i$, ${}^{E}R_j$ are part of the state vector for the total system and, consequently, this equation will be viewed as a measurement equation for both ${}^{E}R_i$ and ${}^{E}R_j$.

For both the least square solution and the Kalman filter design, we would like to linearize the above measurement equation (and as we have done many times in this book), we have

$$\delta d_{ij} = d_{ij} - \hat{d}_{ij} = \frac{\left(\hat{R}_i - {}^{E}\hat{R}_j\right)^T}{\left\| {}^{E}\hat{R}_i - {}^{E}\hat{R}_j \right\|}\left(\delta R_i - \delta R_j\right) + v_{ij} \tag{13.86}$$

If we define

$$^{E}u_{ij} = \frac{\left({}^{E}R_i - {}^{E}R_j\right)}{\left\|{}^{E}R_i - {}^{E}R_j\right\|} \tag{13.87}$$

which is the unit vector pointing from node j to node i.

The linearized equation can then be written as

$$\delta d_{ij} = {}^{E}\hat{u}_{ij}^{T}\delta\,{}^{E}R_i - {}^{E}\hat{u}_{ij}^{T}\delta\,{}^{E}R_j + v_{ij} \tag{13.88}$$

For the ITNS approach, the above equations are directly used.

For the LWIN approach, the equations are viewed for node i as

$$\delta d_{iij} \equiv \delta d_{ij} + {}^{E}\hat{u}_{ij}^{T}\delta\,{}^{E}R_j = {}^{E}\hat{u}_{ij}^{T}\delta\,{}^{E}R_i + v_{iij} \tag{13.89}$$

where $v_{iij} = v_{ij} + {}^{E}u_{ij}^{T}\delta\,{}^{E}R_j$.

v_{iij} is the combination of measurement error and the knowledge error of the position of node j.

Similarly, the equation can viewed for node j as

$$\delta d_{jij} \equiv \delta d_{ij} - {}^{E}\hat{u}_{ij}^{T}\delta\,{}^{E}R_i = -{}^{E}\hat{u}_{ij}^{T}\delta\,{}^{E}R_j + v_{jij} \tag{13.90}$$

where $v_{jij} = v_{ij} - {}^{E}\hat{u}_{ij}^{T}\delta\,{}^{E}R_i$.

v_{jij} is the combination of the measurement error and the knowledge error of the position of node i.

13.5.2 The ITNS Approach

We will illustrate how this is done for a simple network that consists of only two nodes, node i and node j, and this can be easily extended to bigger systems.

For the same setup of the INS for node i and node j as in the last section, in the ITNS approach, there will be two instances of inertial navigation algorithm running:

$$\dot{X}_{Ii} = f\left(X_{Ii}\right) + W_{Ii} \text{ and } \dot{X}_{Ij} = f\left(X_{Ij}\right) + W_{Ij}$$

The state vector of the system is the combination of two state vectors and is given by

$$X_I = \begin{bmatrix} X_{Ii} \\ X_{Ij} \end{bmatrix} \tag{13.91}$$

The linearized model state vector is then given by

$$\delta X_I = \begin{bmatrix} \delta X_{Ii} \\ \delta X_{Ij} \end{bmatrix} \tag{13.92}$$

The linearized error dynamic model is

$$\delta \dot{X}_I = \begin{bmatrix} \delta \dot{X}_{Ii} \\ \delta \dot{X}_{Ij} \end{bmatrix} = \begin{bmatrix} A_{IIi} & 0 \\ 0 & A_{IIj} \end{bmatrix} \begin{bmatrix} \delta \dot{X}_{Ii} \\ \delta \dot{X}_{Ij} \end{bmatrix} + \begin{bmatrix} W_{Ii} \\ W_{Ij} \end{bmatrix} \tag{13.93}$$

The discrete version is given by

$$\begin{bmatrix} \delta X_{Ii}(k+1) \\ \delta X_{Ij}(k+1) \end{bmatrix} = \begin{bmatrix} \Phi_{Ii}(k+1,k) & 0 \\ 0 & \Phi_{Ii}(k+1,k)_{Ij} \end{bmatrix} \begin{bmatrix} \delta \dot{X}_{Ii} \\ \delta \dot{X}_{Ij} \end{bmatrix} + \begin{bmatrix} W_{dIi} \\ W_{dIj} \end{bmatrix} \tag{13.94}$$

where $\Phi_{Ii}(k+1,k)$ and $\Phi_{Ij}(k+1,k)$ are state transition matrices derived from A_{IIi} and A_{IIj} and W_{dIi} and W_{dIj} are the discrete noise processes derived from W_{Ii} and W_{Ij}.

This equation is the basis for the Kalman filter covariance propagation.

Although both the Φ-matrix and the Q-matrix are block-diagonal, the covariance matrix P is not, and the covariance propagation will need to be done for the whole system.

The nonlinear measurement equation for the range measurement between i and j is

$$d_{ij} = \left\| {}^E R_i - {}^E R_j \right\| + v_{ij} \tag{13.95}$$

The linearized measurement error equation will be

$$\delta d_{ij} = {}^E\hat{u}_{ij}^T \delta\, {}^E R_i - {}^E\hat{u}_{ij}^T \delta\, {}^E R_j + v_{ij} \tag{13.96}$$

Consequently, we have the following H-matrix

$$H_{zij} = \begin{bmatrix} 0 & \cdots & 0 & \hat{u}_{ij}^T & 0 & \cdots & 0 & -\hat{u}_{ij}^T \end{bmatrix} \tag{13.97}$$

The nonzero elements are corresponding to node i position error and node j position error.

13.5.3 The LWIN Approach

In this case, we only need to describe one of the local Kalman filters.

The inertial navigation algorithm will be given by

$$\dot{X}_{Ii} = f\left(X_{Ii}\right) + W_{Ii} \tag{13.98}$$

and the linearized error equation is given by

$$\delta \dot{X}_{Ii} = A_{IIi} \delta X_{Ii} + W_{Ii} \tag{13.99}$$

The measurement equations are given by

$$d_{ij} = \left\| {}^E R_i - {}^E R_j \right\| + v_{ij} \tag{13.100}$$

The linearized equations are given by

$$\delta d_{iij} \equiv \delta d_{ij} + {}^{E}\hat{u}_{ij}^{T}\delta {}^{E}R_{j} = {}^{E}\hat{u}_{ij}^{T}\delta {}^{E}R_{i} + v_{iij} \tag{13.101}$$

The corresponding H-matrix is simply

$$H_{ziij} = \begin{bmatrix} 0 & \cdots & 0 & {}^{E}\hat{u}_{ij}^{T} \end{bmatrix} \tag{13.102}$$

13.5.4 Range-Only Least Square Solution

First, let us restate the following important conclusions from Chapter 4:

1. In a 3-D space, a network of range-only collaborators can be anchored by three non-coplanar nodes. In other words, if the geometric shape and size of the network are already defined by the relative distances between the nodes, the position of every node in the network will be determined if the positions of at least three non-coplanar nodes are determined. For a 2-D space, this is true if the positions of two of the nodes are determined.
2. If a node has known ranges to three non-coplanar nodes with known positions, the position of the node can be determined in 3-D space. Only two nodes are required if in 2-D space.

Based on these two principles, we can first determine the position of tier 1 nodes, which are defined to have connections (which means having a known range) to at least three anchoring nodes (i.e., nodes that have known positions without using the position and/or range knowledge of other nodes).

Once all the tier 1 nodes have determined positions, we can continue to determine the positions of tier 2 nodes, which are defined to be the nodes that do not have enough connections to the anchoring nodes but have at least three connections to either anchoring nodes or tier 1 nodes combined.

Once all the tier 2 nodes have determined positions, we can continue to tier 3 nodes, which are defined to be nodes that do not have sufficient connections to anchoring nodes and tier 1 nodes but have at least three connections to tier 2, or tier 1, or anchoring nodes combined.

We can keep doing this recursive process to tier 4, tier 5, and so on until all the positions that can be determined are determined.

As an example, Figure 13.15 illustrates the anchoring nodes, tier 1, tier 2, tier 3, and tier 4 nodes of a network in 3-D space (i.e., three connections are required to determine the position of a node). In this example, nodes 1 to 3 are anchoring nodes. Node 5 has direct connections to the three anchoring nodes (i.e., having range information available for each of the three nodes) and it is therefore a tier 1 node. Nodes 4 and 6, each only having two connections to anchoring nodes, must rely on node 5, a tier 1 node, to determine its position. Consequently, they are tier 2 nodes. Node 8 has to depend on two tier 2 nodes (nodes 4 and 5) and one tier 1 node (node 5) to determine its position. Consequently, node 8 is a tier 3 node.

Nodes 7 and 9 must use a combination of tier 1, tier 2, and tier 3 (node 8) nodes to determine their positions, and they are at tier 4. Obviously, the three-connection criterion drops to two if we are constrained to 2-D space.

If we follow the above approach, we will be able to determine the positions of all the nodes that have sufficient connections. During each of the steps, we only solve for the position of one node at a time. In other words, we do not need to solve the stacked-together equations. We only need to solve equations with three unknowns at a time. This approach is fundamentally the LWIN approach.

When there is no noise and no computational errors, there will be just one accurate solution for the set of equations that we need to solve. In this case, the LWIN approach (solving for the position of one node at a time) will provide the same solution as the ITNS approach (solving for all the unknown positions of the nodes at the same time).

When there are noises and computational errors (as always) and there is redundant information available, then ITNS approach, in which we will stack all the equations and constraints together, will be able to produce a better performance.

Specifically, for every range measurement, we have the following nonlinear equation

$$d_{ij} = \left\| {}^{E}R_i - {}^{E}R_j \right\| + v_{ij}$$

With an initial guess of the solution, the above equation can be linearized to be

$$\delta d_{ij} = {}^{E}\hat{u}_{ij}^{T} \delta {}^{E}R_i - {}^{E}\hat{u}_{ij}^{T} \delta {}^{E}R_j + v_{ij}$$

This equation is the basic equation for the ITNS approach.

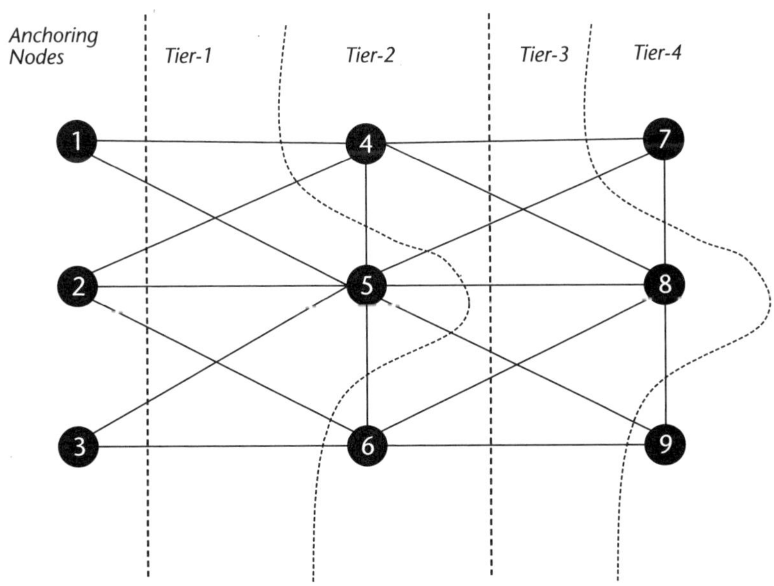

Figure 13.15 Example of anchoring nodes, tier 1, tier 2, tier 3, and tier 4 nodes.

13.5.5 ITNS Approach

We will assume that that we have n nodes with $3n$ unknowns. Without a loss of generality, we also assume that the first l nodes are anchoring nodes. This means that we can write the following 1 vector equations:

$$^{E}R_i = \bar{R}_i \ (i = 1,2,\ldots,l) \tag{13.103}$$

If we have total m connections between the nodes in the network, we will have m equations of the following form

$$^{E}\hat{u}_{ij}^{T}\delta{}^{E}R_i - {}^{E}\hat{u}_{ij}^{T}\delta{}^{E}R_j = \delta d_{ij} - v_{ij} \tag{13.104}$$

The total number of scalar equations will then be

$$3l + m$$

Obviously, in order to come up with a solution, we must have

$$3l + m > 3n \tag{13.105}$$

Solving the above set of equations with proper weighting based on measurement noise and initial knowledge uncertainty, we will be able to provide an optimal solution. This approach is fundamentally the ITNS approach.

13.5.6 LWIN Approach

The problem becomes much easier if we take the LWIN approach and solve the problem recursively. At each recursion, we only solve for one unknown position vector.

Let the unknown position vector be given by R_i and then for every connection of node i to a node with known position, we can write the following equation

$$^{E}\hat{u}_{ij_k}^{T}\delta{}^{E}R_i - {}^{E}\hat{u}_{ij_k}^{T}\delta{}^{E}R_{j_k} = \delta d_{ij_k} - v_{ij_k} \quad j_k = j_k, j_2,\ldots,j_q \tag{13.106}$$

which can be written as

$$^{E}\hat{u}_{ij_k}^{T}\delta{}^{E}R_i = \delta d_{ij_k} + {}^{E}\hat{u}_{ij_k}^{T}\delta{}^{E}R_{j_k} - v_{ij_k} \triangleq \delta\bar{d}_{i_k j_k} + w_{ij_k} \quad j_k = j_1, j_2,\ldots,j_q \tag{13.107}$$

Stacking the equations together, we have

$$\begin{bmatrix} ^{E}\hat{u}_{ij_1}^{T} \\ ^{E}\hat{u}_{ij_2}^{T} \\ \vdots \\ ^{E}\hat{u}_{ij_q}^{T} \end{bmatrix} \delta{}^{E}R_i = \begin{bmatrix} \delta\bar{d}_{i_k j_1} \\ \delta\bar{d}_{i_k j_2} \\ \vdots \\ \delta\bar{d}_{i_k j_q} \end{bmatrix} + \begin{bmatrix} w_{ij_1} \\ w_{ij_2} \\ \vdots \\ w_{ij_{1q}} \end{bmatrix} \tag{13.108}$$

This is an easy least square problem that we can solve for the solution.

References

[1] Gao, C., G. Zhao, and H. Fourati, (eds.), *Cooperative Localization and Navigation: Theory, Research, and Practice*, Boca Raton, FL: CRC Press, 2020.

[2] Sivaneri, V. O., and J. N. Gross, "Cooperative Navigation Between a Ground Vehicle and an Unmanned Aerial Vehicle in GNSS-Challenged Environments," *Computer Science*, 2016.

[3] Hardy, J., et al., "Unmanned Aerial Vehicle Relative Navigation in GPS Denied Environments," *2016 IEEE/ION Position, Location and Navigation Symposium (PLANS)*, 2016, pp. 344–352.

[4] Kealy, A., et al., "Collaborative Navigation as a Solution for PNT Applications in GNSS Challenged Environments—Report on Field Trials of a Joint FIG/IAG Working Group," *Journal of Applied Geodesy*, Vol. 9, No. 4, 2015, pp. 244–263.

[5] Vetrella, A. R., et al., "Cooperative Navigation in GPS-Challenging Environments Exploiting Position Broadcast and Vision-Based Tracking," *2016 International Conference on Unmanned Aircraft Systems (ICUAS)*, Arlington, VA, 2016, pp. 447–456.

[6] Ko, H., B. Kim, and S. -H. Kong, "GNSS Multipath-Resistant Cooperative Navigation in Urban Vehicular Networks," *IEEE Transactions on Vehicular Technology*, Vol. 64, No. 12, 2015, pp. 5450–5463.

[7] Gao, Y., et al., "UWB/GNSS-Based Cooperative Positioning Method for V2X Applications," *Proc. of 27th International Technical Meeting of the ION Satellite Division*, 2014, pp. 3212–3221.

[8] Ducatelle, F., et al., "Cooperative Navigation in Robotic Swarms," *Swarm Intell.*, Vol. 8, November 23, 2013, pp. 1–33.

[9] Yao, Y., "Cooperative Navigation System for Multiple Unmanned Underwater Vehicles," *IFAC Proceedings*, Vol. 46, No. 20, 2013, pp. 719–723.

[10] Lee, J. K., et al., "Network-Based Collaborative Navigation in GPS-Denied Environment," *The Journal of Navigation*, Vol. 65, 2012, pp. 445–457.

[11] Fankhauser, P., "Collaborative Navigation for Flying and Walking Robots," *2016 IEEE/RSJ International Conference on Intelligent Robots and Systems (IROS)*, Daejeon, Korea, October 9–14, 2016.

[12] Berefelt, F., et al., "Collaborative GPS/INS Navigation in Urban Environment," *Proc. of the 2004 National Technical Meeting of The Institute of Navigation*, San Diego, CA, 2004.

[13] Grejner-Brzezinska, D. A., et al., "Positioning in GPS-Challenged Environments: Dynamic Sensor Network with Distributed GPS Aperture and Inter-Nodal Ranging Signals," *Proc. of 22nd International Technical Meeting of the Satellite Division of the Institute of Navigation (ION GNSS 2009)*, Savannah, GA, 2009.

[14] Kaba, J., and S. Wu, "Distributed Multi-Sensor Fusion for Improved Collaborative GPS-Denied Navigation," ION International Technical Meeting, Anaheim, Ca, January 2009.

[15] Sanderson, A. C. "A Distributed Algorithm for Cooperative Navigation Among Multiple Mobile Robots," *Advanced Robotics*, Vol. 12, 1997, pp. 335–349.

[16] Brown, A., and J. Nordlie, "Integrated GPS/TOA Navigation Using a Positioning and Communication Software Defined Radio," *Proc. of PLANS 2006*, San Diego, CA, 2006.

[17] Whelan, D., G. Gutt, and P. Enge, "Boeing Timing & Location: An Indoor Capable Time Transfer and Geolocation System," *Proc. of 5th Stanford University, Symposium on Position, Navigation, and Timing*, Menlo Park, CA, 2011.

[18] Tan, Z., et al., "New Method for Positioning Using IRIDIUM Satellite Signals of Opportunity," *IEEE Access*, Vol. 7, June 24, 2019, pp. 83412–83423.

[19] Chowdhary, G., et al., "GPS-Denied Indoor and Outdoor Monocular Vision Aided Navigation and Control of Unmanned Aircraft," *Journal of Field Robotics*, Vol. 30, No. 3, 2013, pp. 415–438.

[20] https://en.wikipedia.org/wiki/Network_Time_Protocol.
[21] https://en.wikipedia.org/wiki/Precision_Time_Protocol.
[22] Eidson, J., "IEEE 1588 Standard Version 2—A Tutorial," National Institute of Standards and Technology (NIST), 2006, https://web.archive.org/web/20100331083017/http://ieee1588.nist.gov/2006%20IEEE1588%20Agenda/Eidson_1588_Version%202.pdf.

CHAPTER 14

Signal of Opportunity PNT

14.1 Introduction

Radio-based GNSS systems are almost perfect solutions for PNT. GNSS systems provide very low-cost and very low SWAP user equipment (e.g., the GNSS chip inside our cell phone probably costs less than $1 for each phone). GNSS services are widely available, and the performance is typically very good.

The only issues with GNSS are that GNSS can be easily jammed due to its low signal power, the civil GNSS signals can be easily spoofed, and, in addition, the GNSS infrastructure can be the target of attack during the war and consequently become unavailable.

Most of the advantages (and disadvantages) of GNSS systems are because GNSS is an RF-based system. The signal of opportunity (SOOP)-based PNT [1–9], also RF-based, attempts to provide PNT services by leveraging the large amount of RF signals that already exist in our environment.

SOOP has the hope to come close to the affordability, the availability, and even the performance of GNSS and, in the meantime, provides the opportunity to be less vulnerable to jamming and attacks by taking advantages of the much higher power of most of the signals and the greatly increased diversity.

The first major advantage of SOOP is that there are large amount of radio sources available, and, consequently, there are opportunities for greatly improved availability by using many of these sources, and for improved integrity by cross-checking between these sources. Specifically:

1. There are about 27,000 TV stations in the world and the number is growing.
2. There are about 44,000 radio stations in the world and the number is growing.
3. There are more than 4 million cell phone towers and other type of communication system towers in the world, and the number is growing rapidly.
4. There are at least 2,000 communication satellites orbiting the Earth, and many more constellations are being planned and deployed for the future. It is worth mentioning that the SpaceX's Starlink constellation being developed will undoubtedly have a major impact.
5. There are billions of WiFi, Bluetooth, and cell phones generating RF signals.

The second major advantage of SOOP is that most of these signals are significantly more powerful than GPS. Table 14.1 summarizes some of the typical power levels as received by the user:

Table 14.1 Typical Power Levels as Received by the User of Various SOOP sources

	Nominal Power at Receiver (dBm)	*Power Ratio to GPS*
GPS	−128	1.E+00
Typical LEO mobile phone (e.g., Iridium)	−98	1.E+03
Typical satellite radio (e.g., Sirius/XM)	−86	2.E+04
TV	−45	2.E+08
Radio station	−50	6.E+07
Cell phone	−61	5.E+06
WiFi	−64	3.E+06

It would be very hard to jam, spoof, or destroy all these signal sources, especially those with very high power. However, it would be also very difficult to be able to receive and use all of these signals efficiently. In any case, there are a significant amount of opportunities to leverage these signals and it will take some careful planning and development of the total system architecture and efficient and low-cost receivers to bring this technology to practical use.

There are many different ways how the SOOP signals can be used. The following is a list of typical types of measurements that can be generated by a SOOP signal receiver and can be used for PNT purposes.

1. Received signal strength (RSS);
2. Angle of arrival (AOA);
3. Time of arrival (TOA);
4. Frequency of arrival (FOA);
5. Time difference of arrival (TDOA);
6. Frequency difference of arrival (FDOA).

14.1.1 Received Signal Strength (RSS)

The RSS is one of the simplest but also lowest-quality measurements that we can use from a SOOP receiver. The RSS can be transformed to become a measurement of range (typically with large errors). With RSS measurements from three or more non-coplanar sources with known locations, we can determine the 3-D position of the receiver.

Mathematically, we already solved the range-based positioning problem in Chapter 4.

Figure 14.1 illustrates the concept of RSS-based positioning.

14.1.2 AOA

The AOA or, more correctly, the direction of arrival (DOA), is another type of measurement that can be produced by a SOOP receiver.

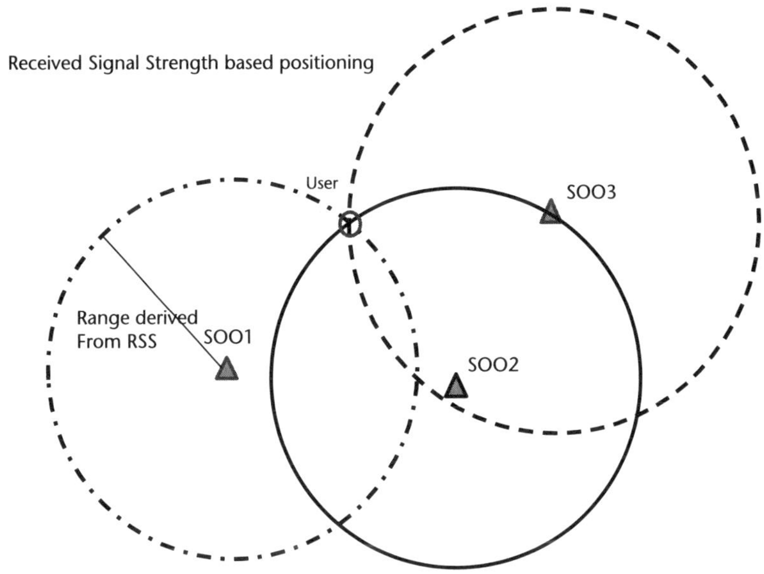

Figure 14.1 RSS-based positioning.

AOA typically requires phased-array antennas to determine the direction along which the signal is coming from with respect to a sensor reference frame. For each signal source, there are two angles or a unit vector that can be measured.

There are at least three scenarios how AOA can be used:

1. If the sensor frame to the Earth attitude is known, then AOA measurements from two or more sources of known locations should be able to determine the position of the receiver.
2. If the position of the receiver is known, the AOA measurements from two or more sources of known locations should be able to determine the attitude of the receiver.
3. If neither position nor the attitude is known, AOA measurements from three or more sources will be required to determine both the position and the attitude.

Mathematically, we already solved each of the above three problems in Chapters 4, 7, and 10.

Figure 14.2 illustrates the concept of one scenario of AOA-based positioning.

14.1.3 TOA

TOA is more correctly called pseudo-TOA (PTOA) since, in reality, most of the TOA measurements are based on at least two time references, the time reference of the sources and the time reference of user. Position and time can be determined if all the sources have the same time reference and this is exactly the same problem of the GPS-based PNT. PTOA is exactly the same thing as pseudo-range.

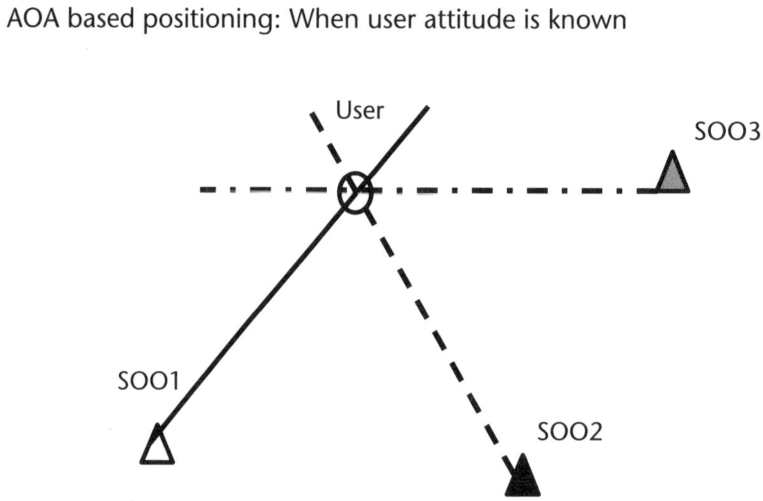

Figure 14.2 One scenario of AOA-based positioning.

There are a number of satellite systems that do use synchronized time between satellites; for example, both the Iridium constellation and the Global Star constellation use time references that are synchronized across the satellites. Consequently, it is possible to use, for example, message preambles, as a time mark, and measure the TOA of the time mark as the PTOA.

Once we have the measurements, the solution approach is the same as GPS. Figure 14.3 illustrates PTOA-based positioning.

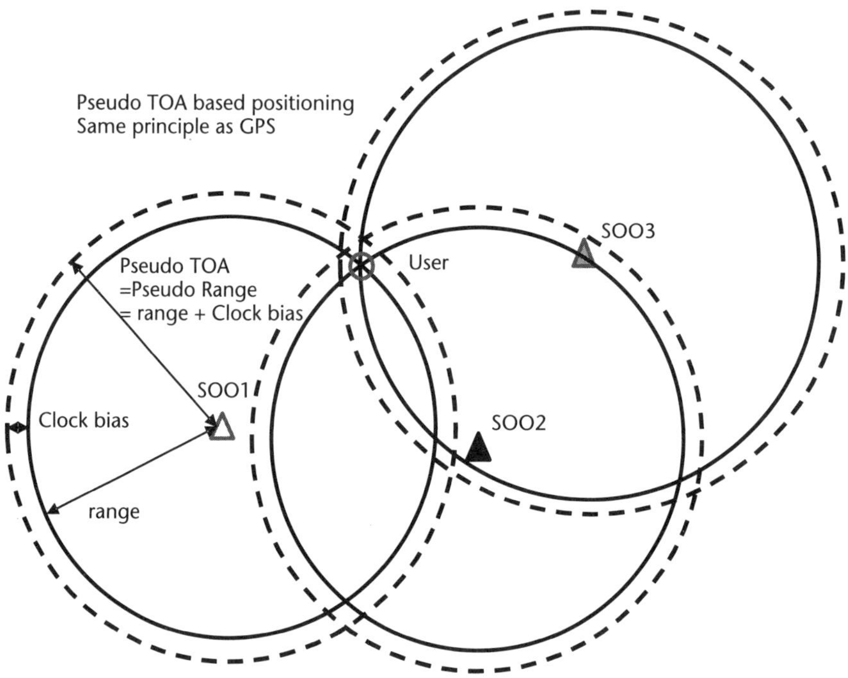

Figure 14.3 PTOA-based positioning.

14.1.4 FOA

FOA should be more correctly described as pseudo-FOA (PFOA). The primary reason that FOA or PFOA is useful is because of the Doppler effect: the frequency changes when the relative LOS velocity changes. There are at least two scenarios where PFOA is useful. First, the PFOA can be used in equations to solve for velocity and clock rate errors. Second, when the signal source is a LEO satellite with a very predictable velocity vector, the measured pseudo Doppler actually represents the projection of the velocity vector on the LOS vector. Since the LOS vector is a function of the current position, the Doppler measurement, in this case, provides partial observability of the position. Figure 14.4 illustrates the concept of a PFOA measurement of a LEO satellite.

14.1.5 TDOA and FDOA

TDOA and FDOA are among the most practical measurements that we can use for SOOP-based PNT systems. This is because:

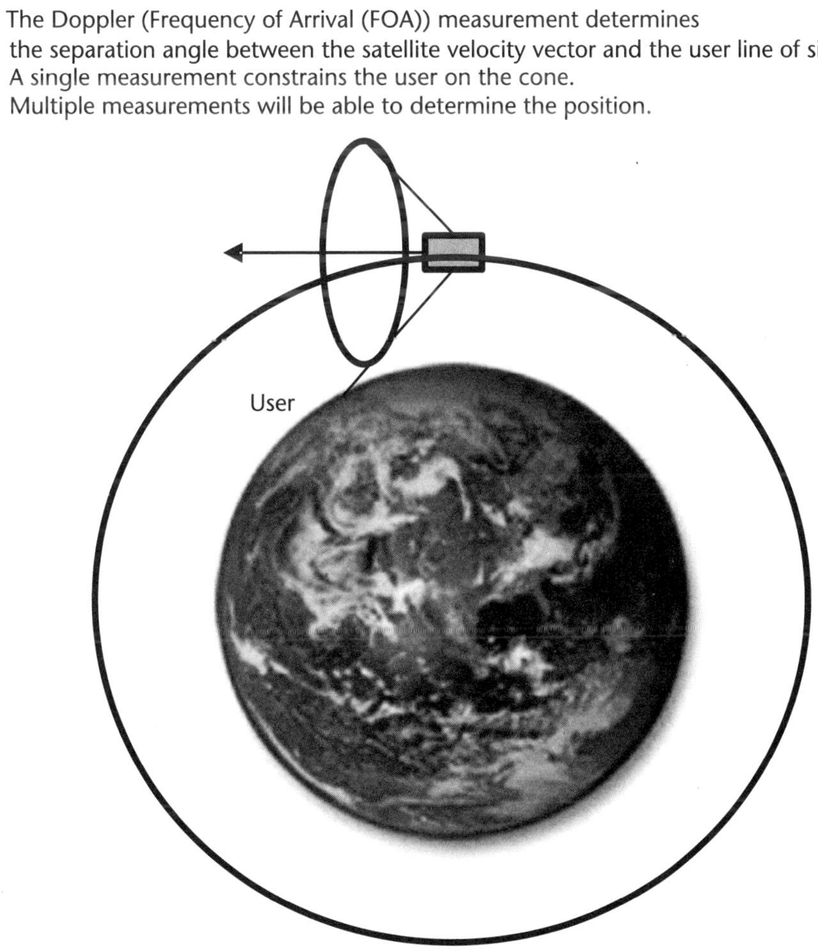

Figure 14.4 Positioning using the FOA (Doppler) measurement of a LEO satellite.

1. The expected performance is comparable to the typical GNSS or differential GNSS-based system.
2. The assumptions are most practical: we do not assume that the sources are synchronized to the same clock.

As hinted in Chapter 4, we can measure the TOA using two receivers, one is called the reference receiver (Ref receiver) and the other is the navigator's receiver (Nav receiver, or sometimes referred to as a user receiver). The difference between the TOAs as measured by the two receivers can be called the difference of TOA (DTOA), which is fundamentally the same as the TDOA. The TDOA can be measured directly by correlating the two signals of interest. Again, the more correct term for the DTOA and TDOA are actually pseudo DTOA and pseudo TDOA, as they typically contain the TOA and the relative cock bias between the Ref receiver and the Nav receiver. In the rest of this chapter, we will simply use the TDOA and FDOA to refer to both the DTOA and TDOA as well as the DFOA and FDOA.

When the sources are space-based and the distance between the Nav receiver and the Ref receiver are not too far (e.g., <100 km), the differential nature of the TDOA has an additional benefit. That is the cancellation, or at least the partial cancellation, of signal propagation errors such as ionosphere delay and troposphere delay errors in the same exact manner as differential GPS.

The differential Doppler measurement, or the pseudo-FDOA measurement, can be viewed as the difference of the projections of the velocity vector on the range vectors in a similar way as illustrated in Figure 14.4.

Before we explain more details of TDOA-based positioning in our context, let us first describe the classical TDOA-based geolocation problem, which is the reverse problem of what we have.

For simplicity, we will discuss the classical TDOA-based 2-D geolocation problem, which is what LORAN (Long Range Navigation) is based upon. Given three signal sources that are synchronized, the TOA for the signals to arrive at a user can be measured, and we can form two independent TDOAs, the differences of TOA from two of the three sources. Each of the TDOAs constrains the Nav receiver on a pair of hyperbola in 2-D space (and hyperboloid surfaces in 3-D space). Two TDOA measurements constrain the Nav receiver position on the intersection of the hyperbola pairs. There could be 1 to 8 intersection points and typically it is not hard to determine which one is the correct solution based on other knowledge that we have. Figure 14.5 illustrates the classical TDOA-based positioning.

The problem we have is a reverse problem of the classical problem. In our setup, the signal sources and the user receiver have swapped roles. Specifically, as illustrated in Figure 14.6, the user receiver and the reference receiver are at the two focal points of a pair of the hyperbola, and the signal sources are on the hyperbola curve. Instead of using the knowledge of the focal points and the TDOA to determine the position of a point on the hyperbola curve, we use the points on the hyperbola curve, and the knowledge of one of the focal points (the reference receiver), and the TDOA measurements to determine the position of the other focal point, the user receiver.

Since the mathematical essence of most of the modalities of the SOOP-based PNT has been discussed in Chapter 4, we will focus on TDOA and FDOA-based methods in this chapter.

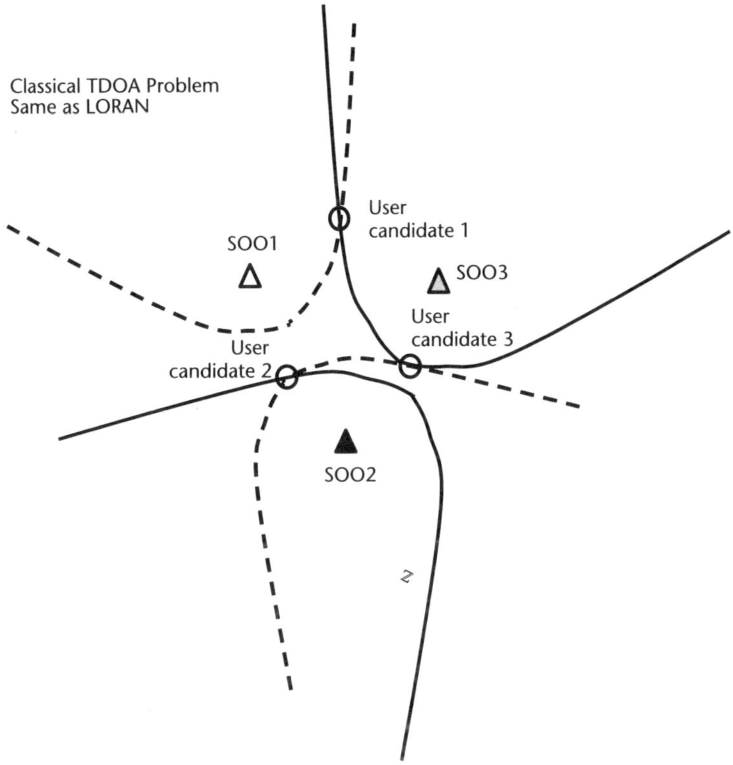

Figure 14.5 Classical TDOA-based positioning (LORAN-like).

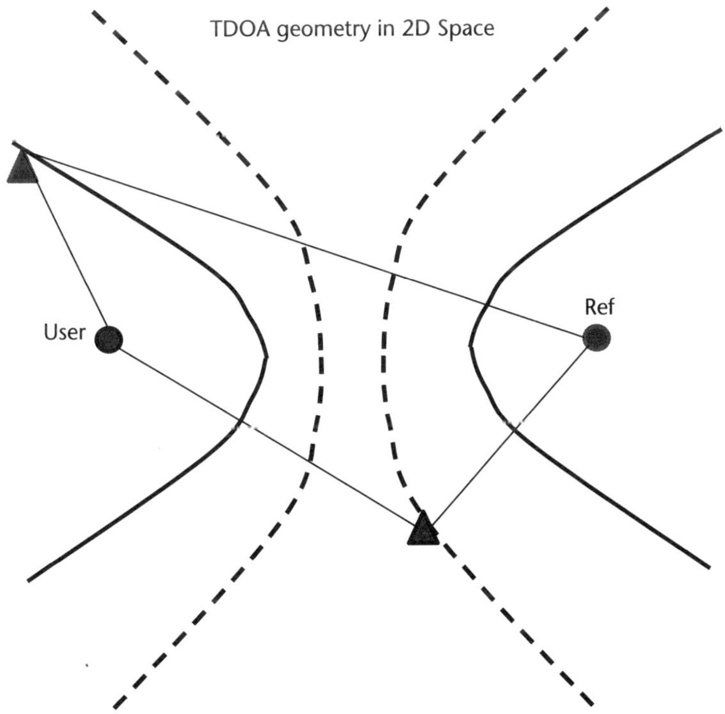

Figure 14.6 Relationships of the Ref receiver, the user (Nav) receiver, and the signal sources in a TDOA position problem.

14.2 Architecture of TDOA and FDOA SOOP PNT System

There are at least three important variations of the architecture:

1. TDOA/FDOA-only PNT system (i.e., an SOOP receiver similar to a GPS receiver);
2. TDOA/FDOA integrated into an All Source PNT system (and, most importantly, INS);
3. For both of the above architectures, the TDOA and FDOA can be derived in two different ways:
   a. TDOA is obtained as the DTOA and the TOA is the measured TOA of a time mark such as a message preamble of a demodulated digital signal.
   b. TDOA is obtained by directly correlating the raw I/Q data collected from the Nav receiver and the Ref receiver.

Figure 14.7 illustrates the architecture of a TDOA/FDOA-only PNT system where the TDOA/FDOA is represented by a box with no internal details.

Figure 14.8 illustrates the architecture of the TDOA and FDOA integrated with an INS.

Figure 14.9 illustrates the two types of the TDOA and FDOA determination approach.

Although not expected to provide the best performance, a TDOA/FDOA-only system has the advantages of being simple, low SWAP, and inexpensive. It saves all the complexity, cost, and SWAP associated with the integrated solution. For this type of architecture, it is important to use a sufficient number of signal sources so that:

1. At any moment, there are a sufficient number of sources (four or more) that can be used to determine a PNT point solution to ensure continuous availability and access.

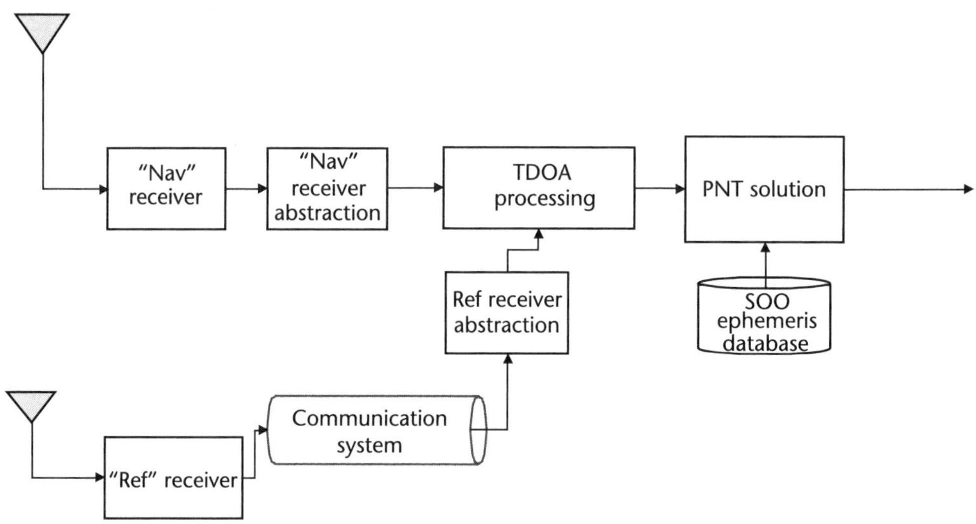

Figure 14.7 Architecture of a TDOA/FDOA-only PNT system.

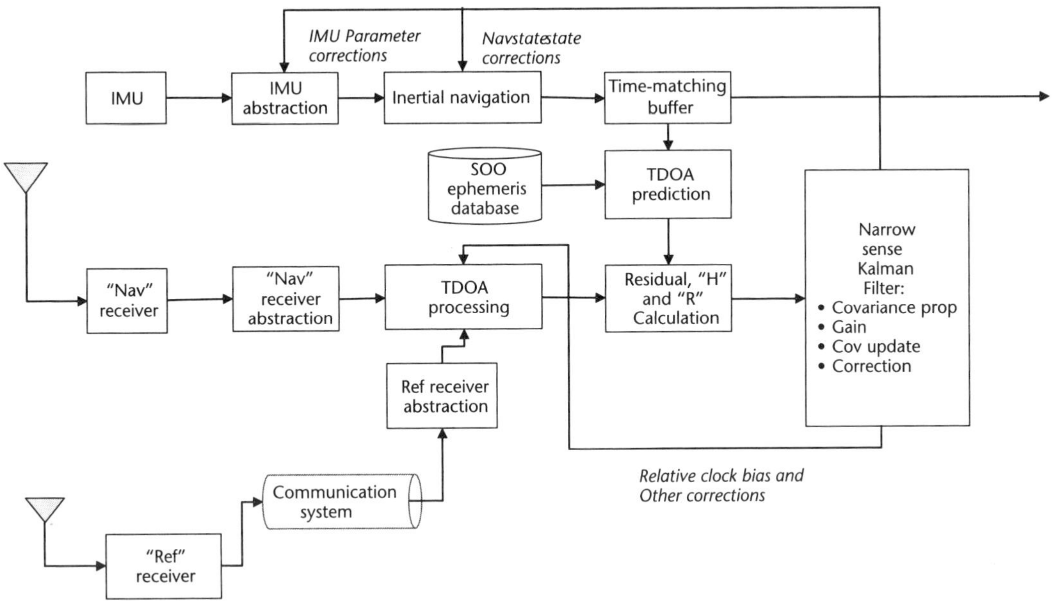

Figure 14.8 Architecture of TDOA/FDOA integrated with inertial navigation and All Source PNT.

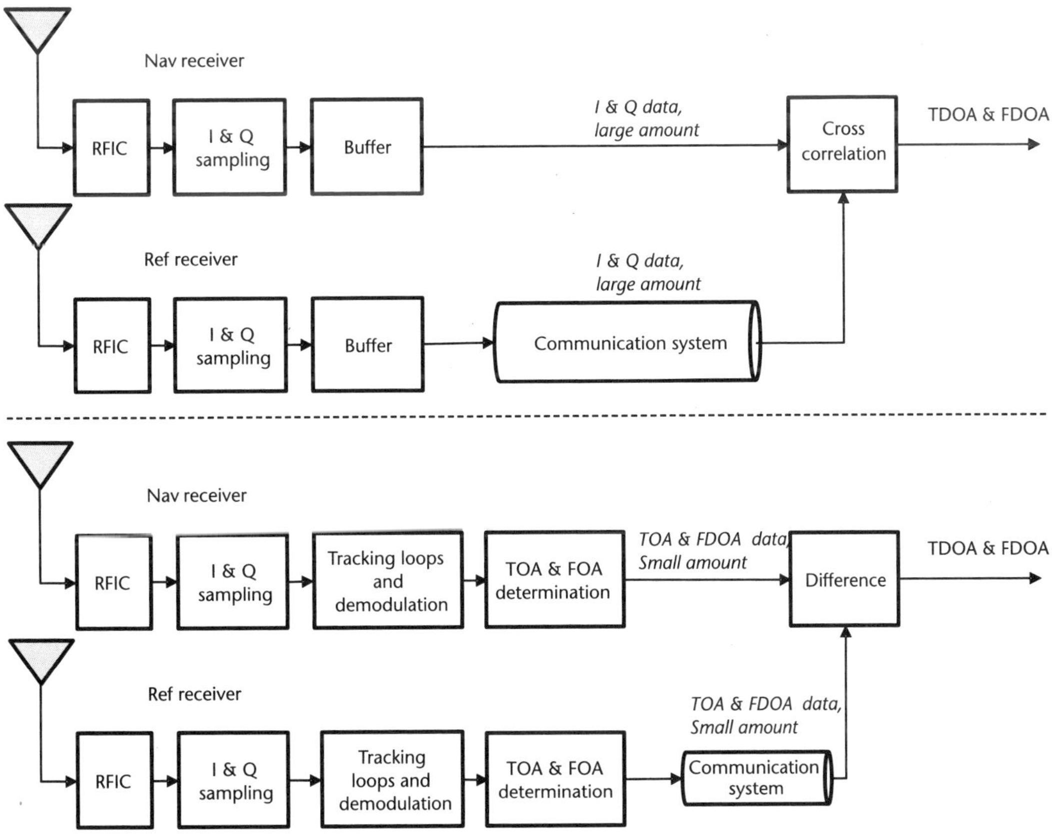

Figure 14.9 Two types of the TDOA and FDOA determination approach.

2. There are a sufficient number of sources that can be used to perform integrity monitoring or cross-checking to ensure the integrity of the solution.

The integration with the INS or other sensor sources become necessary when any one or both of the two conditions above cannot be satisfied. For example, if we only use LEO satellite signals such as Iridium, we typically only see one or two satellites at a time. We will need to rely on the motion of the satellite to provide the observability and improve the dilution of the precision over time. As the result, we must bridge the solution while the receiver and the satellites are moving and the geometry is changing. Inertial navigation will be one example of bridging (or flywheeling) that can be done.

When we have the luxury to know the format of the data messages modulated on a signal, we have the opportunity to measure the TOA of a message pattern such as the header or preamble of a message, and, consequently, we can get a pseudo-range measurement in a way that is similar to GPS. The TOA or pseudo-range obtained this way contains the clock bias of the source and the clock bias of the user receiver.

Similarly, the TOA or pseudo-range can be measured by the Ref receiver, which contains the clock error of the source and the clock error of the Ref receiver.

When the difference of the two TOAs is calculated, the error due to the clock error of the source signal is canceled out, and what is left is the relative clock error between the Ref receiver and the Nav receiver.

The advantage of this approach is that the communication burden from the Ref receiver to the Nav receiver is very light: we only have to send the TOA and pseudo-ranges.

A more general approach, which does not depend on any detailed knowledge about the data modulated on the signal, is simply to take the raw I and Q data from the two receivers and perform cross-correlation between the two sampled signals. The delta time corresponding to the cross-correlation peak provides a measurement of the TDOA. This approach only relies on the fact that the same signal is received by both receivers with different delays and possibly different Doppler. It does not need to look for a recognizable time mark; it does not even need to use a digitally modulated signal.

Since the SOOP-based PNT system relies on many different types of signals of opportunity, a software-defined radio (SDR) often makes sense. SDR provides the flexibility of receiving and processing many different types of signals using the same hardware. Figure 14.10 illustrates the typical architecture of an SDR system.

An SDR uses an RFIC as the low SWAP solution of the RF front end, which is responsible for downconverting the signal to the intermediate frequency (IF) band where the A/D sampling is feasible and then digitizing the signal by A/D. The trend is such that the A/D sampling rate is getting increasingly higher and, in many cases, the RF signal can be directly sampled without analog downconversion.

High data rate processing is typically done by the FPGA where heavy-duty data processing can be done in parallel. Low rate data processing and receiver management are typically done by a processor where traditional program languages are used to program the processor. Signal tracking loops are typically done by the processor.

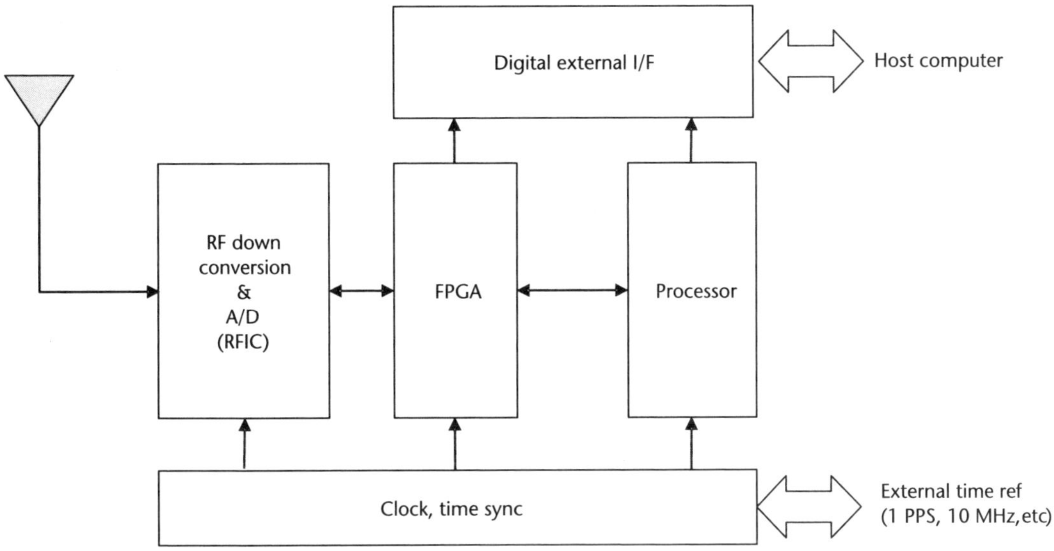

Figure 14.10 Typical architecture of an SDR receiver.

The rapid development of the RF System-On-Chip (RFSOC) technology can put a broadband SDR almost entirely on a single chip. The Xilinx Zynq UltraScale+ RFSOC is a good example of this technology [10].

14.3 TDOA and FDOA Measurement by Cross-Correlation

TODA and FDOA measurements by cross-correlation are a topic that has been well addressed since the 1970s [11–17] and serve as the foundation for TDOA/FDOA-based positioning and geolocation methods.

Let $x(t)$ and $y(t)$ be the signal received by two receivers from the same source $s(t)$

$$\begin{aligned} x(t) &= s(t) + n_x(t) \\ y(t) &= A_r s(t-d)\exp\left(-j2\pi f_d t\right) + n_y(t) \end{aligned} \tag{14.1}$$

where d is the TDOA and f_d is the FDOA.

However, the reality is that there are often multiple signals present in the band of interest. As a result, we have

$$\begin{aligned} x(t) &= s(t) + \sum_{k=1}^{n_s} s_k(t) + n_x(t) \\ y(t) &= A_r s(t-d)\exp\left(-j2\pi f_d t\right) + \sum_{k=1}^{n_s} a_k s\left(t-d_k\right)\exp\left(-j2\pi f_{d_k} t\right) + n_y(t) \end{aligned} \tag{14.2}$$

Let us first consider the simplest case, when the difference of Doppler is 0 or small enough to be ignored. In this case, we have

$$\begin{aligned} x(t) &= s(t) + n_x(t) \\ y(t) &= A_r s(t-d) + n_y(t) \end{aligned} \tag{14.3}$$

Let the self-correlation and cross-correlation of the signals be given by

$$\begin{aligned} R_x(\tau) &= E\left(x(t)x^*(t-\tau)\right) = R_s(\tau) + R_{n_x}(\tau) \\ R_y(\tau) &= E\left(y(t)y^*(t-\tau)\right) = \left|A_r\right|^2 R_s(\tau) + R_{n_y}(\tau) \\ R_{yx}(\tau) &= E\left(y(t)x^*(t-\tau)\right) = A_r R_s(\tau - d) + R_{n_y n_x}(\tau) \end{aligned} \tag{14.4}$$

In the frequency domain, the above equations can be written as

$$\begin{aligned} S_x(f) &= S_s(f) + S_{n_x}(f) \\ S_y(f) &= \left|A_r\right|^2 S_s(f) + S_{n_y}(f) \\ S_{yx}(\tau) &= A_r S_s(f)\exp(-j2\pi d) + S_{n_y n_x}(f) \end{aligned} \tag{14.5}$$

If we assume that the signal-to-noise ratio (SNR) is high enough that we can ignore the noise, we have

$$\frac{S_{yx}(f)}{S_x(f)} = A_r \exp(-j2\pi d) \;\; \text{for } f_0 - \frac{B_s}{2} \le f \le f_0 + \frac{B_s}{2} \tag{14.6}$$

where B_s is the bandwidth of the signal of interest.

The inverse Fourier transformation will give us

$$\begin{aligned} r_{yx}(\tau) &= \text{inverse}FT\left(\frac{S_{yx}(f)}{S_x(f)}\right) = \int_{f_0 - B_s/2}^{f_0 + B_s/2} \frac{S_{yx}(f)}{S_x(f)} \exp(j2\pi f t)\, df \\ &= \frac{A_r \sin\left(\pi B_s(\tau - d)\right)}{\pi(\tau - d)} \exp\left(j2\pi f_0(\tau - d)\right) \end{aligned} \tag{14.7}$$

The amplitude of the complex sinusoidal is a sinc function and its width (i.e., how wide the sinc pulse is) depends on the bandwidth B_s.

If we determine d by looking for the peak of the function, $r_{yx}(\tau)$, the performance of TDOA determination is estimated by the following equation [14, 15]:

$$\sigma\left(\hat{d} - d\right) = \frac{\sqrt{3}}{2\pi}\left(\frac{1}{\sqrt{T}}\right)\left(\frac{1}{B_s\sqrt{B_s}}\right)\left(\frac{1}{\sqrt{\gamma}}\right) = \frac{0.2757}{B_s}\frac{1}{\sqrt{B_s T \gamma}} \tag{14.8}$$

where T is the observation time and γ is the effective receiver SNR that satisfies the following equation:

$$\frac{1}{\gamma} = \frac{1}{2}\left(\frac{1}{\gamma_1} + \frac{1}{\gamma_2} + \frac{1}{\gamma_1\gamma_2}\right) \tag{14.9}$$

γ_1, γ_2 are the SNR of the first and the second receivers.

In practical implementations, the correlations are typically done using FFT and inverse FFT and the following provides a step-by-step algorithm for such processing.

1. Calculate the FFT of the two signals:

$$\begin{aligned} X(f) &= FFT\left(x(t)\right) \\ Y(f) &= FFT\left(y(t)\right) \end{aligned} \tag{14.10}$$

2. At this moment, while the signal is in the frequency domain, we have the opportunity to easily filter out the frequency contents in which we are not interested by simply setting certain undesired frequency contents to 0.
3. Calculate the power spectral density functions:

$$\begin{aligned} S_x(f) &= X(f)X^*(f) \\ S_{xy}(f) &= X(f)Y^*(f) \end{aligned} \tag{14.11}$$

4. Calculate the weighed cross-correlation:

$$s_{xy}(f) = \frac{S_{xy}(f)}{S_x(f)} = \frac{Y^*(f)}{X^*(f)} \tag{14.12}$$

   Note that the second part of the equation above provides significant savings computationally. We can skip step 3 and simply calculate the ratio of the conjugate of the FFT of the two signals.

   The choice among the unweighted cross-correlation and other different types of weighted cross-correlation depends on the practical situation. Whichever gives us the cleaner and sharper correlation peak, whichever leads to better measurement of the TDOA, will be the right choice.
5. Perform inverse FFT:

$$r_{xy}(\tau) = IFFT\left(s_{xy}(f)\right) \tag{14.13}$$

When the relative Doppler cannot be ignored, we will need to use a natural extension of the approach by going back to the signal model with relative Doppler shift:

$$\begin{aligned} x(t) &= s(t) + n_x(t) \\ y(t) &= A_r s(t-d)\exp\left(-j2\pi f_d t\right) + n_y(t) \end{aligned} \tag{14.14}$$

We can form the cross-ambiguity function (CAF) as

$$A(\tau,f) = \int_0^T y(t)x^*(t-\tau)\exp(2\pi f t)\,dt \tag{14.15}$$

We will then find such τ, f to maximize the above function.
If we let $f = 0$, then the above reduces to classical cross-correlation

$$A(\tau,0) = \int_0^T y(t)x^*(t-\tau)\,dt = R_{yx}(\tau) = C(y,x,\tau) \tag{14.16}$$

Using the cross-correlation symbol $C(y, x, \tau)$, CAF can be written as cross-correlation of the Doppler shifted function and

$$A(\tau,f)= \int_0^T y(t)x^*(t-\tau)\exp(2\pi ft)\,dt = C\left(y(t)\exp(2\pi ft),x(t),\tau\right) \tag{14.17}$$

What this means is that we will do a 2-D search, for each candidate relative Doppler frequency f, we will do a cross-correlation function calculation.

If we use FFT, it is important to realize that the FFT of $x(t)$ only has to be calculated once which remains the same for each of the Doppler search steps. The FFT of $y(t)\exp(2\pi ft)$ can be calculated efficiently by shifting the FFT of $y(t)$ because the following well-known property of the Fourier transformation:

$$FT\left(y(t)\exp\left(2\pi f_d t\right)\right) = Y\left(f - f_d\right) \tag{14.18}$$

In other words, we only have to calculate the FFT of $y(t)$ once as well. The computational savings created by recognizing that we only have to do each of the two FFTs once is significant.

It is also important to realize that, when we perform the FFTs, we may be able to determine the Doppler frequency of both signals and in that case, we will be able to do is to simply shift one of the FFT results such that the Doppler between the two signals match without doing a 2-D search.

The following example illustrates this point. Figures 14.11 and 14.12 show the FFTs of two signals $a(t)$ and $b(t)$.

Because of the relative Doppler, the direct correlation of these two signals without relative Doppler compensation shows no correlation peaks as illustrated in Figure 14.13.

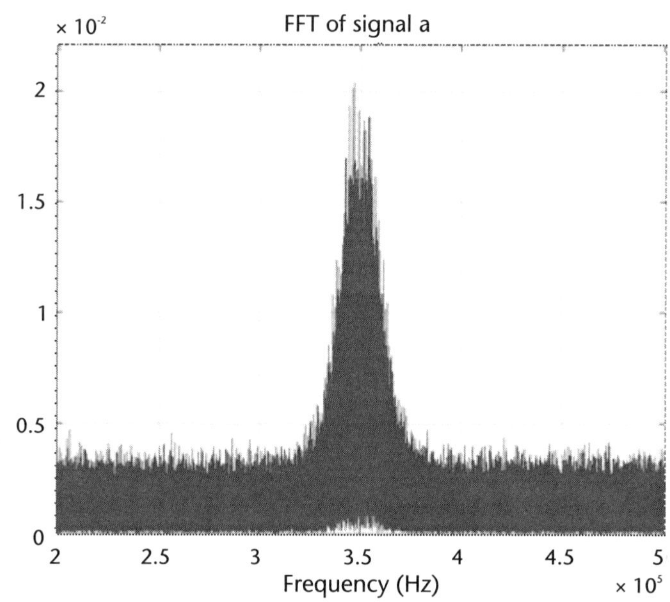

Figure 14.11 FFT of example signal $a(t)$.

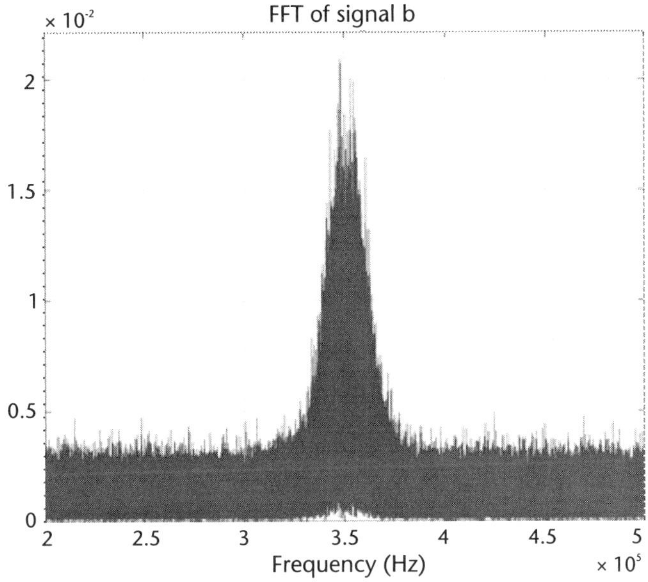

Figure 14.12 FFT of example signal $b(t)$.

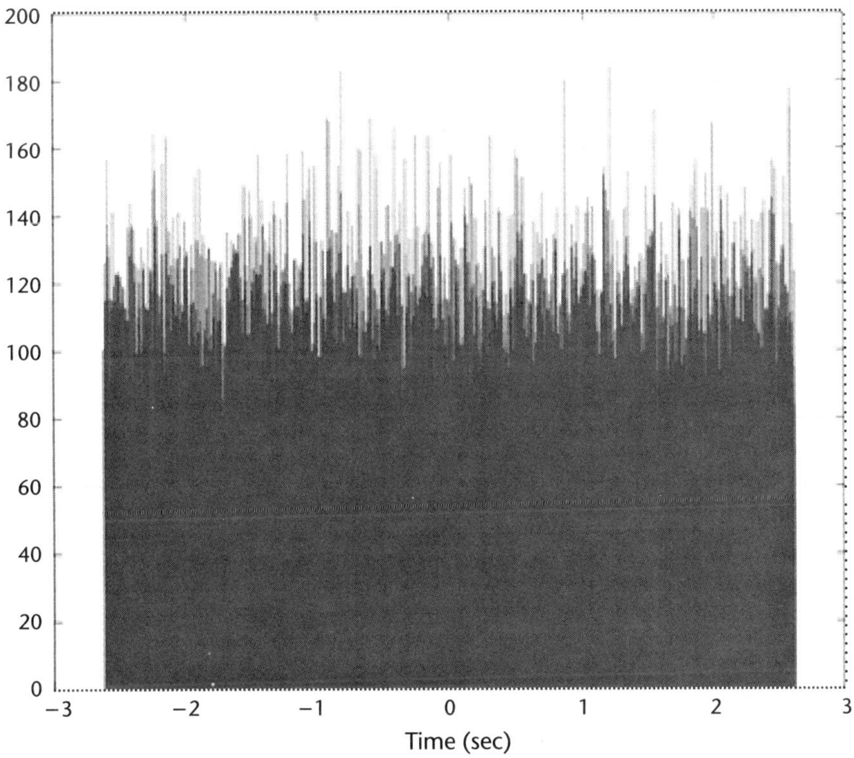

Figure 14.13 Correlation of $a(t)$ and $b(t)$ without relative Doppler compensation. No obvious peaks.

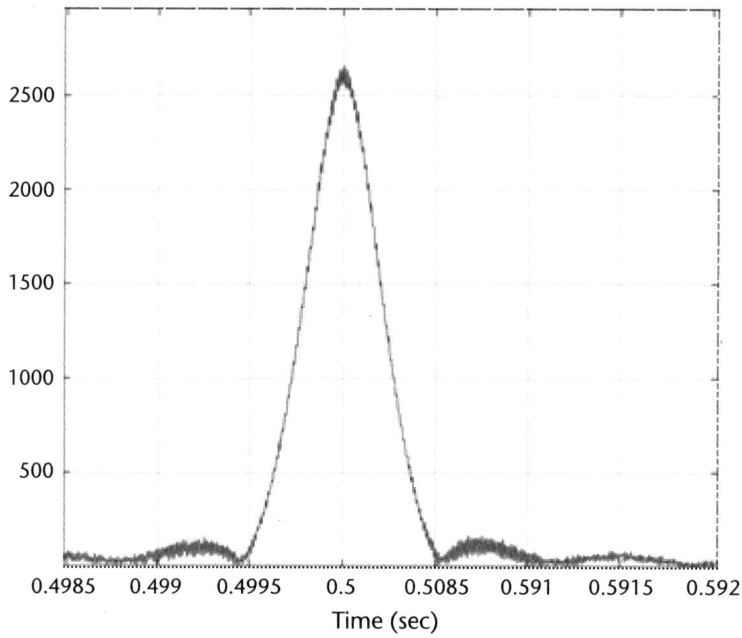

Figure 14.14 Correlation peak shown after relative Doppler frequency compensation.

If we shift the FFT of signal $b(t)$ by aligning the center frequencies of the two signals, the correlation shows a clear peak as illustrated in Figure 14.14.

14.4 Receiver PNT Using TOA Pseudo-Range Measurement Without a Reference Receiver

PNT can be done without the reference receiver if the following two conditions are satisfied:

1. If all of the SOOP signal sources to be used have synchronized clocks or we have determined a clock correction model and the associated parameters such that the time references for each of the sources can be corrected and therefore become synchronized.
2. It is acceptable that signal propagation error cancellation by the difference of the TOA or pseudo-range will not be done, or the signal propagation delay errors can be modeled and corrected using other sources.

In this section, we will develop the equations for the PNT problem using both pseudo-range (TOA) and Doppler (FOA) when the SOOP signal sources are synchronized. However, these equations are important not only because they provide the solution for the problem that we have in this section, but also important because the SOOP PNT problems using asynchronous sources can also be transformed to be mathematically equivalent when a reference receiver is used. In addition, these equations can be used to form the least-square solution problem to derive instantaneous PNT solutions using a sufficient number of SOOP sources. They can also be used as the measurement equations for a Kalman filter-based integrated PNT system using SOOP measurements as the aiding source.

Given n signal sources with known positions located in a coordinate frame with positions given as

$$r_k = \begin{bmatrix} x_k \\ y_k \\ z_k \end{bmatrix} \quad (k = 1,2,\cdots,n)$$

and a navigator that wants to find its own position

$$r = \begin{bmatrix} x \\ y \\ z \end{bmatrix}$$

by measuring the TOA or pseudo-range to each one of the source nodes.

Assuming that each of the source nodes uses the same time reference (synchronized source nodes), we have the following equation:

$$\rho_k = \left\| r - r_k \right\| + b + v_k = \sqrt{(x - x_k)^2 + (y - y_k)^2 + (z - z_k)^2} + b + v_k \tag{14.19}$$

where b is the clock bias measured in distance unit (e.g., meters) and v_k is the range measurement error.

The same clock bias b is assumed for all the signal sources. This is because all the source nodes are assumed to be synchronized, and the only time offset is the offset between the navigator's time and the time of the source nodes.

If the signal source is a LEO satellite (such as Iridium or GlobalStar), the Doppler shift due to the satellite motion is large and also changes with geometry. As a result, the Doppler shift or the FOA is often considered a good source of information, especially when there is difficulty measuring the TOA or pseudo-ranges.

Since the user velocity is often much smaller than the LEO satellite velocity, we typically use the following equation to model the Doppler measurement (and, when needed, the user velocity vector can be easily included in the model):

$$f_{dk} = f_k \frac{v_{\text{LOS}} + b_r}{c} + v_f = f_k \frac{-(v_k^T l) + b_r}{c} + v_f = \left(\frac{f_k}{c}\right)\left(-v_k^T \frac{(r_k - r)}{\left\| r_k - r \right\|} + b_r\right) + v_f \tag{14.20}$$

where f_k is the carrier frequency for the source signal k, v_{LOS} is the velocity along the LOS vector between the satellite and the user receiver, $l = (r_k - r)/\| r_k - r \|$ is the LOS unit vector from the user to the satellite, b_r is the clock rate error in the unit of velocity (meter/second, that is, after scaling by speed of light), v_f is the Doppler measurement noise, v_k is the velocity vector of the satellite k, and r_k is the position vector of the satellite k.

It will be useful to linearize the above equations around an approximate solution

$$ {}_0r = \begin{bmatrix} {}_0x \\ {}_0y \\ {}_0z \end{bmatrix} $$

For TOA or pseudo-range, we have seen many times the following equations in slightly different forms

$$ \Delta\rho_k\left({}_0r\right) = \rho_k - \left\|\left({}_0r - r_k\right)\right\| = -u_{\text{LOS}}\left({}_0r\right)_k^T\left(r - {}_0r\right) + b + v_k = -u_{\text{LOS}}\left({}_0r\right)_k^T \Delta r + b + v_k $$

where

$$ u_{\text{LOS}} = u_{\text{LOS}}(r_0)_k = (r_k - {}_0r)/\|r_k - {}_0r\| $$

is the unit vector from the user to the satellite.

Let r_{u0} be an initial guess of the position; then an initial guess of the Doppler is given by

$$ {}_0f_{dk} = \left(\frac{f_k}{c}\right)\left(-v_k^T \frac{\left(r_k - {}_0r\right)}{\left\|r_k - {}_0r\right\|}\right) $$

The linearized mode for the deviation of the Doppler as the result of the deviation of position is derived here:

$$ \begin{aligned} \Delta f_{dk} &= f_{dk} - f_{d0} + \left(\frac{f_k}{c}\right)b_r + v_f = -\left(\frac{f_k}{c}\right)v_k^T\left(\frac{\left(r_k - r\right)}{\left\|r_k - r\right\|} - \frac{\left(r_k - {}_0r\right)}{\left\|r_k - {}_0r\right\|}\right) + \left(\frac{f_k}{c}\right)b_r + v_f \\ &\approx \left(\frac{f_k}{c}\right)v_k^T\left(\frac{I_{3x3}}{\left\|r_k - {}_0r\right\|} - \frac{\left({}_0r - r_k\right)\left(r - r_k\right)^T}{\left\|r_k - {}_0r\right\|^3}\right)\left(r - {}_0r\right) + \left(\frac{f_k}{c}\right)b_r + v_f \\ &\approx \left(\frac{f_k}{c}\right)v_k^T \frac{1}{\left|\left\|r_k - {}_0r\right\|\right.}\left(I_{3x3} - \frac{\left({}_0r - r_k\right)\left({}_0r - r_k\right)^T}{\left\|r_k - {}_0r_k\right\|^2}\right)\Delta r + \left(\frac{f_k}{c}\right)b_r + v_f \\ &= \left(\frac{f_k}{c}\right)v_k^T \frac{1}{\left\|r_k - {}_0r\right\|}\left(I_{3x3} - \frac{\left(r_k - {}_0r\right)\left(r_k - {}_0r\right)^T}{\left\|r_k - {}_0r\right\|^2}\right)\Delta r + \left(\frac{f_k}{c}\right)b_r + v_f \\ &= \left(\frac{f_k}{c}\right)v_k^T \frac{1}{\left\|r_k - {}_0r\right\|}\left(I_{3x3} - u_{\text{LOS}}\left({}_0r\right)_k u_{\text{LOS}}\left({}_0r\right)_k^T\right)\Delta r + \left(\frac{f_k}{c}\right)b_r + v_f \end{aligned} $$

In summary, for each source signal k, we have the following linearized measurement equations

$$ \Delta\rho_k\left({}_0r\right) = -\left(u_{\text{LOS}}\left({}_0r\right)_k\right)^T \Delta r + b + v_k \tag{14.21} $$

$$\Delta f_{dk} = \left(\frac{f_k}{c}\right) v_k^T \frac{1}{\left\|{}_0r - r_k\right\|}\left(I_{3x3} - u_{\text{LOS}}\left({}_0r\right)_k u_{\text{LOS}}\left({}_0r\right)_k^T\right)\Delta r + \left(\frac{f_k}{c}\right) b_r + v_f \quad (14.22)$$

Introducing the simplified symbols:

$$h_{pp}(k) = -\left(u_{\text{LOS}}\left({}_0r\right)_k\right)^T \quad (14.23)$$

and

$$h_{fp}(k) = \left(\frac{f_k}{c}\right) v_k^T \frac{1}{\left\|{}_0r - r_k\right\|}\left(I_{3x3} - u_{\text{LOS}}\left({}_0r\right)_k u_{\text{LOS}}\left({}_0r\right)_k^T\right) \quad (14.24)$$

Equations (14.21) and (14.22) can then be rewritten as

$$\Delta\rho_k\left({}_0r\right) = h_{pp}(k)\Delta r + b + v_k \quad (14.25)$$

and

$$\Delta f_{dk} = h_{fp}(k)\Delta r + \left(\frac{f_k}{c}\right) b_r + v_f \quad (14.26)$$

These equations can be directly used in the extended Kalman filter design when we integrate the measurement solution with the propagation solution (such as inertial navigation). At this moment, we will use these equations to form a least square solution for Δr and, consequently, r can be estimated as

$$r = {}_0r + \Delta r \quad (14.27)$$

If we only use the pseudo-range measurements, we have

$$\Delta P = \begin{bmatrix} \Delta\rho_1\left({}_0r\right) \\ \Delta\rho_2\left({}_0r\right) \\ \vdots \\ \Delta\rho_n\left({}_0r\right) \end{bmatrix} = \begin{bmatrix} h_{pp}(1) & 1 \\ h_{pp}(2) & 1 \\ \vdots & \vdots \\ h_{pp}(n) & 1 \end{bmatrix} \begin{bmatrix} \Delta r \\ b \end{bmatrix} + \begin{bmatrix} v_1 \\ v_2 \\ \vdots \\ v_n \end{bmatrix} \triangleq H_p \begin{bmatrix} \Delta r \\ b \end{bmatrix} + v_p \quad (14.28)$$

If we just use the Doppler measurements, we have

$$\Delta F = \begin{bmatrix} \Delta f_{d1}\left({}_0r\right) \\ \Delta f_{d2}\left({}_0r\right) \\ \vdots \\ \Delta f_{dn}\left({}_0r\right) \end{bmatrix} = \begin{bmatrix} h_{fp}(1) & \frac{f_1}{c} \\ h_{fp}(2) & \frac{f_2}{c} \\ \vdots & \vdots \\ h_{fp}(n) & \frac{f_n}{c} \end{bmatrix} \begin{bmatrix} \Delta r \\ b_r \end{bmatrix} + \begin{bmatrix} v_{f1} \\ v_{f2} \\ \vdots \\ v_{fn} \end{bmatrix} \triangleq H_f \begin{bmatrix} \Delta r \\ b_r \end{bmatrix} + v_f \quad (14.29)$$

If we use both the pseudo-range and the Doppler, we have

$$\begin{bmatrix} \Delta \mathrm{P} \\ \Delta F \end{bmatrix} = \begin{bmatrix} \Delta\rho_1({}_0r) \\ \Delta\rho_2({}_0r) \\ \vdots \\ \Delta\rho_n({}_0r) \\ \Delta f_1({}_0r) \\ \Delta f_2({}_0r) \\ \vdots \\ \Delta f_n({}_0r) \end{bmatrix} = \begin{bmatrix} h_{pp}(1) & 1 & 0 \\ h_{pp}(2) & 1 & 0 \\ \vdots & \vdots & 0 \\ h_{pp}(n) & 1 & 0 \\ h_{fp}(1) & 0 & \frac{f_1}{c} \\ h_{fp}(2) & 0 & \frac{f_2}{c} \\ \vdots & \vdots & \vdots \\ h_{fp}(n) & 0 & \frac{f_n}{c} \end{bmatrix} \begin{bmatrix} \Delta r \\ b \\ b_r \end{bmatrix} + \begin{bmatrix} v_p \\ v_f \end{bmatrix} = H_{pf} \begin{bmatrix} \Delta r \\ b \\ b_r \end{bmatrix} + \begin{bmatrix} v_p \\ v_f \end{bmatrix} \tag{14.30}$$

Each of the above problems can be easily solved as unweighted or weighted least square problems that have been done extensively in this book.

14.5 SOOP PNT with a Reference Receiver

When the SOOP signal sources are not synchronized, the introduction of the reference receiver becomes a significant contributor in the following two ways:

1. By differencing the TOA or pseudo-range measurements derived from the user (navigator) receiver and the reference receiver, we will be able to cancel out the clock errors of the signal sources, consequently converting the PNT problem mathematically into a form that is the same as the PNT problem when the signal sources are synchronized.
2. By differencing the TOA or pseudo-range measurements derived from the user (navigator) receiver and the reference receiver, we will be able to achieve the same effect as differential GPS in the sense that significant amount of propagation errors such as ionosphere and troposphere delays will be canceled out if the SOOP signal sources are from space.

Using the same setup as the previous section, we can write the TOA or pseudo-range measurement equation of the navigator with respect to the SOOP source k as

$$\rho_{uk} = \|r - r_k\| + b_u + b_k + v_{uk} = \sqrt{(x - x_k)^2 + (y - y_k)^2 + (z - z_k)^2} + b_u + b_k + v_{uk} \tag{14.31}$$

where b_u is the user receiver clock bias measured in distance units (e.g., meters), b_k is the source k clock bias measured in distance unit (e.g., meters), and v_{uk} is the user pseudo-range measurement error.

Similarly, we can also write the Doppler measurement equation as

$$f_{duk} = \left(\frac{f_k}{c}\right)\left(-v_k^T \frac{(r_k - r)}{\|r_k - r\|} + b_{ru} + b_{rk}\right) + v_{fuk} \tag{14.32}$$

where b_{ru} is the user receiver clock rate bias measured in velocity unit (e.g., meters/

sec), b_{rk} is the source k clock rate bias measured in velocity unit (e.g., meters/sec), and v_{fuk} is the user Doppler measurement error.

Similar equations can be written for the reference receiver whose position is known:

$$\rho_{rk} = \left\| r_r - r_k \right\| + b_r + b_k + v_{rk} = \sqrt{\left(x_r - x_k\right)^2 + \left(y_r - y_k\right)^2 + \left(z_r - z_k\right)^2} + b_r + b_k + v_{rk} \quad (14.33)$$

where b_r is the reference receiver clock bias measured in distance units (e.g., meters) and v_{rk} is the reference pseudo-range measurement error.

The Doppler measurement model is given by

$$f_{drk} = \left(\frac{f_k}{c}\right)\left(-v_k^T \frac{\left(r_k - r_r\right)}{\left\| r_k - r_r \right\|} + b_{rr} + b_{rk}\right) + v_{frk} \quad (14.34)$$

where b_{rr} is the reference receiver clock rate bias measured in velocity units (e.g., meters/sec) and v_{frk} is the reference Doppler measurement error.

If we perform a single difference between the user pseudo-range and the reference pseudo-range, we have

$$SD\rho_{urk} = \rho_{uk} - \rho_{rk} = \left\| r - r_k \right\| - \left\| r_r - r_k \right\| + \left(b_u - b_r\right) + \left(v_{uk} - v_{rk}\right) \quad (14.35)$$

$SD\rho_{urk}$ is the single difference pseudo-range, which is our TDOA measurement. Let

$$\xi_k = SD\rho_{urk} + \left\| r_r - r_k \right\|$$

$b_{ur} = b_u - b_r$ be the relative clock offset between the user node and the reference node.

$v_{urk} = v_{uk} - v_{rk}$ be the differences of the pseudo-range measurement errors.

We arrive at the same form of equations as the standard synchronized pseudo-range-based positioning problem:

$$\xi_{urk} = \left\| r - r_k \right\| + b_{ur} + v_{urk} \quad (14.36)$$

Similarly, we can perform the single difference between the two Doppler measurement equations and we have

$$SDf_{duk} = f_{duk} - f_{drk} = -\left(\frac{f_k}{c}\right)\left(v_k^T \frac{\left(r_k - r\right)}{\left\| r_k - r \right\|} - v_k^T \frac{\left(r_k - r_r\right)}{\left\| r_k - r_r \right\|}\right) + \left(\frac{f_k}{c}\right)\left(b_{ru} - b_{rr}\right) + \left(v_{fuk} - v_{frk}\right) \quad (14.37)$$

Let

$$\eta_k = SDf_{urk} - \left(\frac{f_k}{c}\right) v_k^T \frac{\left(r_k - r_r\right)}{\left\| r - r_r \right\|} \quad (14.38)$$

If we define $b_{rur} = b_{ru} - b_{rr}$ to be the relative clock rate bias between the user receiver and the reference receiver and $v_{furk} = v_{fuk} - v_{frk}$ to be the difference of Doppler measurement errors between the user receiver and the reference receiver, we have

$$\eta = \left(\frac{f_k}{c}\right)\left(-v_k^{\ T}\frac{(r_k - r)}{\|r_k - r\|} + b_{rur}\right) + v_{furk} \tag{14.39}$$

The implications of the above equations are:

1. Most importantly, the source-specific clock bias and clock rate bias terms are canceled out.
2. The measurement equations, both the pseudo-range and the Doppler, are in the same form as the synchronized SOOP problem. That means that all the equations on the solution methods and the error analysis methods can be reused and therefore not repeated here.
3. The clock bias that the solution will be able to solve for is the relative clock bias between the user receivers and the reference receiver. In other words, we are achieving time synchronization and time transfer between the reference node and the user nodes.
4. The pseudo-range errors in the single-difference equations are the difference of the errors for the user receiver and reference receiver. This has two implications:
   a. The receiver noises, typically high-frequency and filterable, are increased.
   b. The signal propagation errors are partially canceled.
5. The same can be said about the Doppler measurements.

Simulation Examples:

Example 14.1: Positioning Using a Single Iridium Satellite Using TOA and Doppler Measurements

In this case, as we can imagine, the geometric dilution of precision will be very bad at the beginning and will start to improve as the satellite moves across the sky.

Figure 14.15 shows the simulation results of positioning performance using a single Iridium satellite over time. The user position is fixed during the process. In this case, both the TOA and Doppler are used.

The dotted lines show the 2 sigma upper and lower bounds for the error. The line in the middle is a single instance of Monte Carlo simulation result. The positioning does not start until 1 minute after the run. The first 1 minute of data is used to create the initial solution using the batch least square approach.

Example 14.2: Positioning Using a Single Iridium Satellite Using TOA Measurements Only

Figure 14.16 shows the performance of the same scenario except that, in this run, only the TOA is used. The performance is only slightly worse.

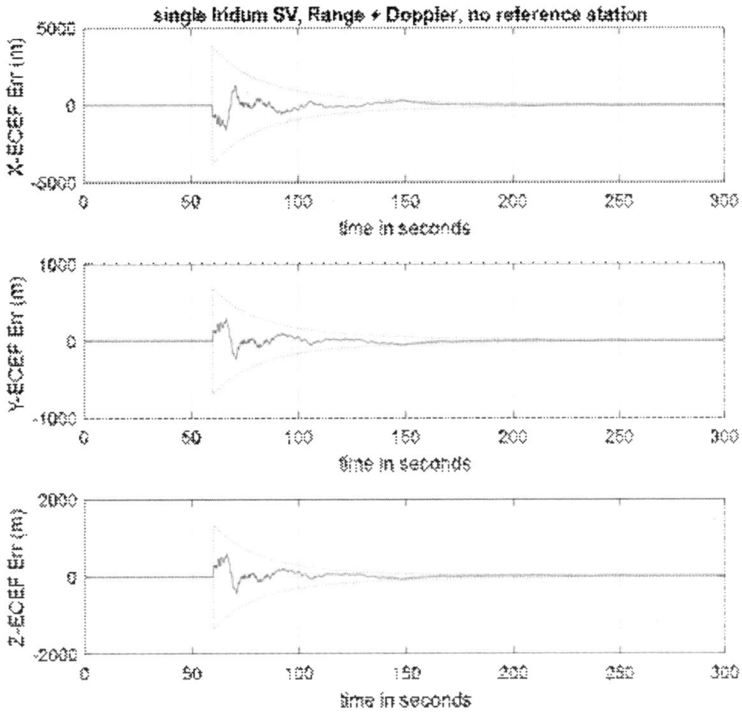

Figure 14.15 Positioning using a single Iridium satellite with TOA and Doppler measurements.

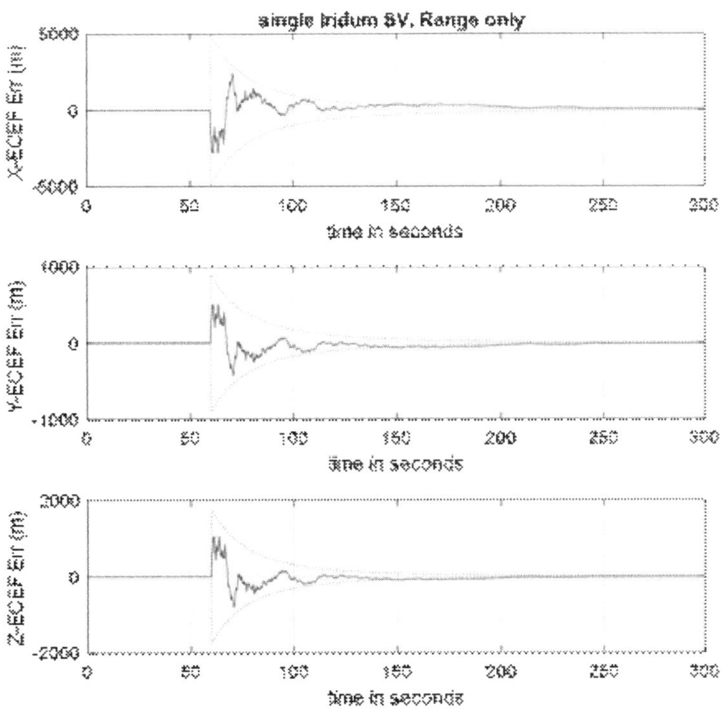

Figure 14.16 Positioning using a single Iridium satellite with TOA only.

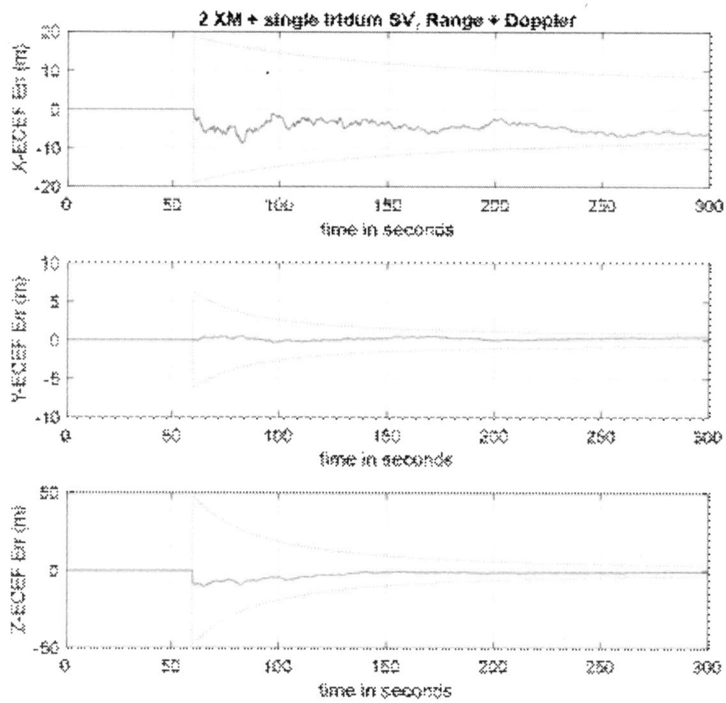

Figure 14.17 Positioning using two XM radio satellites and one Iridium satellite.

Example 14.3: Positioning Using Two XM GEO Satellites and One Iridium Satellite

Figure 14.17 shows the performance when two XM radio signals and one Iridium satellite are used to determine the position. In this case, the geometric dilution of precision becomes much better.

14.6 SOOP PNT Integrated with Inertial and All Source PNT

With all the equations that we have developed, the integration of the SOOP measurements with the INS (and, consequently, into the All Source PNT system) is very easy. First, we have the INS nonlinear algorithms repeated from Chapter 3.

14.6.1 INS Nonlinear Algorithm

The following is the INS nonlinear algorithm:

1. The attitude propagation equation in quaternion form:

$$\dot{q}_B^E = \frac{1}{2}\left(q_B^E \otimes {}^B\omega_I^B - {}^E\omega_I^E \otimes q_B^E\right) \tag{14.40}$$

2. The DCM from the quaternion:

$$C_B^E = q2DCM\left(q_B^E\right) \tag{14.41}$$

3. The velocity propagation equation:

$$\begin{aligned}\frac{d}{dt}{}^{E}v_{E}^{B} &= C_{B}^{E}\,{}^{B}f_{I}^{B} + {}^{E}g - 2\,{}^{E}\omega_{I}^{E} \times {}^{E}v_{E}^{B} - \left[{}^{E}\omega_{I}^{E}\times\right]\left[{}^{E}\omega_{I}^{E}\times\right]{}^{E}r_{E}^{B} \\ &= C_{B}^{E}\,{}^{B}f_{I}^{B} + {}^{E}g_{c} - 2\,{}^{E}\omega_{I}^{E} \times {}^{E}v_{E}^{B}\end{aligned} \tag{14.42}$$

where

$${}^{E}g_{c} = {}^{E}g - \left[{}^{E}\omega_{I}^{E}\times\right]\left[{}^{E}\omega_{I}^{E}\times\right]{}^{E}r_{E}^{B}$$

4. The position propagation equation:

$$\frac{d}{dt}{}^{E}r_{E}^{B} = {}^{E}v_{E}^{B} \tag{14.43}$$

For the covariance propagation of the Kalman filter, we will need the linearized error models for attitude, velocity, and position repeated from Chapter 3 (all symbols were defined there).

The error state vector contains the INS attitude, velocity, and position errors together, all the IMU-related error parameters. The system noises are the gyro and accelerometer noises and parameter random walks:

$$\begin{aligned}\delta X_{I}^{\,T} &= \left[\delta\,{}^{E}\dot{\theta}_{B}^{E}, \delta\,{}^{B}b_{G}^{\,T}, \delta\,{}^{G}s^{T}, \delta\,{}^{B}\beta_{G}^{T}, \delta\,{}^{B}\theta_{G}^{T}, \left(\delta\,{}^{E}v_{E}^{B}\right)^{T}, \delta\,{}^{B}b_{A}^{T}, \delta\,{}^{A}s^{T}, \delta\,{}^{B}\beta_{A}^{T}, \left(\delta\,{}^{E}r_{E}^{B}\right)^{T}\right] \\ W_{I}^{\,T} &= \left[-\left(C_{G}^{E}ARW\right)^{T}, RRW^{T}, SRW_{G}^{T}, \beta RW_{G}^{T}, \theta RW_{G}^{T}, \left(C_{A}^{E}VRW\right)^{T}, ABRW^{T}, SRW_{A}^{T}, \beta RW_{A}{}^{,T}\,0\right]\end{aligned} \tag{14.44}$$

We then have the following 30-state linearized error propagation model

$$\delta\dot{X}_{I} = A_{II}\delta X_{I} + W_{I}$$

$$A_{II} = \begin{bmatrix} [{}^{E}\omega_{I}^{E}\times] & -C_{B}^{E} & -C_{G}^{E}\mathrm{diag}({}^{G}\omega_{I}^{B}) & -C_{B}^{E}J({}^{B}\omega_{I}^{B}) & C_{B}^{E}[{}^{B}\omega_{I}^{B}\times] & 0_{3x3} & 0_{3x3} & 0_{3x3} & 0_{3x3} & 0_{3x3} \\ 0_{3x3} & 0_{3x3} & 0_{3x3} & 0_{3x3} & 0_{3x3} & 0_{3x3} & 0_{3x3} & 0_{3x3} & 0_{3x3} & 0_{3x3} \\ 0_{3x3} & 0_{3x3} & 0_{3x3} & 0_{3x3} & 0_{3x3} & 0_{3x3} & 0_{3x3} & 0_{3x3} & 0_{3x3} & 0_{3x3} \\ 0_{3x3} & 0_{3x3} & 0_{3x3} & 0_{3x3} & 0_{3x3} & 0_{3x3} & 0_{3x3} & 0_{3x3} & 0_{3x3} & 0_{3x3} \\ 0_{3x3} & 0_{3x3} & 0_{3x3} & 0_{3x3} & 0_{3x3} & 0_{3x3} & 0_{3x3} & 0_{3x3} & 0_{3x3} & 0_{3x3} \\ [(C_{B}^{E}m\,{}^{B}f_{I}^{B})\times] & 0_{3x3} & 0_{3x3} & 0_{3x3} & 0_{3x3} & -2[({}^{E}\omega_{I}^{E})\times] & C_{B}^{E} & C_{A}^{E}\mathrm{diag}({}^{A}f_{I}^{B}) & C_{B}^{E}J({}^{B}f_{I}^{B}) & \left[\dfrac{\partial\,{}^{E}g}{\partial\,{}^{E}r_{E}^{B}}\right] \\ 0_{3x3} & 0_{3x3} & 0_{3x3} & 0_{3x3} & 0_{3x3} & 0_{3x3} & 0_{3x3} & 0_{3x3} & 0_{3x3} & 0_{3x3} \\ 0_{3x3} & 0_{3x3} & 0_{3x3} & 0_{3x3} & 0_{3x3} & 0_{3x3} & 0_{3x3} & 0_{3x3} & 0_{3x3} & 0_{3x3} \\ 0_{3x3} & 0_{3x3} & 0_{3x3} & 0_{3x3} & 0_{3x3} & 0_{3x3} & 0_{3x3} & 0_{3x3} & 0_{3x3} & 0_{3x3} \\ 0_{3x3} & 0_{3x3} & 0_{3x3} & 0_{3x3} & 0_{3x3} & I_{3x3} & 0_{3x3} & 0_{3x3} & 0_{3x3} & 0_{3x3} \end{bmatrix} \tag{14.45}$$

14.6.2 Measurement Equations and Error Models

With the symbols defined in the previous section, the measurement equation for the TDOA or single-difference pseudo-range measurement is given by

$$\xi_{urk} = \left\|{}^{E}r_{E}^{B} - r_{k}\right\| + b_{ur} + v_{urk} \tag{14.46}$$

The single-difference FDOA or Doppler measurement is given by

$$\eta = \left(\frac{f_k}{c}\right)\left(-v_k^T \frac{\left(r_k - {}^E r_E^B\right)}{\left\|r_k - {}^E r_E^B\right\|} + b_{rur}\right) + v_{furk} \tag{14.47}$$

The predicted measurement is simply generated by

$$\hat{\xi}_{urk} = \left\|{}^E \hat{r}_E^B - r_k\right\| + \hat{b}_{ur} \tag{14.48}$$

and

$$\hat{\eta} = \left(\frac{f_k}{c}\right)\left(-v_k{}^T u_{los,k}\left({}^E \hat{r}_E^B\right) + \hat{b}_{rur}\right) \tag{14.49}$$

and the linearized model used for covariance updates are given by

$$\begin{aligned}\Delta\xi_k &= \left(-u_{\text{LOS}}\left({}^E \hat{r}_E^B\right)_k\right)^T \delta\, {}^E r_E^B + b_{ur} + v_{pk} \\ &= h_{pp}(k)\delta\, {}^E r_E^B + b_{ur} + v_{pk}\end{aligned} \tag{14.50}$$

and

$$\begin{aligned}\Delta\eta &= \left(\frac{f_k}{c}\right) v_k^T \frac{1}{\left\|{}^E \hat{r}_E^B - r_k\right\|}\left(I_{3x3} - u_{los,k}\left({}^E \hat{r}_E^B\right) u_{los,k}\left({}^E \hat{r}_E^B\right)^T\right)\delta\, {}^E r_E^B + \left(\frac{f_k}{c}\right) b_{rur} + v_{fk} \\ &= h_{fp}(k)\delta\, {}^E r_E^B + \left(\frac{f_k}{c}\right) b_{rur} + v_{fk}\end{aligned} \tag{14.51}$$

In the above equations, the following symbols for the unit vectors are used interchangeably

$$\hat{u}_{los,k} = u_{\text{LOS}}\left({}^E \hat{r}_E^B\right)_k = \frac{\left(r_k - {}^E \hat{r}_E^B\right)}{\left\|r_k - {}^E \hat{r}_E^B\right\|}$$

The Kalman state vector consists of the INS states plus the two clock error states:

$$X^T = \left[X_I^T, b_{ur}, b_{rur}\right] \tag{14.52}$$

The linearized propagation error model is then given by

$$\frac{d}{dt}(\delta X) = \begin{bmatrix} A_{II} & 0_{30x1} & 0_{30x1} \\ 0_{1x30} & 0 & 0 \\ 0_{1x30} & 0 & 0 \end{bmatrix}\delta X + \begin{bmatrix} W_I \\ brw \\ brrw \end{bmatrix} \tag{14.53}$$

The linearized measurement equation can be written as

$$\begin{bmatrix} \delta\xi_k \\ \delta\eta_k \end{bmatrix} = \begin{bmatrix} H_p(k) & 1 & 0 \\ H_f(k) & 0 & \frac{f_k}{c} \end{bmatrix} \delta X + \begin{bmatrix} v_{urk} \\ v_{rurk} \end{bmatrix} \tag{14.54}$$

where

$$H_{pp}(k) = \left[0,0,\ldots,0,h_{pp}(k)\right] \tag{14.55}$$

$$H_{fp}(k) = \left[0,0,\ldots,0,h_{fp}(k)\right] \tag{14.56}$$

References

[1] Jones, M., "Signals of Opportunity: Holy Grail or a Waste of Time?" *GPS World*, February 2018, https://www.gpsworld.com/signals-of-opportunity-holy-grail-or-a-waste-of-time/.

[2] Zheng, L., and R. Xue, "Signals of Opportunity Navigation Methods for Complex Lower Airspace Flight," *2011 IEEE 5th International Conference on Cybernetics and Intelligent Systems*, 2011.

[3] El Gemayel, N., et al., "A Low Cost TDOA Localization System: Setup, Challenges and Results," *2013 10th Workshop on Positioning, Navigation and Communication (WPNC)*, Dresden, 2013, pp. 1–4.

[4] Tan, Z., et al., "New Method for Positioning Using IRIDIUM Satellite Signals of Opportunity," *IEEE Access*, 2019.

[5] Lawrence, D., et al., "Test Results from a LEO-Satellite-Based Assured Time and Location Solution," *Proc. ION GNSS*, Monterey, CA, January 2016, pp. 125–129.

[6] Reid, T. G., et al., "Leveraging Commercial Broadband LEO Constellations for Navigating," *Proc. ION GNSS*, Portland, OR, September 2016, pp. 2300–2314.

[7] Per, E., et al., "Orbital Diversity for Satellite Navigation," *Proc. ION GNSS*, Nashville, TN, September 2012, pp. 3834–3846.

[8] Hill, J., "The Principle of a Snapshot Navigation Solution Based on Doppler Shift," *Proc. ION GPS*, Salt Lake City, UT, September 2001, pp. 3044–3051.

[9] Joerger, M., et al., "Analysis of Iridium-Augmented GPS for Floating Carrier Phase Positioning" *Journal of The Institute of Navigation*, Vol. 57, No. 2, Summer 2010, pp. 137–216.

[10] https://www.xilinx.com/products/silicon-devices/soc/rfsoc.html.

[11] Ohm, D. R., "Kinematic and Cyclostationary Parameter Estimation for Co-Channel Emitter Location Applications," Ph.D. dissertation, Oregon State University, 2007.

[12] Johnson, J. J., "Implementing the Cross Ambiguity Function and Generating Geometry-Specific Signals," Master's Thesis, Naval Postgraduate School, Monterey, CA, 2001.

[13] Ahmed, H. I., et al., "Estimation of Time Difference of Arrival (TDOA) for the Source Radiates BPSK Signal," *IJCSI International Journal of Computer Science Issues*, Vol. 10, Issue 3, No. 2, May 2013.

[14] Carter, G. C., "Coherence and Time Delay Estimation," *Proceedings of the IEEE*, Vol. 75, February 1987, pp. 236–255.

[15] Knapp, C. H., and G. C. Carter, "The Generalized Correlation Method for Estimation of Time Delay," *IEEE Transactions on Acoustics, Speech, and Signal Processing*, Vol. ASSP-24, No. 4, August 1976.

[16] Azaria, M., and D. Hertz, "Time-Delay Estimation by Generalized Cross-Correlation Methods," *IEEE Transactions on Acoustics, Speech, and Signal Processing*, Vol. ASSP-32, No. 2, April 1984.

[17] Stein, S. S., "Algorithms for Ambiguity Function Processing," *IEEE Transactions on Acoustics, Speech, and Signal Processing*, Vol. ASSP-29, No. 3, June 1981.

CHAPTER 15

Real-Time Systems and the Software Development Approach and Applications to PNT Systems

15.1 Introduction

In the earlier chapters, we focused on what we normally call the PNT application software, including the architecture, the algorithms, and the software itself. In addition, we have also provided some discussions on some of the sensors for PNT systems.

In this chapter, we will address other aspects of the systems that have not yet received proper coverage in the earlier chapters. In particular, we cover the following topics:

1. The computing system including both the computer hardware and the infrastructure software, which provides the support for the application software to achieve the purposes of a PNT system.
2. System and software architecture including, most importantly, the architecture standards.
3. System and software development process and quality standards.

Most of the discussions in this chapter are going to be generally applicable to many types of systems including and beyond PNT systems. We will stay general when it is beneficial and will discuss specifics to PNT systems when necessary.

15.2 The Architecture of Real-Time Systems

15.2.1 What Is Architecture All About?

Simply speaking, architecture is about how a system is decomposed into components, and how the components are integrated to achieve the purposes of the system. Consequently, decomposition, components, and integration are some of the most important aspects of an architecture.

The architecture is such that the system can achieve two objectives:

1. The system provides the required functional capabilities.
2. The system achieves certain quality attributes related to the development, construction, test, maintenance, and operations of the system.

The functional aspect is simple. For example, a PNT system must provide positioning, navigation, and timing capabilities to support systems that are communicating with it using some specified interface hardware, software, and data protocols. It is the minimum requirement that an architecture is designed such that the system can achieve its functional requirements.

The quality attributes are what make the differences between a good architecture and a bad one. For example, a good architecture of a system can help to achieve some or all of the following objectives:

1. Allow complicated systems to become manageable.
2. Allow different parts of a team to work on various parts of the system.
3. Allow people of different skills to work on various parts of the system where the skills are most useful.
4. Allow components developed in the past to be reused.
5. Allow components or subsystems to be acquired separately, if the components are specified to have standard interfaces. Consequently, the components or subsystems become interoperable.
6. Allow the system integration to be done in an organized and manageable fashion.
7. Allow the development project to minimize cost and follow a manageable and predictable schedule.
8. Allow the system to be easily maintained by being able to repair, replace, and upgrade components.
9. Allow the system to be easily integrated into larger systems, within which other systems may or may not comply with certain standards.

It is therefore extremely important to ensure both the functional requirements and the quality attribute requirements are elicited, addressed, and evaluated in the architecture development process.

15.2.2 System Architecture Development Approach

As mentioned earlier, the development of system or software architecture is driven by two types of input: (1) the functional and usage requirements that are often provided as use cases or as functional requirements, and (2) the quality attributes of the architecture that are often provided as quality attribute scenarios.

The functional and usage requirements drive the capabilities provided by the system for which the architecture is to be developed. The quality attributes drive the architecture decisions that are made not only to address functional requirements but to also to address developmental and operational concerns such as how easily the design can be changed to accommodate changes of the requirements, how easily the system can be operated, and how the design reuses the capabilities that the organization developed in the past.

As an example, for the PNT system architecture design, we are typically driven by the following quality attributes:

1. Different PNT sensors or subsystems created by different vendors can be easily integrated and work together collaboratively to achieve PNT capabilities (interoperability and compatibility).
2. Products from different vendors can be used interchangeably as long as they comply with the same standards to promote multisourcing and competitive pricing (multisourcing or competition).
3. The hardware and software components can be reused to create different products (hardware and software component reuse).
4. Commercial technology, especially mobile phone technology, receives much higher investment than military technology and enjoys a huge market and can or must be leveraged to support PNT applications. In other words, the architecture shall not create roadblocks to hinder the application of commercial technologies but to facilitate such opportunities (leverage commercial technology).
5. Systems can be rapidly developed using existing hardware and software components. Integration and tests are easy and efficient (rapid and efficient development and integration).
6. Systems, software, and hardware are easily maintainable. Design changes can be easily implemented with minimal cost and schedule impact (maintainability).
7. System, software, and hardware designs are easily understood by new engineers who can therefore start to contribute after a short training or learning period (understandability).

There are many different methodologies available to derive the architecture from the above-mentioned two types of input. However, it is a good practice to go through the following most important steps and each of the key steps can be augmented or elaborated using various methodology and tools:

1. Derive black box functional requirement by usage and requirement analysis. The black box requirements are the external requirements on the system that are independent of the internal design. It is most important to decide first what the black box requirements are to allow enough freedom of the internal design and the optimizations of the architecture. This step can be done informally without the help of an architecture model or can be derived from use cases (i.e., the usage scenarios using traditional IDEF0 functional analysis or using the more modern Unified Modeling Language (UML) and System Modeling Language (SysML)-based approach).
2. Propose candidate logical and/or physical system decompositions (i.e., decompose the system into subsystems or components). This step can be done in many different ways. Regardless of what approach is used, strong domain knowledge is the key to deciding on how the system can be decomposed into smaller systems and how the smaller systems can work together to achieve the capabilities of the total system while still trying to achieve good-quality attribute response. The following approaches are recommended:
   a. Motivate the logical or physical decomposition by functional

decomposition. By decomposing the main capabilities of the system (not every capability), it often provides enough hints as to how the logical system or physical system can be decomposed.

b. Use design patterns. Design patterns are architecture patterns that have been proven by past projects with known advantages and disadvantages. The use of design patterns greatly improves the probability of success. Most of the architecture examples presented in this chapter offer design patterns that can be utilized by the user.

c. Use design tactics and principles.

3. Flow the requirements down to subsystems or components. This step can again be done in many different ways with or without using an architecture model. When an architecture model is used, IDEF0 and UML/SysML-based approaches can be used by elaborating how the system black box capabilities are achieved by the collaboration of the subsystem or component capabilities.
4. Evaluate and improve the architecture by ensuring that the functional capabilities are supported by the architecture and by comparing the quality attribute responses between different candidate architectures or individual architecture decisions.
5. For each of the subsystems or components, steps 2, 3, and 4 can be repeated to hierarchically elaborate the architecture to a level that the lowest level of components is readily available or readily developed. It is important to note that this readiness and the depth of the architecture elaboration vary with the specific purposes and scenarios and the concerns to be addressed.

The architecture development and approach are illustrated by Figure 15.1. It is important to mention that the architecture can be developed with or without using an architecture model. It is highly recommended that the architecture is developed with the help of architecture models that can be visualized. The use of architecture modeling languages is recommended. For example, graphical modeling languages such as IDEF0, UML, and SysML can help significantly to elicit, document, and

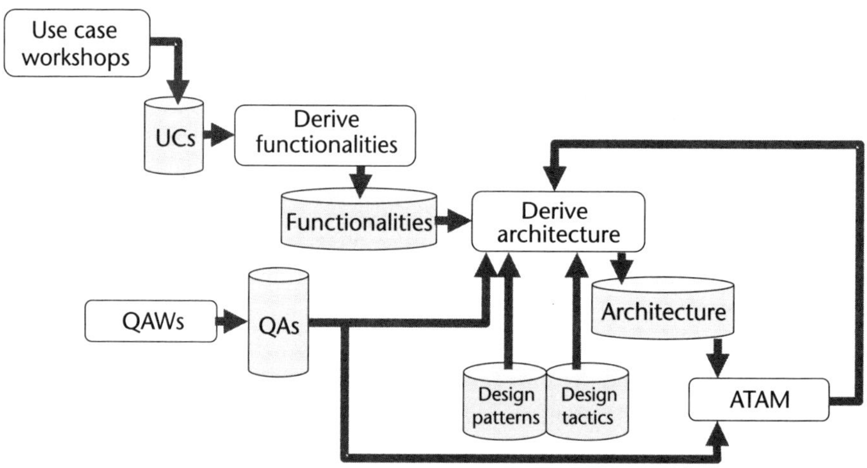

Figure 15.1 Architecture development process.

communicate the architecture. The use of traditional system block diagrams also helps significantly the description of the architecture.

15.2.3 Architecture Design Patterns, Tactics, and Examples

15.2.3.1 Introduction

One of the most difficult things in the architecture development process as illustrated in Figure 15.1 is the step labeled as Derive Architecture. That is where architecture development becomes closer to an art than to science. The key to this step is to rely on the design patterns and the design tactics and principles that are developed and learned in the past to be effective. By using these design patterns and design tactics and principles, combined with architecture assessment and evaluation steps, there is much better hope in developing architecture in a scientific way. In the rest of this section, we will provide a number of example design patterns and will discuss some of the tactics and principles embedded in these examples [1].

As mentioned before, the concept of component is fundamental to architecture. This concept has been natural for the hardware and system world but has not been as clear for the software world because the early software developers often consider software as a program in the sense that it is a sequence of actions, instead of an object. It is not until object-oriented programing becomes popular and widely adopted that software component becomes a recognized concept.

A software component is a software element that is encapsulated with well-defined services and interfaces, can be reused, is not specific to the context where the component is used, and can be independently deployed. Typically, a component is constructed using smaller components.

Component-based architecture decomposes a system into a set of components that interact with each other using a structured approach determined by the architecture. Software components often take the form of objects or collections of objects (such as a C++ or Java Class) in some binary or textual form, adhering to some interface description language (such as IDL used by CORBA) so that the component may exist autonomously from other components in a computer.

Component-based architecture facilitates the planned reuse of software (product-line approach). It also allows independent vendors to be in the business of producing components that can be marketed to different high-level integrators.

It is important to recognize that large grain components (i.e., components that integrate a significant portion of the total system capability) allow new systems to be developed in a very rapid fashion and lead to significant cost and scheduling savings.

15.2.3.2 Typical Sensor, Computer, Actuator, and User Interface (UI) Logical Architecture

Many different real-time systems found in the industry exhibit the basic (almost trivial) pattern of sensor, computer, actuator, and UI architecture as illustrated in Figure 15.2.

It is advantageous to view this architecture as a logical architecture rather than a physical architecture. When viewed as a logical architecture, the sensor or actuator

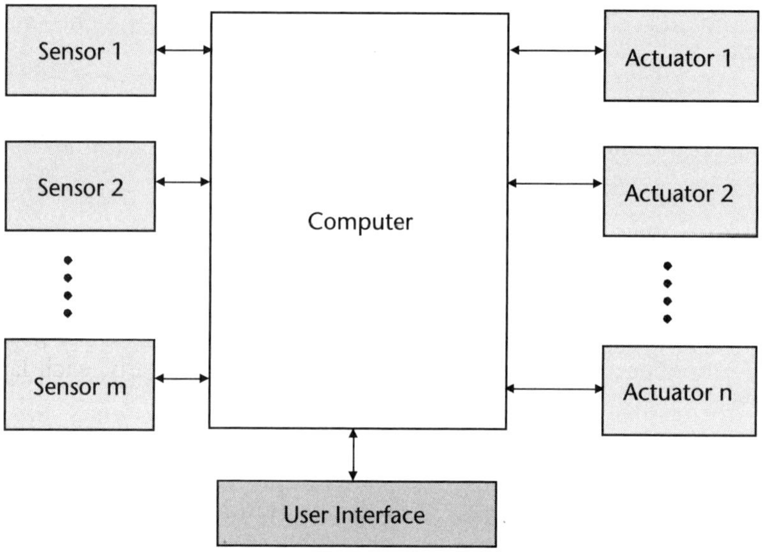

Figure 15.2 Generic sensor, computer, actuator, and UI architecture.

elements are logically providing the capabilities of the sensors or the capabilities of the actuator. The physical implementation of the sensor or actuator itself can contain computers and other electronics. Similarly, the computer in this architecture is a logical element that provides the capabilities of a computer. Physically, the logical computer can be implemented by a single computer in the modern sense, a cluster of computers, or even an ancient mechanical or analog computer.

An important key consequence of this simple architecture is that most of the system design and development efforts focus on what happens inside the computer, that is, the logical software that is typically realized by a real computer or several computers or even by FPGAs or ASICs.

15.2.3.3 Layered System Architecture

Layered architecture is another very frequently used design pattern. The basic design principle or tactics used to design a layered architecture are the separation of concern, which is to use different layers of components to address different concerns of the system.

Layered architecture is a particular way of managing the dependency and, consequently, the complexities within the system. The basic idea applies to both systems and software. In a layered architecture, the system or software is divided into groups of components, with each group called a layer.

A strictly layered architecture requires that the layers form a serial dependency stack. In other words, if the layers are numbered from bottom to top to be layer 0, layer 1, ..., and layer N, then layer k can only depend on layer $k - 1$.

Relaxed layered architecture allows the upper layers to depend on all the lower layers.

In reality, the architect and developer can define a specific layered architecture with dependency rules between the strictly layered architecture and relaxed layered architecture to provide the best compromise between complexity management, performance, and efficiency.

The restricted dependency in a layered architecture means that a change of a layer's interface can only impact the immediate layer above in a strictly layered architecture or layers above for the relaxed layered architecture.

Layers are typically selected according to the separation of concern principle, with the upper layers taking care of the higher-level and more abstract concerns and the lower layers taking care of the lower-level and more specific concerns.

Figure 15.3 provides a view of a layered architecture for the system. It is important to point out that, in a layered architecture, the system is organized into groups of components with each group called a layer. Typically, each layer is desired to address a different level of concerns and requires different skills and knowledge bases to define, design, and implement.

As shown in Figure 15.3, the bottom layer is the sensor, actuator, and UI hardware that provides the input information and actions and output information and actions between the system and the outside world. In other words, it is this layer of hardware devices that interacts with the outside world.

The next layer, which includes the sensor, actuator, and UI device interface hardware, provides the hardware interface between the computer hardware and the actual devices. If standard interfaces are used, such as standard A/D (Analog-to-Digital), standard D/A (Digital-to-Analog), and standard serial data bus, this layer can be constructed using the standard hardware and can be used for systems constructed for very different purposes. Some examples of standard serial data buses are MIL-STD-1553, ARINC 429, ARINC 629, ARINC664, Ethernet, RS232, RS422, RS 485, USB, Camera-Link, IEEE 1394, and Spacewire. Some of these standard interfaces are commercial standards and some of these are specially designed for a narrower class of applications. It is important to recognize that today's technology development, especially electronics technology, is more likely to be led by the commercial or consumer industry rather than by the defense and space industry. In

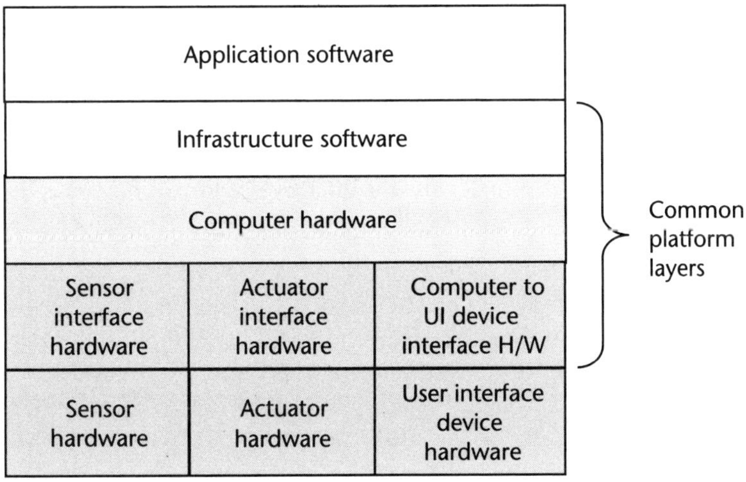

Figure 15.3 Layered system architecture.

particular, the mobile phone industry has been the primary driving force for low-cost, low SWAP, and high-performance computing and communication devices that have a great impact on the rest of industry. It is therefore important for the industry to steer (like it or not) towards standards that are adopted by the mobile phone industry in order to leverage these development with minimum cost and minimum time delays.

The computer hardware layer provides the capability to execute software that performs computation and implements logics and algorithms that process the input information that comes from the sensor, actuator, user hardware interface layer and produce the output information for that layer as well.

The infrastructure software layer provides a software abstraction of the computation and input/output capabilities provided from the hardware layers. This software abstraction greatly simplifies the task of developing the application software.

Finally, the application software implements the system's detailed logics and algorithms that are executed by the computer hardware. The fact that the application software sits on the top of the infrastructure software indicates that the development of application software does not have to be concerned with many details of the hardware but only has to be concerned with a software abstraction of the hardware, and this is enabled by the infrastructure software.

The infrastructure software layer, the computer hardware layer, and the sensor, actuator, UI device hardware interface layer form a system that can be designed in such a way where the same design can be used for a large class of systems. We call these three layers a platform, which can be reused for many different applications when it is hooked up with different sensor, actuator, and UI devices and different application software.

Consequently, both the top and bottom layers are specific to a particular application system and the middle three layers are for general purposes. The bottom hardware layer provides the specific sensor, actuator, and UI devices that are required for the specific application. The application software implements the specific logics and algorithms for the specific application. The key requirements of this system architecture are therefore that the application software talks to the infrastructure software through a standard interface and that the sensor, actuator, and UI devices talk to the interface hardware through standard interfaces.

This approach has been consciously and unconsciously applied to the architecture design of various systems in the industry, and there has been a trend that the platform layers are becoming more general. Furthermore, the application software and sensor and actuator hardware devices are becoming more independent of the platform layers. Integrated Modular Avionics (IMA), which has been implemented for the modern Boeing 787 and Airbus A380 aircraft, is a good example of mature implementations of this approach.

Figure 15.4 shows another typical example of the layered architecture of a real-time system with an emphasis on software and more refined layering. In this example, the layers are organized again to address different levels of concerns.

The top layer, "Mission and system autonomy layer," is concerned with achieving the mission objectives by operating the system using the core control system's abstract capabilities (interfaces). This layer is not concerned with how the specific operations are achieved.

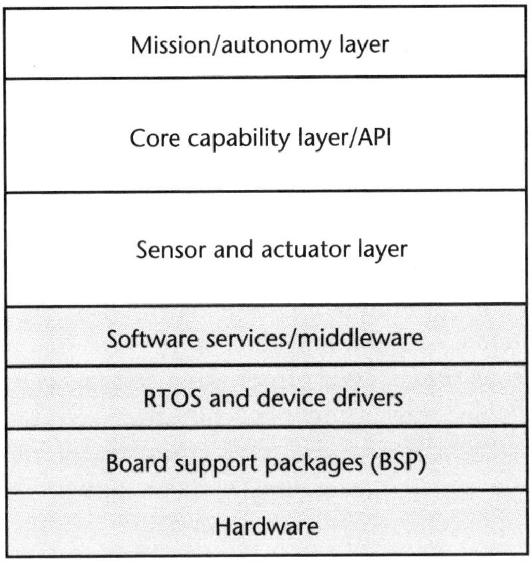

Figure 15.4 Layers of typical aerospace real-time software.

The next layer down is the core capability layer, which is concerned with how to provide the abstracted capabilities using various algorithms. To the upper layer, this layer is abstracted by its API (Application Programming Interface).

The next layer down is the sensor and actuator layer, which handles the specific details of individual sensor and actuators. To the upper layer, the abstract sensor and actuator interfaces are device-independent.

The above three layers are called the application layers.

Below the application layers is the software service or middleware layer, which provides services such as dispatching, communication, and data management.

Below the middleware layer is the real-time operating system (RTOS) and the device driver layer. The operating system (OS) is shielded from the hardware by the board support package (BSP) layer.

The lower three layers are sometimes called the infrastructure layers. The infrastructure layers typically consist of three layers: the middleware layer, the OS and the device driver layer, and the BSP layer.

The infrastructure software shields the computer hardware and the input and output device specifics from the application software (however, it does not shield the high-level characteristics of sensor and actuators) by providing the application layers with abstract services that do not have to change when the computer hardware, the I/O (input and output) devices, or the infrastructure software itself change for other reasons.

Typically, the services provided by the infrastructure software are:

- Multitasking scheduling or dispatching services;
- Timing services;
- Communication services;
- Data management services.

These services are typically provided through the middleware, which augments the generic real-time operating system (RTOS) capabilities to provide services that the application software can use with minimum effort.

Almost all the RTOS used by real-time systems comply or conform to the Portable Operating System Interfaces (POSIX) standard. Using a POSIX-compliant or conformant RTOS means that the RTOS can be easily exchanged to be replaced by another vendor's products if needed without causing a problem to the middleware and the application software.

For aerospace systems, as mentioned before, ARINC 653 defines a standard for partitioned RTOS capabilities to the application software. In an ARINC 653-compliant system, the throughput (time) and memory are partitioned into independent resources that appear to be independent virtual computers. Failures and faults that happen to a particular partition do not propagate into other partitions. ARINC 653 allows independent application software to run on the same computer. To some extent, these applications can be independently qualified and certified. Figure 15.5 illustrates this concept.

The application layers implement the functionalities of the real-time system on the top of the platform provided by the infrastructure layers.

Typical application software has at least three layers: the sensor and actuator layer, the core algorithm layer, and the operation and autonomy layer.

The sensor and actuator abstraction layer on the bottom provides an abstraction of the sensors and actuators. Although the sensor and actuator layer itself depends on specifics of the sensors and actuators, it is typically a design goal that the abstraction provided by this layer is independent of the specifics of the devices and, consequently, the change of sensors and actuators does not impact on the layers above.

The core algorithm layer implements the detailed algorithms for the system. It is typically a design goal to make this layer independent of the hardware devices and the implementation focused on the abstract algorithms.

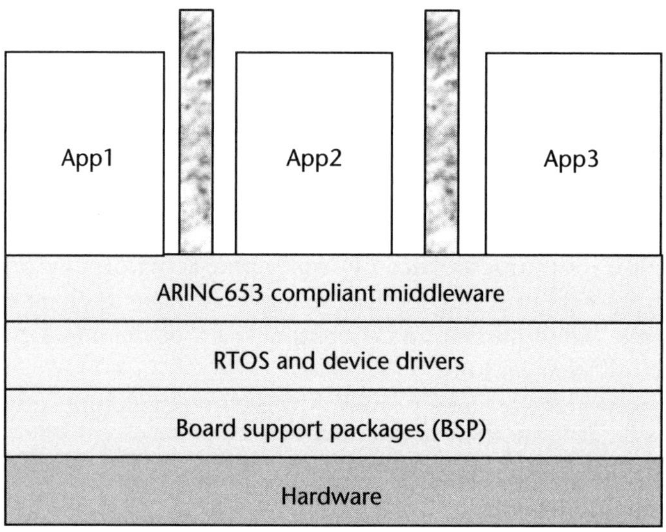

Figure 15.5 Concept of independent applications running on the same hardware through the ARINC 653-compliant partitioned RTOS.

The operation and autonomy layer provides the operational and autonomy capabilities. These capabilities are concerned with operating the system using the capabilities provided by the core algorithm layer.

It is a good practice for each of the layers to be built using components.

15.2.3.4 Distributed Versus Centralized Physical Architecture

Distributed and centralized physical architectures are two important variations of the architecture when the logical system is realized by physical systems. When the logical computer is implemented by a physically centralized computing engine, the physical architecture is centralized. When the logical computer is implemented by physically distributed computing engines, the architecture is distributed. These two variations have their advantages and disadvantages and the choice of a particular architecture always requires careful evaluation and trade study for the specific application and specific objectives.

For PNT systems, we are often concerned with three levels of architecture:

1. The network or system-of-systems level;
2. The vehicle level;
3. The location of interest level.

At the network or system-of-systems level, we are concerned with the large system that consists of various PNT infrastructure systems such as GPS, GLONASS, Galileo, and Beidou, various augmentation systems such as the wide area augmentation system (WAAS), local area augmentation systems such as pseudo-lite systems, various SOOP systems such as radio and TV stations, and various user systems or devices. As the result, in term of the physical architecture, the PNT system-of-systems has no choice but to be distributed.

At the vehicle level, there could be several locations of interest where sensors can be mounted, and the location of sensors are constrained by many things such as FOV, thermal and mechanical constraints, thermal deformations, and structural bending. The physical architecture has to be distributed at least from sensor point of view. The computing architecture can be centralized or distributed and needs to be traded and optimized for specific applications.

A logically centralized PNT system can be designed at the location of interest level or at the vehicle level depending on the requirements for specific applications and the computing architecture can be physically distributed or centralized.

15.2.3.5 Integrated Modular Avionics (IMA) and Convergence of Systems or Devices

IMA [2–5] is a significant advance in the aerospace avionics system architecture. IMA integrates systems that are traditionally implemented by separate physical boxes on aircraft by a uniform general-purpose network of computers interconnected by high-speed communication data buses such as the modern ARINC 664 PART 7 (Avionics Full-Duplex Switched Ethernet) or the older ARINC 429 data bus. The sensor and actuator hardware devices are connected to this network and are accessible by any one of the computing engines across the network. The software modules that were traditionally developed individually for each of the boxes are now application software modules running on the computer network on the top of common hardware and low-level software resources.

IMA offers significant advantages in terms of reducing the development cost, hardware cost, weight, and the maintenance cost. For example:

1. Using a common API to access the hardware and network resources greatly simplifies the hardware and software integration effort.
2. IMA allows the application developers to focus on the application layer and, consequently, enables much more efficient development and simpler integration and test.
3. IMA reuses extensively tested hardware and lower-level software of the platform layers over time and across projects and even the community.
4. IMA allows application software to be relocated and reconfigured on spare computing engines if the hosting computing engine is detected to be faulty during operations. This reduces the cost of redundancy and fault management in the system.

IMA has been adopted by F22-Raptor, Airbus A380, Boeing 787, and several other important avionics systems.

ARINC 653 (Avionics Application Standard Software Interface) is a software specification for space and time partitioning. It defines an API for software of avionics, following the architecture of Integrated Modular Avionics. IMA implementations rely on ARINC 653-compliant RTOS or middleware to allow the memory and time available from a computer to be partitioned in a way as if they were multiple hardware computers.

What is called IMA in the aerospace industry is called the convergence of electronics or the convergence of systems in other communities. Traditional systems, often implemented in multiple individual boxes, are combined into a converged system. The computing systems are integrated and general-purpose. The purposes of the different traditional systems are implemented by multiple application software components. These software components are supported by multiple types of shared sensors and actuators.

One example of such a converged system is a cell phone. Today's cell phone is the convergence of many traditional individual devices including phone, clock or watch, calendar, camera, camcorder, TV, video recorder or player, audio recorder or player, book, heart rate monitor, step counter, GPS receiver, portable computer, game machine, and many more. All these different capabilities are realized on the top of the same cell phone hardware and infrastructure software by individual application software components that we call apps.

Example 15.1: IMA for Boeing 787 [2]

Due to the many benefits of IMA, both the Boeing 787 and Airbus 380 have adopted IMA-based avionics architecture. Boeing 787's IMA is based on what is called the Common Core System (CCS), which was developed by Smith Industry (now General Electric). Figure 15.6 is picture of the 787 CCS.

Figure 15.6 The 787 CCS.

CCS is the 787's central nervous system. The CCS is a computing platform that includes dual common computing resource (CCR) cabinets that host processing and power control modules and network switches. Application-specific modules (ASMs) can be installed in the cabinets.

The CCS platform runs an ARINC 653-partitioned operating environment with an Avionics Full Duplex Switched Ethernet (ARINC 664 PART 7) network backbone. The CCS provides shared system platform resources to host airplane functional systems such as avionics, environmental control, electrical, mechanical, hydraulic, auxiliary power unit, cabin services, flight controls, health management, fuel, payloads, and propulsion.

The CCS has a common data network (CDN), which includes network switches inside the CCR cabinets and externally mounted throughout the aircraft. CDN is a fiber optic Ethernet that connects all the systems that need to communicate with the CCS and with each other.

The CCS is designed to utilize remote data concentrators (RDC) to consolidate inputs from aircraft systems and aircraft sensors that include analog and digital interfaces, such as ARINC 429 and controller area network (CAN) bus.

The CCS replaces multiple computers and hosts up to 80 avionics and utility functions. More than 100 different line replaceable units (LRUs) were eliminated.

Example 15.2: All Source PNT System on Converged Avionics

In a converged avionics system (note that we are following the industries of the abuse of the term avionics, to mean real-time vehicle electronics systems, not in the traditional narrow sense of aviation electronics), the sensor and actuator access is provided to the converged avionics system where a number of separate applications can be running using and producing data from and to the sensor and actuators and generate output that are useful for different purposes. In comparison, traditional systems consist of multiple physical boxes, each of which provides a different service by interfacing with the sensors and actuators associated with that particular function and performing processing for that function.

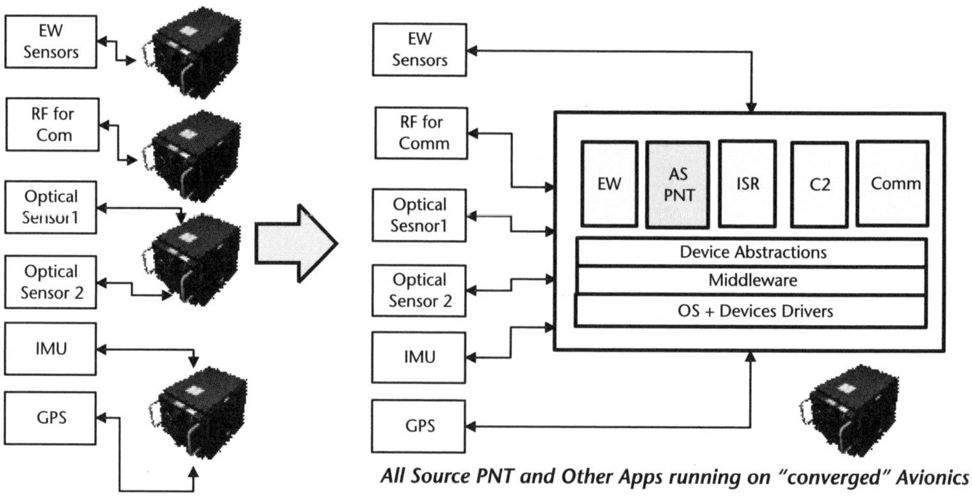

Figure 15.7 All Source PNT system and other applications on converged avionics.

Although All Source PNT can be implemented on almost any types of physical architecture, however, it is especially well suited to be implemented on a Converged Avionics architecture. The fundamental reason for this is because All Source PNT system requires the integration of many different types of sensors that are available on a platform (e.g., a car, an aircraft, a ship). Each of these sensors may or can be used for other purposes and, consequently, it is profitable to collect the sensor data into a converged avionics system and make them available for different types of applications. In particular, it provides the opportunity to reduce development and recurring cost, to reduce SWAP, and to optimize performance. Figure 15.7 illustrates this concept.

15.3 Open Architecture and Open Architecture Standard

15.3.1 Overview

As mentioned before, architecture is all about decomposing the system into components and integrating the components into systems.

Open architecture requires the components to comply to publicly available standard so that:

1. Components can be developed and produced by different vendors and other sources.
2. Components from different sources can be used in an interchangeable fashion.

It is important to realize that:

1. Just component-based architecture is not necessarily open architecture.
2. Open source architecture is not necessarily open architecture.

3. COTS (Commercial Off-The-Shelf) architecture is not necessarily open architecture.

An open system architecture is an architecture in which the interrelationships of the components are defined by interface standards, and the architectural principles and guidelines are consistent with an open systems approach [6].

An open system is an integrated collection of interacting components designed to satisfy stated needs in which the implementations of components are conformant to an standard interface specification; and the interface specification of components are fully defined and available to the public and are maintained according to group consensus.

Consequently, an open architecture approach will lead to the following quality attribute responses:

1. Development-related quality attributes:
   a. Easy development and integration: Systems can be developed and integrated using the preexisting, fully tested, and fully certified components.
   b. Reusability: Components and subsystems developed by different vendors can be easily reused.
2. Operation-related quality attributes:
   a. Operational flexibility: Architecture allows configuration and reconfiguration of available assets to meet rapidly changing operational requirements.
   b. Interoperability: Components and subsystems developed by different vendors can operate together and integrate together seamlessly. Severable software and hardware modules can be changed independently.
3. Maintenance and service-related quality attributes:
   a. Upgradability and technology innovation: Delivery of new capabilities or replacement technology without changing all the components in the entire system.
   b. Maintainability and technology refreshability: Systems can be easily repaired or refreshed by replacement and refresh of components.
4. Competition, cost savings, and cost avoidance:
   a. Enabling of competitive ecosystem: Open architecture with severable modules allows components to be openly competed.
   b. Cost savings and cost avoidance: Reuse of technology, modules, and/or components from any supplier across the acquisition life cycle.

Because of the potential benefits of open architecture approach, the U.S. Department of Defense (DoD) has officially adopted the Modular Open System Approach (MOSA) as an integrated business and technical strategy for the assessment and implementation of open systems.

In January 2019, the top U.S. military leaders reinforced their commitment to open-system standards in a tri-service memorandum signed by the secretaries of the U.S. Navy, Army, and Air Force [7]. The memo directed the Pentagon's service acquisition executives and program executive officers to use MOSA and the existing and emerging open-system standards including:

- Sensor Open Systems Architecture (SOSA);
- Future Airborne Capability Environment (FACE) [8, 9];
- Vehicular Integration for C4ISR/EW Interoperability (VICTORY) [10];
- Open Mission Systems/Universal Command and Control Interface (OMS/UCI).

These existing and emerging standards cover various aspects of systems that are important for the U.S. DoD but, by no means, provide a comprehensive and consistent coverage of all types of systems.

SOSA leverages the OpenVPX embedded computing standard and focuses on single-board computers and how they can be integrated into sensor platforms.

Of particular interest to PNT system developers, both FACE and VICTORY play a major role in the future of PNT systems.

FACE is a comprehensive open avionics standard to enable developers to create and deploy applications across military aviation systems through a common operating environment. It seeks to increase capability, security, safety, and agility while also reducing costs. The main technical objective of FACE is to create an environment that supports a software component that can be used in an open environment. In that sense, the applicability of FACE is definitely beyond avionics systems.

VICTORY is an Army standard specifically allowing different systems from various vendors on an Army vehicle to be integrated and interoperational through an open vehicle data bus or network system. The main technical objective of VICTORY is to create an in-vehicle data bus standard and the hardware and software layers of services to support the data bus to allow systems (boxes) to be easily integrated.

It is the intention of the DoD that future programs will use an open system approach. It has been written into law, 10 USC 2446, that a major defense acquisition program that receives Milestone A or Milestone B approval after January 1, 2019, shall be designed and developed, to the maximum extent practicable, with a modular open system approach to enable incremental development and enhance competition, innovation, and interoperability.

MOSA requires a major defense acquisition program to use an integrated business and technical strategy that:

1. Employs a modular design that uses major system interfaces between a major system platform and a major system component, between major system components, or between major system platforms.
2. Is subjected to verification to ensure that major system interfaces comply with, if available and suitable, widely supported and consensus-based standards.
3. Uses a system architecture that allows severable major system components at the appropriate level to be incrementally added, removed, or replaced throughout the life cycle of a major system platform.

Each of the existing and emerging standards is related to PNT in some way. Table 15.1 provides a rough summary of the coverage of the PNT system architecture by FACE and VICTORY, which are the most relevant.

Table 15.1 Coverage of FACE and VICTORY Versus Desired PNT Architecture Standard Contents

	Desired Architecture Standard Contents	*FACE*	*VICTORY*
PNT system of systems and network architecture	1. How PNT systems collaborate to achieve network level objectives 2. Interface and communication standard 3. Standard for time synchronization and time transfer, ranging, and LOS measurements	Very limited coverage for this level of architecture	Very limited coverage for this level of architecture
Vehicle-level PNT system architecture	1. How subsystems share sensor information to achieve multiple operational purposes 2. Standard for sensor information sharing 3. Standard for portable hardware and software components 4. Standard for time synchronization 5. Standard for alignment of references 6. Standard for communication messages	Good coverage of vehicle-level PNT architecture; focus on portable software component and FACE computing environment, allow one broad type of architecture paradigm	Good coverage of vehicle level PNT architecture using a data bus-centric approach; near complete coverage for one specific architecture paradigm
Location of interest level PNT architecture	All the above still apply	Good coverage	Less concerned about what is happening inside the box; more concerned with the communication, collaboration, and resource sharing between the nodes

15.3.2 FACE

FACE is one of the best examples of the standards created to support MOSA. The main objective of FACE is to establish an open standard under which avionics software components are used, reused, and deployed.

The FACE technical standard is an open avionics standard for making military computing operations more robust, interoperable, portable, and secure. The standard enables developers to create and deploy a wide catalog of applications for use across the entire spectrum of military aviation systems through a common operating environment. The latest edition of the standard further promotes application interoperability and portability with enhanced requirements for exchanging data among FACE components and emphasis on defining common language requirements for the standard.

Figure 15.8 illustrates the context of FACE. Specifically, FACE defines:

1. The standard FACE computing environment that can be supported on existing and future avionics hardware and communication networks;

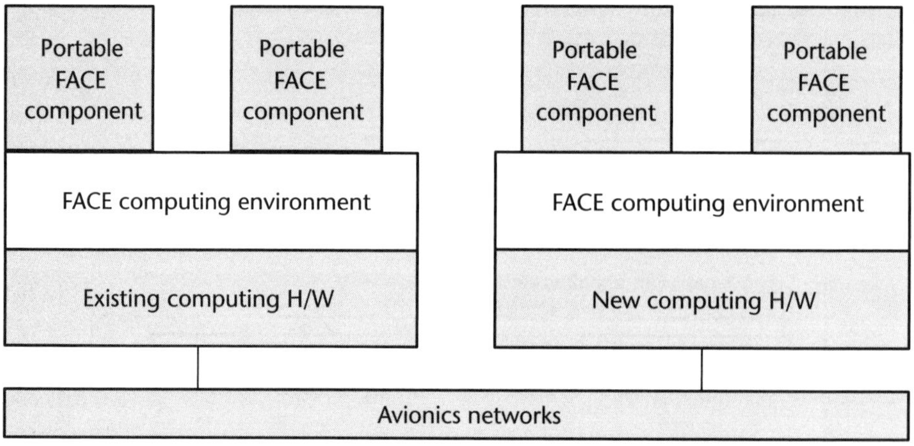

Figure 15.8 FACE is to provide a standard to allow portable FACE software components to be created and deployed across a spectrum of systems through the support provided by the FACE computing environment.

2. The portable FACE component standard, under which a component of application software that is conformant to the standard can be reused across many different types of avionics systems.

FACE removes the barrier for portable components of traditional tightly coupled components by moving the tightly coupled concerns out of what is now called the FACE portable component and into the FACE computing environment. Adapters will be added to the FACE computing platform that would handle the variability of platform device ICDs.

FACE uses a five-layer architecture and each of the layers is called a segment in the FACE standard. The five segments of the FACE reference architecture are:

1. Operating system segment (OSS);
2. I/O services segment (IOSS);
3. Platform-specific services segment (PSSS);
4. Transport services segment (TSS);
5. Portable components segment (PCS).

Figure 15.9 illustrates the layers of the FACE reference architecture.

Each of the layers is briefly described below according to [8, 9].

The OSS is described on Figure 15.9 as being underneath all the other segments and vendor-supplied code. It hosts various OSs, run times, and framework interfaces and provides the vertical interface to the software in all the other segments.

The IOSS layer is an adaption layer that adapts the device drivers from various vendors' device drivers for the interface card. This segment shields the changes of the vendor-specific hardware and device drivers from the all the layers above. In other words, when the hardware and device driver change, only the IOSS segment has

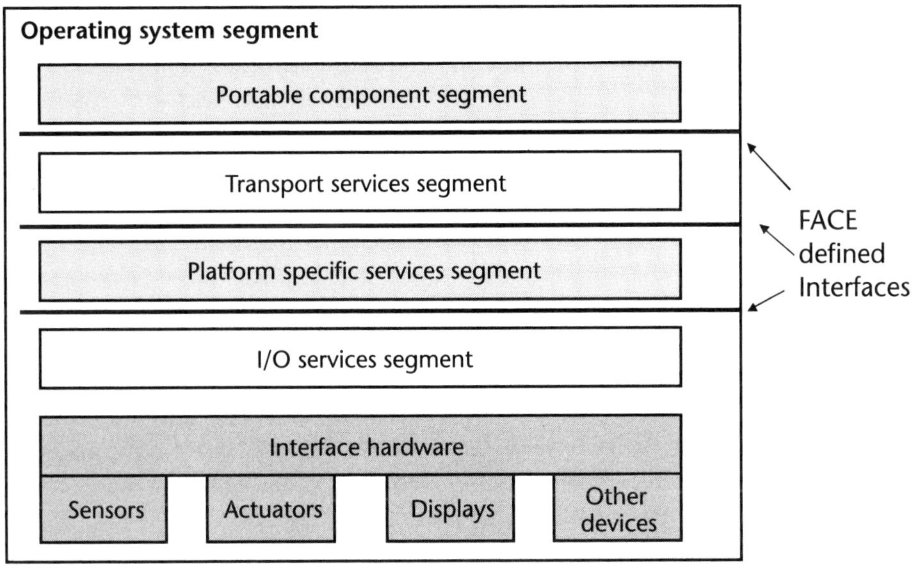

Figure 15.9 FACE reference architecture.

to change. As long as the interface between the PSS and the IOSS does not change, the layers above do not need to change.

The PSSS layer provides an abstraction of platform-unique devices. The FACE standard interface shields the variations of device ICDs from the layers above. When a new device is introduced, as long as the capabilities, data, and operational approach can be abstracted to use the same FACE standard interface for such kind of devices, the layers above do not need to change.

The TSS is a layer that supports transport of data. There are two types of data transportations: (1) the transportation of data between the portable components achieved by the TSS layer alone, and (2) the transportation of data between the portable components to lower layers (which will eventually reach hardware devices) through the PSSS layer. For this purpose, data communication and distribution services such as data distribution service (DDS) middleware, for example, are used to implement this layer.

The PCS is a layer consisting of a collection of portable FACE components and common services that provide the desired platform-level capabilities when executing together. Because of the services provided by the TSS layer, the abstraction of the devices in PSSS layer, and the abstraction of device drivers in the IOSS layer, the PCS layer is supposed to enjoy portability that is independent of:

1. Specific underlying OS used such as Linux, VxWorks, Integrity, ARINC 653-compliant OS, and even windows that supports POSIX;
2. Specific platform devices that provide certain functional capabilities such as a sensor, actuator, display, and other devices such as an IMU, a GPS receiver, and control and display unit (CDU);
3. Specific hardware interface devices and their device drivers such as Ethernet, 1553, RS422, RS485, and ARIC429.

15.3.3 VICTORY

VICTORY was initiated to develop a framework for the integration of electronic mission equipment including C4ISR and electronic warfare systems on ground platforms. The framework includes an architecture, a standard specification, and reference designs. The architecture includes definitions of common terminology, systems, components, and interfaces. The specification provides technical specifications for the systems, components, and interfaces identified in the architecture.

If we disregard the aviation purpose of FACE and the ground vehicle purpose of VICTORY, it is possible to apply both FACE and VICTORY to the same system. Despite that, FACE and VICTORY use very different underlying assumptions for the total system architecture.

Specifically, FACE assumes single or multiple computer hardware that is connected by some kind of network above which there will be FACE-portable software components supported by the FACE computing environment.

VICTORY is built on the idea that the system on the vehicle is largely boxes, instead of software modules, connected by a data bus. It does not assume that all end points can run middleware or even software. Figure 15.10 illustrates an example of the basic underlying architecture for which VICTORY is designed.

Like FACE, VICTORY is another example of MOSA with the purpose of creating modularity and interoperability.

A number of fundamental principles, called VICTORY Tenets in the VICTORY standard document, forms the core of the VICTORY architecture. These tenets assert that electronics system designs on vehicle platforms should (repeated verbatim from the VICTORY Standard public document with no copyright notice [10]):

1. Utilize a data bus-centric design.
2. Leverage on-the-wire network interfaces.

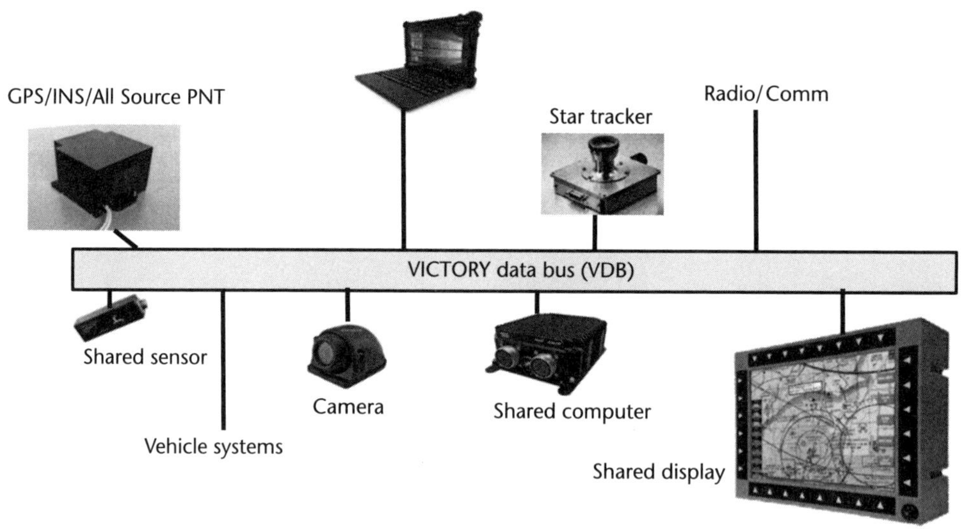

Figure 15.10 The data bus-centric design of the VICTORY architecture.

3. Provide hardware components that can be shared so that software components can be deployed without having to add hardware.
4. Use only open standard physical and logical interfaces between systems and between C4ISR/EW components.
5. Utilize a set of shared data services.
6. Provide shared hardware and software IA (Information Assurance) component types to enable system integrators to build defense in depth security designs that adequately protect platform resources.

15.3.3.1 Data Bus-Centric Design

The foundation of the VICTORY architecture is the vehicle data bus (VDB). The VDB provides the core integration mechanism (i.e., the integration of the systems on the vehicle are achieved by using VDB). As the result, the network interfaces become the most important interfaces in the architecture.

The use of the data bus, instead of point-to-point communication, helps to keep the architecture simple and helps to reduce the impact of the change of individual components in the system.

In order to support this basic architecture design principle, each system that is compliant with VICTORY component types must provide VDB interfaces that include the interfaces for data, configuration, control, status reporting, and fault management according to the VICTORY standard specification.

VICTORY will use the existing standard whenever possible. As the result, VDB is based on IP and uses Ethernet as the standard for the physical layer (PHY), the media access control (MAC), and the data link layer.

The following are some of the key attributes of the data bus-centric design:

1. Data from sensors, the platform, and systems are published on the VDB. As the result, the data are accessible by nodes that are attached to the VDB.
2. Tight coupling of sensors and their processing functionality are avoided unless there is a performance requirement that would otherwise be unattainable.
3. Components attached to the VDB must provide VDB interfaces for configuration, control, status reporting, and fault management. Applications should access data, control, configuration, status reporting, and fault management interfaces via the VDB interfaces.
4. Message formats and protocols for all data, control, configuration, and status interfaces must be open.

15.3.3.2 On-the-Wire Network Interfaces

VICTORY promotes interoperability between components through the definition of network interfaces that have on-the-wire specifications. The key concept driving interoperability between components implementing the VICTORY component type specifications is that the message formats and protocols that they use to communicate are standardized.

The primary goal of VICTORY is to promote interoperability between components and systems to allow the government the choice in who provides systems and components, while achieving interoperability between them. A secondary goal of

VICTORY is to reduce the technical risk (cost and schedule) of moving software components between computing platforms to allow the government to procure software components instead of pre-integrated boxes. The on-the-wire network interfaces are key to VICTORY.

15.3.3.3 Shared Hardware Components

Another core tenet of the VICTORY architecture is the concept of shared hardware components. The purpose of shared hardware components is to reduce the number of hardware components on a platform.

The VICTORY architecture promotes the idea that processing, display, and user input hardware should be shared across multiple systems. To realize this goal, the VICTORY architecture promotes the use of processing hardware that is able to host third-party software applications that are developed to the appropriate application programming interface (within performance limits of the platform). The architecture also promotes the use of display devices that provide for third-party Warfight Machine Interface (WMI) and input devices that work with third-party applications running on shared processing units (SPUs). The VICTORY standard specifications define standards to be used for sharing computing, display, and user input devices with multiple applications.

15.3.3.4 Standard Open Interfaces and APIs

VICTORY architecture promotes the idea that not only the data and management interfaces be implemented on the VDB, but that those interfaces utilize standard open interfaces and APIs. This approach further improves the interoperability and reuse of the components in the system.

15.3.3.5 Shared PNT Services

Of particular interest is that VICTORY recognizes that PNT information is required by almost every C4ISR/EW system attached to the VDB. As a result, VICTORY identifies PNT as one of the most important shared services. By making PNT services available for the entire vehicle, it reduces the need for duplicated processing and duplicated hardware devices.

One of the most important PNT services is time synchronization. The system clock of each hardware component attached to the VDB should be synchronized so that applications running on the distributed processing units have the same notion of time.

15.3.3.6 Shared Information Assurance (IA) Components

In VICTORY, IA support is provided as part of the VDB architecture that identifies the components and architectural patterns that are necessary for implementing a wide array of IA solutions. This enables system designers to include IA solutions during the design phase.

Figure 15.11 is a diagram copied from the official VICTORY architecture document (a public document), which provides another view of VICTORY, depicting its data bus-centric approach and its capability to enable C4IRS/EW systems and access or interface to the platform systems.

15.3.4 The DoD PNT Architecture Standard (PNTA)

The PNTA is yet another effort in MOSA. It is to standardize the interfaces and operational environment for PNT systems to promote interoperability, reusability, and reduced cost and schedule. PNTA is still in the early stages of development. However, before PNTA is officially released and adopted. FACE and VICTORY, described earlier, provide a significant amount of guidance to PNT system developers.

15.3.5 A Proprietary Vendor's Approach to Support Open Systems and the Open Standard

For a provider of a system, subsystem, or component that is to be integrated into a larger system designed based on open system architecture standard, it important to realize that the most important difficulty is the problem of progressive adoption. In other words, during the time period when the standard is gradually adopted, a vendor will need to provide a design supporting the open system standard and, in the meantime, supporting other non-open system architecture and interfaces since not everyone has adopted the open standard yet.

This is exactly the same problem if one is to define and promote a unified language for the world. The unified language could be beautiful and would have made everything in the world easier. However, before the unified language is fully adopted by everybody, life would be harder instead of easier, because one needs to

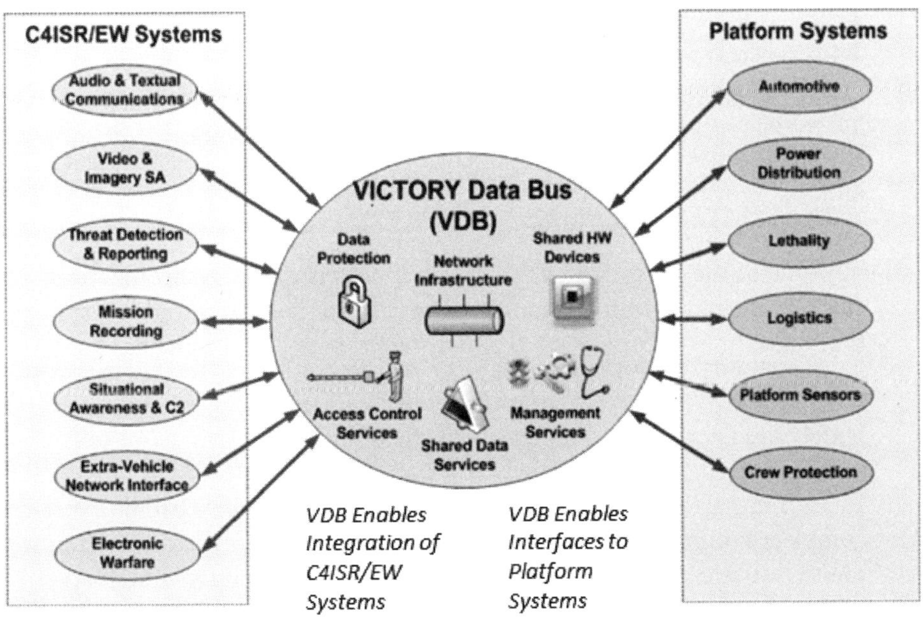

Figure 15.11 VICTORY enables the integration of C4ISR/EW systems and provides the access and interfaces to platform systems. (*Source:* [10].)

master not only the different languages currently in use but also the new unified language. As a result, an open system architecture system or component provider will need to develop a strategy that supports both open system standard and proprietary interfaces.

The solutions to the problem are fortunately simple:

1. The vendor will need to use a good internal architecture to protect or shield most of the designs from changes of external interfaces by the use of abstraction layers.
2. When one has to comply with a different interface requirement, including the open standard interface, one can use adapters in the abstraction layer to adapt the same internal design to different interfaces including open standard interfaces.

In other words, we treat the open standard interface as another interface that we have to support and we use good architecture practices to allow the support of many different interface standards. Figure 15.12 illustrates this concept.

As illustrated in Figure 15.12, a good internal architecture will keep most of the designs and the code unchanged except the adaption layer. The adaption layer is used to deal with the imperfect world: the coexistence of multiple, yet not necessarily consistent, and evolving standards, together with the nonstandard proprietary interfaces that have to be supported.

15.4 Real-Time System Quality and Development Process Standard

15.4.1 Overview

There are a number of software process standards that are followed in the industry. The choice of standard is driven by the market that the system is to address.

For military applications, there has been a history of MIL-STD-2167, MIL-STD-498, and IEEE 12207 [11–13]. These processes typically do not provide sufficient flexibility in terms of adopting modern software development methodology and, consequently, are being used less than they were in the 1980s and 1990s.

For the commercial aviation market, due to the need of FAA certification, virtually all aircraft-related software development follows the DO-178B/C standard [14, 15] created by RTCA. DO-178B/C is objective-driven and consequently provides flexibility for different software development methodologies to be used.

Capability Maturity Model (CMM) Integration (CMMI) [16], developed and advocated by Carnegie Mellon University's Software Engineering Institute (SEI), is widely used across various sectors of the industry. CMMI is not only applicable to software engineering; it applies to system engineering as well.

It is important to mention that, due to large cost and schedule overruns, or even complete failures of a large number of high-profile software development efforts, the industry is moving more towards a risk-driven, iterative development approach rather than the traditional waterfall approach. IBM's Rational Unified Process (RUP) [17–19] and the open-source version of a similar process called the Unified

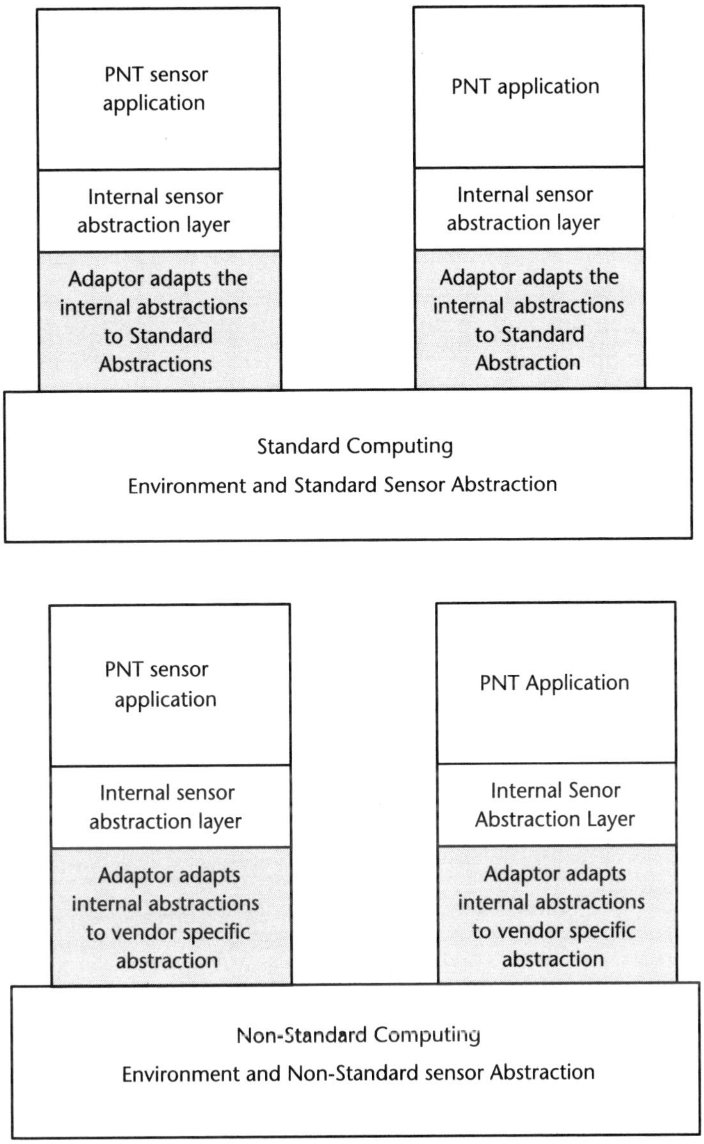

Figure 15.12 Use of internal good architecture and adaptors to deal with the coexistence of evolving, multiple open standards and proprietary interfaces and environments.

Process (UP) and the Open Unified Process (OpenUP) are among the popular ones that are adopted in the industry.

The iterative development processes such as RUP are allowable under DO-178B/C and CMMI. It will be difficult, although not impossible, to adopt an iterative approach for the MIL-STD-2167 type of processes.

15.4.2 The DO-178 (DO-178B and DO-178C) Standard

The DO-178 Software Considerations in Airborne Systems and Equipment Certification provides guidance for software development. DO-178 is published by RTCA and was jointly developed by RTCA and EUROCAE. DO-178B and DO-178C are the recent versions and are widely adopted.

DO178-B and DO-178C have become a de facto standard, and the FAA's Advisory Circular AC20-115B and AC20-115C established DO-178B and DO-178C as the accepted means of certifying all new aviation software.

DO178-B/C is primarily concerned with the development processes. As a result, certification to DO178-B/C requires delivery of multiple supporting documents and records. The quantity of items needed for DO178-B/C certification and the amount of information that they must contain are determined by the level of certification.

The Design Assurance Levels (DAL) are used to classify the criticality of the aerospace system software and the different levels of test and validation objectives with different levels of independence are defined for each of the levels. Table 15.2 summarizes the definitions of the DALs.

DO-178B/C is objective-driven and, consequently, there is a lot of flexibility in terms of the software development life-cycle process. Both the traditional waterfall process and more modern iterative processes can be allowed under DO-178 B/C. However, it does provide the following key activities of the development effort in terms of the objectives, what has to be done, and what documents are produced:

- Planning;
- Development;
- Verification;
- Configuration management;
- Quality assurance;
- Certification liaison.

The following is a list of DO-178B/C required documents and records. Not all are required at all certification levels.

DO-178B/C Documents Related to Planning

- Plan for software aspects of certification (PSAC)l
- Software development plan (SDP);

Table 15.2 Definitions of DALs

DAL	*Failure Consequence*	*Description of Failure Consequence*
A	Catastrophic	May cause a crash
B	Hazardous	Large impact on performance, safety, or crew operation; serious or fatal injuries among passengers
C	Major	Major impact on performance, safety, or crew operation; leads to passenger discomfort
D	Minor	Noticeable impact; leads to passenger inconvenience
E	No effect	No impact to performance and safety

- Software verification plan (SVP);
- Software configuration management plan (SCMP);
- Software quality assurance plan (SQAP);
- Software requirement standards (SRS);
- Software design standards (SDS);
- Software code standards (SCS).

DO-178B/C Documents Related to Development

- Software requirement data (SRD);
- Software design description (SDD);
- The actual software code and images.

DO-178B/C Documents and Records Related to Verification

- Software verification cases and procedures (SVCP);
- Software configuration index (SCI);
- Software accomplishment summary (SAS);
- Software verification results (SVR);
- Problem reports.

DO-178B/C Documents or Records Related to Configuration Management

- Software configuration index (SCI);
- Software life-cycle environment configuration index (SECI);
- Software configuration management records;
- Software quality assurance records.

DO-178B/C Documents or Records Related to Quality Assurance

- Software quality assurance records (SQAR);
- Software conformity review (SCR);
- Software accomplishment summary (SAS).

15.4.3 Mil-STD-2167/Mil-STD-498/IEEE 12207 Standards

Military Standard 498 (Mil-STD-498) was a U.S. military standard with the purpose of software development and documentation. It was released in 1994, in replacement of DOD-STD-2167A and a few other related standards. Mil-STD-498 was intended to be an interim standard and was canceled in 1998 and then replaced by J-STD-016/IEEE 12207. However, these standards were developed by the same key technical personnel and follow the same philosophy and the essential contents are all the same.

These standards provide a rigorous guideline as to how software shall be developed with quality ensured. The process included guidelines on requirement analysis, architecture, and detailed design, coding, unit test, integration test, qualification test, configuration management, and maintenance. These standards specify the following documents to be produced:

- Software development plan (SDP);
- Software test plan (STP);
- Software installation plan (SIP);
- Software transition plan (STrP);
- Operational concept description (OCD);
- System and subsystem specification (SSS);
- Software requirement specification (SRS);
- Interface requirement specification (IRS);
- System and subsystem design description (SSDD);
- Software design description (SDD);
- Interface design description (IDD);
- Database design description (DBDD);
- Software test description (STD);
- Software test report (STR);
- Software product specification (SPS);
- Software version description (SVD);
- Software user manual (SUM);
- Software input and output manual (SIOM);
- Software center operator manual (SCOM);
- Computer operation manual (COM);
- Computer programming manual (CPM);
- Firmware support manual (FSM).

15.4.4 CMMI

CMMI is a process improvement approach that provides companies and organizations with the essential elements of effective processes. CMMI is the successor of the software CMM. The software CMM was developed from 1987 until 1997. CMMI version 1.1 was released in 2002 followed by version 1.2 in 2006, version 1.3 in 2010, and version 2.0 in 2018. The goal of the CMMI project is to improve the usability of maturity models by integrating many different models into one framework. It was created by members of industry, government, and the SEI. The main sponsors included the Office of the Secretary of Defense (OSD) and the National Defense Industrial Association.

CMMI best practices are published in documents called models. A process model is a structured collection of practices that describe the characteristics of effective processes. The practices are those proven by experience to be effective. A process model is used to help to ensure stable, capable, and mature processes and is used as a guide to improve the processes.

The CMMI model is not a process but rather describes the characteristics of the processes. Consequently, very different processes and system and software development methodology can be consistent with practices identified by CMMI. For example, both waterfall development processes and iterative development processes can be consistent with CMMI practices.

There are currently two areas of interest covered by CMMI models: development and acquisition. Only the development area will be discussed in this section.

The CMMI-DEV model provides best practices in the following 22 process areas:

- Causal Analysis and Resolution;
- Configuration Management;
- Decision Analysis and Resolution;
- Integrated Project Management;
- Measurement and Analysis;
- Organizational Innovation and Deployment;
- Organizational Process Definition;
- Organizational Process Focus;
- Organizational Process Performance;
- Organizational Training;
- Project Monitoring and Control;
- Project Planning;
- Process and Product Quality Assurance;
- Product Integration;
- Quantitative Project Management;
- Requirements Management;
- Requirements Development;
- Risk Management;
- Supplier Agreement Management;
- Technical Solution;
- Validation;
- Verification.

The maturity of a company or organization's process and practices can be appraised and rated at one of the following five levels:

- Level 1: Initial;
- Level 2: Repeatable;
- Level 3: Defined;
- Level 4: Managed;
- Level 5: Optimized.

Figure 15.13 illustrates the five maturity levels.

15.4.5 Waterfall Versus Iterative Development Process

It is important to note that, for system and software development, there are two very different ways of development. The first way is often referred to as the waterfall process, which features the sequential execution of system and software development tasks. A car product line is an example of a waterfall process. In a waterfall process, a working product is produced only at the end of the process. The second way is often referred to as iterative process, which features growing or maturing the product through iterations. At each iteration, there is a working product that has partial or immature capabilities. A good example of iterative development will be the growth of life.

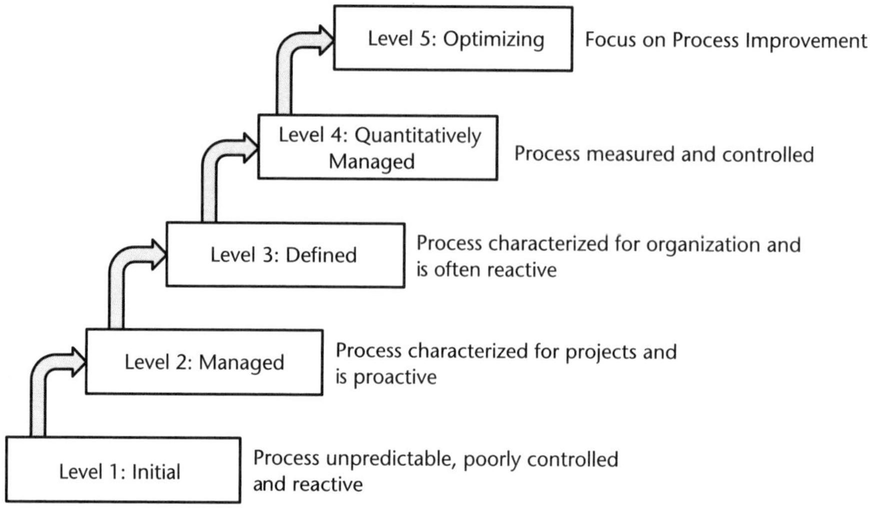

Figure 15.13 Five maturity levels and their characteristics.

The waterfall process is very efficient when the requirement and design are mature and it is simply a matter of implementation or mass production. The disadvantage of a waterfall process is that it does not tolerate changes. A change in the requirement or design can cause major cost and schedule issues in a waterfall process. Consequently, a waterfall process will be a good approach for mass production of a well-designed product. However, a waterfall approach may not be a good process to mature the design of a product.

Iterative development grows the product from a very immature prototype to a fully functional and fully qualified product through iterations. In each iteration, one will do a bit of everything including requirement analysis, architecture design, detailed design, implementation, integration, and test. Iterative development is appropriate when the requirement and design is highly likely to change during the development. Iterative development is efficient if the product developed is not for mass production. (Note that software is never for mass production; we ignored the production in the trivial case of making copies or distributing through a web site.)

One of the key benefits of the iterative development effort is that it exposes and reduces risks early and often and, consequently, in most cases, iterative development process is better than the traditional waterfall process for software development.

Among many variations of the iterative development process, the IBM RUP represents one of the most successful processes using the iterative development approach. Figure 15.14 illustrates the RUP.

As illustrated in Figure 15.14, the life cycle is divided into four sequential phases (waterfall in nature), and each phase consists of a number of iterations (iterative in nature). For each iteration, one does a little bit of business modeling, requirements, analysis and design, implementation, test, and deployment. The purpose of inception is to reduce the business risk. The purpose of the elaboration is to reduce the architecture risk. The purpose of construction phase is to construct the product and reduce the implementation risk and the purpose of the transition phase is to reduce the deployment risk.

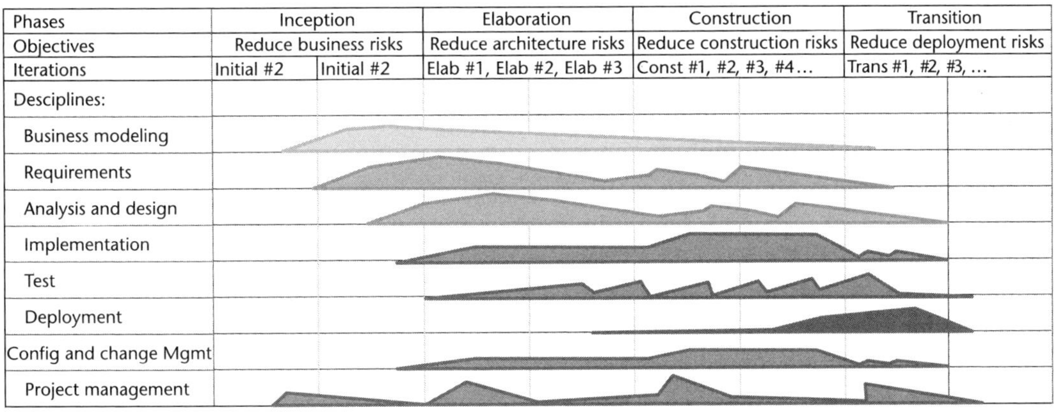

Figure 15.14 The RUP.

15.5 Simulation, Test, and Verification and Validation (V&V) Approaches

15.5.1 Overview

For a real-time system development, simulation plays a very important role in the life cycle of system development.

During the inception phase of the development efforts, the simulation is used to provide validation of the system concept and to provide the sizing, cost, and schedule estimate.

During the (traditional) preliminary design and critical design phases, similar to the elaboration phase for RUP, the simulation is then used to verify that: (1) the requirement of system is consistent with the use cases or usage scenarios, (2) the architecture of the system supports all the required capabilities, and (3) the detailed design works and meets the requirements.

During the implementation phase, or the construction phase in the RUP-like process, the software-in-the-loop (SITL) simulation can be used as a software development platform to integrate and test the implemented system or software. The SITL simulation allows most of the application software bugs to be detected and removed early and easily in a desktop environment.

Beyond the fidelity provided by the SITL simulation, a processor-in-the-loop (PITL) simulation will run the truth models on a general-purpose computer and the flight software on a target processor. The PITL simulation is typically performed in real time and consequently allows the risks associated with the processor and real-time execution retired early in an easy and convenient environment.

The highest-fidelity simulation is provided by the hardware-in-the-loop (HITL) simulation environment where a test version of the real-time computer is integrated with I/O hardware driven by real or simulated hardware devices. The HITL provides the realistic environment for the real-time software to be tested. Human-in-the-loop simulation is a special type of HITL simulation where the human interface hardware devices are included in the simulation and consequently allow humans to interact with the system.

With the exception of the simulation for the concept development, it is usually more productive if the commonality of various simulations mentioned above, including the simulations for the SITL test, simulations for the PITL test, and simulations for the HITL test, are maximized and are natural extension of lower-fidelity simulations.

15.5.2 Simulation for Concept Development and V&V

During the concept development of a design of a real-time system, simulations play an important role of verifying that the concept is feasible and that the objectives of the system can be achieved. The simulation also helps to develop an early estimate of the scope and size of the effort, as well as an estimate of the cost and schedule.

Simulation for concept development typically requires that the simulation be simple and fast and could be easily tailored to analyze different scenarios in a short time. Consequently simulations for this purpose are typically coded in languages such as MATLAB and Simulink or in the form of a spreadsheet. It will be too much overhead at this phase to try to use simulations that have much higher fidelity with the exception where the risks for the projects are identified to be implementation, cost, and schedule risks instead of technology and design risks.

If the risks are on the implementation side, early versions of higher-fidelity simulation such as SITL, PITL and even HITL simulations may be necessary.

15.5.3 SITL Simulation and V&V

Once a detailed design is developed, it becomes necessary to develop and maintain a high-fidelity simulation for the purposes of design V&V, performance analysis, implementation, and risk reduction.

It is a trade-off, but most of the time, it is productive for the high-fidelity simulation to be a SITL simulation. The SITL simulation has the advantage of simultaneously validating system architecture and system performance, validating the algorithm design, and validating the software implementation.

Modern SITL simulations typically are available on the developer's desktop. The SITL simulations allow the designers, analysts, and implementers to quickly implement and validate design changes in an efficient manner and consequently are very important factors for efficient software and system development processes. This is especially true for iterative development process where early and frequent testing and fast turnaround are essential for efficient execution of iterations.

Typical SITLs run faster than real time to allow efficient testing and debugging. The SITL can also be configured to run in real time when risks associated with real-time processes and multitasking are to be reduced early in the development process.

15.5.4 PITL and V&V

PITL simulations provide the next level of fidelity of simulation for the real-time systems. Some PITL platforms just have the target processor, which runs the software, and the simulation truth models, which run on a general-purpose processor, and they communicate with one another though standard interfaces such as Ethernet and TCP/IP. Some PITL simulations are high-fidelity in such a way that all the hardware interfaces appear to be real to the processor and, in this case, hardware-related, low-level software can be integrated and tested in this environment.

Typically, PITL platforms are more expensive and less available and take more effort to run test cases. Consequently, PITL should be used less frequently than SITL. Most of the debugging and testing should take place in the SITL environment.

PITL allows risks associated the specific target processor to be reduced by testing and debugging in the PITL environment. With the advancement of modern infrastructure software, such risk rarely happens to the application software. PITL is most useful for infrastructure software risks reduction.

15.5.5 HITL and V&V

HITL simulations provide the highest-fidelity simulation where engineering development units will be connected together to conduct integration test. The HITL platform typically only runs in real time and it uses the engineering development model of real-time computers, sensors, and actuators to perform tests where hardware interfaces are important.

In the HITL platform, depending on what interfaces or integration issues are considered as risks, not all hardware will be included in the simulation; only those devices deemed risky for the successful development of the system are included.

Clearly, HITL has the most fidelity and it is also most expensive and least available. Consequently, HITL is only used to perform test and validation that cannot be done elsewhere, such as the hardware and software interface, interactions with sensors and actuators, and tests that are designed to test the entire system such as test cases used as part of the formal qualification test (FQT).

15.6 Integrated System and Software Engineering and Model-Driven/Model-Based Development

Unlike software development in other industries, the software development for the aerospace industry has been always separated into two sets of activities. The first set includes the definition and validation of the algorithms, the development of a simulation that validates the algorithm design and system performance, and the development of an algorithm design document to communicate the design to the engineers who will be writing the code. The second set of activities includes the development, testing, integration, and qualification of the code.

This traditional approach has suffered for many years from a fundamental flaw: the domain knowledge is separate from the code, testing, and integration activities. Consequently, the second sets of activities become very inefficient and are often the area of risk for the projects.

Indeed, software cost and schedule are identified as a key risk in almost all aerospace real-time system development projects. While there are many reasons for this fact, the separation of domain expertise from software coding, debugging, and testing is one of the key problems in the traditional aerospace software development process.

This problem is becoming increasingly more apparent due to the following reasons:

- The increase of the complexity of the algorithms has made it necessary for the architecture of these algorithms and the software to be done by individuals who have the domain knowledge. Combined system domain expertise and system and software architecture expertise have proven to be critical for the efficient and successful development of complex, performance-demanding, safety-critical aerospace systems and software.
- The increase of the complexity has made it necessary for the person who writes and debugs the software that implements the algorithm to have sufficient domain knowledge to be efficient.
- The modern iterative development approach (such as IBM's RUP), which has proven to be a superior approach to the traditional waterfall process for software development, demands the capability for fast iterations that consist of tightly coupled requirement, architecture, detailed design, implementation, integration, and test activities. The traditional approach of system engineering throwing something over the wall to software engineering simply does not fit the iterative development approach.

The Integrated Algorithm, Simulation, and Software Development (IASSD) is an approach that integrates the two sets of activities mentioned earlier into a single set of activities to be performed by an integrated team of engineers who develop the algorithms, simulation, and flight software at the same time. The IASSD is typically performed in an iterative development process such as RUP.

The following are a number of approaches in which IASSD can be practiced.

The IASSD can be done by integrating the domain and software team developing the algorithms, simulation, and flight software using the traditional hand-code approach. For these purposes, domain experts are trained with software development skills and software engineers are trained with domain knowledge.

The IASSD can be done using a technique called auto-code. When auto-code is used, the algorithms, simulation, and flight software are defined and implemented in modern graphical languages such as Simulink and Matrix-X, which are commonly used by control engineers. The flight code is then automatically generated from the higher-level graphical description of the design. This approach is often referred to as model-based design.

Auto-code can also be done from an UML-based architecture model, and this approach is often associated with model-driven design.

Regardless of the approach taken, a critical factor in successful IASSD-based development is the development and the use of flight software in the loop simulations throughout the life cycle. For a hand code-based development approach, the hand-coded flight software is included in the simulation. For the model-based approach, the flight software in the loop simulation is really just the graphical model-based simulation itself.

15.7 Software Reuse and Software Product Lines (SPL)

As mentioned before, cost and schedule problems for real-time software systems are persistent across the industry. Better process, better approach, better tools, and better designs provide better productivity and consequently reduce the cost and schedule.

It is important to realize that no productivity improvement saves more cost and schedule than eliminating certain development completely. Software reuse and software product lines are ways to reduce cost and schedule along this line.

It is also important to note that reuse means much more than just the reuse of code. The concept, the requirement, the architecture, the architecture model, the design document, the code, and the test cases can all be reused.

Software reuse is a broader term than software product line. Software reuse includes any form of reuse. However, the product-line approach only refers to planned and managed reuse in a prescribed way. Because of the planning and development of reusable assets in a product-line approach, the software product-line approach often achieves significantly better results than traditional plain reuse.

Over the years, software reuse has evolved from the reuse of a functions or subroutines to the reuse of modules, classes, and objects that contain data and functions to components that provide a significant amount of capabilities for a software system.

In terms of the size of the reused elements, it has grown from the small grain functions or objects to large grain components. As we discussed earlier, large grain components allow rapid assembly, integration, and test of software systems using just a few large components.

A software product-line approach goes further from the advancement of the reuse strategy into a fully strategic reuse. That means developing many software system products with some common managed set of features from a common set of core assets in a prescribed way.

The establishment of a software product line consequently is more complicated than the establishment of a code library.

By carefully studying successful and unsuccessful examples of software product-line practices in the industry, SEI of CMU has developed a fairly mature framework for software product-line practices, which is called the SPL framework [20, 21]. This framework identifies three essential activities including core asset development, product development, and management. This framework identifies and describes 29 practice areas (PAs) that fall into three categories: software engineering, technical management, and organizational management.

The software engineering practice areas are:

1. Architecture definition;
2. Architecture evaluation;
3. Component development;
4. Mining existing assets;
5. Requirement engineering;
6. Software system integration;
7. Testing;

8. Understanding relevant domains;
9. Using externally available software.

The technical management practice areas are:

1. Configuration management;
2. Make/buy/mine/commission analysis;
3. Measurement and tracking;
4. Process discipline;
5. Scoping;
6. Technical planning;
7. Technical risk management;
8. Tool support.

The organizational management practice areas are:

1. Building a business case;
2. Customer interface management;
3. Developing an acquisition strategy;
4. Funding;
5. Launching and institutionalizing;
6. Market analysis;
7. Operations;
8. Organizational planning;
9. Organizational risk management;
10. Structuring the organization;
11. Technology forecasting;
12. Training.

For aerospace systems, there are a lot of similarities and common capabilities for different systems used by aircraft, missile, launch vehicles, and spacecraft. Capabilities such as navigation, attitude determination, orbit determination, thermal control are very similar across many different types of vehicles. For flight control, similarity exists among smaller classes of vehicles. In addition, virtually for all real-time aerospace systems, the RTOS can be very similar. These similarities allow reuse at various levels. By careful planning, software product lines can be established to go beyond traditionally simple reuse into strategic reuse through the software product line.

The successful deployment of the product line can lead to large-scale productivity gains, it improves time to market, it helps to maintain a market presence, it helps to sustain growth, it reduces demand for manpower, and it improves product quality and improves customer satisfaction.

Some of the key factors that contribute to the successful development of a software product line are:

1. A compelling business case;
2. Deep domain knowledge;

3. A rich legacy base;
4. A dedicated champion;
5. Organizational cohesion;
6. Courage to try new engineering approaches.

References

[1] Li, R., and M. Santina, "Aerospace Real-Time Control System and Software," Ch. 6 in *The Control Handbook: Control System Applications*, 2nd ed., W. S. Levine, (ed.), Boca Raton, FL: CRC Press, 2018.

[2] New Item: "GE Aviation Common Core System for the 787 Dreamliner Marks Debut of Open System Architecture for the Type of Commercial Aviation System," https://www.greencarcongress.com/2011/10/geccs-20111028.html.

[3] https://en.wikipedia.org/wiki/Integrated_modular_avionics.

[4] Watkins, C. B., and R. Walter, "Transitioning from Federated Avionics Architectures to Integrated Modular Avionics," *26th Digital Avionics Systems Conference (DASC)*, Dallas, TX, October 2007.

[5] Littlefield-Lawwill, J., and R. Viswanathan, "Advancing Open Standards in Integrated Modular Avionics: An Industry Analysis," *26th Digital Avionics Systems Conference (DASC)*, Dallas, TX, October 2007.

[6] https://www.theopensystem.org/the-idea.

[7] https://www.militaryaerospace.com/computers/article/16722050/pentagon-reinforces-mandate-for-electronics-design-opensystems-standards-like-sosa-face-and-victory.

[8] The Open Group, "Technical Standard for Future Airborne Capability Environment (FACE™)," Edition 2.1, 2014.

[9] https://www.opengroup.org/face.

[10] VICTORY Standards Support Office, "Vehicular Integration for C4ISR/EW Interoperability (VICTORY) Architecture," Version A, April 2, 2019.

[11] Sorensen, R., "MIL-STD-498, J-STD-016, and the U.S. Commercial Standard," *CrossTalk Magazine*, December 16, 2004.

[12] https://en.wikipedia.org/wiki/MIL-STD-498.

[13] https://en.wikipedia.org/wiki/DOD-STD-2167A.

[14] https://en.wikipedia.org/wiki/DO-178B.

[15] https://en.wikipedia.org/wiki/DO-178C.

[16] https://en.wikipedia.org/wiki/Capability_Maturity_Model_Integration.

[17] https://en.wikipedia.org/wiki/Rational_Unified_Process.

[18] Jacobson, I., G. Booch, and J. Rumbaugh, *The Unified Software Development Process*, Reading, MA: Addison-Wesley Professional, 1999.

[19] Kroll, P., and P. Kruchten, *Rational Unified Process Made Easy: A Practitioner's Guide to the RUP*, Reading, MA: Addison-Wesley Professional, 2003.

[20] Northrop, L. M., et al., "Framework for Software Product Line Practice, Version 5.0," White Paper, Software Engineering Institute, December 2012.

[21] https://en.wikipedia.org/wiki/Software_product_line.

List of Acronyms

1-D	One-dimensional
2-D	Two-dimensional
3-D	Three-dimensional
AGC	Automatic gain control
AHE	Adaptive histogram equalization
AHRS	Attitude or Heading Reference System
AKAZE	Accelerated KAZE
AOA	Angle of arrival
ARINC	Aeronautical Radio, Incorporated
ASIC	Application-specific integrated circuit
ASPN	All Source Position and Navigation
BCCE	Balanced CLAHE (contrast-limited adaptive histogram equalization) and contrast enhancement
BF&DRP	Bilateral filter and dynamic range partitioning
BRISK	Binary robust invariant scalable key points
BSP	Board support package
C4ISR	Command, Control, Communications, Computers, Intelligence, Surveillance, and Reconnaissance
C5ISR	Command, Control, Communications, Computers, Cyber, Intelligence, Surveillance, and Reconnaissance
CAF	Cross-ambiguity function
CLAHE	Contrast-limited adaptive histogram equalization
CMM	Capability Maturity Model
CMMI	Capability Maturity Model Integration
CONOPS	Concept of operations
CSAC	Chip-scale atomic clock
CCS	Common Core System
CSWAP	Cost, size, weight, and power
CTE	Charge transfer efficiency
CXO	Crystal oscillator
DCM	Directional cosine matrix

DGPS	Differential Global Positioning System
DOA	Directional of arrival
DoD	U.S. Department of Defense
DTOA	Difference of time of arrival, differential time of arrival
ECEF	Earth centered, Earth fixed
ECI	Earth-centered inertial
EGM	Earth gravity model
EGNOS	European Geostationary Navigation Overlay Service
EKF	Extended Kalman filter
ENU	East North up
EO/IR	Electro-optical/infrared
EW	Electronic warfare
FACE	Future Airborne Capability Environment
FDCIR	Fault Detection, Containment, Isolation, and Response
FDO	Finite distance object
FOV	Field of view
FFOV	Full field of view
FT	Fourier transformation
FFT	Fast Fourier transformation
FPGA	Field programmable gate array
GARN	Gravity anomaly referenced navigation
GEO	Geostationary Earth orbit
GGARN	Gravity gradient anomaly referenced navigation
GGI	Gravity gradient instrument
GLONASS	GLObal NAvigation Satellite System
GNSS	Global Navigation Satellite System
GPS	Global Positioning System
GRAPHIC	Group and Phase Ionosphere Correction
GTR	Generic terrain reference
HE	Histogram equalization
HEO	Highly elliptical orbit
HIL	Hardware-in-the-loop
HSF	High spatial frequency
IA	Information assurance
IASSD	Integrated Algorithm, Simulation and Software Development
ICD	Interface Control Document
IDEF0	Icam DEFinition for Function Modeling
IERS	International Earth Rotation and Reference Systems Service

I/F	Interface
IF	Intermediate frequency
IFT	Inverse Fourier transformation
IFFT	Inverse fast Fourier transformation
IGRF	International Geomagnetic Reference Field
IMA	Integrated modular avionics
IMU	Inertial measurement unit
INS	Inertial Navigation System
IRP	IERS Reference Pole
I/Q	In-phase/quadrature-phase
ITNS	Integrated total network solution
JGM	Joint Gravity Model
KAZE	Not an acronym, a Japanese word that means wind
LEO	Low Earth orbit
LIDAR	Light Detection and Ranging
LIS	Lost-in-space
LLA	Latitude, longitude, altitude
LORAN	LOng RAnge Navigation
LOS	Line of sight
LSF	Low spatial frequency
LWIR	Long-wave infrared
LWIN	Local with immediate neighbor
MARN	Magnetic anomaly referenced navigation
MBL	Map-based localization
MEO	Medium Earth orbit
MI	Mutual information
MOSA	Modular open system approach
NAMAD	North American Magnetic Anomaly Database
NCPNT	Network collaborative PNT
NCC	Normalized cross-correlation
NED	North East down
NTP	Network Timing Protocol
OCXO	Oven-controlled crystal oscillator
ORB	Oriented FAST and Rotated BRIEF (ORB)
OMS/UCI	Open Mission Systems/Universal Command and Control Interface
PFA	Probability of false alarm
PHE	Plateau histogram equalization

PIL	Processor-in-the-loop
PMD	Probability of missed detection
PNT	Position, navigation, and timing
PR	Pseudo-range
PROM	Programmable read-only memory
PSD	Power spectral density
PTP	Precision Timing Protocol
QZSS	Quasi-Zenith Satellite System
RAM	Random access memory
REF	Reference
RF	Radio frequency
RFIC	Radio frequency integrated circuit
RFSOC	Radio frequency system-on-chip
RSO	Resident space objects
RTOS	Real-time operating system
SAA	Stellar attitude acquisition
SAR	Synthetic aperture radar
SDR	Software-defined radio
SIFT	Scale invariant feature transform
SIL	System Integration Lab
SITAN	Sandia Inertial Terrain Aided Navigation
SITL	Software-in-the-loop
SLAM	Simultaneous localization and mapping
SNR	Signal-to-noise ratio
SOC	System-on-chip
SOOP	Signals of opportunity
SOSA	Sensor open systems architecture
SPL	Software product line
SRTM	Shuttle Radar Topography Mission
STL	Satellite timing and location
SURF	Speeded-up robust features
SWAP	Size, weight, and power
SWIR	Short-wave infrared
SysML	System Modeling Language
TCXO	Temperature-compensated crystal oscillator
TDOA	Time difference of arrival
TOA	Time of arrival
TOT	Time of transit

TRN	Terrain referenced navigation
UI	User interface
UKF	Unscented Kalman filter
UML	Unified Modeling Language
V&V	Verification and validation
VICTORY	Vehicular Integration for C4ISR/EW Interoperability
VFOV	Variable fields of view
WAAS	Wide area augmentation system
WDMAM	World Digital Magnetic Anomaly Map
WGS	World Geodetic System
WMI	Warfight Machine Interface
WMM	World Magnetic Model
XO	Oscillator

About the Author

Dr. Rongsheng (Ken) Li is a Senior Technical Fellow at Boeing. He is one of the industry's leading innovators and experts in positioning navigation and timing, guidance navigation and control, and several other related technical fields. He has more than 50 issued or pending patents and a number of publications in his fields.

Dr. Li is currently working for Boeing Research & Technology (BR&T) and is responsible for strategizing, creating, defining, and developing advanced guidance, navigation, and control technologies for Boeing. He is also responsible for developing the next generation of technical leaders through mentoring and teaching.

Prior to BR&T, Dr. Li worked at both Boeing Network & Space Systems and Hughes Space & Communication for 19 years where he had been one of the key developers of Stellar Inertial Attitude Determination and Control Systems for several U.S. key space assets and spacecraft product lines.

Before joining Boeing and Hughes, Dr. Li was the system analysis department manager at BAE (British Aero Equipment) Lear Astronics Corporation, where he was responsible for the modeling, simulation, design, and analysis of aircraft flight control, navigation, guidance, and flight management systems. He has been the primary system architect, system analyst, and software developer for several important navigation, flight control, and flight management systems in the industry.

In 2016, Dr. Li was awarded the Asian American Engineer of the Year (AAEOY) award, a prestigious, national-level recognition in the United States. Dr. Li teaches courses on spacecraft attitude control, Kalman filtering and applications, system and software architecture, and GPS and applications at the University of California at Los Angeles. He received a Ph.D. in electrical engineering from the University of Southern California in 1989 and an M.S. in 1984 and a B.S. in 1981 in EE/AE (Electrical/Aerospace Engineering) in China.

Dr. Li grew up in China and he spent only 9 years in elementary school through high school before he started attending college when he was 15 years old. Dr. Li was fortunate enough to be part of the first generation of university students after China ended its Cultural Revolution and restored the higher education system in 1977.

Dr. Li and his wife live in Southern California. They are proud parents of two sons, one a graduate of and the other a current student at Harvard Medical School. Dr. Li enjoys Bible teaching, singing, and photography.

Index

K

L

M

N

O

P

The GNSS Technology and Applications Series

Elliott Kaplan and Christopher Hegarty, Series Editors

A-GPS: Assisted GPS, GNSS, and SBAS, Frank van Diggelen

All Source Positioning, Navigation, and Timing, Rongsheng (Ken) Li

Applied Satellite Navigation Using GPS, GALILEO, and Augmentation Systems, Ramjee Prasad and Marina Ruggieri

Digital Terrain Modeling: Acquisition, Manipulation, and Applications, Naser El-Sheimy, Caterina Valeo, and Ayman Habib

Geographical Information Systems Demystified, Stephen R. Galati

GNSS Applications and Methods, Scott Gleason and Demoz Gebre-Egziabher

GNSS Interference Threats and Countermeasures, Fabio Dovis, editor

GNSS Markets and Applications, Len Jacobson

GNSS Receivers for Weak Signals, Nesreen I. Ziedan

GNSS for Vehicle Control, David M. Bevly and Stewart Cobb

GPS/GNSS Antennas, B. Rama Rao, W. Kunysz, R. Fante, and K. McDonald

Implementing e-Navigation, John Erik Hagen

Introduction to GPS: The Global Positioning System, Second Edition, Ahmed El-Rabbany

Location-Based Services in Cellular Networks: From GSM to 5G NR, Adrián Caldaldo García, Stefan Maier, Abhay Phillips

MEMS-Based Integrated Navigation, Priyanka Aggarwal, Zainab Syed, Aboelmagd Noureldin, and Naser El-Sheimy

Navigation Signal Processing for GNSS Software Receivers, Thomas Pany

Principles of GNSS, Inertial, and Multisensor Integrated Navigation Systems, Second Edition, Paul D. Groves

RF Positioning: Fundamentals, Applications, and Tools, Rafael Saraiva Campos, and Lisandro Lovisolo

Server-Side GPS and Assisted-GPS in Java™, Neil Harper

Spread Spectrum Systems for GNSS and Wireless Communications, Jack K. Holmes

Understanding GPS/GNSS: Principles and Applications, Third Edition, Elliott Kaplan and Christopher Hegarty, editors

Ubiquitous Positioning, Robin Mannings

Wireless Positioning Technologies and Applications, Second Edition, Alan Bensky

For further information on these and other Artech House titles, including previously considered out-of-print books now available through our In-Print-Forever® (IPF®) program, contact:

Artech House Publishers
685 Canton Street
Norwood, MA 02062
Phone: 781-769-9750
Fax: 781-769-6334
e-mail: artech@artechhouse.com

Artech House Books
16 Sussex Street
London SW1V 4RW UK
Phone: +44 (0)20 7596 8750
Fax: +44 (0)20 7630 0166
e-mail: artech-uk@artechhouse.com

Find us on the World Wide Web at: www.artechhouse.com